Steel: Design, Properties and Applications

Steel: Design, Properties and Applications

Editor: Felix Sampson

MURPHY & MOORE
www.murphy-moorepublishing.com

Published by Murphy & Moore Publishing
1 Rockefeller Plaza,
New York City, NY 10020, USA
www.murphy-moorepublishing.com

Steel: Design, Properties and Applications
Edited by Felix Sampson

International Standard Book Number: 978-1-63987-506-1 (Hardback)

Cataloging-in-Publication Data

Steel : design, properties and applications / edited by Felix Sampson.
p. cm
Includes bibliographical references and index.
ISBN 978-1-63987-506-1
1. Steel. 2. Steel--Analysis. 3. Steel--Metallurgy. I. Sampson, Felix.
TS305 .S74 2022
669.142--dc23

Contents

Preface

This book has been an outcome of determined endeavour from a group of educationists in the field. The primary objective was to involve a broad spectrum of professionals from diverse cultural background involved in the field for developing new researches. The book not only targets students but also scholars pursuing higher research for further enhancement of the theoretical and practical applications of the subject.

Steel design or structural steel design is a sub discipline of structural engineering that is used to design steel structures. Steel is an alloy which is made up of iron and carbon. The tensile strength of steel is high which makes it hard and strong. Designing and construction of steel structures depend on chemical composition and mechanical properties of steel. Two commonly used methods for designing steel structures are allowable strength design method (ASD) and load and resistance factor design method (LRFD). Steel is widely used in automobile industry, infrastructure business, construction of roads, small and heavy instruments and equipments. This book unravels the recent studies in the field of steel design and its applications. It unfolds the innovative aspects of steel design which will be crucial for the progress of this field in the future. This book attempts to assist those with a goal of delving into this field of study.

It was an honour to edit such a profound book and also a challenging task to compile and examine all the relevant data for accuracy and originality. I wish to acknowledge the efforts of the contributors for submitting such brilliant and diverse chapters in the field and for endlessly working for the completion of the book. Last, but not the least; I thank my family for being a constant source of support in all my research endeavours.

Editor

1

Crack-Depth Prediction in Steel Based on Cooling Rate

M. Rodríguez-Martín,[1,2] S. Lagüela,[1,3] D. González-Aguilera,[1] and P. Rodríguez-Gonzálvez[1]

[1]*Department of Cartographic and Terrain Engineering, University of Salamanca, Polytechnic School of Avila, Hornos Caleros 50, 05003 Ávila, Spain*
[2]*Faculty of Sciences and Arts, Catholic University of Avila (UCAV), C/Canteros s/n, 05005 Ávila, Spain*
[3]*Applied Geotechnologies Research Group, University of Vigo, Campus Lagoas-Marcosende, Rúa Maxwell s/n, 36310 Vigo, Spain*

Correspondence should be addressed to M. Rodríguez-Martín; ingmanuel@usal.es

Academic Editor: Gianluca Percoco

One criterion for the evaluation of surface cracks in steel welds is to analyze the depth of the crack, because it is an effective indicator of its potential risk. This paper proposes a new methodology to obtain an accurate crack-depth prediction model based on the combination of infrared thermography and the 3D reconstruction procedure. In order to do this, a study of the cooling rate of the steel is implemented through active infrared thermography, allowing the study of the differential thermal behavior of the steel in the fissured zone with respect to the nonfissured zone. These cooling rate data are correlated with the real geometry of the crack, which is obtained with the 3D reconstruction of the welds through a macrophotogrammetric procedure. In this way, it is possible to analyze the correlation between cooling rate and depth through the different zones of the crack. The results of the study allow the establishment of an accurate predictive depth model which enables the study of the depth of the crack using only the cooling rate data. In this way, the remote measure of the depth of the surface steel crack based on thermography is possible.

1. Introduction

Welding is the most important joining process for metallic materials but is also an aggressive technique, taking into account the fact that an important heating process is applied in order to meld the material and that high thermomechanical stresses induced by the welding procedure could cause several security problems. Defects present in the material are harmful and dangerous for the integrity of the joints. The most commonly observed welding flaws include the following: lack of fusion, lack of penetration, gas holes, porosity, cracking process, and inclusions, which are all typified in the international quality standards [1]. Some types of defects may occur more frequently than others for a particular welding process. In order to maintain the desired level of structural integrity, welds must be inspected in accordance with the quality standards. The results of the inspection of welds also provide useful information to identify potential problems in the manufacturing process and to improve the welding operations [2]. In the current industry and building practice, welding inspection is the responsibility of inspectors certified according to international standards [3].

In these cases, the cracking process is critical because the quality of the result may influence the failure of an important structural element and, consequently, the full collapse of the structure with drastic consequences. Welds are the origin of structural feebleness in the majority of cases and should be systematically checked in order to ensure the structural integrity of the components. Therefore, establishing a detailed geometric study of the crack is really important, since the knowledge of the size of the crack (measures along surface and depth) presents enormous importance for the evaluation of materials. If the depth of the crack is not known, the rejection of the material will be automatic due to the impossibility of predicting its risk; else, if the depth of the crack is known, this information could be used to calculate the risk of maintaining the material and to assess the possibility of its rejection [4]. Thus, the possibility of measuring accurately the depth of small cracks is highly useful, since they represent the least dangerous factor and will therefore present a more likely probability to be accepted without removal [5]. Moreover, the crack depth would increase with the crack propagation length [6].

The methods most frequently used for the measurement of surface cracks are ultrasounds [5] and radiography [7]. More novel and sophisticated methods are scanning cameras [8], laser [9], 2D stereoimaging [10], and macrophotogrammetry [11]. Among them, the only one that allows the complete dimensioning for crack depth is the radiographic method [7] although an approximation for the prediction of depth of the crack has been raised with thermography using absolute values of temperature [12].

Infrared Thermographic (IRT) methods have also been used in the welding field for the control and metallurgical characterization of the welding process through the study of emissivity [13]. IRT has been used to inspect and measure the adhering slag on the weld during the welding process [14]. Another important use of the thermographic methods is the inspection and evaluation of materials, mainly composite materials. However, their application to inspect steel welds is not as settled as the previous methods. Relevant research shows positive results, ensuring a promising future for the application of the technique in testing engineering. The most extended modality of thermography used to evaluate materials is the one denominated as *active thermography*, which is based on the study of the thermal reaction of the material when external signal stimulation is applied. The signal can be applied in different ways, such as heat pulses [15], one-frequency modulated signal, also known as lock-in method [16], mechanical vibrations [17], ultrasounds [18], and microwaves [19]. In this paper, authors apply a new approach to establish a protocol for the prediction of crack depth from cooling rate data. Cooling rate data are extracted for each pixel from a sequence of thermal images using an algorithm (*pixelwise algorithm for time derivative of temperature*) based on the experimental proposal of [20]. Cooling rate data are integrated with geometrical depth data extracted following the macrophotogrammetric procedure established for welding in [11]. The objective is to establish a prediction model that allows the estimation of depth of the crack using an IR camera and low-temperature long-pulsed heating/cooling of the metal.

This paper is organized as follows: after this introduction, Section 2 presents the equipment used and the testing methodology; Section 3 analyzes the results obtained and the information gathered from the combined thermographic and geometric knowledge of the defects detected with thermography. Last, Section 4 explains the conclusions drawn from the presented study.

2. Materials and Methods

The materials needed and the methods followed to implement the procedure are shown in this section.

2.1. Materials. A welded plaque of low carbon with a thickness of 7.5 mm was used as subject of the study, given the presence of a crack with a little notch in the face of weld (Figure 1).

TABLE 1: Parameters for the geometric calibration of the IR camera NEC TH9260.

Focal length (mm)	14.35 ± 0.44
Format size (mm)	$5.98 \times 4.50\ (\pm 0.07)$
Pixel size (mm)	$9.34 \cdot 10^{-3} \times 9.37 \cdot 10^{-3}$
Principal point (X_{PP}) (mm)	2.99 ± 0.07
Principal point (Y_{PP}) (mm)	2.28 ± 0.09
Lens distortion	
K_1	$-1.32 \cdot 10^{-2}$
K_2	$1.08 \cdot 10^{-3}$
P_1	$-3.43 \cdot 10^{-4}$
P_2	$1.44 \cdot 10^{-3}$

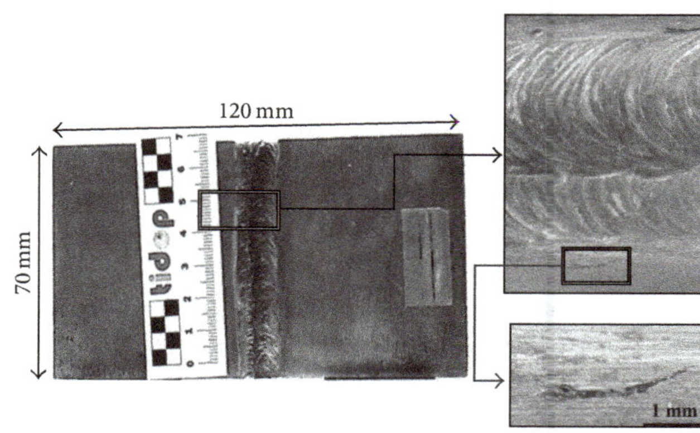

FIGURE 1: Low-carbon steel weld used as specimen. This presents a toe crack close to the limit of the weld (macroimage).

The plaque has been welded with Tungsten Inert Gas (TIG) welding, presenting butt-welding with edge preparation in V. The crack is oriented parallel to the longitudinal axis of the weld and is consequently denominated as *toe crack* according to the quality standard [21].

The thermal analysis is performed with an IR (infrared) camera. The IR camera used for this work is NEC TH9260 with 640 × 480 Uncooled Focal Plane Array Detector (UFPA), with a resolution of 0.06°C and a measurement range from −40°C to 500°C. The camera is geometrically calibrated prior to data acquisition using a calibration grid based on the emissivity difference between the background and the targets, presented in [22]. The calibration parameters of the IR camera in the focus position used for data acquisition during the thermographic tests are shown in Table 1.

Finally, the photogrammetric process has been applied following the methodology established in [11]. For the implementation, a commercial digital single lens reflex (DSLR) camera, Canon EOS 500D, with a Sigma 50 mm macro lens is used. A tripod is used to stabilize the camera during acquisition due to the high exposure times required to get the correct lighting exposition. In order to homogenize and optimize the illumination conditions, two halogen lamps (50 W each) are also used.

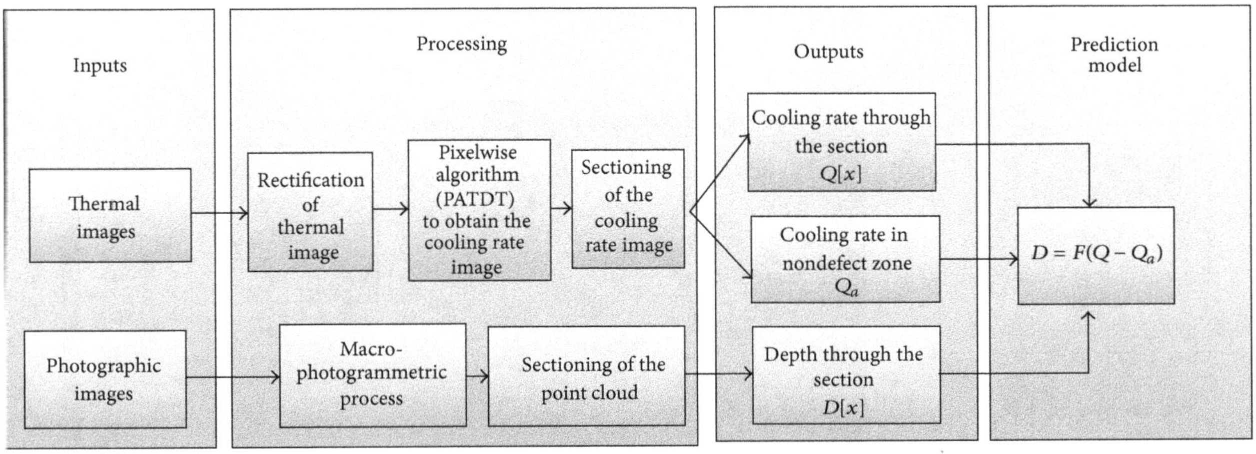

FIGURE 2: Methodology followed to obtain the prediction model. Inputs are the thermal and photographic images. Processing has two parts: the thermal procedure (firstly, the rectification of the image is implemented, after a PATDT algorithm is applied in order to obtain the cooling rate for each pixel of the image; subsequently, sectioning of the image through different lines is implemented in order to get discrete cooling rate functions) and the macrophotogrammetric procedure [10]. Finally, the processing of the inputs allows obtaining the cooling rates and depths through a concrete different section of the crack; with these outputs, a mathematical approach is applied in order to obtain the prediction model.

2.2. Methodology. The methodology applied for this research presents two different parts (Figure 2): on the one hand, a thermal processing is applied to extract the cooling rate for each pixel of the thermal matrix through two different sections; on the other hand, a photogrammetric procedure is applied to obtain the real depth values of the cracks through the two mentioned sections. The objective is to compare the two types of results in order to study the possible relation between them through a linear mathematical approach.

2.2.1. Thermal Data Procedure to Extract the Cooling Rates

(1) Acquisition of Thermal Images. The data acquisition protocol starts with the heating of the weld by a Joule effect heater until 40°C. The superficial temperature is controlled with a contact thermometer TESTO720 with Pt-100, resolution 0.1°C, and accuracy ±0.2°C. The thermometer is held on the surface of the plaque, in order to ensure total contact with the plaque and avoid the interference of the ambient conditions in the measurement. When the desired temperature is obtained, the heater is switched off and the IR camera is switched on. The monitoring of the cooling period is made with 50 thermograms, one every 5 seconds. Finally, the thermal images are extracted in RAW format, which implies one temperature value for each pixel, so that each image constitutes a matrix with the temperature values of the weld in the time of the acquisition.

(2) Rectification of Thermal Images. The objective of the rectification is to scale the thermal images to the real geometry, allowing the establishment of a correlation between temperature and depth values, the latest coming from the macrophotogrammetric procedure. If the thermal images have 2D metric, their different points can be related with the 3D metric dense point cloud obtained with macrophotogrammetry. The first step consists of the extraction of the temperature matrix of each thermographic image: each position in the matrix contains the temperature value of the corresponding part of the object contained in the image pixel. Values are corrected on an emissivity basis, using as reference the temperature values measured at the beginning of the test (prior heating) with the contact thermometer. The emissivity value is calculated using Stefan-Boltzmann's Law, and the correction is applied to the matrix for obtaining the real temperature values.

Once the temperature values are corrected, the infrared image matrix is subjected to a rectification algorithm. The core algorithm of image rectification is the plane projective transformation, ruled by

$$X = \Delta x + \frac{a_0 + a_1 x' + a_2 y'}{c_1 x' + c_2 y' + 1}$$
$$Y = \Delta y + \frac{b_0 + b_1 x' + b_2 y'}{c_1 x' + c_2 y' + 1}, \quad (1)$$

where X and Y are the rectified (real) coordinates of the element, x' and y' are the pixel coordinates in the image, and a_0, a_1, a_2, b_0, b_1, b_2, c_1, and c_2 are the mathematical coefficients of the projective matrix that enclose rotation, scale, translation, and perspective.

Therefore, the knowledge of the coordinates of 4 points in the object is the only requirement for the determination of the projective matrix, as well as the geometric calibration parameters of the camera $(\Delta x, \Delta y)$.

(3) Pixelwise Algorithm for Time Derivative of Temperature (PATDT). Once images have been rectified, the next step is to apply an algorithm based on the approach developed in [21]. In this way, the cooling rate of the heated

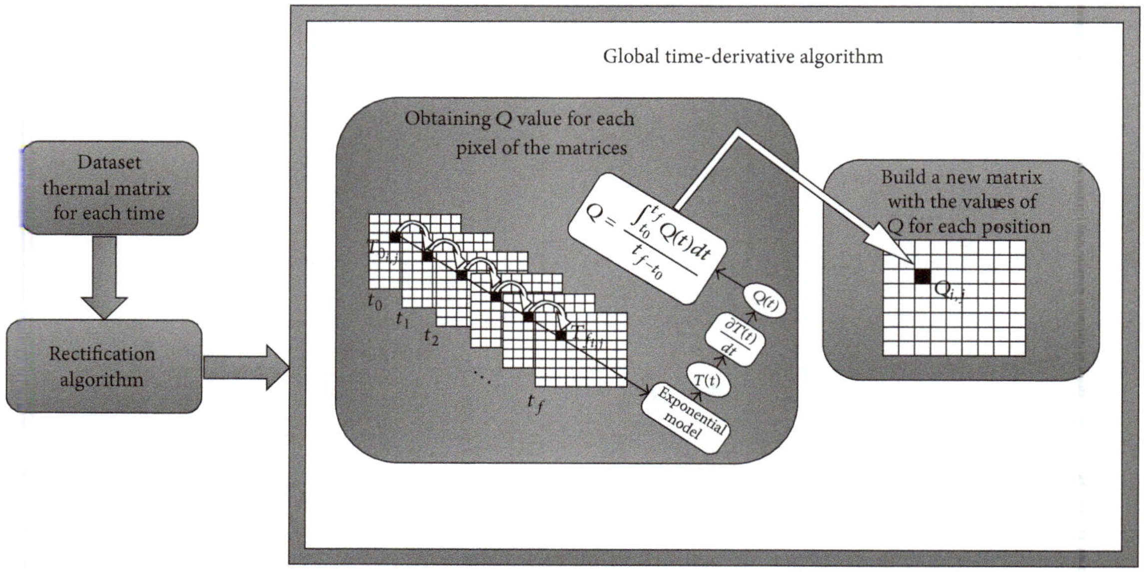

FIGURE 3: Steps followed by the global time-derivative algorithm. After data acquisition and rectification, the exponential model is applied for each (i, j) position with the values of temperature for each time instant. As a result, a temperature function is obtained and the time derivate of the temperature function is the cooling function, which is averaged for each position of the matrix, obtaining a matrix of cooling rates.

steel is analyzed according to Newton's Cooling Law. This law indicates that temperature decreases exponentially. This procedure based on the monitoring of the cooling allows the detection of defects without using complex processing algorithms (Figure 3), like those indicated in Section 1.

The PATDT applies Newton's Cooling Law for each pixel of the thermal image using an exponential fit model for the temperature-time data of each thermogram. Using this approach, a cooling rate function $Q(t)$ is established for each pixel as time derivate of the temperature function, $T(t)$. The sequence followed by the algorithm is the following (Figure 3):

(a) An exponential fit according to Newton Law is established for each temperature value $T_0, T_1, T_2, \ldots, T_j$, associated with each time instant $t_0, t_1, t_2, \ldots, t_f$ and for each pixel position of matrix, obtaining an expression $T(t)$ for each pixel position.

(b) The derivation $\partial T(t)/dt$ is calculated in order to obtain the cooling rate function $Q(t)$ for each pixel position.

(c) An integrated average Q for $Q(t)$ between the initial and final time is calculated for each pixel position.

A new matrix with the same dimensions as the thermal matrix is created. The integrated average cooling rate $Q_{i,j}$ is introduced on each i, j position of the matrix obtaining a cooling rate matrix.

In order to work with relative cooling rate data, each cooling rate value corresponding to the crack (Q_{ij}) is subtracted from the average of immediate exterior pixels Q_a *from a 10-pixel zone placed immediately near the crack*. In this way, the relative cooling rate is obtained for each pixel of the crack and a relative cooling rate matrix is obtained as $Q_{ij} - Q_a$, being i, j the dimensions of the crack submatrix.

2.2.2. Macrophotogrammetric Procedure to Extract Depth Data. The generation of the photogrammetric 3D model of the crack is done following the procedure established in detail by the authors in [11]. The first step is the acquisition of photographic images with a DSLR camera following a semispherical trajectory centered on the object. Once images are acquired, keeping always a constant distance between the lens and the object, two processing steps are applied: first, the automatic determination of the spatial and angular positions of each image, regardless of the order of acquisition and without requiring initial approximations or camera calibration, and second, the automatic computation of a dense 3D point cloud (submillimetric resolution), so that each pixel of the image renders a specific point of the model of the weld (Figure 4).

2.2.3. Correlation between Cooling Rate and Depth Data for Each Section for the Generation of the Model and Validation. When the cooling rate function for each section is obtained following the thermal procedure established in Section 2.2.1 and the depth function for each section is extracted following the macrophotogrammetric procedure explained in Section 2.2.2, the correlation between the points of both functions is established for every two different longitudinal sections in order to design a mathematical lineal model that allows the crack-depth prediction based on the cooling rate (Figure 2). Two transversal sections are used for the external validation of the model. Additionally, the method is applied to other cracks in other welds in order to establish an external validation of the method (Figure 4).

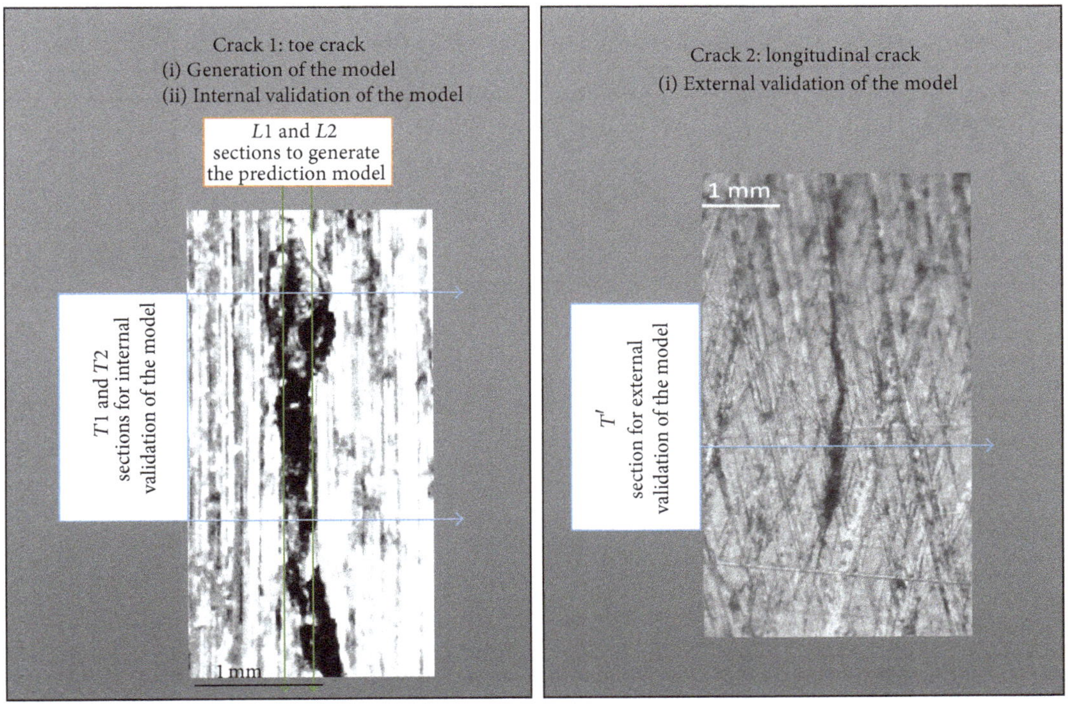

FIGURE 4: Cracks with sections used to generate the model ($L1$ and $L2$), internal validation ($T1$ and $T2$), and external validation (T').

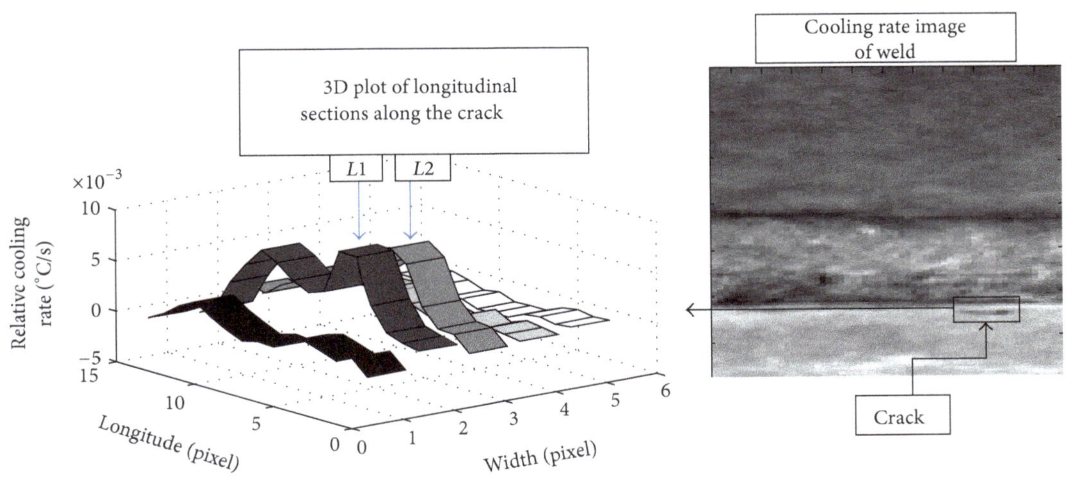

FIGURE 5: 3D representation of the crack zone based on the relative cooling rate matrix obtained following the procedure illustrated by Figure 3. In Figure 6, the two pixel rows of the crack ($L1$ and $L2$) are labeled.

3. Results

3.1. Cooling Rate Results. The graphical representation of the relative cooling rate matrix segmented in the crack region in 3D is shown in Figure 6. The average cooling rate for the adjacent zone to the defect, Q_a, is 0.0149°C/s. The cooling rate through two sections ($L1$ and $L2$) is studied in Figure 6.

3.2. Depth Results. The macrophotogrammetric model for the crack is obtained from the matching of images, which are acquired from convergent positions with respect to the weld, repeating the procedure exposed in [11].

For the computation of depths, the point cloud from the macrophotogrammetric procedure is sectioned in the two longitudinal sections ($L1$ and $L2$) (Figure 4). The section of the point cloud with normal planes $L1$ and $L2$ is applied in order to extract the depth for each pixel (Figure 5). The metrical reference between depths on the dense point cloud and the cooling rate values is made using the distances from the extremes of the crack.

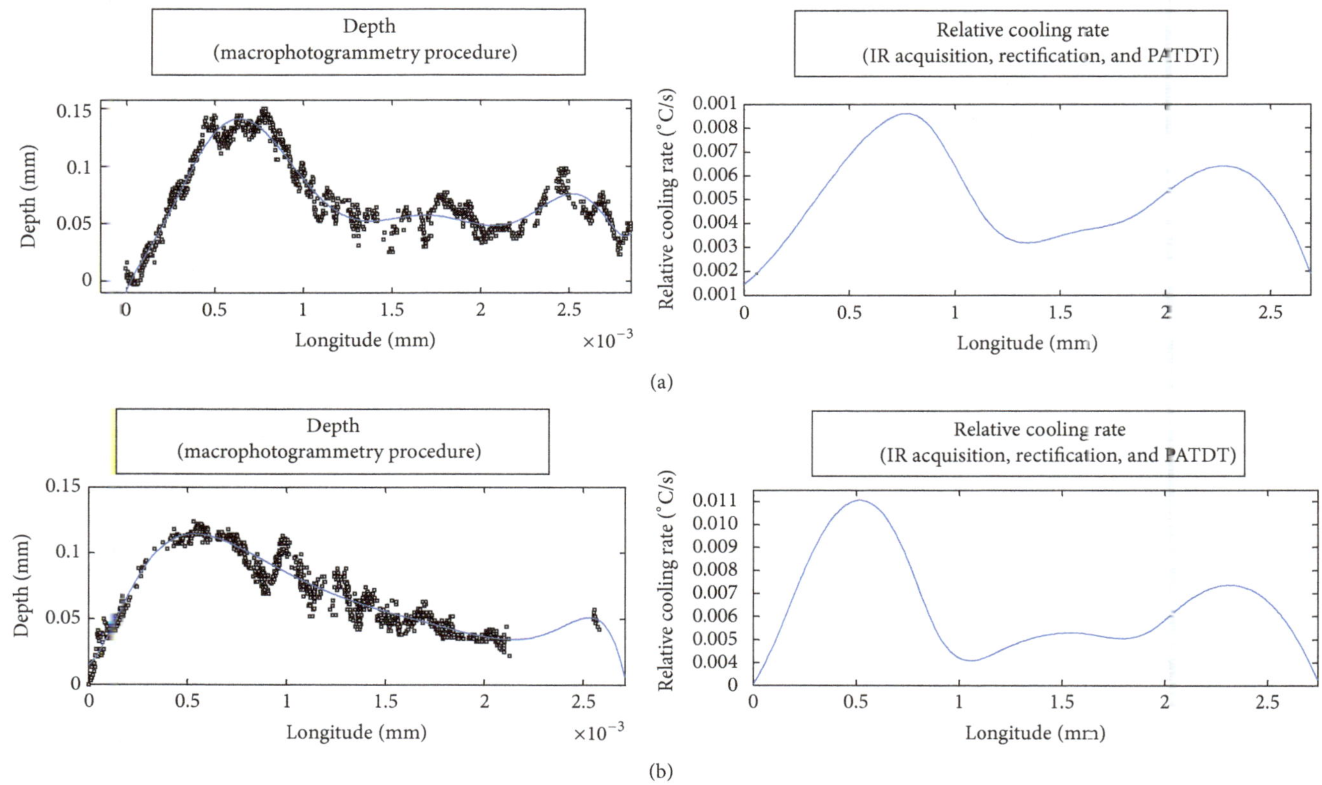

FIGURE 6: Depth function and relative cooling rate function along the longitudinal sections ($L1$ (a) and $L2$ (b)).

The fit is used because the number of points is higher than the number of pixels on the cooling rate image. For this reason, authors have chosen to work with continuous functions in order to ease the correlation of data. The fit model used to obtain a depth function that allows the estimation of the real depth is a 9-degree polynomial fit, which provides an acceptable goodness (Root Mean Square Error (RMSE) of 0.0098 for longitudinal section $L1$ and RMSE of 0.0082 for longitudinal section $L2$).

3.3. Correlation between Cooling and Depth Data. When relative cooling rate and depth data have been obtained through each section (Figure 6), each pair of values (depth-relative cooling rate) is extracted from 20 longitude values and correlated in order to establish a prediction model which allows the estimation of the depth using just the cooling rate value.

The correlation between depth and relative cooling rate $(Q - Q_a)$ (Figure 7) is obtained with a fitting error of 0.86 (Table 2). For this reason, the prediction model is considered as acceptable, which demonstrates the existence of a coherent mathematical relationship between depth (mm) and cooling rate (°C/s).

TABLE 2: Parameters of the lineal fit which correlates relative cooling rate and depth.

$D = a(Q - Q_a) + b$		
Gain (a)	Offset (b)	R^2
15.71	−0.002534	0.857

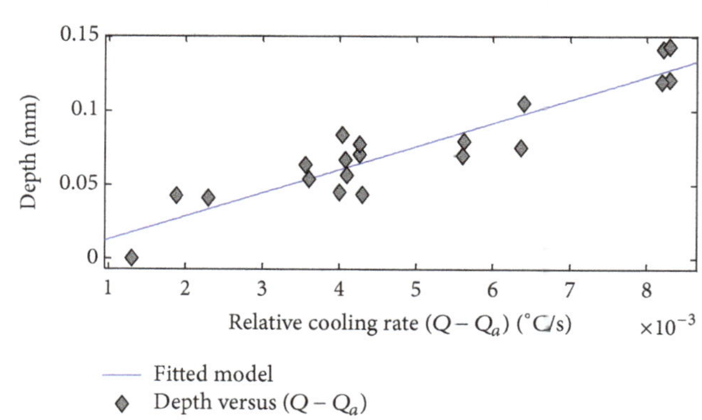

FIGURE 7: Correlation between depth and relative cooling rate for each pixel in the crack.

4. Validation

4.1. Internal Validation of the Prediction Model. In order to test the validity of the model, an internal validation procedure is applied through the analysis of two transversal sections ($T1$ and $T2$) in the same crack (Figure 4). These sections are performed in a direction where the difference of depth between positions is the highest. The aim of this step is

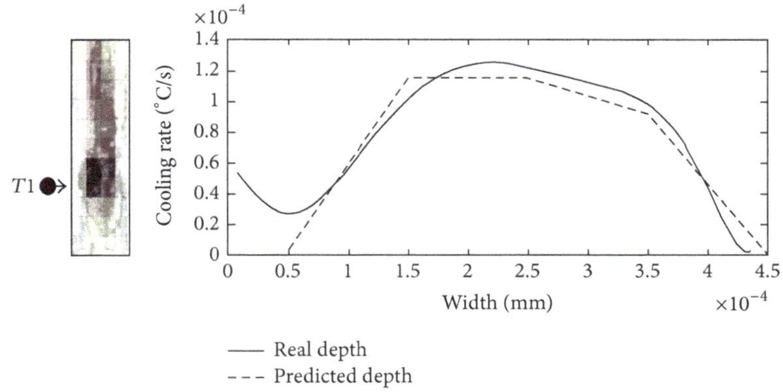

(a)

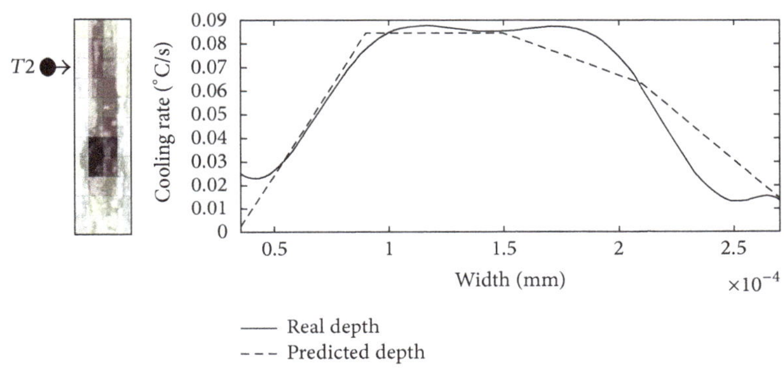

(b)

FIGURE 8: Transversal sections used for the internal validation of the model ($T1$ (a) and $T2$ (b)).

to analyze the correspondence between the predicted depth values $P[x]$ (where x is the width) calculated with the prediction model (Table 2) from the cooling rate data with the real depth values $V[x]$ extracted from the sectioning of the point cloud. These sections are significantly smaller than the longitudinal sections used to build the prediction model (Figure 4) but present a higher contrast of the cooling rate. For this reason, the application of the prediction model to these points belonging to these transversal sections serves to analyze the internal coherence of the model and thus the goodness of the fit.

The prediction model is applied to each transversal section (Figure 8). Then, the prediction model is compared with the real model in order to analyze the accuracy. The error is calculated for each x value with the following discrete expression:

$$\varepsilon = \sum_{x_0}^{x_f} |V[x] - P[x]|, \tag{2}$$

where V is the real value of depth and P is the predictive value (both for the same value of x). The error of the prediction model for the first transversal section, $T1$ (Figure 8), is 0.00169 mm (14%) and the error for the second transversal section, $T2$, is 0.0010 mm (13%).

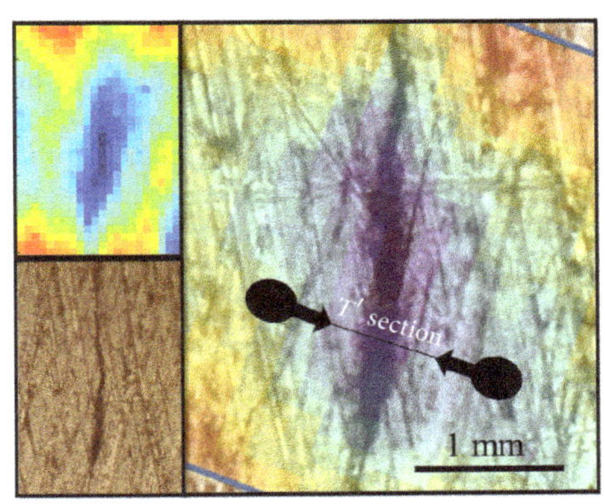

FIGURE 9: Cooling rate matrix of the new crack, with pseudocolor assigned to each cooling rate value (top left) and macrophotography of the crack (bottom left). Overlapping of cooling rate map with macrophotography of the crack in order to correlate the different depths of the crack with the different cooling rate data (right). The transversal section T' used to assess the prediction model has been plotted.

4.2. External Validation of the Prediction Model. In this section, the robustness of the prediction model is tested through its application to the estimation of the depth in a crack open to surface in a different plate, made of the same material (Figure 9). In this case, the crack is a longitudinal crack open to surface placed on the weld (weld cap removed) and generated by the stress provoked during the welding process. This crack is significantly smaller (tenths of mm) than the crack used to build the model, which is the reason for its choice. In this way, authors seek to test the model under stringent requirements.

The macrophotogrammetric procedure is repeated for this new crack in order to obtain accurately the depth distribution of the crack (denominated as *real depth*). Then, the plate is heated at 70°C. This temperature, higher than in the previous case, is necessary in order to obtain a correct visualization of the crack in the thermal matrix, given the small size of this crack. The cooling period is monitored for 250 s (the same period). The cooling rate is calculated from the thermal images for each time instant following the procedure established in Section 2.2.1(3). Subsequently, the average cooling rate for the adjacent zone of the crack, Q_a, is calculated. The result is 0.0516°C/s. Then, the prediction model calculated in Section 3.3 is applied through the values of transversal section of the crack (called section T') and the depth values are estimated for each pixel from the cooling rate data (Figure 9).

At this point, real and prediction models are compared in order to analyze the error of the predicted model regarding the reality. Both models are shown in Figure 10. The resulting error of the prediction model is 0.009 mm (18%). This result is acceptable knowing that the maximum depth in the crack is 0.05 mm.

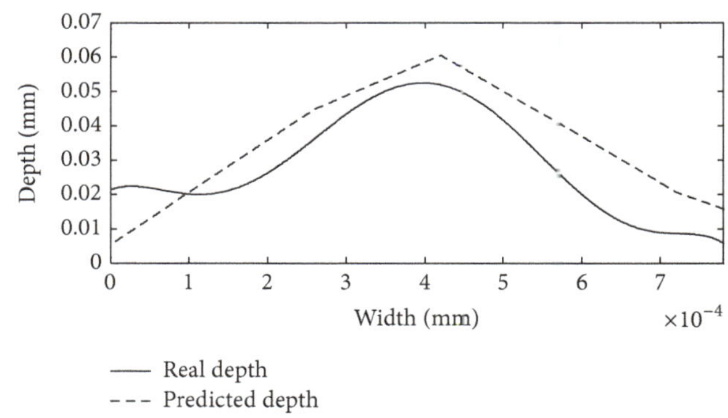

Figure 10: Real and predicted depth through the transversal section T' used for the external validation of the model.

5. Conclusions

A procedure for the prediction of depth of cracks from infrared data has been implemented. This approach represents a methodological innovation in the field of Nondestructive Testing for the evaluation of weld flaws.

The procedure applied allows the accurate prediction of different depths of the crack from thermal images. Low-temperature heating (40°) is applied to the steel and the cooling is monitored for 250 s. Although the heat transfer tends to be homogenized through the material in long cooling times, the values of the cooling rate also depend on the surface properties for each point of the material. Therefore, different surface properties in the defect and the nondefect zone provide a difference in the heat dissipation which provokes a difference in the cooling rates. This difference is clearly visible for each pixel through the sections of thermal images. These thermal data are associated with the real depths obtained with a reliable and experienced macrophotogrammetric procedure [10] using a commercial DSLR camera.

The cooling rate value for each pixel is different for each time moment, since the cooling rate is higher at the beginning of the cooling process than at its end due to the exponential reduction of the temperature. The registration and integration of the cooling rate values from 250-second cooling interval allows the registration of the totality of information. For this reason, the homogenization of the heat transfer through the material is not a limit for the methodology, since results present enough contrast to establish a depth prediction model.

The prediction model established through the correlation between cooling rate and depth maintains statistical consistency allowing an acceptable fit (RMSE between 0.008 and 0.009). Furthermore, the prediction model has been validated following a twofold approach: on the one hand, an internal validation has been implemented, consisting of the application of the prediction model to the crack in two transversal sections (different sections to those used to obtain the prediction model). These transversal sections have been chosen since they present high thermal contrast in a minimal space, subjecting the prediction model to a limit situation. Results demonstrate the good precision of the method to predict depth values of cracks, with 15% of maximum deviation. On the other hand, an external validation has been applied through the analysis of a different crack. In this case, a crack open to surface located in cap removing weld has been used to provide the external validation. With this test, authors analyze not only the validity of the method for other cracks, but also the quality of the prediction model when the heating is done at different temperatures (in this case, higher, 70°C). Results of the prediction model are consistent, with 18% of maximum deviation regarding real depth. The error is mainly due to the interpolation method used for the relation of thermal and depth values. It could be improved using more discrete values and this is a future goal that this work opens.

The procedure has been applied to cracks generated from the stress provoked by the welding process in low-carbon steel. However, the procedure could be extended to the analysis of cracks in other materials. The cooling phenomenon is studied by the Newton Law and depends on the physical property of the materials as conductivity, heat capacity, surface features, and so forth. For this reason, the extension of the technique to other materials will be researched in future works.

Conflict of Interests

The authors declare that there is no conflict of interests regarding the publication of this paper.

Acknowledgment

The authors would like to thank Agencia de Innovación y Financiación Empresarial de Castilla y León (ADE), Ministerio de Economía y Competitividad (Gobierno de España), for the financial support given through human resources grants (FPDI-2013-17516).

References

[1] ISO, "Welding. Fusion-welded joints in steel, nickel, titanium and their alloys (beam welding excluded). Quality levels for imperfection," ISO 5817:2009, European Committee for Standardization, 2009.

[2] T. W. Liao, "Classification of weld flaws with imbalanced class data," *Expert Systems with Applications*, vol. 35, no. 3, pp. 1041–1052, 2008.

[3] IAB, "Minimum Requirements for the Education, Examination and Qualification," Tech. Rep. 252r2-14/SV-00, IAB—International Authorization Board, 2014.

[4] BS, "Guide to methods for assessing the acceptability of flaws in metallic structures," Tech. Rep. BT7910:2013, The British Standards Institution, 2013.

[5] M. V. Felice, A. Velichko, and P. D. Wilcox, "Accurate depth measurement of small surface-breaking cracks using an ultrasonic array post-processing technique," *NDT & E International*, vol. 68, pp. 105–112, 2014.

[6] H.-S. Huang, "Fracture characteristics analysis of pressured pipeline with crack using boundary element method," *Advances in Materials Science and Engineering*, vol. 2015, Article ID 508630, 13 pages, 2015.

[7] E. Maire, J.-Y. Buffière, L. Salvo, J. J. Blandin, W. Ludwig, and J. M. Létang, "On the application of X-ray microtomography in the field of materials science," *Advanced Engineering Materials*, vol. 3, no. 8, pp. 539–546, 2001.

[8] S. Henkel, D. Holländer, M. Wünsche et al., "Crack observation methods, their application and simulation of curved fatigue crack growth," *Engineering Fracture Mechanics*, vol. 77, no. 11, pp. 2077–2090, 2010.

[9] L. Zhang, W. Ke, Q. Ye, and J. Jiao, "A novel laser vision sensor for weld line detection on wall-climbing robot," *Optics and Laser Technology*, vol. 60, pp. 69–79, 2014.

[10] M. Dinham and G. Fang, "Autonomous weld seam identification and localisation using eye-in-hand stereo vision for robotic arc welding," *Robotics and Computer-Integrated Manufacturing*, vol. 29, no. 5, pp. 288–301, 2013.

[11] M. Rodríguez-Martín, S. Lagüela, D. González-Aguilera, and P. Rodríguez-Gonzálvez, "Procedure for quality inspection of welds based on macro-photogrammetric three-dimensional reconstruction," *Optics & Laser Technology*, vol. 73, pp. 54–62, 2015.

[12] M. Rodríguez-Martín, S. Lagüela, D. González-Aguilera, and J. Martinez, "Prediction of depth model for cracks in steel using infrared thermography," *Infrared Physics & Technology*, vol. 71, pp. 492–500, 2015.

[13] R. Frappier, A. Benoit, P. Paillard, T. Baudin, R. Le Gall, and T. Dupuy, "Quantitative infrared analysis of welding processes: temperature measurement during RSW and CMT-MIG welding," *Science and Technology of Welding and Joining*, vol. 19, no. 1, pp. 38–43, 2014.

[14] S. Nagaraju, A. Rajadurai, M. Vasudevan, M. Menaka, and R. Subbaratnam, "Novel method for quantitative assessment of slag detachability in austenitic stainless steels welds made by SMAW," *Science and Technology of Welding and Joining*, vol. 13, no. 8, pp. 739–743, 2008.

[15] X. Maldague, F. Galmiche, and A. Ziadi, "Advances in pulsed phase thermography," *Infrared Physics & Technology*, vol. 43, no. 3–5, pp. 175–181, 2002.

[16] G. Busse, A. Gleiter, and C. Spiessberger, "NDE using lockin-thermography: principle and recent developments," in *Nondestructive Testing of Materials and Structures*, vol. 6 of *RILEM Bookseries*, pp. 627–632, Springer, Dordrecht, The Netherlands, 2013.

[17] A. Castelo, A. Mendioroz, R. Celorrio, and A. Salazar, "Vertical cracks characterization and resolution from lock-in vibrothermography," in *Proceedings of the 12 International Conference on Quantitative Infrared Thermography*, Bordeaux, France, July 2014.

[18] A. Mendioroz, A. Castelo, R. Celorrio, and A. Salazar, "Characterization and spatial resolution of cracks using lock-in vibrothermography," *NDT & E International*, vol. 66, pp. 8–15, 2014.

[19] D. Palumbo, F. Ancona, and U. Galietti, "Quantitative damage evaluation of composite materials with microwave thermographic technique: feasibility and new data analysis," *Meccanica*, vol. 50, no. 2, pp. 443–459, 2014.

[20] M. Rodríguez-Martín, S. Lagüela, D. González-Aguilera, and P. Arias, "Cooling analysis of welded materials for crack detection using infrared thermography," *Infrared Physics and Technology*, vol. 67, pp. 547–554, 2014.

[21] ISO, "Welding and allied processes—classification of geometric imperfections in metallic materials—part 1: fusion welding," EN-ISO 6520-1:2009, European Committee for Standardization, 2009.

[22] S. Lagüela, H. González-Jorge, J. Armesto, and J. Herráez, "High performance grid for the metric calibration of thermographic cameras," *Measurement Science and Technology*, vol. 23, no. 1, Article ID 015402, 2012.

2

Enhancing Ductility of 1045 Nanoeutectic Steel Prepared by Aluminothermic Reaction through Annealing at 873 K

Zhengning Li, Fuan Wei, Peiqing La, Hongding Wang, and Yupeng Wei

State Key Laboratory of Advanced Processing and Recycling of Nonferrous Metals, Lanzhou University of Technology, Lanzhou 730050, China

Correspondence should be addressed to Peiqing La; pqla@lut.cn

Academic Editor: Pavel Lejcek

The 1045 steel with lamellar spacing of pearlite in nanometer was prepared by aluminothermic reaction casting and annealed at 873 K (600°C) with different time. Microstructures of steels were investigated by X-ray diffraction (XRD), scanning electron microscope (SEM), and transmission electron microscope (TEM). Tensile properties of the steels were measured. The results showed that the lamellar spacing of the pearlite increased with the annealing time. It was found that the microstructure of steels consisted of nanocrystalline-ferrite matrix and laminar pearlite phase. The average grain sizes of the ferrite were 26.9, 27.0, 26.1, and 34.9 nm for the cast steel and samples annealed for 2, 4, and 6 h, respectively. As the annealing time increased, the volume fraction of the pearlite almost remained constant, while the laminar spacing of pearlite increased from 146 to 300 nm. The tensile and yield strength varied slightly; the elongation obviously improved. After annealing for 4 h, the elongation increased to be 33%, which was the reported highest value for the steel up to now and about twice of the conventional 1045 steel.

1. Introduction

Due to high strength, considerate ductility, and low production cost of the steel, it occupied about 90% of the world's metals production. Owing to potential large decrease of raw material and energy consumption in using, steels with higher strength, considerable ductility, and lower cost are attracting considerable interests [1, 2].

The 1045 carbon steel as an excellent structure material had been widely used in manufactures matches plate and guide pillar of the mould, moving component of the steam turbine, compressor, and pump, such as bent axle, transmission shaft, gear, and worm. The steel had high strength, good machinability, and low cost; after annealing, the steel had considerable toughness, plasticity, and abrasive resistance [3–6]. The improvement of ductility and tensile strength was the most concerned problem in the development of advanced 1045 carbon steel. In metals and alloys, high ductility was often accompanied by low strength, and they rarely can be enhanced simultaneously [4].

The flow stress of the carbon steels was not only connected with the volume fraction of pearlite but also with the microstructure characteristics of the individual phase constituents (which included lamellar spacing and colony size of the pearlite and grain size of the ferrite). For the case of low carbon steels with C (carbon content) $\leq 0.25\%$, the yield strength was mainly affected by the ferrite grain size; while carbon content increased, the contribution of pearlite starts being relevant. And when $C > 0.25\%$, the lamellar spacing (S) becomes important, but usually there was not a direct linear relationship between YS and UTS with $S^{-1/2}$ [7].

In general, strength of steels significantly increased with carbon content; while ductility largely decreased at the same time, there was no method to obtain high ductility in high carbon steel up to now. In present work, high ductility without loss of strength has been successfully obtained in 1045 carbon steel which has lamellar spacing of pearlite in nanometer without any expensive alloying elements [8–14].

2. Methods

Commercial ferric oxide and aluminum powders were weighted according to stoichiometry of the aluminothermic reaction (1); then carbon and silicon powders that were

TABLE 1: The composition of the raw powder (wt.%).

Raw material	Fe_2O_3	Al	Si	Fe	C
Weight (g)	55.20	18.63	0.16	25.77	0.27

TABLE 2: Chemical composition of the raw powders (wt.%).

Raw material	Purity	Impurity
Fe_2O_3	>99%	Cl^-, SO_4^{-2}, N, Cu
Al	>99%	N, Si, Fe, Cu
Fe	>99.9%	P, Si, Cu
Si	>99.9%	Fe, Al, Ca, Cu
C	>99.9%	SiO_2, Al_2O_3, FeO, CaO, P_2O_5, CuO

designed to be 0.45 and 0.25 wt% without Mn element in the produced steel were added in the reactant powders. The composition of the raw powders and purity was shown in Tables 1 and 2, respectively.

$$Fe_2O_3 + 2Al + 1.33Fe = 3.33Fe + Al_2O_3. \quad (1)$$

The powders were dry-mixed at a speed of 150 rpm (rotation per minute) for 8 h using a planetary ball mill with Al_2O_3 spheres. Then the mixed powders were pressed under a uniaxial pressure of 40 MPa as compacts in a steel tube. The substrate and the tube were ultrasonically cleaned in ethanol before using. The compacts were put on copper substrate in a reactor. A lamellar igniter of about 2 g was put on the surface of the pressed powders, which was given in details elsewhere [7–11]. The substrate, stainless steel jars, steel mould, and inner wall of the reactor were cleaned by ethanol before using, and the copper substrate was kept horizontal in the reactor.

The reactor was purged with argon gas at room temperature and heated to 473 K (200°C); then the reactor was purged again and 5 MPa argon gas was introduced. The reactor was continually heated and the igniter was ignited at 533 K (260°C), and released exothermic heat started aluminothermic reaction of the reactant powders. The reactants transformed to Fe (C, Si) and Al_2O_3 melts in areas where the combustion wave passed compacts. The reaction completed in a few seconds, and the products were kept in the reactor to cool down under the pressure of argon gas. The products were taken out from the reactor at room temperature and surrounding of the steel product was a gray layer of Al_2O_3. Due to the thermal expansion coefficient between the Al_2O_3 and the 1045 carbon steel being large, the binding between the carbon steel and Al_2O_3 phase was very weak, and thus Al_2O_3 phase could be removed by hands. The produced steel ingot was about 110 mm in diameter and 10 mm in height. Samples were cut from the steel ingot using wire cutting machine; they were annealed at 873 K (600°C) in a furnace for 2, 4, and 6 h and then cooled to room temperature in the furnace, respectively.

The cross sections of the samples were polished with sand paper and ultrasonically cleaned in ethanol for ten minutes and then examined by D8 X-ray diffraction (XRD) using Cu Kα radiation. Polished cross section of the steels was etched by 4% nitric acid alcohol for 45 s, subsequently, examined by a JSM-6700F field emission scanning electron microscope (FE-SEM).

Lamellar spacing of the pearlite without and with annealed at different time was measured by Image-Pro plus software from high magnification SEM pictures. We drew a line which is perpendicular with the lamellar pearlite in high magnification SEM picture and measured the length of this line (l_0) and then counted the number of the lamellar pearlites (n) which intersects with the line; lamellar spacing of the pearlite includes one lamellar ferrite and one lamellar cementite; the lamellar spacing of the pearlite was $S = l_0/n$. The lamellar spacing of the pearlite was an average value of ten high magnification SEM pictures; the standard deviation was less than 5%.

Sheet specimens cut from the rolled steels were grounded into thin foils with less than 100 μm in thickness by hands and then punched into plates with a diameter of 3 mm by a puncher. The plates were thinned electrolytically by a twin-jet electropolishing device and were examined on a JEM-2010 transmission electron microscope (TEM) with voltage of 200 kV. The electrolyte was a solution of 2 vol.% perchloric acid in ethanol.

Tensile specimens that were 17 mm in length were cut from the steels. The tensile tests were performed on the Shimadzu AT10 t tester with a crossing speed of 0.2 mm min^{-1} and repeated three times; final value was an average of the three times; the standard deviation was less than 5%.

3. Results

XRD pattern of the 1045 carbon steel without and with annealing at 873 K (600°C) for different time is shown in Figure 1; the peaks correspond to ferrite phase, but the peaks of cementite phase are not detected. The average grain size of the ferrite of the cast 1045 steel and the steel annealing with different time was calculated, which was deduced from the Scherrer equation (2):

$$d = \frac{K\lambda}{B\cos\theta}, \quad (2)$$

where d, λ, and θ represent the average grain size, the wavelength of characteristic X-ray, and the diffraction angle, respectively. K is a constant, and B is the half width of the broadening X-ray diffraction by virtue of grain refinement. The average grain size of the ferrite without and with annealing at 873 K (600°C) for different time is 26.9, 27.0, 26.1, and 34.9 nm, respectively. It remains nearly constant before annealing at 600°C with 6 h and then increases with the annealing 6 h. Typical FE-SEM micrographs of the 1045 carbon steel without and with annealing at 873 K (600°C) for different time are shown in Figures 2–5; the phase that looks like island is the pearlite and the other is the ferrite in the low magnification SEM micrographs (Figures 2(a)–5(a)). It also can be demonstrated by energy dispersive spectrometer (EDS) elements analysis of the cast 1045 steel in Figure 6; the area with higher carbon content is pearlite phase, and the other area is ferrite phase. The volume fraction of the pearlite in the steels without and with annealing at 2, 4,

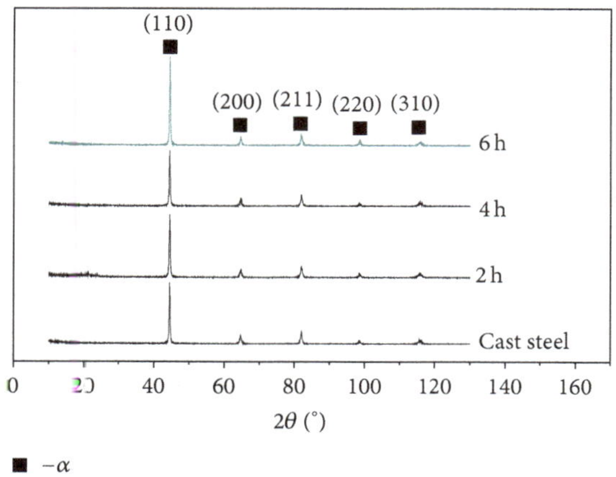

Figure 1: XRD pattern of the 1045 carbon steel without and with annealing at 873 K (600°C) with different time.

and 6 h is calculated by Image-Pro plus software from low magnification SEM pictures are about 22, 20, 22, and 23%; the result showed that the volume fraction of the pearlite of the steels is not changed after annealing. Figures 2(b)–5(b) are higher magnification of the A region in Figures 2(a)–5(a), respectively. Figures 2(c)–5(c) are the higher magnification of the *B* region in Figures 2(b)–5(b), respectively. The measured lamellar spacing of the pearlite in 1045 steels increases with the annealing time, without and with annealing at 2, 4, and 6 h, is about 146, 172, 255, and 300 nm, respectively. Bright field TEM micrographs of the pearlite in steels without and with annealing at 873 K (600°C) with different time are shown in Figure 7; the lamellar pearlites are composed of black stripes and gray matrix; the black stripes are cementite and the gray matrix is the ferrite, which can be demonstrated from SAED (selected area electron diffraction) image of the pearlite in steel with annealing at 873 K (600°C) for 2 h in Figure 8. Figures 8(b) and 8(d) are the SAED image of the red region in Figures 8(a) and 8(c), respectively; the SAED patterns are indexed to be cementite and ferrite, respectively, and the weak diffraction halo indicates the presence of a small amount of noncrystalline phase. The lamellar spacing of the pearlite measured from bright field of TEM micrographs, and the result complies with the value measured from SEM micrographs.

Tensile curves of the 1045 steel without and with annealing at 873 K (600°C) for different time are shown in Figure 9 and summarized in Table 3. The elongation increases from 15% of the cast steel to 33% when the annealing time is 4 h, and the tensile and yield strength of the steels almost are not changed with the annealing time increases; when the annealing time increases to 6 h, the elongation of the steel decreases to 22%; the tensile and yield strength of the steel decrease a little. The steel with annealing at 873 K (600°C) for 4 h has highest elongation of 33%. The elongation is the highest value in 1045 steels, and about twice for conventional 1045 carbon steel but the strength of them is the same [13].

The fracture appearance of 1045 carbon steels without and with annealing at 873 K (600°C) with different time is shown in Figure 10; fracture surface of the cast steel is uneven

Table 3: Tensile properties of the 1045 steel without and with annealing for different time.

	Tensile strength (MPa)	Yield strength (MPa)	Total elongation (%)
The cast 1045 steel	490	265	16
2 h	483	264	25
4 h	488	267	33
6 h	448	255	22

and like lamellar tearing, when the steel is annealed at 873 K (600°C) for 4 h, there are a lot of dimples that appear on the fracture surface of the steel.

4. Discussions

During the process of aluminothermic reaction casting, the temperature of the products was calculated to be about 3040 K (2767°C) and the melts were superheated, and it led to a large super cooling degree in solidification process [15]. The melt was cleaned by the large super heating. Therefore, the austenite nucleation rate was very high in the solidification process. And it was beneficial to get much smaller austenite grains; the interfacial area per unit volume increased significantly. It could provide more pearlite nucleation position during the transformation of austenite to pearlite. Thus, the pearlite lamellar spacing was small in nanometer [16].

The pearlite consists of lamellar ferrite and lamellar cementite; the carbon content of them is largely different; the lamellar spacing of the pearlite was determined by carbon diffuses extent; the longer the annealing time is, the more favorable it is for the diffusion of carbon atoms, and the lamellar spacing of the pearlite increased obviously. Therefore, lamellar spacing of the pearlite increased with the annealing time. Due to the pearlite lamellar spacing of the cast steel prepared by aluminothermic reaction casting method was smaller than the conventional 1045 steels in nanometer; the free energy of grain boundary was high, and there was lack of stability of grains; the carbon diffused more fast and lamellar spacing of the pearlite increased faster than the micrometer lamellar spacing steels [17].

The elongation of the steel was greatly enhanced from 16% to 33% through annealing at 873 K (600°C) in comparison to cast steel. There were three major reasons about the improvement of the ductility. At first, large thermal stress formed in the process of the cast steel solidification owing to the fast cooling rate of melts metal; the thermal expansion coefficient between the ferrite and cementite in the pearlite is different. The stress was released with increase of annealing time, which can be demonstrated by XRD in Figure 2; taking (110) peak, for example, the distance of crystal face (*d*) of the steels without and with annealing at different time and the standard distance of crystal face of the steel without any stress were calculated by software Jade 6; the difference between them was given in Table 4. The smaller the difference is, the more the stress is relieved completely.

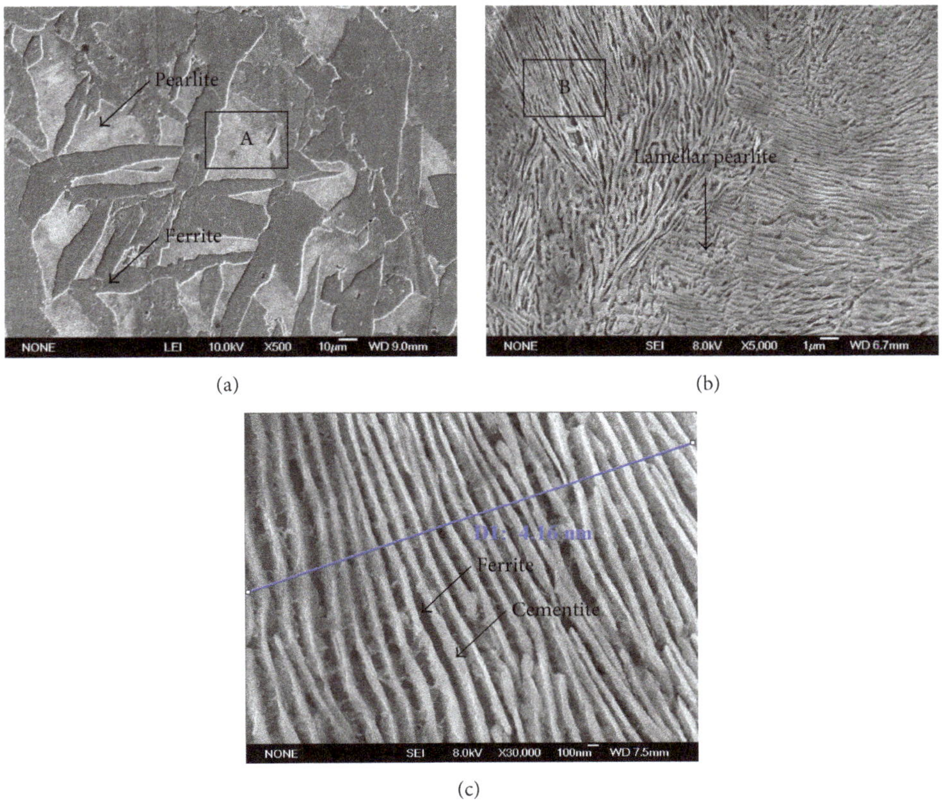

FIGURE 2: Typical FE-SEM micrographs of the cast 1045 carbon steel prepared by the aluminothermic reaction casting ((b) is the higher magnification of the A region in (a); (c) is the higher magnification of the *B* region in (b)).

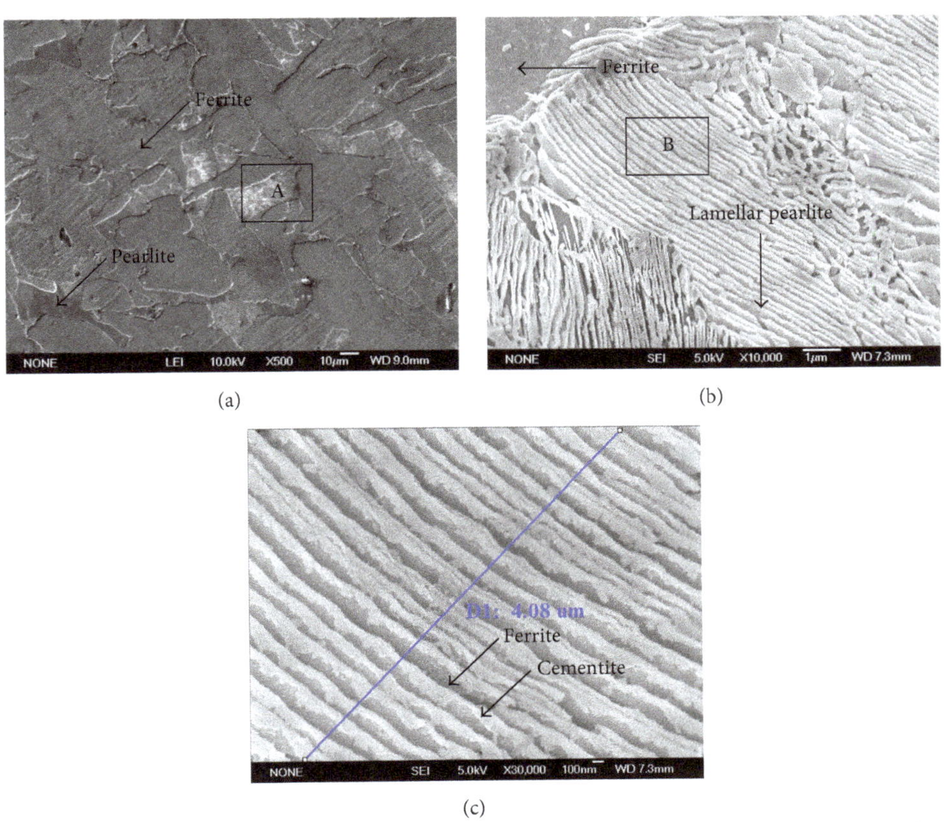

FIGURE 3: Typical FE-SEM micrographs of the 1045 carbon steel after annealing at 873 K (600°C) for 2 h ((b) is the higher magnification of the A region in (a); (c) is the higher magnification of the *B* region in (b)).

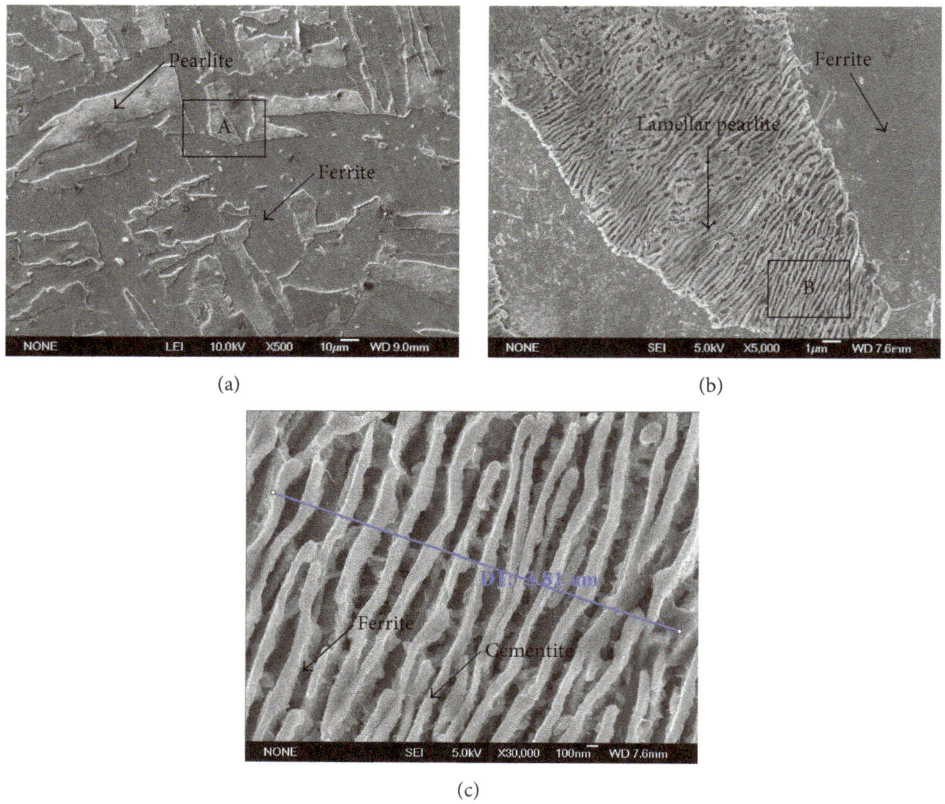

FIGURE 4: Typical FE-SEM micrographs of the 1045 carbon steel after annealing at 873 K (600°C) for 4 h ((b) is the higher magnification of the A region in (a); (c) is the higher magnification of the *B* region in (b)).

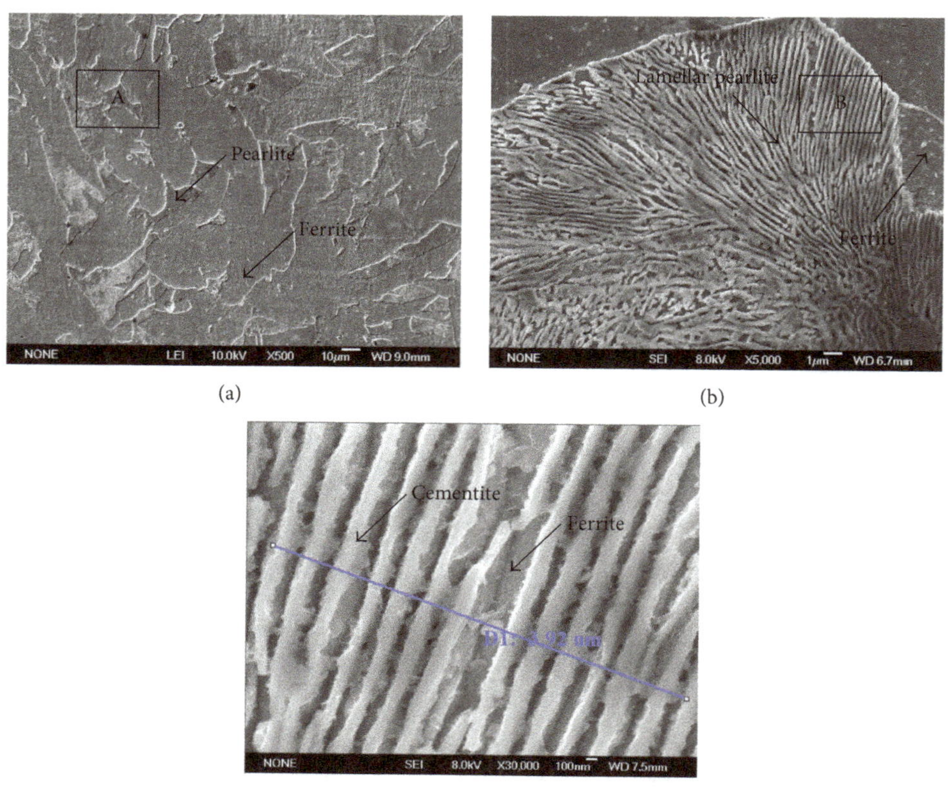

FIGURE 5: Typical FE-SEM micrographs of the 1045 carbon steel after annealing at 873 K (600°C) for 6 h ((b) is the higher magnification of the A region in (a); (c) is the higher magnification of the *B* region in (b)).

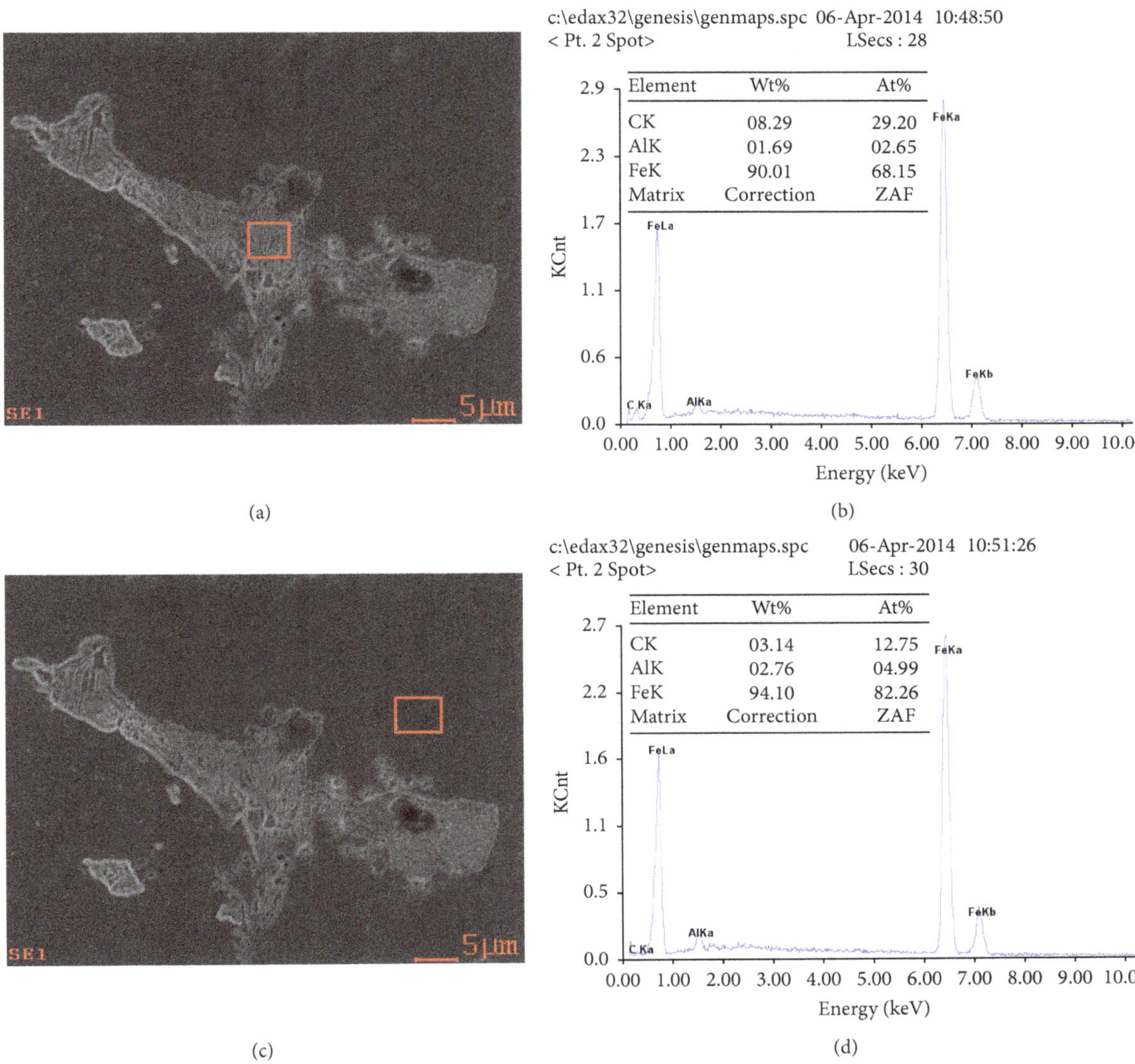

FIGURE 6: EDS elements analysis of the cast 1045 steel ((b) is correspondence of red region in (a); (d) is correspondence of red region in (c)).

Secondly, there is a small amount of the amorphous phase in the cast steel. The amorphous phase transforms into the nanocrystalline phase when the specimen was annealed at 873 K (600°C). Thirdly, nanograined/ultrafine-grained steel exhibited unusual deformation behavior compared with the fine-grained steel, and tensile ductility was is low relatively. For nanograined steel, during deformation, dislocations are trapped at grain boundaries; the time for dislocations to move to grain boundaries is shorter than the time of tensile test. This decrease in dislocation density reduces accumulation of dislocations inside the grains and consequently leads to less work hardening when compared with corresponding steels of coarse grain size. Grain refinement also leads to reduced work hardening capacity. As a result, plastic instability (necking) occurs at an early stage during tensile test, which results in limited uniform elongation in cast steels. The variation of tensile strength, yield strength, and percent elongation with the inverse of the square root of interlamellar spacing (S) was shown in Figure 11. After annealing for 4 h, although the grain size of ferrite remained constant, lamellar spacing of the steel increasing as 255 nm is also smaller than conventional microstructure, and the thermal stress was relieved. The plastic deformation was more uniform and the stress concentration was much decreased compared to the steel with conventional microstructure [6]. In addition, under the action of external stress, cracks were nucleated and grew in the cementite phase which was hard and brittle. The smaller lamellar spacing resulted in more interface area of ferrite and cementite phase in the pearlite, which was resistant to crack propagation owing to the deflection of crack at the interface. Therefore, the elongation of the steels was increased after annealing [4, 7, 18, 19]. For the mechanical properties of the specimens, the yield strength and tensile strength are almost invariable with increase of the annealing time, which is attributed to the few increase in the lamellar spacing of the pearlite and the grain size of

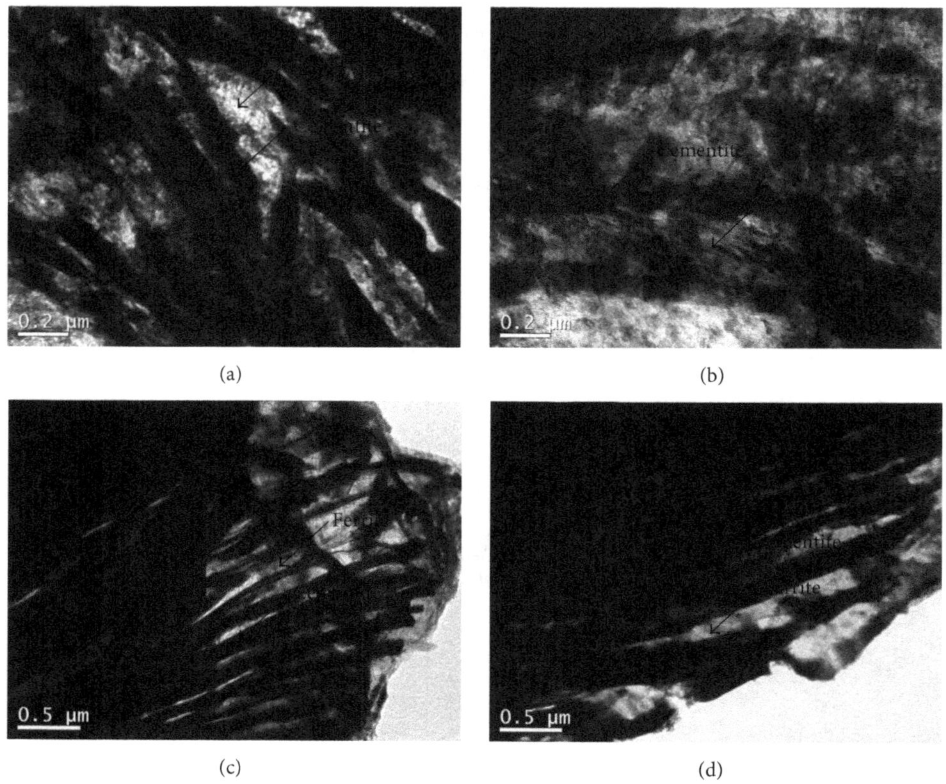

FIGURE 7: Bright field TEM micrographs of the pearlite in 1045 carbon steel without (a) and with annealing at 873 K (600°C) for 2 h (b), 4 h (c), and 6 h (d).

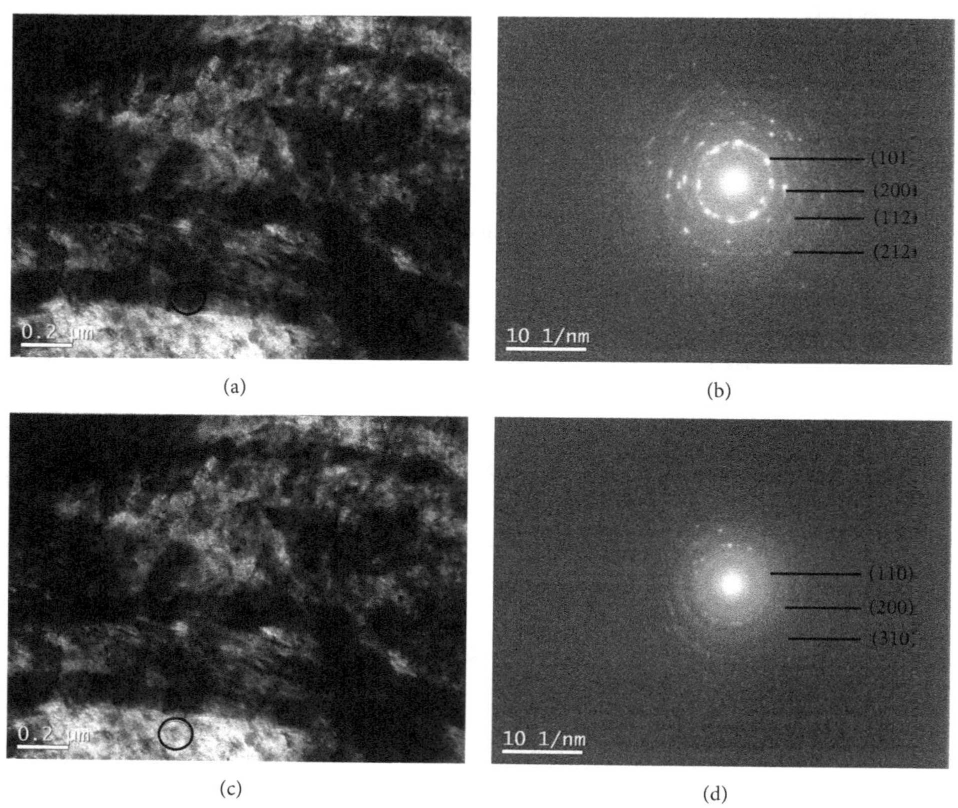

FIGURE 8: SAED image of the pearlite in 1045 carbon steel with annealing at 873 K (600°C) for 2 h.

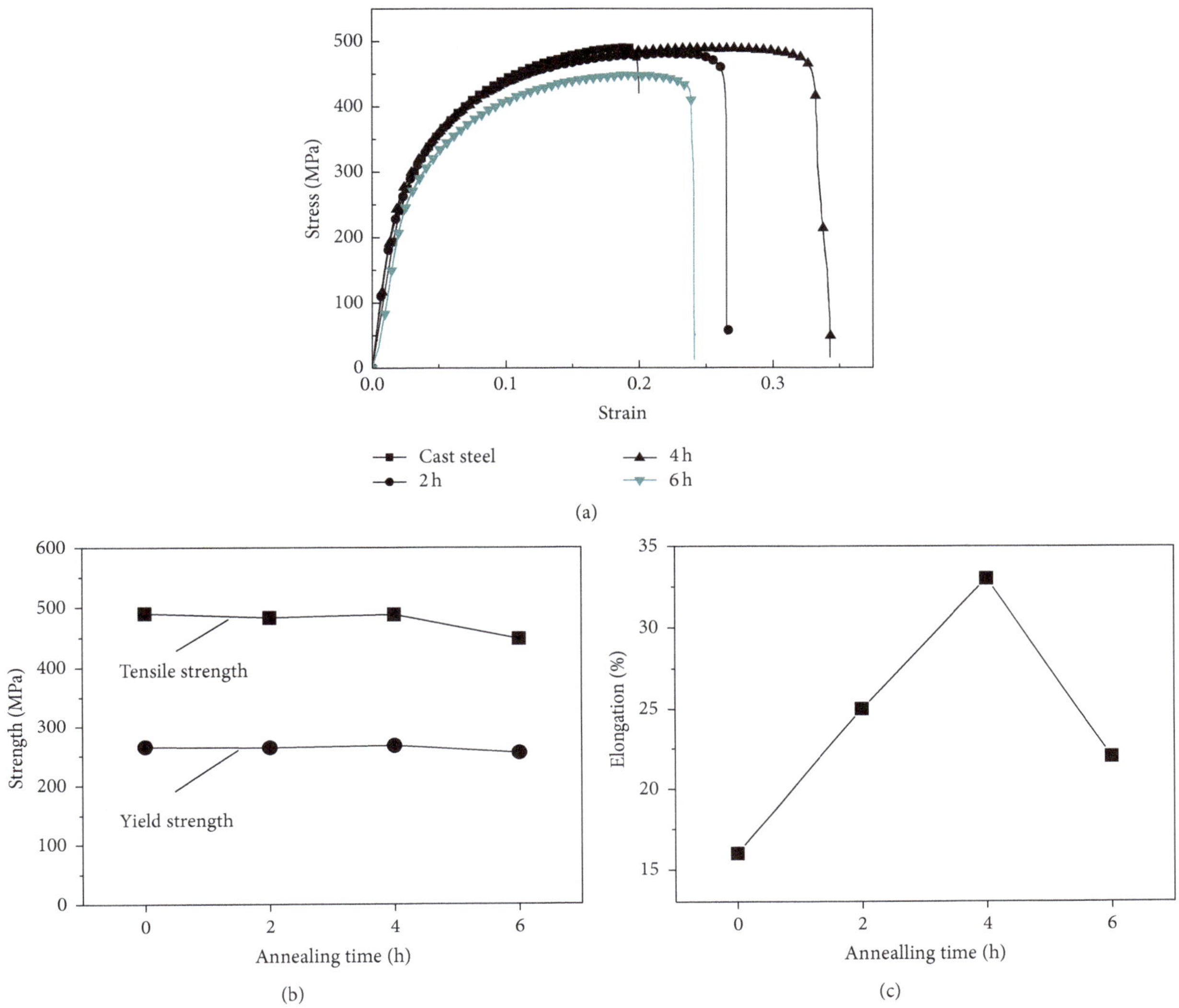

FIGURE 9: Tensile stress-strain curves (a), tensile and yield strength (b), and elongation rate (c) of the 1045 steel without and with annealing at 873 K (600°C) for different time.

ferrite. However, after annealing for 6 h, the elongation of the steel was decreased, the lamellar spacing of the pearlite and the grain size of ferrite increased sharply, the dislocation movement became easier, but the interface between lamellar ferrite and lamellar cementite was reduced, which decreased the resistance to crack propagation by crack deflection at the interface, and the elongation was decreased. The plastic deformation was always associated with the free movement of dislocations. When lamellar spacing of the pearlite was larger than 255 nm and the ferrite in the pearlite was softer than cementite, it was also deformed during the course of pearlite deformation; therefore, dislocations move more freely in the ferrite zone; consequently, it results in tensile strength decreasing observably [4, 8, 20, 21].

TABLE 4: The d values (distance of crystal face) of the (110) peak of the steels without and with annealing at different time, the standard d value of (110) peak, and the difference between them.

	d (10^{-1} nm)	Standard d (10^{-1} nm)	$\lvert\Delta d\rvert$ (10^{-1} nm)
The cast 1045 steel	2.0370	2.0262	0.0108
2 h	2.0379	2.0262	0.0117
4 h	2.0352	2.0262	0.0090
6 h	2.0351	2.0262	0.0089

5. Conclusions

The lamellar spacing of pearlite increasing with the annealing time, without and with annealing at 2, 4, and 6 h was about 146, 172, 255, and 300 nm. When the annealing time is less than or equal to 4 h, the tensile and yield strength of the steel almost were not changed with the annealing time increasing; after annealing at 6 h, the tensile and yield strength of the steel decreased a little. The elongation of the steel increased first and then decreased; after annealing at 4 h, the elongation

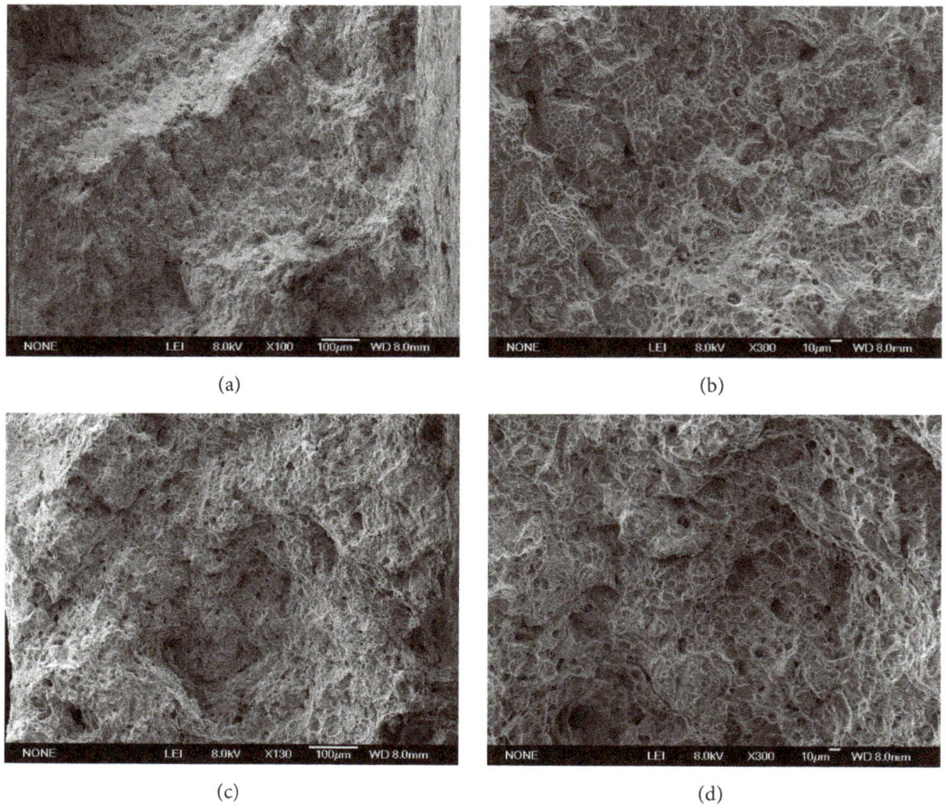

FIGURE 10: The fracture appearance of 1045 carbon steel without and with annealing at 873 K (600°C) for different time: macrograph (a) and microscopy (b) of cast steel; macrograph (c) and microscopy (d) of the steel annealed at 873 K (600°C) for 4 h.

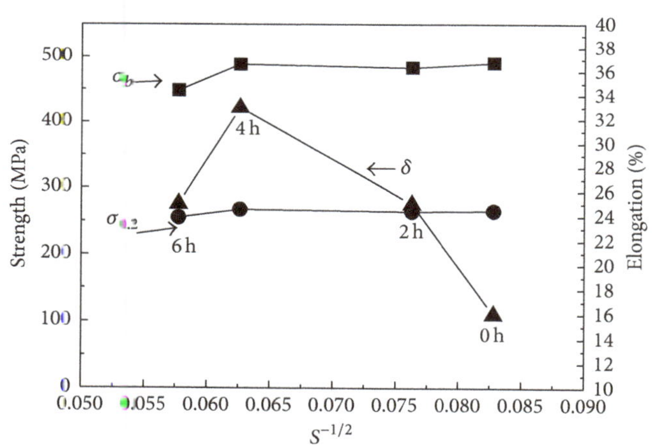

FIGURE 11: The variation of tensile strength, yield strength, and percent elongation with the inverse of the square root of interlamellar spacing, S.

increased from 15% of the cast steel to 33%. The steel annealed at 873 K (600°C) for 4 h has the highest elongation of 33% and the tensile strength was about 488 MPa; the elongation value was about twice of the conventional 1045 steel whereas the strength remained constant. Therefore, this is a better route to enhance the ductility of 1045 nanoeutectic steel prepared by aluminothermic reaction, but it does not lose strength.

Competing Interests

The authors declare that there is no conflict of interests regarding the publication of this paper.

Acknowledgments

This work is supported by the National Natural Science Foundation of China (51164022) and Gansu Provincial Science and Technology Support Program (1304GKCA027).

References

[1] G. Buzzichelli and E. Anelli, "Present status and perspectives of European research in the field of advanced structural steels," *ISIJ International*, vol. 42, no. 12, pp. 1354–1363, 2002.

[2] R. Song, D. Ponge, D. Raabe, J. G. Speer, and D. K. Matlock, "Overview of processing, microstructure and mechanical properties of ultrafine grained bcc steels," *Materials Science and Engineering A*, vol. 441, no. 1-2, pp. 1–17, 2006.

[3] H. W. Song, S. R. Guo, and Z. Q. Hu, "Coherent polycrystal model for the inverse Hall-Petch relation in nanocrystalline materials," *Nanostructured Materials*, vol. 11, no. 2, pp. 203–210, 1999.

[4] K. S. Kumar, H. Van Swygenhoven, and S. Suresh, "Mechanical behavior of nanocrystalline metals and alloys," *Acta Materialia*, vol. 51, no. 19, pp. 5743–5774, 2003.

[5] Y. Zhao, T. Topping, J. F. Bingert et al., "High tensile ductility and strength in bulk nanostructured nickel," *Advanced Materials*, vol. 20, no. 16, pp. 3028–3033, 2008.

[6] S. Ramtani, G. Dirras, and H. Q. Bui, "A bimodal bulk ultra-fine-grained nickel: experimental and micromechanical investigations," *Mechanics of Materials*, vol. 42, no. 5, pp. 522–536, 2010.

[7] M.-C. Zhao, T. Hanamura, F. Yin, H. Qiu, and K. Nagai, "Formation of bimodal-sized structure and its tensile properties in a warm-rolled and annealed ultrafine-grained ferrite/cementite steel," *Metallurgical and Materials Transactions A*, vol. 39, no. 7, pp. 1691–1701, 2008.

[8] F. Khodabakhshi, M. Kazeminezhad, and A. H. Kokabi, "Constrained groove pressing of low carbon steel: nano-structure and mechanical properties," *Materials Science and Engineering A*, vol. 527, no. 16-17, pp. 4043–4049, 2010.

[9] E. Ghassemali, A. Kermanpur, and A. Najafizadeh, "Microstructural evolution in a low carbon steel during cold rolling and subsequent annealing," *Journal of Nanoscience and Nanotechnology*, vol. 10, no. 9, pp. 6177–6181, 2010.

[10] S. M. Hosseini, A. Najafizadeh, and A. Kermanpur, "Producing the nano/ultrafine grained low carbon steel by martensite process using plane strain compression," *Journal of Materials Processing Technology*, vol. 211, no. 2, pp. 230–236, 2011.

[11] K.-T. Park, Y.-S. Kim, J. G. Lee, and D. H. Shin, "Thermal stability and mechanical properties of ultrafine grained low carbon steel," *Materials Science and Engineering A*, vol. 293, no. 1-2, pp. 165–172, 2000.

[12] C. Mangler, C. Gammer, H. P. Karnthaler, and C. Rentenberger, "Structural modifications during heating of bulk nanocrystalline FeAl produced by high-pressure torsion," *Acta Materialia*, vol. 58, no. 17, pp. 5631–5638, 2010.

[13] V. L. Tellkamp, E. J. Lavernia, and A. Melmed, "Mechanical behavior and microstructure of a thermally stable bulk nanostructured Al alloy," *Metallurgical and Materials Transactions A*, vol. 32, no. 9, pp. 2335–2343, 2001.

[14] T. B. Holland, I. A. Ovid'ko, H. Wang, and A. K. Mukherjee, "Elevated temperature deformation behavior of spark plasma sintered nanometric nickel with varied grain size distributions," *Materials Science and Engineering A*, vol. 528, no. 2, pp. 663–671, 2010.

[15] P. La, F. Wei, S. Hu, C. Li, and Y. Wei, "White cast iron with a nano-eutectic microstructure and high tensile strength and considerable ductility prepared by an aluminothermic reaction casting," *Materials Science and Engineering A*, vol. 561, pp. 317–320, 2013.

[16] H. Wang, D. Zhu, C. Zou, and Z. Wei, "Effect of high pressure on the lamellar spacing and mechanical properties of Ti-48Al alloy," *Rare Metal Materials and Engineering*, vol. 41, no. 1, pp. 42–44, 2012.

[17] F. L. Lian, H. J. Liu, J. J. Sun et al., "Ultrafine grain effect on pearlitic transformation in hypereutectoid steel," *Journal of Materials Research*, vol. 28, no. 5, pp. 757–765, 2013.

[18] F. Khodabakhshi and M. Kazeminezhad, "The annealing phenomena and thermal stability of severely deformed steel sheet," *Materials Science and Engineering A*, vol. 528, no. 15, pp. 5212–5218, 2011.

[19] C. C. Koch, R. O. Scattergood, M. Saber, and H. Kotan, "High temperature stabilization of nanocrystalline grain size: thermodynamic versus kinetic strategies," *Journal of Materials Research*, vol. 28, no. 13, pp. 1785–1791, 2013.

[20] T. S. Wang, Z. Li, B. Zhang, X. J. Zhang, J. M. Deng, and F. C. Zhang, "High tensile ductility and high strength in ultrafine-grained low-carbon steel," *Materials Science and Engineering A*, vol. 527, no. 10-11, pp. 2798–2801, 2010.

[21] P. Zhang, S. X. Li, and Z. F. Zhang, "General relationship between strength and hardness," *Materials Science and Engineering A*, vol. 529, no. 1, pp. 62–73, 2011.

3

A Stress Measurement Method for Steel Strands Based on LC Oscillation

Dongjun Chen,[1,2] Benniu Zhang,[1] Xingxing Li,[1] Chong Tu,[1] Can Yuan,[1] Wenzhao Li,[1] Zhixiang Zhou,[1] and Zongbao Liang[1]

[1] *Chongqing Jiaotong University, Chongqing 400074, China*
[2] *CSCEC Xinjiang Construction & Engineering (Group) Co., Ltd., Urumchi 830000, China*

Correspondence should be addressed to Benniu Zhang; benniuzhang@gmail.com

Academic Editor: Antonino Squillace

The prestress loss is one of the main factors affecting the safety of prestressed concrete structure. While the detecting signals like sound and light are difficult to spread in steel strands, there is no effective method for prestress detection of the bonded prestressed steel strands in existing structures yet. In this paper, taking into consideration that the electromagnetic oscillation characteristic can make the signal propagate effectively on the bonded prestressed steel strands, a nondestructive prestress detection method based on the electromagnetic effect to detect oscillation frequency is proposed. In a detection circuit, the steel strands are simulated as an inductance component, in which an induced electromagnetic signal passes through the steel strands to form resonance. And then, a frequency meter is used to detect the oscillation frequency of the resonant circuit. The oscillation frequency is supposed to have relationship with the prestress loading on the steel strands. A section of steel strands with a length of 1.2 m is adopted to test the correlation of stress and oscillation frequency. Both the theoretical and experimental results show that the resonant frequency of the circuit decreases with the increase of the stress of the strand and is linear in a certain range.

1. Introduction

Prestressed concrete structures are widely used in civil engineering and hydraulic engineering because of their advantages of reducing deformation, improving bearing capacity and cracking resistance. However, the stress relaxation of prestressing tendons, the occurrence of cracks and shrinkage of concrete, and the deformation of anchorages result in loss of prestress. This will decrease the service life of the structures. Therefore, it is important to detect the health status of the prestressed structure at regular intervals.

The expression of the new analytical model [1, 2] is established to estimate the stress distribution of the prestressing tendon, which may have a very important influence on nondestructive detection. A series of studies [3–8] show that the performance of prestressed concrete structures will be affected by the prestressed reinforcement corrosion and crack propagation. Under the cyclic loading, the cracks of the prestressed structure will further expand. And it will reduce the stiffness of the concrete structure. Saiedi et al. [9] have tested the performance of three 13-year-old concrete T-beams in the low-temperature environment. Their results show that the prestressed structures which were exposed to low temperature may reduce the strength by about 19%. Hou et al. [10] have done a series of studies on the mechanical performance of the prestressing tendons at 20–800°C. In their findings, the strength and stiffness loss of prestressed tendons are slower than those of traditional prestressed steel strands. It can be concluded that the temperature has a great influence on the structure and the steel strand. Therefore, it is necessary to consider the influence of temperature on the prestress detection. The influence of time-dependent prestressed concrete members [11] puts forward an interesting conclusion that the change in the length of the transmission exists in some prestressed loading conditions, and the length of the transmission increases as time goes on. In other words, the prestress will decrease as the time goes by.

Accordingly, the measurement of the steel strand is representative, and detection of the structure in real time is of great significance. Acoustic emission (AE) [4–6] is also an excellent method to monitor the materials' prestressing, and it uses the characteristics of acoustic information to detect the location and size of an acoustic source. Other studies [12–15] such as the application of a new material (unbonded carbon fiber) and technologies (novel optico-acoustic and sensors on a single optical fiber) in the prestressed concrete have also made some progress.

The vibrating wire strain gauge (VWSG) [16, 17] and crack reopening method [18, 19] are two kinds of methods to measure the existing prestressed structures. VWSG is a nondestructive technique; however, crack reopening method to detect existing stress is a semidestructive method. And an embedded vibrating beam strain gauge (VBSG) [18] is used in T-beam to discover the existing stresses in prestressing strands. The first natural frequency, static displacement and strain of the concrete are used to calculate the existing stresses. However, the force of the concrete structure is influenced by the embedded gauge, and VWSG and VBSG are easily to be affected by environmental factors such as electromagnetic waves and the vibration of the structure. Owing to the fact that both VWSG and VBSG are embedded in concrete structures, they are consumables and cannot be reused. The data acquisition systems and frequency measurement instruments are also much cheaper than the systems used on VWSG and VBSG. What is more, the test equipment which we design can be reused, and it is convenient to carry and can improve the efficiency of testing personnel.

In practical engineering, the length of the prestressed steel strand is up to tens of meters, while the length of the strand used in the test is only 1.2 m. The reason for this is that the purpose of the experiment is to verify the relationship between the stress and magnetism characteristic in Section 2. The steel strand specimen is equivalent to the force model of the steel strand under the tensile force of the universal testing machine. The model which is used in the test can simulate the real stress state well.

The aim of this study is to measure the stress of existing prestressed structures. Accordingly, an electromagnetic oscillation (EMO) method is proposed. To verify the effectiveness of this method, a series of stress tests by electromagnetic oscillation loading on two ends of the steel strand are implemented in laboratory. The form of the steel strand is similar to the inductance in an LC oscillation circuit. Meanwhile, an EMO circuit is created. It is based on the magnetoelastic effect to measure the electromagnetic oscillation frequency and calculate the stress of prestressed structures. The theory which is deduced in Section 2 can get a comparative result, and the most important is that the experimental data match theoretical derivation well.

2. Theory and Model

2.1. Formula Derivation. As shown in Figure 1, the inductance L in the electromagnetic oscillating circuit is parallel to the capacitance C. When a parallel resonance occurs, the voltage and current are in phase, as shown in Figure 2.

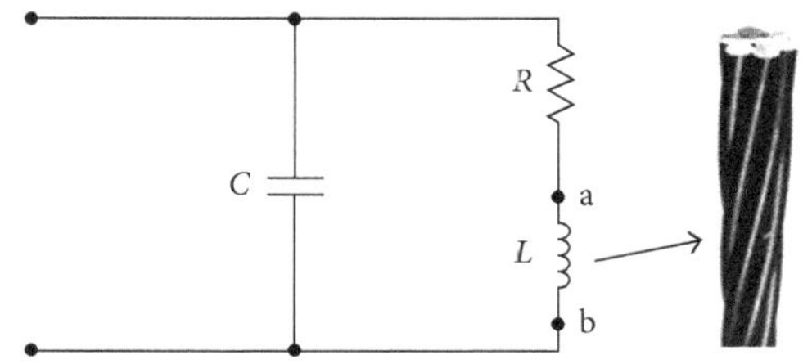

FIGURE 1: Circuit schematic.

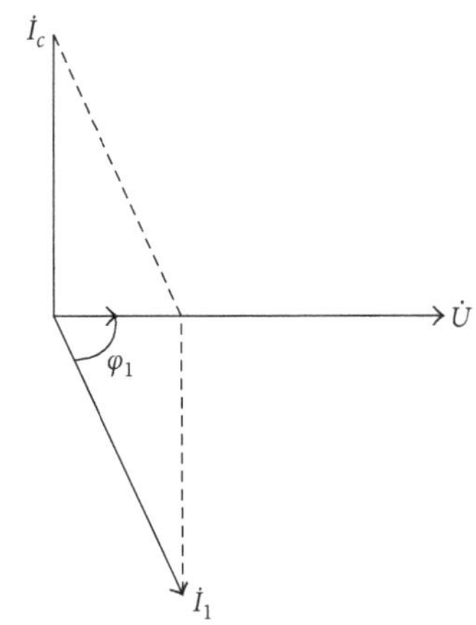

FIGURE 2: Vector model.

And the vector diagram can be written as

$$I_1 \sin\varphi_1 = I_c. \tag{1}$$

And it can be drawn from circuit diagram 1 that

$$I_1 = \frac{U}{\sqrt{R^2 + X_L^2}} = \frac{U}{\sqrt{R^2 + (2\pi f L)^2}},$$

$$\sin\varphi_1 = \frac{X_L}{\sqrt{R^2 + X_L^2}} = \frac{2\pi f L}{\sqrt{R^2 + (2\pi f L)^2}}, \tag{2}$$

$$I_c = \frac{U}{X_C} = 2\pi f C U.$$

Under parallel resonance situation, there is no pure resistance in the circuit, and $\varphi_1 = 90^\circ$. Therefore, according to the above formula (1) to (2), the resonant frequency is obtained as follows:

$$f \approx \frac{1}{2\pi\sqrt{LC}}, \tag{3}$$

where L is the inductance value of the steel strand, C is the capacitance value of the oscillator, and f is the oscillation frequency.

For conductors, the conductor generates a magnetic field inside and around it when the current passes through it. The magnetic lines of force are concentric closed rings, and the direction is determined by the right-hand rule. As the

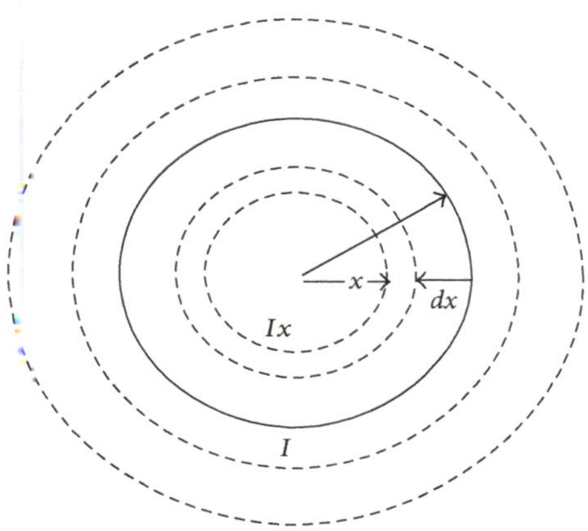

FIGURE 3: Magnetic field of concentric.

current changes, the flux changes. Assuming ψ = total flux chain, ϕ = flu, and N = turn, where the total flux chain $\psi = N\phi = LI$, the inductance can be expressed as

$$L = \frac{\psi}{I}. \tag{4}$$

As shown in Figure 3, assuming r is the radius of the steel strand intersecting surface, I is the current value passing through it, and a constant H_x is the magnetic field intensity of the concentric circular whose radius is x, the direction is tangent to the circle.

Assume I_x is the electric current of circular cross section with radius x. According to the Ampere rules, it can be obtained as

$$\int_0^{2\pi x} H_x \cdot dl = I_x, \tag{5}$$

where

$$H_x = \frac{I_x}{2\pi x}. \tag{6}$$

The inductance of the steel strand is the result of the interaction between the internal flux linkage and the external flux linkage. For the inductance generated by the internal flux linkage, the skin effect is ignored here, and assuming that the current is evenly distributed inside the steel strand, then the current per unit area is equal and can be simply expressed as

$$\frac{I}{\pi r^2} = \frac{I_x}{\pi x^2}. \tag{7}$$

Substituting I_x in (6) leads to

$$H_x = \frac{1}{2\pi r^2} x. \tag{8}$$

For magnetic wires, the magnetic permeability is μ, and the magnetic flux density can be expressed as

$$B_x = \mu H_x = \frac{\mu I}{2\pi r^2} x. \tag{9}$$

It can be assumed that the strand is uniform in thickness, for the strand thickness is d_x, length is l, and the magnetic flux differential $d\phi_x$ can be expressed as

$$d\phi_x = B_x d_x \cdot l = \frac{\mu I l}{2\pi r^2} x dx. \tag{10}$$

This part of the magnetic flux is not around the whole strand but rather part of its cross-sectional area πx^2. Thus, the current density is consistent, and the $\pi x^2/\pi r^2$ has a magnetic flux linkage that can be written as

$$d\Psi_x = \left(\frac{x^2}{r^2}\right) d\phi_x = \frac{\mu I l}{2\pi r^4} x^3 dx. \tag{11}$$

The internal magnetic chain inside the entire steel strand can be obtained by integrating dx from 0 to r:

$$\Psi_{\text{in}} = \frac{\mu I l}{2\pi r^4} \int_0^r x^3 dx = \frac{\mu I l}{8\pi}. \tag{12}$$

The inductance produced by the internal flux chain of (4) is

$$L_{\text{in}} = \frac{\Psi_{\text{in}}}{I} = \frac{\mu l}{8\pi} \tag{13}$$

For the inductance generated by the external magnetic flux, as shown in Figure 4, it is assumed that the magnetic field strength of the concentric circular magnetic circuit is H_x at the external radius $x > r$ of the wire. Since the circle of radius x surrounds the entire current, the current flowing through the circular cross section of radius x is $I_x = I$. For nonmagnetic wires, it is assumed that the magnetic permeability is μ_0. According to formula (6), the magnetic flux density at radius x can be expressed as

$$B_x = \mu_0 H_x = \frac{\mu_0 I}{2\pi x}. \tag{14}$$

Since this part of the magnetic flux surrounds the entire current, the magnetic chain is numerically equal to the magnetic flux. The differential $d\phi_x$ of the magnetic flux of the wire whose thickness is d_x and length is l can be expressed as

$$d\Psi_x = d\phi_x = B_x \cdot d_x \cdot l = \frac{u_0 I l}{2\pi r^2} x dx. \tag{15}$$

The external magnetic chain between $D1$ and $D2$ can be obtained by integrating $d\Psi_x$ from $D1$ to $D2$:

$$\Psi_{\text{out}} = \frac{\mu_0 I l}{2\pi} \int_{D1}^{D2} \frac{1}{x} dx = \frac{\mu_0 I l}{2\pi} \ln \frac{D2}{D1}. \tag{16}$$

The inductance between the two external points of the wire can be written as

$$L_{\text{out}} = \frac{\Psi_{\text{out}}}{I} = \frac{\mu_0 l}{2\pi} \ln \frac{D2}{D1}. \tag{17}$$

Thus, the total inductance can be expressed as

$$L = L_{\text{in}} + L_{\text{out}} = \frac{\mu l}{8\pi} + \frac{\mu_0 l}{2\pi} \ln \frac{D2}{D1}. \tag{18}$$

Since the strand is ferromagnetic, the magnetic permeability is $\mu \gg \mu_0$ and the above equation is simplified as

$$L = \frac{\mu l}{8\pi} = \frac{\mu (l_0 + \Delta l)}{8\pi}, \tag{19}$$

where l_0 is the original length and Δl is the length increment.

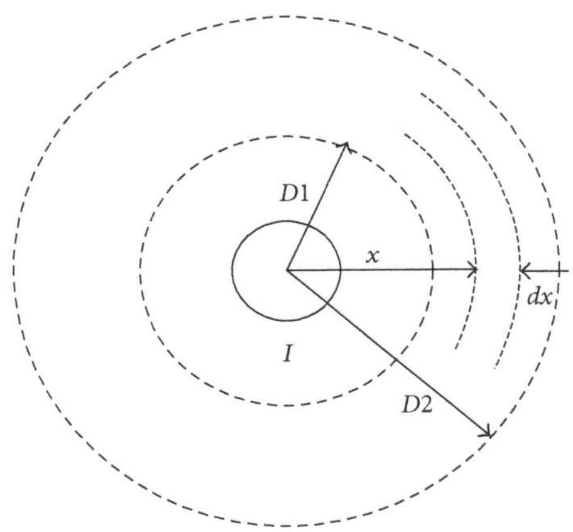

Figure 4: Concentric magnetic field of the external magnetic flux linkage.

According to the simplified constitutive relation of the linear elastic material, the stress value of linear elastic material can be expressed as

$$\sigma = E\varepsilon = E\frac{\Delta l}{l} \tag{20}$$

where E is the elastic modulus of steel strand and ε is the strain.

It can be assumed that if H is the magnetic field intensity and B is the magnetic induction intensity, then the permeability is defined as

$$\mu = \mu_r\mu_0 = \frac{B}{H}. \tag{21}$$

For the magnetic field strength H, the steel strand is equivalent to a long direct current wire, and the magnetic field strength can be expressed as

$$H = \frac{I}{2\pi r}. \tag{22}$$

As the length changes of the steel strand stretching process are much larger than the cross-sectional area changes, the current size I and radius r are almost unchanged; hence, magnetic field strength can be regarded as unchanged.

The magnetic induction intensity B can be defined as

$$B = \mu_0 (H + M), \tag{23}$$

where M is the magnetization and μ_0 is the permeability of vacuum.

For the magnetization M, in the volume V of the magnet, a large number of magnetic dipoles are contained within this volume, and they all have magnetic moments. It is assumed that the magnetic moment of all the atoms of the steel strand is arranged along the direction of the steel strand; if the magnetic moment of each atom is m_0, then the size of the total magnetic dipole can be expressed as

$$m = Nm_0 = \frac{m}{M_0}N_A m_0, \tag{24}$$

where ρ is the average density of the steel strand, M_0 is the molar mass, and N_A is Avogadro's number, and the magnetization can be expressed as

$$M = \frac{(m/M_0)N_A m_0}{V}. \tag{25}$$

Under the effect of stress, when a small length change occurs, the cross-section radius r is almost unchanged, or even it can be ignored, where we assume that the steel strand deformation is Δl, so the magnetization M size becomes

$$M = \frac{(m/M_0)N_A m_0}{V} = \frac{(m/M_0)N_A m_0}{\pi r^2 \cdot l} = \frac{(m/M_0)N_A m_0}{\pi r^2 \cdot (l_0 + \Delta l)}. \tag{26}$$

Combining (22), (23), and (26), it can be obtained that

$$\begin{aligned} \mu &= \frac{\mu_0 \left((I/2\pi r) + ((m/M_0)N_A m_0/\pi r^2 (l_0 + \Delta l)) \right)}{I/2\pi r} \\ &= \mu_0 \left(1 + \frac{(2m/M_0)N_A m_0}{Ir\left(l_0 + (\sigma l_0/E)\right)} \right), \end{aligned} \tag{27}$$

$$L = \frac{\mu_0 \left(1 + (2(m/M_0)N_A m_0/Ir\left(l_0 + (\sigma l_0/E)\right))\right)\left(l_0 + (\sigma l_0/E)\right)}{8\pi}. \tag{28}$$

Then combining (3) and (28), the result of formula derivation can be obtained as follows:

$$f = \frac{1}{2\pi\sqrt{\left(\mu_0 \left(1 + (2(m/M_0)N_A m_0/Ir\left(l_0 + (\sigma l_0/E)\right))\right)\left(l_0 + (\sigma l_0/E)\right)/8\pi\right) \cdot C}}. \tag{29}$$

2.2. Simulation Result. Substituting $\mu_0 = 4\pi \times 10^{-7}\ \mathrm{H} \times \mathrm{m}^{-1}$, $E = 1.95 \times 10^5$ MPa, $C = 0.1$ uF, $l_0 = 19.5$ m, $r = 0.005$ m, $I = 0.1$ A, $m = 10$ kg, $M_0 = 55.8 \times 10^{-3}\ \mathrm{kg} \times \mathrm{mol}^{-1}$, $N_A = 6.02 \times 10^{23}\ \mathrm{mol}^{-1}$, and $m_0 = 1.2488 \times 10^{-27}\ \mathrm{A} \times \mathrm{m}^2$ in (28), it can be obtained that

$$f = \frac{1000}{2\pi\sqrt{2.792 + 5 \times 10^{-7}\sigma}}. \tag{30}$$

The results are shown in Figure 5.

3. Experiments

According to the working principle of the *LC* oscillating circuit and the limitation of the piston stroke of the universal testing machine, a steel strand specimen whose both ends were insulated and derusted was used to verify the theoretical derivation. Both ends of the prestressed steel strand were clamped in the jaws of the universal testing machine to simulate the anchor zone, and then the axial force is

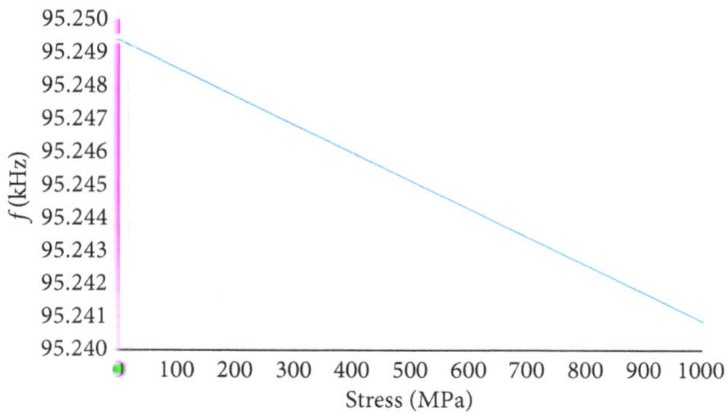

FIGURE 5: Relationship diagram between frequency and stress.

applied to simulate the prestressing of the steel strand after stretching. The deformation of the steel strand, change of oscillation circuit frequency, and stress-strain curve can be observed by applying the force.

3.1. Manufacture of Steel Strand Sample. The steel strand used in the test is a common 1×7 strand structure whose parameters are shown in Table 1. The test specimens were made of steel strand with a length of 1.2 m, and both ends of the steel strand were polished with sandpaper. Then, the wire and steel strand were welded by soldering, and it is essential that both ends of the wire's direction should be consistent. After the steel strand firmly welded the whit wire, both ends of the steel strand should be winded with the insulation adhesive tape without leaving gaps. The insulating electrical tape's winding length should be greater than the depth of the universal testing machine to avoid short circuit caused by direct contact. Finally, as the applied load is relatively large, only the electrical tape may cause the steel strand and the universal testing machine to slip which could lead to errors to the experimental data. In order to avoid this situation, both ends of the steel strand were wrapped with coarse insulating electrical tape to ensure the stability of the circuit and the steel strand.

3.2. Brief Introduction of Circuit. An *LC* oscillator can be built by using the integrated circuit and some peripheral circuits, and the schematic is shown in Figure 6. The *LC* oscillator module uses MC1648 as the oscillator chip, and the voltage comparator chip uses LM393. In addition, the signal conditioning module uses an LM393 chip. In the schematic, *U*1 is the oscillator chip, *C*1 is a resonant capacitor, *C*3 and *C*4 are power filter capacitors, and *C*2 and *C*5 are the input filter. Besides, *L*1 is the sum of the inductance of the steel strand and its series, capacitor *R*1 is the matching impedance, and the measuring range of the digital frequency meter module is between 20 kHz and 100 kHz.

3.3. The Process of Experiment. During the experiments, the steel strand is preloaded to 2 kN, and the final load is set to 8 kN. Such small stress is chosen based on the following reasons: (1) we hope the deformation of the steel strand can be limited in elastic range to test multiple equal cycles; (2) due to the clip limitation of loading, much larger tension may cause relative slippage between the test machine and the strand.

Before the test, the universal testing machine was used to clamp the ends of the insulated steel strand. The wire, which was leaded from both ends of the steel strand, is oriented perpendicular to the surface of the paper. The jaw clamps the 2/3 position of the end of the insulated steel strand to confirm that the steel strand and wire are not directly contacted with the universal testing machine and the joints were twined by insulating tape. And the structure of the experimental system is shown in Figure 7.

The test procedure is as follows. Firstly, turning on the power supply of the oscillation circuit, the universal testing machine is used to preload the steel strand to 2 kN. After three times of loading, it is confirmed that the applied load and the deformation of the steel strand tend to be stable. Finally, the loading rate of 10 mm/min is loaded to 8 kN. In this period, the frequency meter collects the frequency data once per second, and the collected data are transmitted to the mobile device via bluetooth, and a total of 1 set of test data is collected during the loading process. After the data collection, the tension is unloaded to zero, and the change trend of the frequency is observed during unloading. At the end of a trial, the stress-strain data and corresponding curves are derived from the software which is used in the test. In this way, the same steel strand specimen is loaded 6 times, and a total of 6 sets of data are collected during these processes.

4. Experimental Results and Data Analysis

4.1. Experimental Results. Under the same experimental conditions, the 6 loading tests have been completed. The effective data and the calculated data are shown in Table 2. By analyzing the stress-strain curves corresponding to the frequency data in the same time, the data which are corresponding to the nonlinear part of the curve are removed. The median frequency, frequency repeatability error, and frequency standard deviation are calculated by the effective frequency data.

4.2. Data Analysis. In the experiment, the steel strand is in the linear elastic stage. So, the stress of the steel strand increases with the increase of the applied load of the universal testing machine. And the increase of stress will lead to the increase of strain. In general, the load, stress, and strain of the steel strand increase with time.

As shown in Figure 8, the median frequency calculated in Table 2 is in linear regression with time. The stress-strain curve is shown in Figure 9; the blue line indicates the stress-strain relationship between the stretching jaw and the steel strand without sliding, and the red line indicates that, under the same stress, the stress-strain relationship occurs in the case of relative slip.

The following conclusions can be drawn from the analysis:

(1) As can be seen from Loading 1, 2, 5, and 6, the oscillation frequency decreases with the increase of

Table 1: Parameters of the steel strand.

Structure of steel strand	Nominal diameter of steel strand, D (mm)	Nominal area of steel strand, S (mm^2)	Ultimate tensile strength, R_m (MPa) No less than	Maximum tension, F_m (kN) No less than	Nonproportional extension force, $F_{P0.2}$ (kN) No less than	Maximum elongation, A_{GT} (%) No less than	Stress relaxation rate after 1000 h, r (%) No more than
1×7	15.2	139	1720	241	217	3.5	4.5

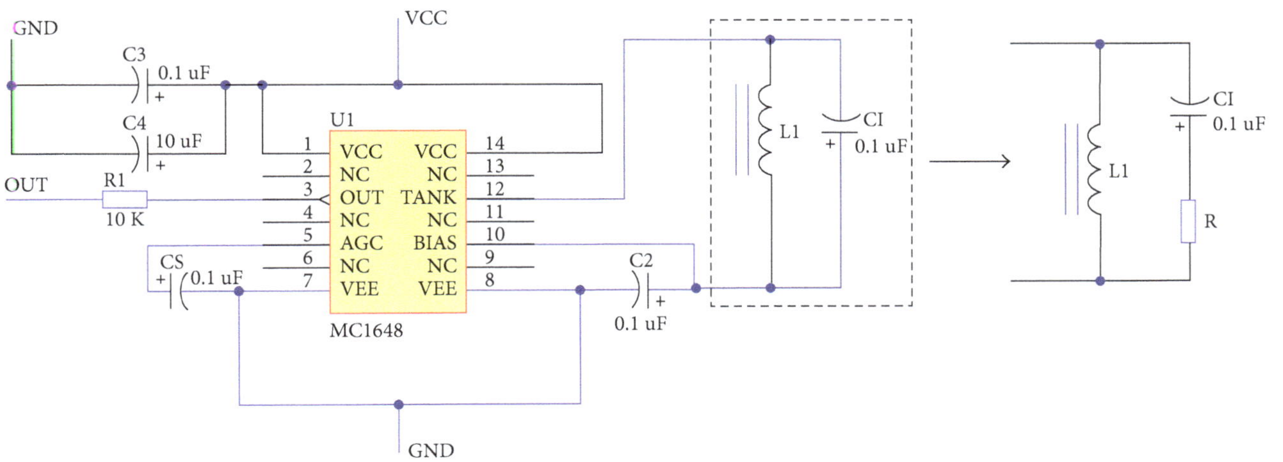

Figure 6: Circuit principle diagram.

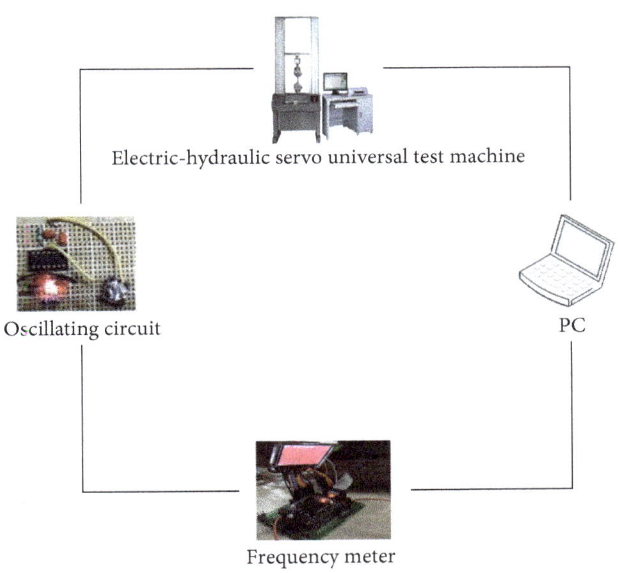

Figure 7: The experiment structure diagram.

time, while the stress increases with the increase of time. It can be concluded that the oscillation frequency decreases with the increase of stress and has a linear relationship within a certain range. This is consistent with the theoretical derivation of the second part of the article.

(2) As shown in Loading 3 and 4 of Table 2, in the first 6 seconds, the stress and the oscillation frequency showed a good linear relationship. And with the increase of stress, the oscillation frequency decreased correspondingly. The subsequent 4 seconds have shown the opposite relationship; that is, the oscillation frequency increases with the increase of stress. The reason is that the friction force between the two ends of the universal testing machine and the steel strand is smaller than the tensile force, which results in the relative sliding between the steel strand and the jaw of the universal testing machine.

(3) Figure 9 also shows that, under the same stress, the strain of the steel strand is greater than that of the case without sliding. The tensile force of the steel strand after sliding decreases, and the oscillation frequency increases, which proves that the stress is inversely proportional to the frequency of oscillation. From the data of the frequency meter, the oscillation frequency increases with the unloading of stress. And it is also consistent with the theoretical derivation.

(4) It can be concluded from the analysis of the relationship between the median frequency and time in Figure 8 that, with the increase of the stress in the steel strand, the oscillation frequency decreases and the linear relationship exists in a certain range. It is consistent with the theoretical derivation as well.

(5) The relationships shown in Figure 8 show that the gradient of frequency variation is small in fixed time, and there is a certain relationship with the force rate of the steel strand. The bigger the test force rate is, the bigger the change frequency of the oscillation frequency is and the more obvious the frequency changes are.

TABLE 2: The experimental data.

Median strain (10^{-4})	Median stress (MPa)	Measurement times Loading 1 *f* (kHz)	Loading 2 *f* (kHz)	Loading 3 *f* (kHz)	Loading 4 *f* (kHz)	Loading 5 *f* (kHz)	Loading 6 *f* (kHz)	Repeatability error	Standard deviation	Median frequency (kHz)
0.7379	14.3885	74.3009	74.2940	74.2838	74.2754	74.2748	74.2677	0.0170%	0.01266	74.28277
1.0146	19.7842	74.3018	74.2926	74.2838	74.2751	74.2746	74.2669	0.0174%	0.01294	74.28247
1.2987	25.3237	74.3018	74.2911	74.2828	74.2750	74.2740	74.2656	0.0176%	0.01309	74.28172
1.6012	31.2230	74.3010	74.2900	74.2817	74.2748	74.2733	74.2647	0.0175%	0.01300	74.28092
1.9111	37.2662	74.2995	74.2889	74.2809	74.2747	74.2720	74.2639	0.0172%	0.01274	74.27998
2.162	42.1583	74.2995	74.2883	74.2804	74.2742	74.2720	74.2632	0.0173%	0.01287	74.27960
2.4682	46.6906	74.2984	74.2881	74.2807	74.2736	74.2712	74.2625	0.0173%	0.01284	74.27908
2.7449	50.9353	74.2972	74.2875	74.2809	74.2734	74.2705	74.2619	0.0171%	0.01266	74.27857
3.0142	54.1727	74.2951	74.2869	74.2809	74.2738	74.2696	74.2619	0.0162%	0.01206	74.27803
3.2835	57.5540	74.2951	74.2855	74.2804	74.2745	74.2687	74.2617	0.0161%	0.01198	74.27765

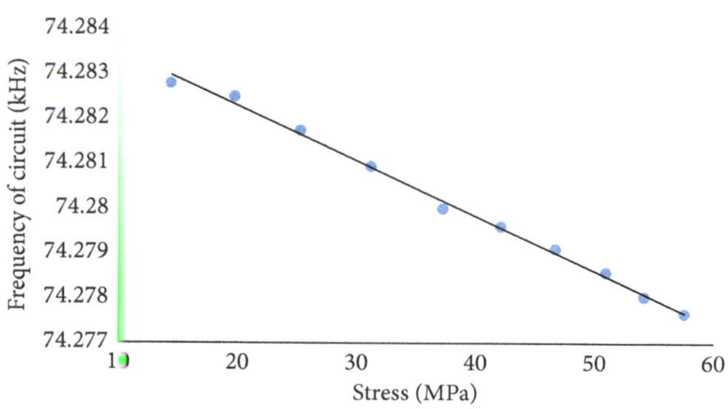

FIGURE 8: Relationship diagram between median frequency and median stress.

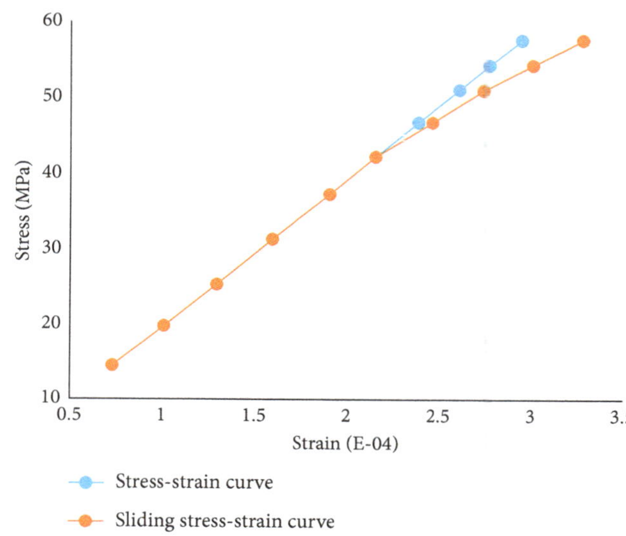

FIGURE 9: Relationship diagram between stress and strain.

5. Conclusion

It is a challenging problem to detect the prestressing force of the structural steel strand in service concrete. The deformation of concrete, the relaxation of the prestressed steel strand, the corrosion of the steel strand, and the creep of concrete will cause the loss of prestress. However, the prestress loss has not been solved well, so it is important to detect the prestressed concrete structure effectively. The traditional method for detecting the prestressing force of the steel strand in the existing structure is complex, and the equipment that is used to obtain the prestressing force is expensive and inconvenient to carry. The vibration frequency of the oscillating circuit is used to detect the loss of prestress in the steel strand, which avoids the damage to the structure caused by the prestress detection and provides a timely and accurate method for detecting prestress.

The idea of the whole system is to connect the steel strand to the oscillating circuit, and the steel strand is equivalent to inductance in the circuit. The vibration frequency of the circuit is obtained by vibration, and the prestressing force in the steel strand can be obtained by calculation. A series of experimental data show that there is a certain relationship between the prestress of the steel strand and the frequency of oscillation. The inverse proportion relationship between the stress and frequency is obtained by data analysis, which is consistent with the theoretical derivation of this paper. It is proved that the method of detecting the prestressing force in the steel strand by the oscillation frequency of the circuit is practicable. It is also verified the previous conjecture that there is a certain relationship between the electromagnetic force and the strand force.

Adopting the EMO circuit, the experiments have been greatly improved in measuring accuracy and range. The measuring force of the steel strand is closer to the actual situation. The frequency meter used in the experiment is in the range of 20–200 kHz, whose measuring range is larger than the original one. And the measurement accuracy is 0.001 kHz.

Compared with the traditional method, it is more convenient and feasible to measure the prestressing force of the steel strand by the oscillating circuit, and it can reflect the force of the steel strand without any damage to prestressed structures.

Nevertheless, in the process of the steel strand, the experiment has failed to solve the rotation and frequency drift well. These are the problems that we have to deal with in the next step.

Conflicts of Interest

The authors declare that there are no conflicts of interest regarding the publication of this paper.

Acknowledgments

This work is jointly supported by the National Key Research and Development Program of China (2017YFC0806007), the Science and Technology Construction Project of Ministry of Transport of China (2015319814020), the Technology Innovation Project of Chongqing Social and People's Livelihood (CSTC2016SHMSZX30026), the Program for Innovation Team Building at Institutions of Higher Education in Chongqing (CXTDG201602013), the Urumqi Science and Technology Plan (Y161320008), the Scientific and Technological Research Program of Chongqing Municipal Education Commission (KJ1400321 and KJ1605202), the National Science Fund for Distinguished Young Scholars (51425801), and the National Natural Science Foundation of China (11372366 and 51508059).

References

[1] A. B. M. Abdullah, J. A. Rice, H. R. Hamilton, and G. R. Consolazio Gary, "An investigation on stressing and breakage response of a prestressing strand using an efficient finite element model," *Engineering Structures*, vol. 123, pp. 213–224, 2016.

[2] A. O. Abdelatif, J. S. Owen, and M. F. M. Hussein, "Modelling the prestress transfer in pre-tensioned concrete elements," *Finite Elements in Analysis and Design*, vol. 94, pp. 47–63, 2015.

[3] L. Z. Dai, L. Wang, J. R. Zhang, and X. H. Zhang, "A global model for corrosion-induced cracking in prestressed concrete structures," *Engineering Failure Analysis*, vol. 62, pp. 263–275, 2016.

[4] W. Velez, F. Matta, and P. Ziehl, "Acoustic emission monitoring of early corrosion in prestressed concrete piles concrete piles," *Structural Control and Health Monitoring*, vol. 22, no. 5, pp. 873–887, 2015.

[5] A. Appalla, M. K. ElBatanouny, W. Velez, and P Ziehl, "Assessing corrosion damage in posttensioned concrete structures using acoustic emission," *Journal of Materials in Civil Engineering*, vol. 28, no. 2, p. 04015128, 2016.

[6] J. Mangual, M. ElBatanouny, P. Ziehl, and F. Matta, "Corrosion damage quantification of prestressing strands using acoustic emission," *Journal of Materials in Civil Engineering*, vol. 25, no. 9, pp. 1326–1334, 2013.

[7] W. P. Zhang, X. G. Liu, and X. L. Gu, "Fatigue behavior of corroded prestressed concrete beams," *Construction and Building Materials*, vol. 106, pp. 198–208, 2016.

[8] B. Fernandes, M. Titus, D. K. Nims, A. Ghorbanpoor, and V. Devabhaktuni, "Field test of magnetic methods for corrosion detection in prestressing strands in adjacent box-beam bridges," *Journal of Bridge Engineering*, vol. 17, no. 6, pp. 984–988, 2012.

[9] R. Saiedi, M. F. Green, and A. Fam, "Behavior of CFRP-prestressed concrete beams under sustained load at low temperature," *Journal of Cold Regions Engineering*, vol. 27, no. 1, pp. 1–15, 2013.

[10] X. M. Hou, W. H. Zheng, V. Kodur, and H. Y. Sun, "Effect of temperature on mechanical properties of prestressing bars," *Construction and Building Materials*, vol. 61, pp. 24–32, 2014.

[11] L. A. Caro, J. R. Marti-Vargas, and P. Serna, "Time-dependent evolution of strand transfer length in pretensioned prestressed concrete members," *Mechanics of Time-Dependent Materials*, vol. 17, no. 4, pp. 501–527, 2013.

[12] X. Yang, P. Zohrevand, A. Mirmiran, M. Arockiasamy, and W. Potter, "Comparative study of unbonded carbon fiber and steel strands in posttensioned pier caps," *Journal of Composites for Construction*, vol. 20, no. 1, p. 04015036, 2016.

[13] C. G. Lan, Z. Zhou, and J. P. Ou, "Monitoring of structural prestress loss in RC beams by inner distributed Brillouin and fiber Bragg grating sensors on a single optical fiber," *Structural Control and Health Monitoring*, vol. 21, no. 3, pp. 317–330, 2014.

[14] N. F. Grace, S. K. Rout, K. Ushijima, and M. Bebawy, "Performance of carbon-fiber-reinforced polymer stirrups in prestressed-decked bulb T-beams," *Journal of Composites for Construction*, vol. 19, no. 3, p. 04014061, 2015.

[15] P. A. Vanniamparambil, F. Khan, K. Hazeli et al., "Novel optico-acoustic nondestructive testing for wire break detection in cables," *Structural Control and Health Monitoring*, vol. 20, no. 11, pp. 1339–1350, 2013.

[16] H. M. Lee, S. W. Choi, D. J. Jung, and H. S. Park, "Analytical model for estimation of maximum normal stress in steel beam-columns based on wireless measurement of average strains from vibrating wire strain gages," *Computer-Aided Civil and Infrastructure Engineering*, vol. 28, no. 9, pp. 707–717, 2013.

[17] H. S. Park, H. Y. Lee, S. W. Choi, and Y. Kim, "A practical monitoring system for the structural safety of mega-trusses using wireless vibrating wire strain gauges," *Sensors*, vol. 13, no. 12, pp. 17346–17361, 2013.

[18] S. Biswal and A. Ramaswamy, "Measurement of existing prestressing force in concrete structures through an embedded vibrating beam strain gauge," *Measurement*, vol. 83, no. 10, pp. 10–19, 2016.

[19] J. A. Paredes, S. Oller, and A. H. Barbat, "New tension-compression damage model for complex analysis of concrete structures," *Journal of Engineering Mechanics*, vol. 142, no. 10, 2016.

4

Strengthening of Steel Columns under Load: Torsional-Flexural Buckling

Martin Vild and Miroslav Bajer

Faculty of Civil Engineering, Institute of Metal and Timber Structures, Brno University of Technology, Veveří 331/95, 602 00 Brno, Czech Republic

Correspondence should be addressed to Martin Vild; vild.m@fce.vutbr.cz

Academic Editor: Ana S. Guimaraes

The paper presents experimental and numerical research into the strengthening of steel columns under load using welded plates. So far, the experimental research in this field has been limited mostly to flexural buckling of columns and the preload had low effect on the column load resistance. This paper focuses on the local buckling and torsional-flexural buckling of columns. Three sets of three columns each were tested. Two sets corresponding to the base section (D) and strengthened section (E) were tested without preloading and were used for comparison. Columns from set (F) were first preloaded to the load corresponding to the half of the load resistance of the base section (D). Then the columns were strengthened and after they cooled, they were loaded to failure. The columns strengthened under load (F) had similar average resistance as the columns welded without preloading (E), meaning the preload affects even members susceptible to local buckling and torsional-flexural buckling only slightly. This is the same behaviour as of the tested columns from previous research into flexural buckling. The study includes results gained from finite element models of the problem created in ANSYS software. The results obtained from the experiments and numerical simulations were compared.

1. Introduction

Strengthening is a type of retrofit works, in which a material is added to a base cross-section in order to transfer additional load. The use of welded plates to strengthen steel members is a common practice because it is fast and cheap. Unloading the structure before strengthening may often be almost unfeasible or economically inconvenient, in which case strengthening under load is carried out (for case study examples see [1–3]). Although there are several rules of thumb, current normative documents do not refer to this problem and no reliable analytical solution has been developed to date [4]. In most of the experiments detailed in literature (e.g., [5–7]), the load resistance of columns strengthened under load was very similar to that of reference columns strengthened without preloading. However, there is no consensus among researchers. Some (Brown [8], Ricker [9], Spal [10], and Unterweger [11]) believe that load resistance is decreased by preloading, others (Tall [12], O'Sullivan [5], Tide [13], Wu and Grondin [14], and Bhowmick and Grondin [15]) claim that well designed and conducted column strengthening under load does not decrease its load resistance.

The paper presents experimental and numerical research into the strengthening of columns under load using welded plates. Experimental research of other investigators (e.g., [6, 7, 16]) was focused on flexural buckling of columns. Also other works of authors investigated flexural buckling of wide flange columns strengthened under load with intermittent welds [17]. All the experimental research into flexural buckling showed that the preload had low effect on the load resistance of the column strengthened under load. Therefore, columns susceptible to local and torsional-flexural buckling were selected. Additionally, the theoretical axis of loading did not pass the column centre of gravity which caused a bending moment. The experiments were conducted at the laboratory of the Institute of Metal and Timber Structures at Brno University of Technology in April 2015. The purpose of this paper and the presented experiments is to contribute

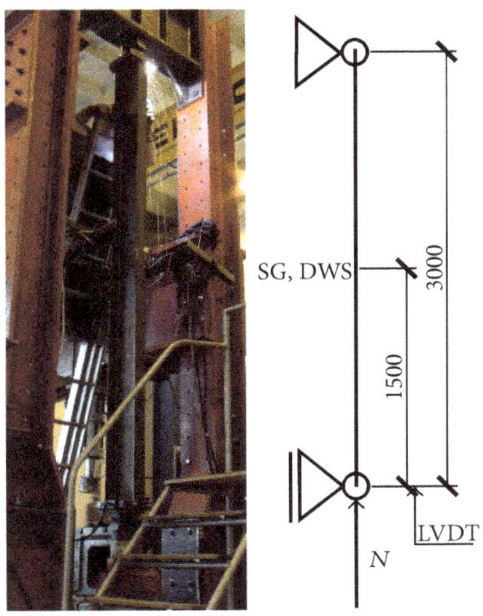

FIGURE 1: Column specimen strengthened under load in the test set-up.

to the existing body of knowledge concerning strengthening under load that aims to make retrofit design both safe and economical.

2. Materials and Methods

Three sets of columns were selected for the research. The experimental sets of columns comprise three columns each. Set (D) includes columns labelled D1, D2 and D3, set (E) columns E1, E2 and E3, and set (F) columns F1, F2 and F3. Numerical models of columns are labelled D, E and F. The columns were subjected to a compressive force (see Figure 1). The axis of loading passing through axis y' did not pass the centre of gravity (intersection of axes y and z) which caused additional bending moment. The cross-sections are presented in Figure 2. All columns were welded with continuous welds; the throat thickness of which was 4 mm. Set (D) included columns with a T-shaped cross-section welded from the flange with the dimensions 140 × 8.1 mm and from the web with the dimensions 200 × 5.4 mm. The columns from set (D) were class 4 according to the classification of cross-sections in EN 1993-1-1 [18]. Set (E) comprised columns with a monosymmetric I-shaped cross-section. The flanges had the dimensions 140 × 8.1 mm and 80 × 7.9 mm, and the web was 200 × 5.4 mm in size. Sets (D) and (E) were necessary for the comparison of behaviour and resistance. Set (F) contained columns with a T-shaped cross-section with the same dimensions as the columns in set (D). The columns from set (F) were first loaded to 70 kN and then strengthened under this load with a second flange. The preload magnitude was chosen as the half of the average experimental load resistance of set (D). The preload ratio (ratio of the preload magnitude and the base section load resistance) 0.5 is considered in Czech technical recommendations [19] as the maximum limit value for strengthening under load. The resulting cross-section was the same as that of the columns from set (E). The columns from sets (E) and (F) were class 3.

Buckling resistance of investigated beam-columns was determined using Cl. 6.3.4 general method for lateral and lateral torsional buckling of structural components in Eurocode 1993-1-1 [18]. Values of $\alpha_{\text{ult},k}$ and $\alpha_{\text{cr,op}}$ were determined numerically using average measured thicknesses of plates and average yield stress determined from coupon tests.

2.1. Experimental Research. Each set comprised three columns (labelled 1, 2, and 3) selected for use in experimental research. All columns were 3 m long and loaded by loading cylinder. The load was transferred through knife-edge bearings (see Figure 3), which provided pinned boundary conditions perpendicular to stronger axis y' and fixed perpendicular to weaker axis z. The load acted on axis y' which does not align with the axis y passing through the centre of gravity. The distances between axes y' and y are 53.5 mm and 18.4 mm in cases of base section and strengthened section, respectively. Hence, the axial force causes also bending moment. Both sets (D) and (E) were loaded monotonically until collapse occurred. The columns from set (F) were first inserted into the loading set-up and loaded to 70 kN. The force was manually held roughly constant (±5 kN) while the second flange with the dimensions 80 × 7.9 mm was welded to the web under the load. The shielded metal arc welding method with electrode wire thickness 2.3 mm was used. The force rose by around 1 kN per 12 s while the steel was being heated by the welding process and dropped at a similar rate while the process was paused and the column cooled. The load and column elongation and shortening of specimen F3 can be seen in Figure 4. The temperature of the column was monitored with a handheld infrared thermometer. After the welding process was finished and the specimen had cooled, columns F1 and F2 were unloaded to 10 kN to determine the change of initial imperfections as suggested by Unterweger [11] and then loaded to failure. Column F3 was loaded to failure without unloading.

Tension coupon tests were conducted to determine steel mechanical properties. Three coupons were machined from webs, three from longer flanges, and three from shorter flanges. The precise thickness of the plates was measured with callipers at several spots and the average value was used for numerical analysis. 1-LY11/6/350 strain gauges (SG) were used (see Figure 2) to measure the strain, which was recalculated into stress using the true stress-strain diagram obtained from the coupon tests. The vertical displacement and force at the bottom of the column and the horizontal deflection and strains at midheight were measured for all specimens. A Southwell plot [20] was used to determine the initial imperfections.

2.2. Numerical Modelling. Numerical simulations were performed to complement the values which had not been measured in the experiment. ANSYS software [21] was used for finite element analyses. The temperature load from thermal analyses was used to model the effects of the continuous weld with a throat thickness of 4 mm. The temperature load corresponded with the deformations measured in these

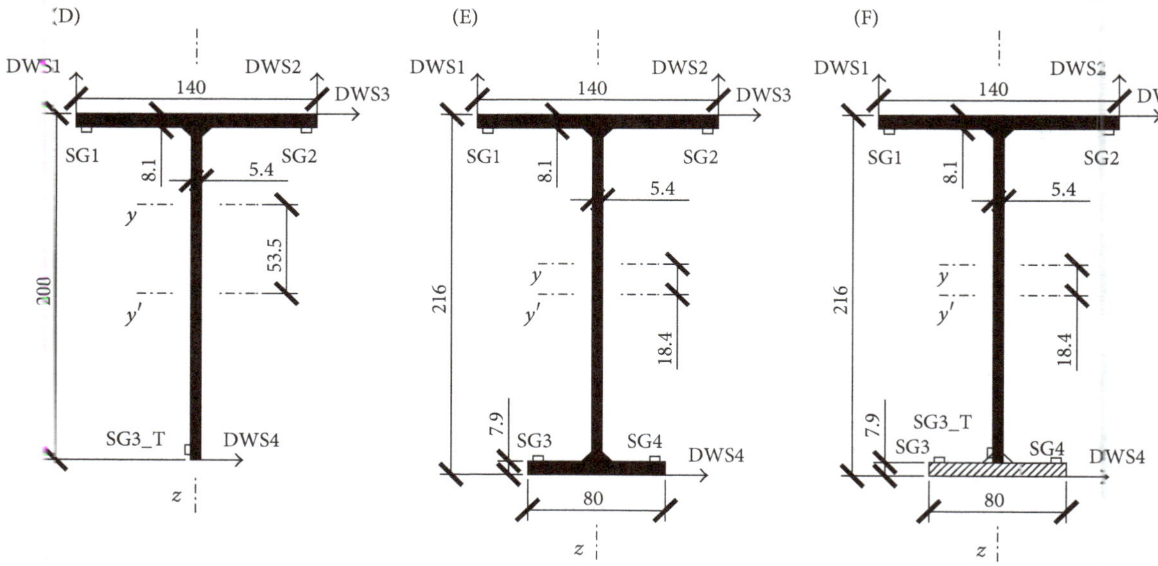

FIGURE 2: Column sets with dimensions, axes, strain gauge (SG) positions, and draw wire sensors (DWS) positions.

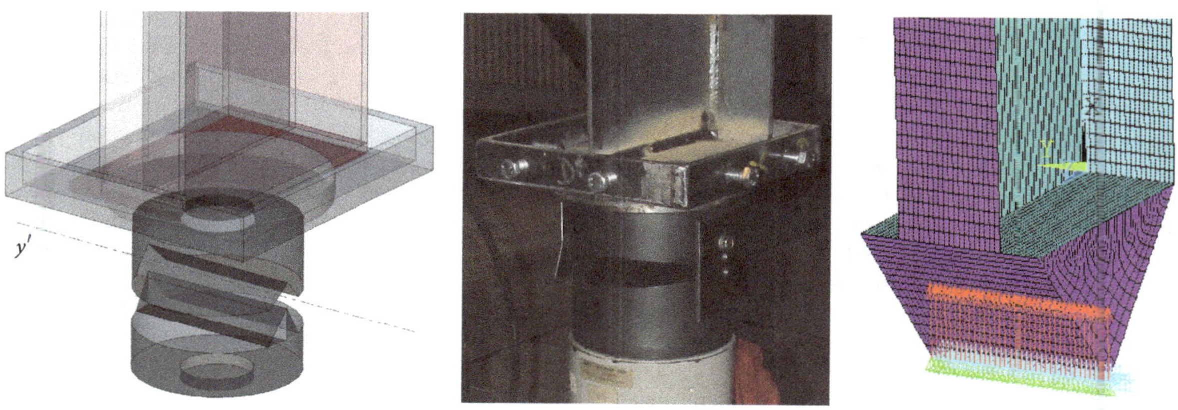

FIGURE 3: Knife-edge boundary conditions.

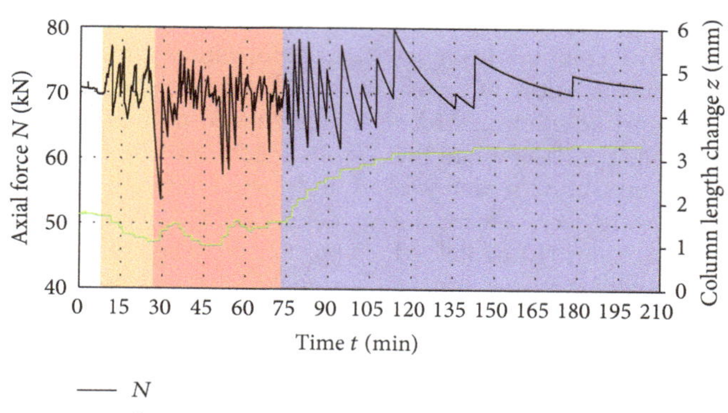

FIGURE 4: Axial force and column length changes during the spot welding (orange background colour), welding process (red colour), and cooling (blue colour).

experiments and similar experiment performed by authors [22] and the longitudinal stresses available in the literature (e.g., [23–25]). The residual stress of bare plates was neglected. The SHELL 181 element type was used for all plates. It is a four-node structural shell element with six degrees of freedom at each node which is well suited for large strain nonlinear applications [21]. The mesh had to be relatively dense (element size 4×10 mm) due to high stress gradients caused by welding simulation. Mapped mesh was used all over the column and welds; the degenerate triangular elements were used only on few plates of knife-edge bearings which were modelled with several plates (see Figure 3). The material model of the knife-edge bearings was elastic (modulus of elasticity $E = 210$ GPa) and the material model of the plates was multilinear isotropic with the average true stress-strain diagram obtained from the coupon tests. The columns in the numerical analyses, labelled D, E, and F, correspond to sets of specimens (D), (E), and (F) used in the experimental research.

The column strengthened under load F was modelled using the following procedure. First, the ideal geometry was created. Pinned boundary conditions around the axis y' and node coupling in longitudinal direction x were used at the position of the knife-edge bearings. Then, the model was updated with imperfections using Block Lanczos Eigen Buckling analyses. The elements of the strengthening plate and fillet welds were deactivated and the base column with a T-shaped cross-section was updated with the first mode of Eigen Buckling analysis. This mode corresponded to the local buckling of the web and caused web undulation. The maximum deflection was chosen as 1 mm, which corresponded to 1/200 of the web width. The strengthening elements were then activated and another Eigen Buckling analysis was performed. The first mode, corresponding to global torsional-flexural buckling, was selected with a maximum deflection of 3.75 mm, which was equal to 1/800 of the column length that is the limit geometrical tolerance stated in EN 1090-2 [26] and roughly coincided with the average results of the Southwell plot. Additionally, other three modes were added with decreasing amplitude (3.75 mm, 1.875 mm, 0.938 mm, and 0.469 mm for the first, second, third, and fourth mode, resp.). The elements were then switched to SHELL 131 and two thermal analyses were conducted. The first was for the residual stresses distribution of the base welded T-shaped cross-section, and the second was for deflections and residual stresses caused by the welding of the strengthening flange under load (see Figure 9). After this, the elements were changed to structural again and five consecutive static analyses were performed. First, the elements of the strengthening plate were deactivated and the residual stresses for the base T-shaped cross-section were implemented. Second, the column was loaded to 70 kN. Third, the elements of the strengthening plate were reactivated and residual stresses and deflections caused by the welding process were implemented. Fourth, the column was unloaded to determine the change in the initial imperfections of the column strengthened under load. Finally, the column was loaded with force until the solver stopped converging, meaning collapse had been reached.

For the purpose of comparison, the models of columns D and E were also created using a similar procedure. The geometry of columns D and E updated the same imperfections as column F. The same program and similar procedure were used by Liu and Gannon [27, 28] and Wang et al. [29] in their recent work on beams strengthened under load.

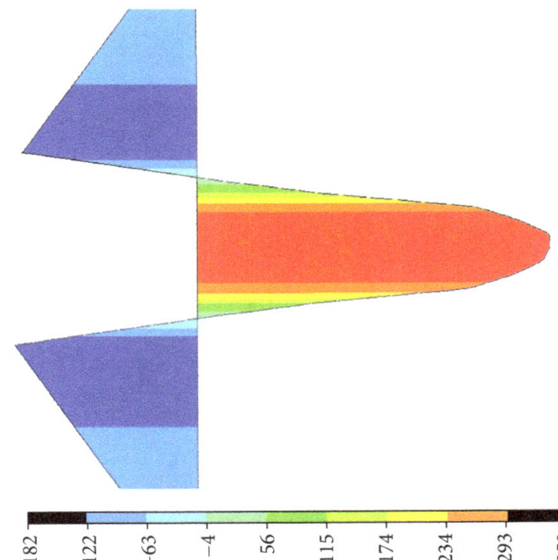

FIGURE 5: Stress distribution in longitudinal direction σ_x [MPa] caused by welding in longer flange used in finite element analysis.

The effect of residual stress was investigated in greater detail because it is the source of unknown variables. The temperature load that was determined to correspond to the measured deformations of specimens (throat thickness 4 mm, area of the weld $A_w = 32\,\text{mm}^2$; see Figures 5 and 9) and then temperature load higher by 33% and lower by 33% was applied to the models of columns D, E, and F. These temperature loads are expected to correspond to the weld with throat thickness 3.3 mm (area of the weld $A_w = 21\,\text{mm}^2$) and 4.6 mm (area of the weld $A_w = 43\,\text{mm}^2$) for temperature load 33% lower and 33% higher, respectively. A little difference in throat thickness leads to a relatively big difference in the area of the weld. Other variables are the travel speed of the welding arc, number and length of pauses, and so forth. These variables are impossible to account for in field welding. The results of this parametric study are described further.

3. Results and Discussion

3.1. Experimental Research. The results of mechanical properties were averaged. Young's modulus of elasticity, yield strength, and ultimate strength were 210 GPa, 316 ± 5 MPa ($\overline{\rho} = 1.57\%$), and 454 ± 5 MPa ($\overline{\rho} = 1.01\%$), respectively.

Stresses recalculated from strain gauges and deflections obtained from draw wire sensors (DWS) at the moment of collapse are plotted in Figure 6; graphs of deflections dependent on the axial force are shown in Figure 8. Note that the deflections caused by welding at the load 70 kN are included. The specimens from sets (E) and (F) failed via torsional-flexural buckling and the specimens from set (D)

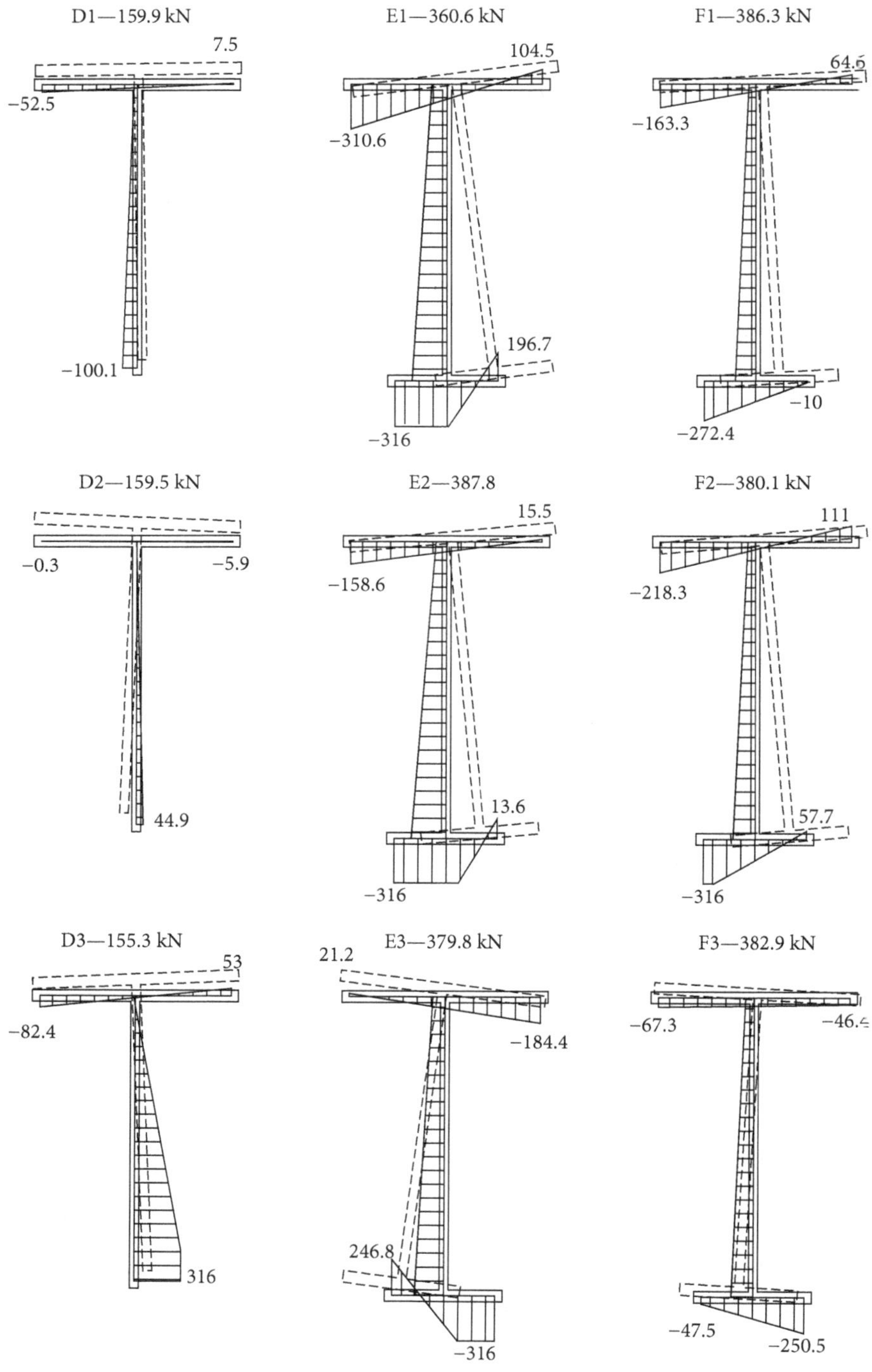

Figure 6: Experimental results at the moment of collapse: load resistance, stresses recalculated from strain gauges [MPa], and deflections.

failed via local buckling of the web, as expected (see Figure 7). In the case of specimens from set (D), strain gauges and draw wire sensors were positioned at midheight and the wave of local buckling occurred at a different location. This is the reason for big differences in the direction of axis y, rotation θ, and stress recalculated from strain gauges.

The welding process caused deformation of the flanges. The weld caused shrinkage and the flanges inclined by approximately 3° in the direction towards the web. The average stress on the flange of the base T-shaped cross-section (the average of SG1 and SG2) at 70 kN of load was only −13.4 MPa (compression). The stress after the welding process even switched into tension, to an average value of 13.9 MPa (tension). The stress on the web of the base T-shaped cross-section (SG3) was −85.9 MPa, which is still a low value compared to the yield stress. Unfortunately, the

FIGURE 7: Local buckling of column D2 (before collapse occurred) and torsional-flexural buckling of columns E3 and F3 (after collapse).

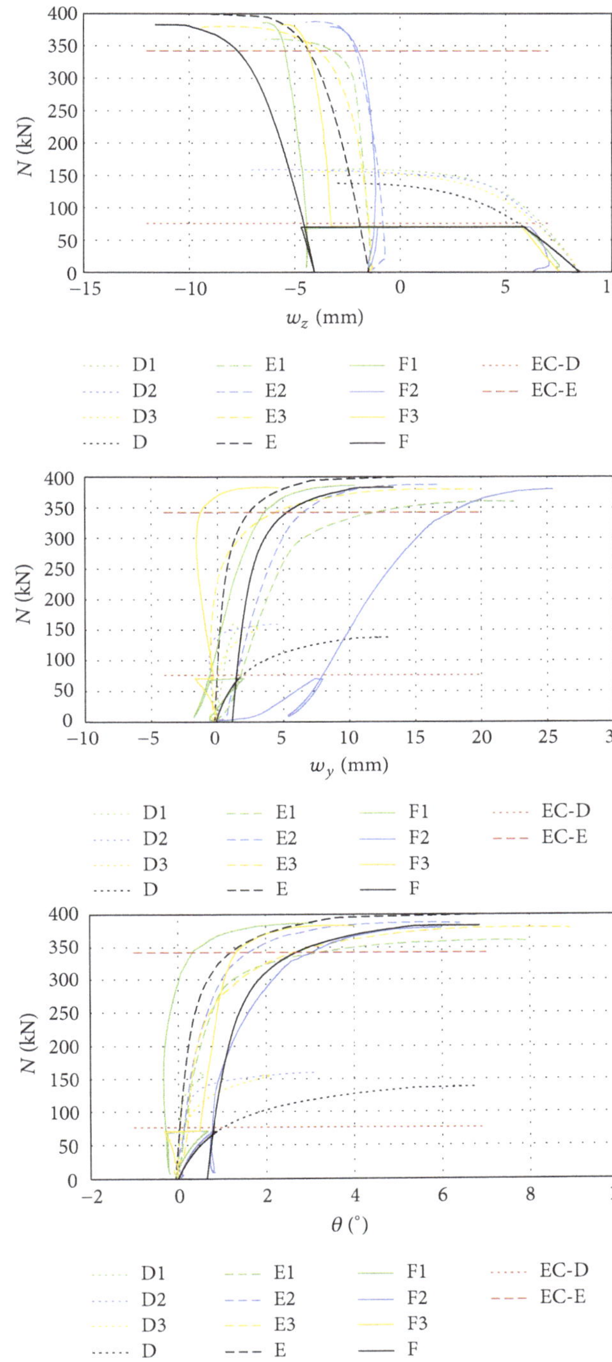

FIGURE 8: Graphs of deflections in the middle of the column in the direction of principal axes y and z and rotation θ, deflections caused by welding are included; numerical results are in black; calculation according to EC 3 is in red.

welding process near SG3_T destroyed the strain gauge but the weld clearly caused shrinkage. The specimens from set (F) were deflected at midheight by welding by an average value of −8.63 mm in the direction of axis z. This value corresponds well with the calculation by Blodgett [24] and Huenersen et al. [25], −7.59 mm and −9.03 mm, respectively.

The average load resistance of set (D) was 157 kN, which is much larger value than the one calculated from EC 3 [18]—76 kN. This issue is known and is handled by researchers, for example, [30]. The average load resistance of set (E) was 376 kN and for set (F) it was 383 kN; both values slightly exceeded resistance calculated according to EC 3—342 kN. In the case of these six specimens, the average resistance of the column strengthened under load (F) was similar and even slightly increased compared to the column with an I-shaped cross-section welded without preloading (E). This could be explained by the effect of residual stresses and deflections caused by welding, which caused the tension in the most compressed base section flange. Furthermore, the stress in the shorter flange, which is decisive for the column resistance, is slightly lower at the moment of collapse in the case of set (F) than in the case of set (E) (see Figure 6). In the case of specimens from set (E) at 70 kN, the average stress on the short flange (average of SG3 and SG4 which were glued after welding and therefore they did not take into account residual stresses) was −29.4 MPa. Using a combination of numerical (only to determine residual stress) and experimental results, it seems that the shorter flange of specimens from set (F) started to yield in average at 21 kN higher load than in the case of specimens from set (E). This behaviour and a slight increase in load resistance were also observed in previous experimental research by the authors [22] and in the literature, for example, by Nagaraja Rao and Tall [6]. However, it seems that specimen E1 had higher initial nonintended eccentricity in supports which resulted in the lower load resistance. Additionally, column E1 was delivered slightly damaged by the supplier.

The unloading and loading load-displacement curves of columns F1 and F2 followed the same path and the load resistances of both specimens were the same as of column F3. Therefore, it is presumed the unloading to 10 kN did not

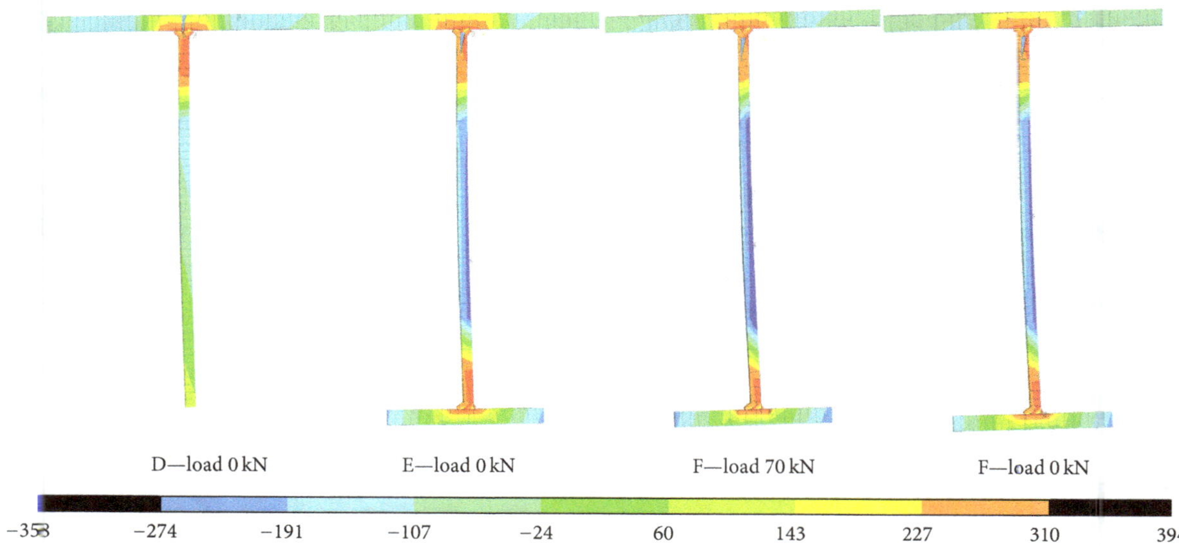

FIGURE 9: Residual stresses in the longitudinal direction σ_x [MPa] caused by welding at midheight.

affect the column behaviour. From the graphs in Figure 8, it can be seen that the strengthened column is stiffer and its load-displacement curve is more vertical. This confirms the theory of increased imperfection by Unterweger [11]. However, the welding process caused predictable deflection in the direction of axis z which was much more serious and random deflection in the direction of axis y and rotation θ which were of the similar magnitude as the increased imperfection.

3.2. Numerical Modelling. The goal was to achieve similar distortion and stress distribution as described in Section 3.1. Then, values which were unclear and hard to measure in the experiment could be read from the finite element models. The results of deflections are included in Figure 8 for comparison with the experimental results. The numerical results correspond closely to those from the experiment. Differences may be accounted for by the different shapes of the initial imperfections. A large difference only appeared in the case of set (D), which failed via local buckling. Unlike in the case of the finite element analysis, the decisive buckling wave that occurred during the experiment did not occur at the midheight where the measurements were taken and therefore only the deflections in the direction of axis z are comparable. The deflections of column F during the loading were slightly higher than those of column E. This suggests that the assumption made by Unterweger [11] that there are higher initial imperfections in the member strengthened under load is correct. However, the experimental results do not show clear pattern due to the unpredictable stresses and deflections caused by welding.

Residual stresses caused by welding are shown in Figure 9. Column F had lower compression stress than column E in the shorter strengthening flange where the most stressed fibres were located when the column later failed via buckling. This might be one of the reasons why the columns

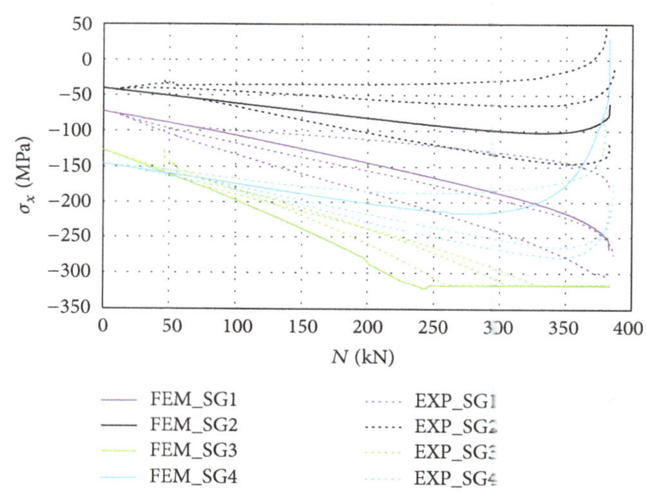

FIGURE 10: Longitudinal stresses on strain gauges during loading, comparison of numerical model and experimental results.

strengthened under load (F) had a slightly higher average load resistance than the regular columns (E) tested in the experiment. Longitudinal stresses σ_x in the position of the strain gauges are compared with the experimental results in Figure 10. The stresses from the experiment were recalculated from measurements obtained from the strain gauges, which unfortunately were not glued to the plain plates, so they did not reflect the residual stresses. For this reason, the origins of experimental stresses (dotted lines) were set to the stresses from the numerical model at zero load. Figure 11 shows the stresses in the longitudinal direction at the moment of collapse. The stress distribution of columns E and F is very similar, with a significant portion of the cross-section passing the yield point in both cases. The load resistance of columns D, E, and F was 137.8 kN, 399.4 kN, and 383.0 kN, respectively.

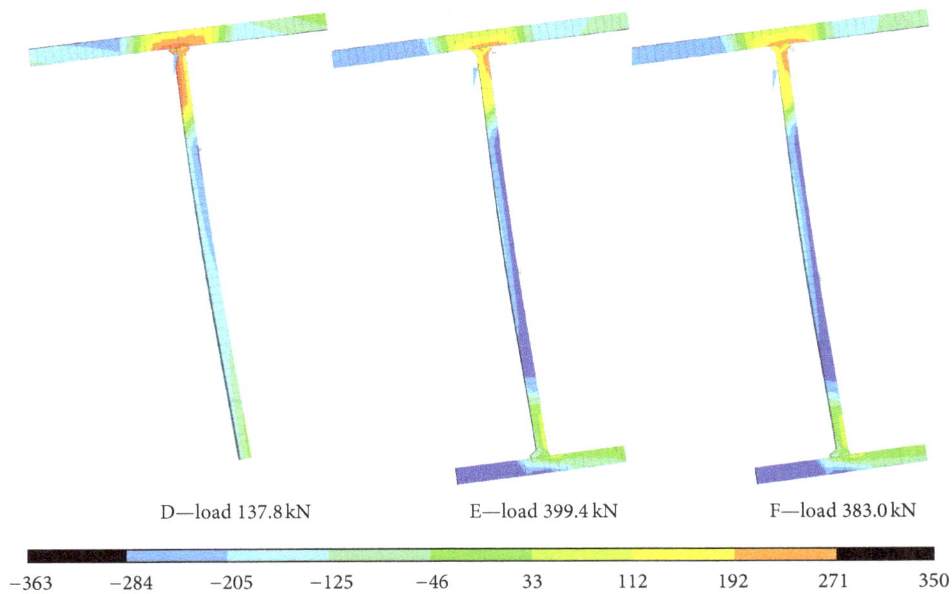

FIGURE 11: Stresses in the longitudinal direction σ_x [MPa] at midheight at collapse.

The member strengthened under load F and under no load E had nearly the same load resistance. The slight increase in load resistance for the set of columns strengthened under load (F) measured in the experiment was therefore not confirmed by these finite element analyses. This suggests that this increase could be caused by smaller initial imperfections of specimens from set (F), the slightly damaged specimen E1, and possible different quality of the strengthening weld, as discussed further.

The results of the parametric study concerning the temperature load causing residual stress distribution are summarised in Table 1. The temperature load corresponding to the average measured deformations of the specimens, the decreased temperature load, and the increased temperature load are labelled exp, 0.67 × exp, and 1.33 × exp, respectively. In the case of column D with the T-shaped cross-section, the load resistance increased with increasing temperature load. This can be explained by the higher deflection in the direction of axis z which decreased the distance between the point of load application and the centre of gravity in the midheight which led to smaller bending moment. In the cases of columns E and F, the load resistance slightly decreased with increasing temperature load. The difference in deflections caused by welding, while big on its own, is small if effects of both welds at longer flange and shorter flange are summed up. The decrease in load resistance can be attributed to the higher compressive residual stress. The decrease in load resistance of the highest and lowest applied temperature load was only 8% but it can be one of the factors contributing to the difference between results of specimens in each set of experimental study and also the difference between numerical models and averages of experimental study. The welder performed welding operations of specimens from sets (D) and (E) and the welds at the longer flange of specimens (F) comfortably on the bench in a horizontal position. The strengthening fillet welds at the shorter flange of specimens (F) were performed in a vertical position, the upper parts from a ladder. Thus, the quality of welds, travel speed of the welding arc, and so forth were affected.

TABLE 1: Results of parametric study: load resistance $N_{b,R}$, deflection in the direction of axis z caused by the fillet weld at longer flange $w_{z,T}$ and at shorter flange $w_{z,I}$

Column	Temperature load	$N_{b,R}$ [kN]	$w_{z,T}$ [mm]	$w_{z,I}$ [mm]
D	0.67 × exp	134	6.3	
	exp	138	8.5	
	1.33 × exp	140	10.3	
E	0.67 × exp	420	6.3	−7.2
	exp	399	8.5	−9.9
	1.33 × exp	387	10.3	−12.4
F	0.67 × exp	405	6.3	−7.4
	exp	383	8.5	−10.5
	1.33 × exp	374	10.3	−13.4

4. Conclusion

The experiments on slender columns susceptible to local and global buckling proved that the method of strengthening T-shaped columns with a second flange under preload magnitude up to 50% of the base section's buckling resistance is safe and feasible. These six experiments support the claim that the load resistance of a column strengthened under load is nearly the same as that of a column welded without preloading. The numerical simulation was managed to determine values that had not been measured in the experiment. The following conclusions can be drawn from the experiments and research involving numerical models:

(1) The force changes rapidly during the shielded metal arc welding process with thick electrode core wire. Temperature changes must be carefully monitored and stress changes considered especially if the column is less stiff than surrounding structural components. Use of thin electrode core wire or intermittent welds could reduce these changes.

(2) The residual stresses and deformations caused by welding are often much higher than the stresses caused by preloading. Especially in cases with slender base sections, the stresses during axial loading rise slowly with increasing load but plummet at the most stressed fibres in the cross-section when the load on the column is nearing its critical load. Therefore, with preload magnitudes up to 50% of the base section's load resistance, the buckling of the base section (both local and global) hardly affects the load resistance of the strengthened section and strengthening plates can prevent further local buckling.

(3) Welding causes shrinkage and tensional residual stresses near the weld. If the strengthening method is designed so that the position of the welds is near the most compressed fibres or causes deformations which have a positive influence, the welds can increase buckling resistance.

(4) Similarly, if possible, strengthening plates should be positioned outside the base cross-section in order to increase the moment of inertia. In the case of columns strengthened under load, these plates reach the yield point later, thus increasing buckling resistance.

The validated numerical model will find application in the behavioural analysis of members strengthened under load in future research in this area. Such research is currently in progress, and other experiments have already been planned to verify the above-mentioned conclusions.

Competing Interests

The authors declare that there is no conflict of interests regarding the publication of this paper.

Acknowledgments

The financial support of Project no. LO1408 "AdMaS UP-Advanced Materials, Structures and Technologies," supported by the Ministry of Education, Youth and Sports under "National Sustainability Programme I" is gratefully acknowledged.

References

[1] H. A. Gordy, "South Shore Plaza," Modern Steel Construction, AISC, 1997.

[2] S. Punch and S. S. Wall, "Constant growth," Modern Steel Construction, AISC, 1994.

[3] K. Horikawa and H. Suzuki, "Studies on repair welding in Japan," *Transactions of JWRI*, pp. 175–183, 1985.

[4] R. L. Brockenbrough, "AISC rehabilitation and retrofit guide: a reference for historic shapes and specifications," *APT Bulletin*, vol. 34, no. 1, p. 53, 2003.

[5] T. P. O'Sullivan, "Strengthening of steel structures under load," *Proceedings of the Institution of Civil Engineers*, vol. 2, no. 1, pp. 76–95, 1953.

[6] N. R. Nagaraja Rao and L. Tall, "Columns reinforced under load," *Welding Journal*, vol. 42, p. 54, 1963.

[7] H. Marzouk and S. Mohan, "Strengthening of wide-flange columns under load," *Canadian Journal of Civil Engineering*, vol. 17, no. 5, pp. 835–843, 1990.

[8] J. H. Brown, "Reinforcing loaded steel compression members," *AISC Engineering Journal*, pp. 161–168, 1988.

[9] D. T. Ricker, "Field welding to existing steel structures," *AISC Engineering Journal*, pp. 1–16, 1988.

[10] L. Spal, *Rekonstrukce Ocelových Konstrukcí*, STNL, Prague, Czechoslovakia, 1968.

[11] H. Unterweger, "Ultimate load capacity of columns strengthened under preload," *Advances in Steel Structures*, vol. 1, pp. 117–124, 1999.

[12] L. Tall, "Reinforcement of steel columns," *AISC Engineering Journal*, pp. 33–39, 1989.

[13] R. H. R. Tide, "Reinforcing steel members and the effects of welding," *AISC Engineering Journal*, vol. 27, no. 4, pp. 129–131, 1991.

[14] Z. Wu and G. Y. Grondin, "Behaviour of steel columns reinforced with welded steel plates," Structural Engineers Report 250, University of Alberta, Edmonton, Canada, 2002.

[15] A. K. Bhowmick and G. Y. Grondin, "Limit state design of steel columns reinforced with welded steel plates," *Engineering Structures*, vol. 114, pp. 48–60, 2016.

[16] V. M. Kolesnikov, "O normách po rasčotu stalnych konstrukcij, usiljennych pod nagruzkoj," *Promyšlennoje Strojitělstvo*, vol. 10, pp. 46–50, 1962.

[17] M. Vild and M. Bajer, "Strengthening under load: the effect of preload magnitudes," *Procedia Engineering*, vol. 161, pp. 343–348, 2016.

[18] CEN, "Eurocode 3: design of steel structures—part 1-1: general rules and rules for buildings," EN 1993-1-1:2005, 2009.

[19] S. R. O. Pontex, *TP 42—Opravy, Obnovy a Přestavby Ocelových Nosných Konstrukcí Mostů*, Ministerstvo dopravy, Praha, Czech Republic, 2014.

[20] R. V. Southwell, "On the analysis of experimental observations in problems of elastic stability," *Proceedings of the Royal Society of London. Series A*, vol. 135, no. 828, pp. 601–616, 1932.

[21] ANSYS®, *Academic Research, Release 15.2, Help System, ANSYS Parametric Design Language Guide*, ANSYS, 2015.

[22] M. Vild and M. Bajer, "Strengthening under load: experimental and numerical research," *IOP Conference Series: Materials Science and Engineering*, vol. 96, no. 1, pp. 1–10, 2015.

[23] L. Gannon, Y. Liu, N. Pegg, and M. Smith, "Effect of welding sequence on residual stress and distortion in flat-bar stiffened plates," *Marine Structures*, vol. 23, no. 3, pp. 385–404, 2010.

[24] O. W. Blodgett, *Design of Welded Structures*, James F. Lincoln Arc Welding Foundation, Cleveland, Ohio, USA, 1966.

[25] G. Huenersen, H. Haensch, and J. Augustyn, "Repair welding under load," *Welding in the World*, vol. 28, no. 9, pp. 174–182, 1990.

[26] CEN, "Execution of steel and aluminium structures—part 2: technical requirements for steel structures," EN 1090-2+A1, 2011.

[27] Y. Liu and L. Gannon, "Experimental behavior and strength of steel beams strengthened while under load," *Journal of Constructional Steel Research*, vol. 65, no. 6, pp. 1346–1354, 2009.

[28] Y. Liu and L. Gannon, "Finite element study of steel beams reinforced while under load," *Engineering Structures*, vol. 31, no. 11, pp. 2630–2642, 2009.

[29] Y.-Q. Wang, L. Zong, R.-X. Zhu, X.-Y. Liu, and Y.-J. Shi, "Behavior of I-section steel beam welding reinforced while under load," *Journal of Constructional Steel Research*, vol. 106, pp. 278–288, 2015.

[30] A. Taras, J. P. Tapley, P. Kugler, and H. Unterweger, "On the stability behaviour and design of T-section beam-columns," in *Proceedings of the 7th European Conference on Steel and Composite Structures (Eurosteel '14)*, 2014.

5

Experimental Investigation of Polyurea-Coated Steel Plates at Underwater Explosive Loading

Li-hui Dai, Cheng Wu, Feng-jiang An, and Sha-sha Liao

State Key Laboratory Explosion Science and Technology, Beijing Institute of Technology, Beijing 100081, China

Correspondence should be addressed to Cheng Wu; chengwu@bit.edu.cn

Academic Editor: Yuanshi Li

To improve the survivability of ship structures at underwater explosion, thin steel plates coated with polyurea were used to investigate the blast protection effect. During the experimental tests of bare steel plates at different standoff, an appropriate distance was selected as the reference standoff to perform the tests of coated plates. Experimental tests of different coating locations (front versus back) and coating thickness were carried out to study the influencing factors of blast resistance for metal substrate plates. Compared with the bare steel plates, the polyurea coating was found to reduce the deformation of the test plates at blast tests in both cases of the front and back surface locations of the polyurea layer. An increase in the coating thickness also mitigates substantially the deformation of plates. In addition, the properties of the material and the substrate-coating bond strength may also affect the protective effect of the polyurea coating.

1. Introduction

The threat of terrorism in recent years has prompted research into the importance of structural protection of surface ships. On October 12, 2000, the USS Cole was attacked in Aden Harbor, Yemen [1]. The terrorist attack did serious damage on the ship and caused huge economic loss. After the terrorist attack, the importance of materials used to improve the survivability of ships had been focused on.

A series of studies have been carried out to investigate the effect of polymer coatings on different structures when subjected to blast damage or penetration in recent years. The results show that polymer coatings improve the ballistic resistance and explosive resistance of metal structures and buildings [2–8]. When brick wall or concrete structure collapses under the impact of explosive loading, the debris of bricks crush splash. Once the surface of the wall or concrete structure was coated with polyurea, free flying of the fragments of the back face of such structures after blast was prevented [2–4].

Other studies had found that polymeric coating increases the impact resistance of hard substrates, such as metal or composite structure [6–11]. Amini et al. [6] investigated the effect of a polyurea layer casted on the back of a plate under dynamic shock. It was found that polyurea enhanced energy absorption and mitigated the failure of steel plates. Ackland et al. [11] performed experimental and numerical studies of the blast resistance of mild plates coated with polyurea on the back face under local explosive loading. The results showed the residual deformation increase along with the increase in the coating thickness. Besides, Ackland et al. [10, 12] found reduction in deformation of mild steel plates when polyurea coating casted on their back side. Furthermore, they found that the blast resistance effect of a thicker coating on the back surface was superior than that of a thinner one. Roland et al. [8, 9] applied elastomeric coatings on the front face to enhance the ballistic resistance of steel armor under impact loading generated by a high-speed projectile. Furthermore, polymer-metal laminates were also used to improve the ballistic performance, and they provided superior ballistic protection than uniform polyurea coatings.

When applying polymeric material to enhance the blast resistance of the hard substrate, the location of the polyurea layer on the front or back face is important. Influencing factors should be taken into account [12–15].

Ackland et al. [12] applied commercially produced coatings on the surface of mild steel plates to carry out close-in blast experimental tests in air. It was found that the deformation of the steel plate coated with polyurea on the front face was even larger than that of the bare steel plate. However the polyurea coated on the back face effectively reduced the deformation of the plate at blast tests. This proved that the polyurea layer coated on the back face of the mild plate was more effective in increasing the blast resistance effect than that applied on the front face. Amini et al. [13–15] investigated the effect of the polyurea coating on the front and back faces of steel plates. When the coating was sprayed on the front face of the circular steel plate, the compression of the polyurea under the shock loading increased its stiffness, resulting in a better impedance match with the steel plate. Thus, the coating on the loading face of a plate transfers more energy to the plate promoting the failure of steel plates. Contrary to that, the coating deposited on the back face of steel captures and dissipates some energy of initial shock loading impacted on the steel plate. That resulted in reducing the deformation and avoiding plate fracture.

According to the results of researchers above, polyurea coatings could be considered to reduce the vulnerability of ship structures subjected to underwater blast loading.

In this paper, a series of underwater near-field explosion tests were carried out to investigate the effect of the polyurea coating on thin steel plates. The research was focused on the questions how the location of polyurea coating with respect to the loading direction and the thickness of polyurea coating affect the blast resistance performance of steel plates. In addition, the dependence of the deformation and failure of test plates on the standoff between the charge and the steel plate also discussed in this paper.

2. Experimental Setup

During the tests of the effects of the coating location (front versus back) and the coating thickness, the final deformations of steel plates coated and uncoated with polyurea were compared as the evaluation criteria after blast tests.

2.1. Test Plate

2.1.1. Materials. A3 grade steel was used as the substrate plates due to its availability; its mechanical properties are presented in Table 1.

Polyurea is readily synthesized from aromatic or aliphatic isocyanates with chemical functionality of oligomeric diamines [16], which could be easily mixed and sprayed onto the surface of a metal like steel. The plates had been prepared by grit blasting to remove the surface metal rust and then covered with an epoxy primer to enhance the adhesion between metal and polyurea. The polyurea was sprayed onto plates, cured at room temperature, and stored at room temperature for at least one week to ensure stability of the material. The polyurea samples used in this experiment were provided by Qingdao Shamu International Trade Co. Ltd, China. The physical and mechanical properties of the polyurea are listed in Table 2, and the stress-strain characteristics of the polyurea under high strain rates are plotted in Figure 1.

Table 1: Mechanical properties of steel used in blast tests.

Steel	Yield strength (MPa)	Tensile strength (MPa)
A3	235	375–450

Table 2: Physical and mechanical properties of polyurea used in blast tests.

Sample	Tensile strength (MPa)	Fracture elongation (%)
Polyurea	18.1	184

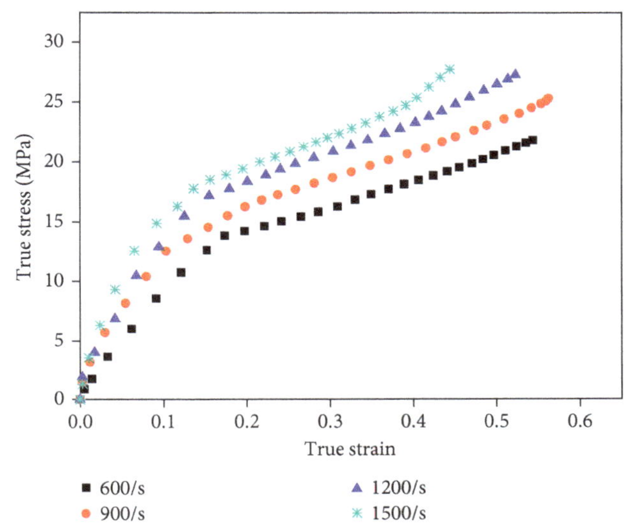

Figure 1: The true tensile stress-strain curves of polyurea under different strain rates.

In Figure 1, the tensile deformation process of polyurea contains three deformation characteristics: an initial linear elastic region corresponding to small deformation; a transitional region of the started yielding; and a viscoplasticity region before fracture. The stress-strain behavior of polyurea at high strain rates exhibits nonlinear and rate dependency. It was also investigated by other researchers [17, 18].

2.1.2. Test Plate Preparation. According to the test requirements, the thickness of the coating polyurea layer on the steel plate was equal to or twice more than the thickness of the substrate plate. In the experimental tests, the thickness of the steel plate was 2 mm; thus, the thickness of polyurea coatings was 2 mm and 4 mm.

It should be noted that, due to spray inhomogeneity, the thickness of the polyurea coating was not as precise as the thickness of steel plates. The average value of the thickness at the center of polyurea coating and at its edges was usually considered as the coating thickness. The calculated areal density of the coatings on the test plates is listed in Table 3.

Table 3: Areal density of different types of test plates.

Type no.	Substrate plate	Coating thickness (mm)	Areal density (kg/m^2)
1	2 mm steel	No	15.6
2	2 mm steel	2	17.7
3	2 mm steel	4	19.8

2.2. Test Setup. The experimental tests were carried out in a water pool, and the experimental test site layout is shown in Figure 2. The length and width of the pool was 2 m × 2 m, and the depth of water was 2 m. The experimental test setup was hanged in the center of the pool, and the charge was 1 m below the free water surface.

The test plate was fixed on the test rig with bolts as shown in Figure 3. The dimensions of test plates were 0.5 m × 0.5 m, with a test area of 350 mm × 350 mm used for the experiment. The details of the plates for blast tests and standoff are listed in Table 4. The long steel rods were used to fix charge on the opposite side of the test plate. The wires tied to the long steel rods could be moved to achieve an appropriate distance between the charge and the test plate.

In each blast test, 10 g cylindrical charge of RDX was suspended in front of the center of the plate to reduce the plate boundary effects. An electronic detonator located centrally in the charge was used for ignition.

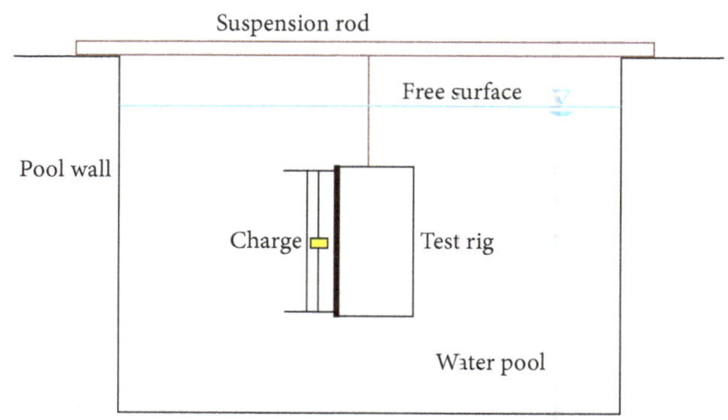

Figure 2: Experimental site layout.

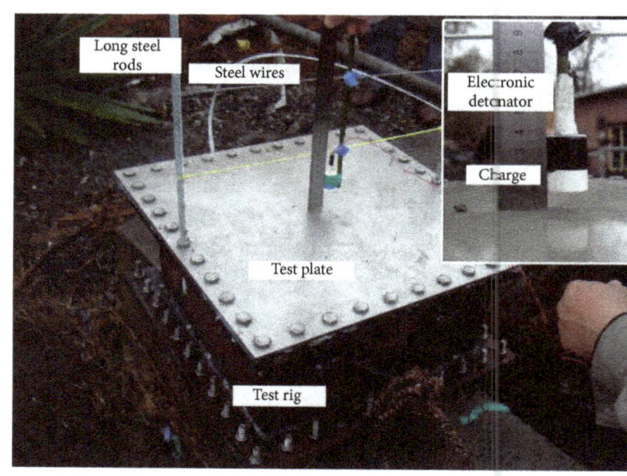

Figure 3: Details of the test setup

3. Results and Discussion

The tasks of the experimental tests were the effect of the standoff distance, coating location (front versus back), and coating thickness on the substrate deformation. The deformations of the substrates were measured after the test plates unbolted from the test rig.

3.1. Bare Steel Plates at Different Standoff. The deformation and damage of bare steel plates at different standoff are presented in Table 5. In addition, the deformations of each test plate along the width direction are plotted in Figure 4. The maximum deformation of the test plate increased as the distance between the charge and the surface of the test plate decreased, and the local variation in deformation increased obviously.

Once the standoff becomes less, much more energy generated by the charge affected on the test plates, and the degree of test plate deformation increased. The test plates even ruptured, just like test no. 6.

As presented in Figure 5(a), test no. 5, at the 10 mm standoff, the plate did not rupture and just have got dishing deformation. While in test nos. 6 and 7, both at 8 mm standoff the test plates ruptured with a crack or broken into three petals as shown in Figures 5(b) and 5(c), respectively.

According to the results of test no. 6, we assumed that the deformation value at this standoff is the deflection limit for this thickness plate, and it may be considered as a reference value. Thus, the 8 mm gap between the charge and the plate, which was in test no. 6, was determined as the appropriate standoff distance in the subsequent tests.

3.2. Steel Plates Coated with Polyurea. Table 6 shows the deformation test results of the substrate plates with and without polyurea. Besides, the data on relative reduction in deformation of the coated plates with respect to bare plates are also presented in Table 6.

The thick (4 mm) and thin (2 mm) polyurea coatings, whether on the front or back surface of steel plates, reduced about 40% and 30% of the deflection deformation, respectively, while the areal density increased by only 27% or 13%.

The bare steel plates undergo elastic-plastic deformation under explosive loading, and they rupture once the deformation exceeds the ductility limit of the material. The coating mitigates the fracture or reduces the deformation of substrate effectively, regardless of whether it is deposited on the front or rear face. It was also found that once the thickness of polyurea coating increased, the maximum

Table 4: Details of plates for blast tests.

Test no.	Substrate plate	Coating thickness (mm)	Surface	Standoff (mm)
1	2 mm steel	No	N/A	30
2	2 mm steel	No	N/A	20
3	2 mm steel	No	N/A	15
4	2 mm steel	No	N/A	12
5	2 mm steel	No	N/A	10
6	2 mm steel	No	N/A	8
7	2 mm steel	No	N/A	8
B3	2 mm steel	4	Front	8
B4	2 mm steel	4	Back	8
B5	2 mm steel	2	Back	8
B6	2 mm steel	2	Front	8

Table 5: Results of bare steel plates under different standoff.

Test no.	Substrate plate	Standoff (mm)	Maximum deflection (mm)
1	2 mm steel	30	59.0
2	2 mm steel	20	67.4
3	2 mm steel	15	68.1
4	2 mm steel	12	78.6
5	2 mm steel	10	79.7
6	2 mm steel	8	113.5 (a crack)
7	2 mm steel	8	Three petals

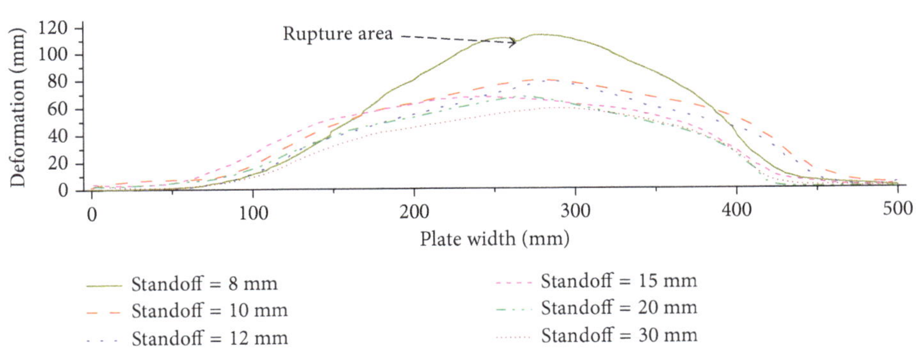

Figure 4: Deformation of bare steel plates after the test along the width direction.

deflection decreased. Therefore, the coating thickness increase is effective for steel plate protection at underwater explosion tests.

As shown in Table 6, in both cases of 2 mm and 4 mm coating thickness, the steel plates with polyurea coated on the front surface performed better than those covered on the rear face at blast tests. The details are discussed below.

3.2.1. Steel Plates with Polyurea Coated on the Front Surface. Figure 6 presents the front faces of two plates coated with different thicknesses of polyurea on the front surfaces. A circular torn hole appeared near the center of each plate. The polyurea coatings disconnected from the steel plate surface and showed a circular debonded area at the center of the plates. The areas surrounded by black dotted lines in Figure 6 are the places where the polyurea layer was totally separated from the plates. The average diameter of the unbonded area was about 260 mm for thin polyurea coating (2 mm), while for thick polyurea coating (4 mm), the value was 210 mm.

The deformation of the steel plate with thin coating (2 mm) was larger than that of the plate coated with thick polyurea (4 mm) as shown in Table 6; thus, the gap between the thin polyurea coating (2 mm) and the steel plate was longer due to the elastic recovery of polyurea. And that may result in a larger debonding area.

(a) (b) (c)

Figure 5: Examples of test plates after test: (a) no rupture (10 mm standoff); (b) a crack (8 mm standoff); (c) three petals (8 mm standoff).

Table 6: Results of test plates coated with polyurea at the same standoff.

Test no.	Coating thickness (mm)	Surface	Maximum deflection (mm)	Areal density increase compared to uncoated plate (%)	Reduction in deformation (%)
6	No	N/A	113.5	—	—
B3	4	Front	66.6	127	41.3
B4	4	Back	68.4	127	39.7
B5	2	Back	81.9	113	27.8
B6	2	Front	78.8	113	30.5

The standoff distance was 8 mm, and the substrate was 2 mm steel plate in each test.

3.2.2. Steel Plates with Polyurea Coated on the Back Surface. Figure 7 presents the back faces of two test plates coated with polyurea of different thicknesses on the back surface. Unlike the damage of polyurea coated on the front surface of steel plates, the polyurea that covered on the back surface of plates not only separated from the substrate plate but also cracked or even fallen off a large-scale area after the blast test. In particular, 4 mm polyurea layer coating completely broken off from the substrate and broke up into several small fragments. In Figure 7(c), there is a hole in the center of the fragment, which may be caused by the initial shock wave.

In general, the fracture of polymer contains brittle fracture and ductile fracture. The brittle fracture usually occurs at the elastic phase, and the fracture cross section is relative smooth without ductile deformation, as shown in Figure 7, whereas the ductile fracture mainly occurs at the plastic or viscous phase, and the cross section shows coarse with obvious ductile deformation.

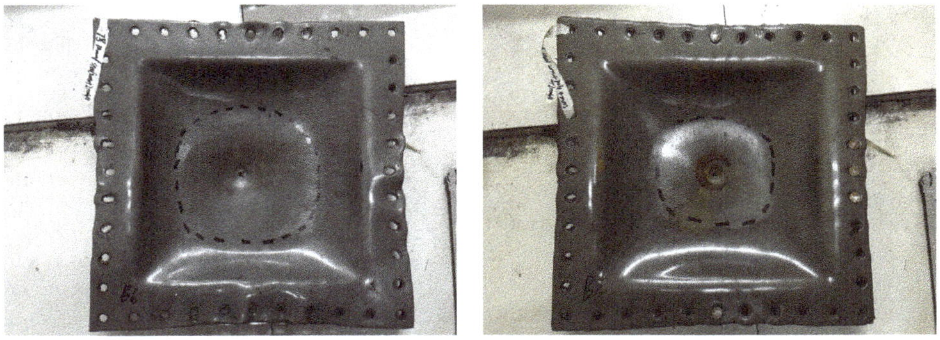

(a) (b)

Figure 6: Front faces of steel plates with the polyurea layer on the front surface after tests. (a) 2 mm polyurea coating (~260 mm diameter of unbonded layer). (b) 4 mm polyurea coating (~210 mm diameter of unbonded layer).

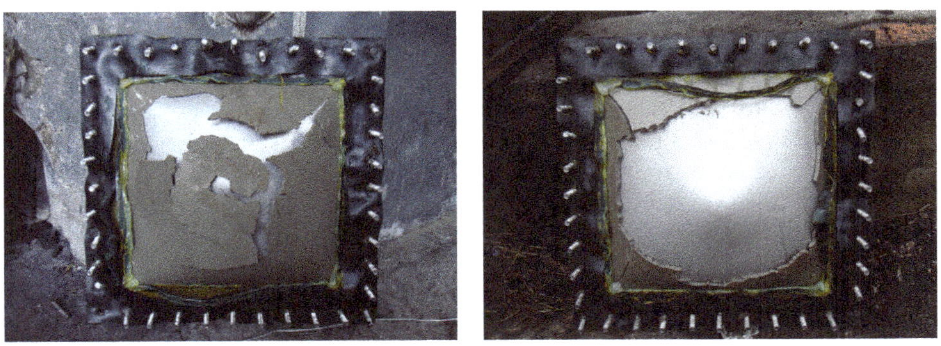

(a) (b)

(c)

Figure 7: After the test, pictures of the back faces of steel plates with polyurea sprayed on the back surface: (a) 2 mm polyurea coating; (b) 4 mm polyurea coating; (c) fragments of 4 mm coating.

There were two main reasons for the debonding effect and the large area broken off after blast tests in Figure 7. The first one was that the bonding strength between the polyurea material and the substrate was not very strong. Once the initial shock wave impacted on the polyurea, the polyurea layer would separate from the back of the substrate easily. The second one was that the elongation of the polyurea was limited. Once the polyurea layer delaminates from the bilayer structure, it could not be compressed by the impact loading; thus, its stiffness would not increase. Under high strain rate loading, the polyurea would crack at its elastic deformation stage because of low stiffness.

The above two reasons could also explain why the protective effect of polyurea coated on the back face of the substrate plate was weaker than that of polyurea coated on the front surface, which was different from other researcher's experimental results [12, 13]. If the interface bonding strength was strong enough, the initial shock wave would not pull the polyurea layer off from the back of the substrate. Instead, the compressive wave could increase its stiffness. The polyurea would not crack easily under high strain rate and could capture and dissipate a large amount of blast energy because of its viscoelasticity [19]. In addition, the compressive polyurea layer could also increase the bilayer structure's tangent modulus, delaying the onset of the necking instability [19, 20]; once the polyurea separated from the back of the substrate because of weak interface bonding strength, it would fracture easily and only dissipate a small part of shock loading energy. Thus, a large amount of energy would impact on the substrate, which resulted in a big deformation.

Besides, when the polyurea was sprayed on the loading direction face, the actual distance from the substrate to the charge was larger due to the presence of the polyurea layer. This may be also a reason for the different deformation of substrate plates.

4. Conclusions

Underwater explosion tests were carried out to investigate the protective effect of thin steel plates coated with the polyurea layer. The deformation or rupture of a bare steel plate is sensitive to the standoff distance. Therefore, it was necessary to select a suitable standoff as a reference in the experimental tests. Whether coated on the front or rear face of the plate, polyurea could provide significant blast resistance protection for the steel substrate with just a small areal density increase. Increasing the thickness of the polyurea layer on thin steel plates is helpful in reducing the deformation of the test plates at blast tests.

In addition, it was found that the effect of the front coating on mitigating the deformation of the plate is slightly better than the back coating, which is different from other researcher's experimental results. The bonding strength between the substrate and coating and the material properties of the polyurea under high strain rate may be the main reasons for different experimental results.

More studies should be carried out in future to investigate the influence of these factors.

Conflicts of Interest

The authors declare that there are no conflicts of interest regarding the publication of this paper.

Acknowledgments

The authors thank the State Key Laboratory Explosion Science and Technology, Beijing Institute of Technology, for providing the test sites and facilities to carry out the experimental tests.

References

[1] A. J. Brown, *AOE 5315 Naval Ship Vulnerability and Underwater Explosion-Introduction to Survivability OnLine Lecture 1*, Virginia Tech Department of Aerospace and Ocean Engineering, Blacksburg, VA, USA, 2004.

[2] S. N. Raman, T. Ngo, P. Mendis et al., "Elastomeric polymers for retrofitting of reinforced concrete structures against the explosive effects of blast," *Advances in Materials Science and Engineering*, vol. 2012, Article ID 754142, 8 pages, 2012.

[3] X. Huang, G. W. Ma, and J. C. Li, "Damage assessment of reinforced concrete structural elements subjected to blast load," *International Journal of Protective Structures*, vol. 1, no. 1, pp. 103–124, 2010.

[4] J. S. Davidson, J. W. Fisher, M. I. Hammons et al., "Failure mechanisms of polymer-reinforced concrete masonry walls subjected to blast," *Journal of Structural Engineering*, vol. 131, no. 8, pp. 1194–1205, 2005.

[5] C. Chen, D. G. Linzell, E. Alpman et al., "Effectiveness of advanced coating systems for mitigating blast effects on steel components," *WIT Transactions on The Built Environment*, vol. 98, pp. 85–94, 2008.

[6] M. R. Amini, J. B. Isaacs, and S. Nemat-Nasser, "Effect of polyurea on the dynamic response of steel plates," in *Proceedings of the 2006 SEM Annual Conference and Exposition on Experimental and Applied Mechanics*, St. Louis, MO, USA, June 2006.

[7] S. A. Tekalur, A. Shukla, and K. Shivakumar, "Blast resistance of polyurea based layered composite materials," *Composite Structures*, vol. 84, no. 3, pp. 271–281, 2008.

[8] C. M. Roland, D. Fragiadakis, and R. M. Gamache, "Elastomer–steel laminate armor," *Composite Structures*, vol. 92, no. 5, pp. 1059–1064, 2010.

[9] R. M. Gamache, C. B. Giller, G. Montella, D. Fragiadakis, and C. M. Roland, "Elastomer-metal laminate armor," *Materials and Design*, vol. 111, pp. 362–368, 2016.

[10] K. Ackland, C. Anderson, and N. St John, "Polymeric coatings for enhanced protection of structures from the explosive effects of blast," *Recent advances in security technology*, vol. 90, 2007.

[11] K. Ackland, C. Anderson, and T. D. Ngo, "Deformation of polyurea-coated steel plates under localised blast loading," *International Journal of Impact Engineering*, vol. 51, pp. 13–22, 2013.

[12] K. Ackland and N. St John, *Experimental and Numerical Investigations into Polymeric Coatings for Blast Protection*, Defense Science and Technology Organization and Maritime Platforms Division, Melbourne, VIC, Australia, 2011.

[13] M. R. Amini, J. Isaacs, and S. Nemat-Nasser, "Investigation of effect of polyurea on response of steel plates to impulsive loads in direct pressure-pulse experiments," *Mechanics of Materials*, vol. 42, no. 6, pp. 628–639, 2010.

[14] M. R. Amini, A. V. Amirkhizi, and S. Nemat-Nasser, "Numerical modeling of response of monolithic and bilayer plates to impulsive loads," *International Journal of Impact Engineering*, vol. 37, no. 1, pp. 90–102, 2010.

[15] M. R. Amini, J. B. Isaacs, and S. Nemat-Nasser, "Experimental investigation of response of monolithic and bilayer plates to impulsive loads," *International Journal of Impact Engineering*, vol. 37, no. 1, pp. 82–89, 2010.

[16] D. Fragiadakis, R. Gamache, R. B. Bogoslovov et al., "Segmental dynamics of polyurea: effect of stoichiometry," *Polymer*, vol. 51, no. 1, pp. 178–184, 2010.

[17] S. N. Raman, T. Ngo, J. Lu et al., "Experimental investigation on the tensile behavior of polyurea at high strain rates," *Materials and Design*, vol. 50, no. 17, pp. 124–129, 2013.

[18] C. M. Roland, J. N. Twigg, Y. Vu, and P. H. Mott, "High strain rate mechanical behavior of polyurea," *Polymer*, vol. 48, no. 2, pp. 574–578, 2007.

[19] M. R. Amini, "Effect of polyurea on dynamic response and fracture resistance of steel plates under impulsive loads," University of California, San Diego, CA, USA, Dissertations & Theses–Gradworks, 2007.

[20] Z. Xue and J. W. Hutchinson, "Neck development in metal/elastomer bilayers under dynamic stretchings," *International Journal of Solids and Structures*, vol. 45, no. 13, pp. 3769–3778, 2008.

6

Multiobjective Optimization of Turning Cutting Parameters for J-Steel Material

Adel T. Abbas,[1] Karim Hamza,[2] Mohamed F. Aly,[3] and Essam A. Al-Bahkali[1]

[1]*Department of Mechanical Engineering, King Saud University, P.O. Box 800, Riyadh 11421, Saudi Arabia*
[2]*Department of Mechanical Engineering, University of Michigan, Ann Arbor, MI 48105, USA*
[3]*Department of Mechanical Design, American University in Cairo, New Cairo, Cairo 11835, Egypt*

Correspondence should be addressed to Adel T. Abbas; atabbas1954@yahoo.com

Academic Editor: Wenbin Yi

This paper presents a multiobjective optimization study of cutting parameters in turning operation for a heat-treated alloy steel material (J-Steel) with Vickers hardness in the range of HV 365–395 using uncoated, unlubricated Tungsten-Carbide tools. The primary aim is to identify proper settings of the cutting parameters (cutting speed, feed rate, and depth of cut) that lead to reasonable compromises between good surface quality and high material removal rate. Thorough exploration of the range of cutting parameters was conducted via a five-level full-factorial experimental matrix of samples and the Pareto trade-off frontier is identified. The trade-off among the objectives was observed to have a "knee" shape, in which certain settings for the cutting parameters can achieve both good surface quality and high material removal rate within certain limits. However, improving one of the objectives beyond these limits can only happen at the expense of a large compromise in the other objective. An alternative approach for identifying the trade-off frontier was also tested via multiobjective implementation of the Efficient Global Optimization (m-EGO) algorithm. The m-EGO algorithm was successful in identifying two points within the good range of the trade-off frontier with 36% fewer experimental samples.

1. Introduction

Heat-treated alloy steels exhibit many attractive properties, such as wear resistance, high strength, and high thermal stability. Examples of uses of heat-treated alloy steels include dies, automotive structures, bearings, and gears. In a highly competitive world, there is much need for optimizing the manufacturing and processing technologies to achieve high-quality products at low cost and high productivity. Focus of this research is on optimizing the cutting parameters of CNC-turning operations of J-Steel material in order to achieve (i) high-quality surface finish (which translates to product quality) and (ii) high material removal rate (which translates to productivity). Cost is not modeled in this study due to it being (at least partially) correlated with productivity and due to several uncertainties that can be site-specific (e.g., cost of labor, energy, and transport).

Examples of notable related work in the literature of machinability of other types of hardened steels include [1–3], and a review of state of the art and identification of active research areas is summarized in [4]. The primary control parameters (also termed "design variables" within an optimization framework) in CNC turning are the machining parameters (cutting speed, feed rate, and depth of cut), material, and geometry of the cutting tools and cutting fluids. Hard-turning operations require no cutting fluids, which is advantageous in terms of cost (purchase and disposal of cutting fluids), environmental impact (elimination of the need to posttreat the cutting fluids for safe disposal), and process tenability (fewer parameters to adjust). By selecting a set of standard Tungsten-Carbide insert tools, thereby considering the tooling as a fixed parameter, this study conducts a thorough examination of the three machining parameters and how they relate to the trade-off between surface quality and material removal rate.

Tuning of the cutting parameters is conducted in this paper within the context of Pareto-optimality in order to find good compromises between surface quality and process

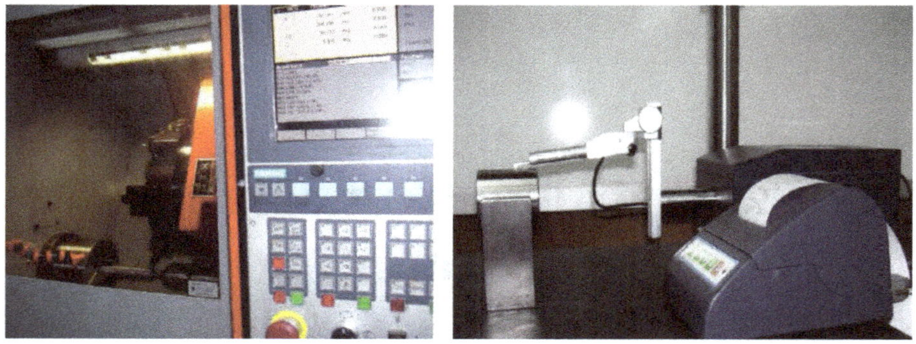

FIGURE 1: Test rig for turning and surface roughness measurement.

productivity. The tuning is done via "dense sampling" (five-level full-factorial design of experiments) as well as multi-objective implementation of an optimization technique from metamodeling based design optimization, "Efficient Global Optimization" (EGO) [5–8]. EGO seeks to locate the optimum of a processes while conducting as few experiments as possible. This is done via incrementally updating a Gaussian process metamodel that approximates the behavior of the actual process. Multiobjective versions of EGO such as Par-EGO [9] and m-EGO, which is some of the authors' earlier work in [10, 11], have been proposed and tested via numerical models. To the best of the authors' knowledge, the present study is the first experimentation with utilizing m-EGO in real machining experiments.

This paper started with a brief introduction and motivation of the performed research. The rest of the paper is organized as follows. Section 2 provides details of the experimental setup and material to be machined. Section 3 showcases the results from a five-level full-factorial experimental matrix (125 samples). Section 4 presents the m-EGO approach for identification of the Pareto trade-off frontier. Section 5 discusses the results and implications from both approaches, and then the paper concludes with a brief summary.

2. Experimental Setup

EMCO Concept Turn 45 CNC lathe equipped with Sinumeric 840-D was used to conduct experimental work. An uncoated Tungsten-Carbide insert was clamped with the tool holder to carry out this work. The specifications for insert and tool holder are TR-V13JBL 2020K and TR-VB1304-F. Clearance angle, cutting edge angle, and nose radius are maintained by 7°, 75°, and 0.4 mm, respectively. After machining, a surface roughness tester TESA was used to measure the surface roughness over a 50 mm length of the machined surface. Photos of the test rig are shown in Figure 1.

Chemical composition of J-Steel material is listed in Table 1. The heat treatment for involved Austenitizing at 850°C for 4-5 hours, quenching in oil, tempered at 600–630°C for 6-7 hours and then air-cooled. Hardness was HV 365–395.

TABLE 1: Chemical composition of the J-Steel material.

Element	C	Si	Mn	Ni	Cr	Mo	V	S	P
Minimum%	0.30	0.10	0.20	2.7	0.7	0.40	<0.25	≤0.015	≤0.015
Maximum%	0.4	0.35	0.7	3.3	1.2	0.7			

TABLE 2: Design variables values in five-level full-factorial matrix.

Level	Cutting speed x_1 (m/min)	Depth of cut x_2 (mm)	Feed rate x_3 (mm/rev)
1	150	0.10	0.050
2	175	0.15	0.075
3	200	0.20	0.100
4	250	0.30	0.150
5	300	0.40	0.200

3. Full-Factorial Sampling

A five-level full-factorial experimental matrix was used to explore various settings of the machining parameters. To frame the experiments within an optimization framework, the three machining parameters of cutting speed, depth of cut, and feed rate are referred to as the design variables x_1, x_2, and x_3, respectively. Additionally, the two performance measures of interest (surface quality and material removal rate) are formulated in a minimization setting, so that the objective is to minimize:

f_1 is the arithmetic average deviation of the assessed machined surface profile (μm)

f_2 is machining time needed to remove a unit volume of the J-Steel material (min/cm^3).

Listing of the five-level settings of the design variables is provided in Table 2. For three design variables, the number of samples corresponding to five levels is $5^3 = 125$. The corresponding objective values are plotted in Figure 2 and full listing of all 125 samples is provided in the appendix.

As the general consensus is that surface roughness in turning operations has high sensitivity to the feed rate [12], Figure 2 was constructed with separate markers for samples

TABLE 3: Listing of undominated samples from full-factorial matrix.

Pareto ID	ID # among 125 samples	x_1 (m/min)	x_2 (mm)	x_3 (mm/rev)	f_1 (μm)	f_2 (min/cm^3)
P1	56	200	0.15	0.050	0.264	0.66667
P2	36	175	0.20	0.050	0.296	0.57143
P3*	66	200	0.30	0.050	0.309	0.33333
P4*	47	175	0.40	0.075	0.381	0.19048
P5*	121	300	0.40	0.050	0.504	0.16667
P6*	98	250	0.40	0.100	0.518	0.10000
P7	123	300	0.40	0.100	0.801	0.08333
P8	99	250	0.40	0.150	1.222	0.06667
P9	124	300	0.40	0.150	1.963	0.05556
P10	125	300	0.40	0.200	2.861	0.04167

*Sample is within the "knee-cap" zone of the Pareto frontier.

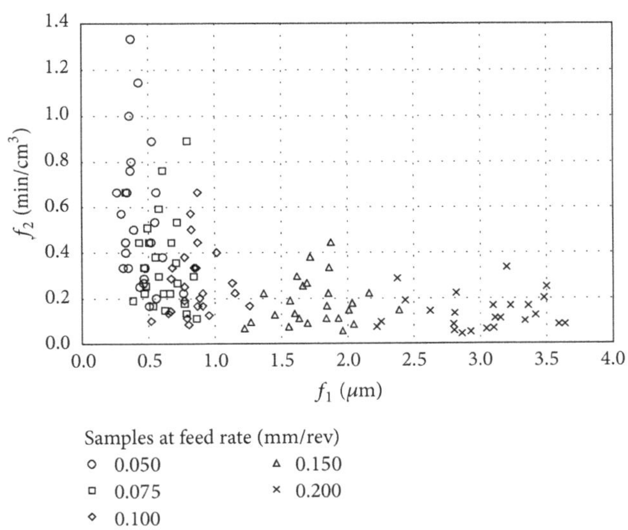

FIGURE 2: Objective values for five-level full-factorial samples.

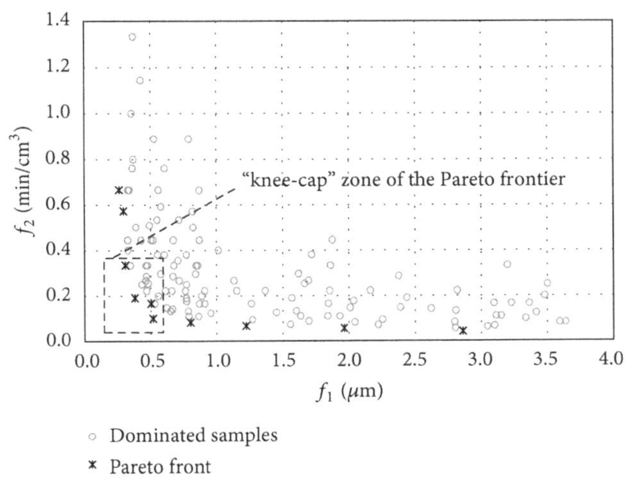

FIGURE 3: Pareto plot for five-level full-factorial samples.

that have different feed rate values. And while this consensus is observed to be mostly true in Figure 2 (samples with the same feed rate tend to cluster together), the following observations are also noted:

(i) For any given feed rate, there is a relatively large range of variation in both surface roughness (f_1) and the machining time (f_2). Thus, identification of the proper combination of cutting speed and depth of cut to match the selected feed rate is of significant importance.

(ii) There is quite a bit of overlap between clusters of same feed rate samples, especially at the submicron range for values of (f_1). This implies that the proper (simultaneous) selection of cutting parameters is of higher significance than simply adjusting the parameter with high sensitivity.

Extracting the "undominated" trade-off frontier results in the Pareto plot shown in Figure 3. In Pareto analysis, a sample is considered "dominated" if another sample exists that has better value in one objective and no-worse values among all the objectives. In two-objective minimization problems, this dominance criterion translates to a sample being dominated if another sample exists towards its lower left in the Pareto plot. Listing of the undominated samples is provided in Table 3.

An important observation in Figure 3 is that the Pareto frontier appears to be "knee-shaped," which is one of the interesting types of multiobjective problems [13]. In these types of problems, the "knee-cap" of the Pareto frontier (highlighted in Figure 3) is where a process planner would likely focus their attention because this is where meaningful trade-off decisions can be made. Excluding sample P5 (because it is near-dominated by P6), there are reasonable gains in machining time at the expense of moderate compromise in surface roughness by going from P3 to P4 to P6. On the other hand, outside of the knee-cap zone, there is very little gain in machining time at the expense of large compromise in surface roughness by going from P6 to P7–P10. Likewise, there is very little gain in surface roughness at large compromise in machining time by going from P3 to P1-P2.

4. Multiobjective Efficient Global Optimization

When full-factorial sampling with large number of levels is feasible in terms of cost and time to conduct the experiments, it is typically the approach of choice since it provides assurance that the entire space of tunable parameters has been explored. When the number of experiments needed becomes large (e.g., exponential growth of the number of samples becomes prohibitive when there are more than three tunable parameters or if the same process needs to be repeated for different tool/inserts), it is often desirable to seek approaches that focus on generating samples in the vicinity of the Pareto frontier instead of exploring the entire space of tunable parameters.

This section explores the "what-if" scenario of how many experimental samples would have been needed to generate a Pareto plot like Figure 3 by utilizing an alternative approach from the literature on metamodeling based design optimization. Metamodeling based design optimization, as described in the review articles in [14, 15], includes approaches that utilize (computationally inexpensive) numerical approximation models to what is otherwise a highly time/cost expensive process to model. The approximate models are then used to "guide" the sampling within the expensive process to what is predicted to be samples with more desirable performance. The approach adopted in this paper is an extension of Jones' "Efficient Global Optimization" (EGO) algorithm [5], which has been further developed and applied in many application studies [6–8]. Multiobjective versions of EGO were also developed, such as the work in [9, 10]. A brief recap of the adopted algorithm is provided in the next subsection. For full details of the most recent version of the algorithm, readers are referred to [11].

4.1. The m-EGO Algorithm. All variations of the EGO algorithm use Kriging [16] metamodels. Like all metamodels, Kriging provides a prediction $\phi(\mathbf{x})$ that approximates time/costly to evaluate process $f(\mathbf{x})$, with $\mathbf{x}$ being a vector of design variables and ϕ and f being the approximate and true process response, respectively. This prediction is done after "training" the metamodel to a number N of known samples $\mathbf{x}^1, \mathbf{x}^2, \ldots, \mathbf{x}^N$ and their corresponding values of $f^1, f^2, \ldots, f^N$. However, what is rather unique in Kriging compared to other metamodels is that it also provides an expected variation $\sigma(\mathbf{x})$, which can also be interpreted as a measure of uncertainty in the predicted values $\phi(\mathbf{x})$. This allows an optimization algorithm to make predictions about the expected limits for the process output (regions with good expected improvement become candidates for exploring new samples) at various points within the search space, as well as identifying which regions of the search space have too low density of samples (and these locations also become candidates for sampling in order to further refine the metamodel). At a high level, the basic steps of an EGO algorithm are as follows:

(1) Train a Kriging model using known samples.

(2) Solve an "Infill Optimization Problem" using the Kriging model predictions $\phi(\mathbf{x})$ & $\sigma(\mathbf{x})$ to generate new sample point(s).

(3) Evaluate the real process $f(\mathbf{x})$ for the new sample points.

(4) Add the new sample points to the set of known samples.

(5) Repeat step 1 unless a termination criterion is met.

(6) Output the best sample point(s).

Primary differences between different implementations of the EGO algorithm are usually in step 2, in particular,

(1) objective(s) for the infill optimization problem,

(2) optimization algorithm to solve the infill problem,

(3) whether new sample points are generated one at a time or in batches.

Other minor differences in implementation of the EGO algorithm can include the initial set of samples (which can be random or structured), as well as the termination criteria (which may be things like reaching a target process output, preset budget on number of samples, or number of iterations without new improved samples).

The m-EGO algorithm (as implemented in [10, 11]) formulates the infill problem as an augmented multiobjective problem that seeks to minimize $\{\phi(\mathbf{x}), -\sigma(\mathbf{x})\}$ and solves it via the evolutionary optimization algorithm SPEA2 [17]. The output of SPEA2 is an "archive" of points distributed on the trade-off frontier of the Kriging model predictions, as well as regions of the search space with high uncertainty. A clustering technique is then applied to the archive of points in order to select a batch of new candidate sample points.

4.2. Implementation and Results. The m-EGO algorithm was tested to see if it would be successful in generating the Pareto trade-off frontier for the same ranges of tunable cutting parameters. Details of m-EGO implementation in this study are as follows:

(i) Structured initial set of samples, using four-level full factorial (64 initial samples). Listing of the design variables at different levels is provided in Table 4.

(ii) Termination condition which is upon reaching a preset budget of 80 experimental samples.

(iii) New samples generated in batches of 8 (thus the termination condition is met after two iterations of the algorithm, since $64 + 2 \times 8 = 80$).

Since this is a "what-if" study, the four levels in the full-factor initial set of samples were selected to coincide with samples that were already experimented in the five-level full-factorial study in Section 3 (relieving the need to conduct

TABLE 4: Design variables values in four-level full-factorial matrix used as initial samples for m-EGO.

Level	Cutting speed x_1 (m/min)	Depth of cut x_2 (mm)	Feed rate x_3 (mm/rev)
1	150	0.10	0.050
2	200	0.20	0.100
3	250	0.30	0.150
4	300	0.40	0.200

TABLE 5: Listing of undominated samples from m-EGO.

Pareto ID	x_1 (m/min)	x_2 (mm)	x_3 (mm/rev)	f_1 (μm)	f_2 (min/cm^3)
E1*	200	0.30	0.050	0.309	0.33333
E2*	223	0.31	0.050	0.374	0.28695
E3*	200	0.40	0.050	0.434	0.25000
E4*	259	0.40	0.050	0.443	0.19279
E5*	300	0.40	0.050	0.504	0.16667
E6*	250	0.40	0.100	0.518	0.10000
E7	300	0.40	0.100	0.801	0.08333
E8	250	0.40	0.150	1.222	0.06667
E9	300	0.40	0.150	1.963	0.05556
E10	300	0.40	0.200	2.861	0.04167

*Sample is within the "knee-cap" zone of the Pareto frontier.

64 new experiments). New experiments were conducted to evaluate the performance of the 16 new samples generated by two iterations of the algorithm and are shown in Figure 4. It is notable in Figure 4 how the newly generated samples are mostly biased towards the Pareto frontier (implying that the m-EGO algorithm does in fact work as intended), with some of the new samples leading to discovery of new undominated samples, two of which within the knee-cap zone of the Pareto frontier. Listing of the undominated samples obtained via the approach is provided in Table 5.

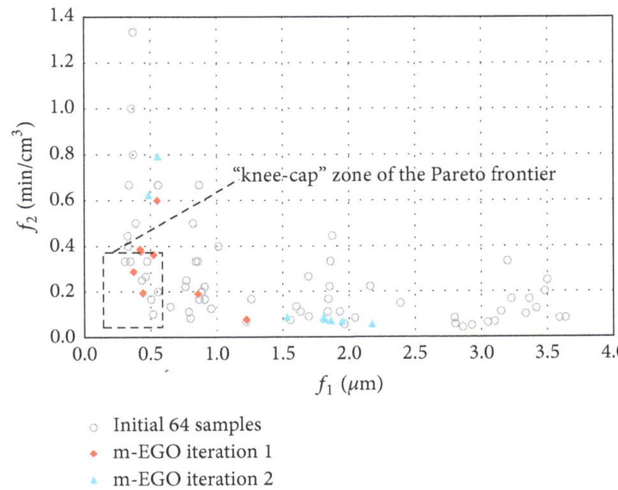

FIGURE 4: Objective values for samples obtained via m-EGO approach.

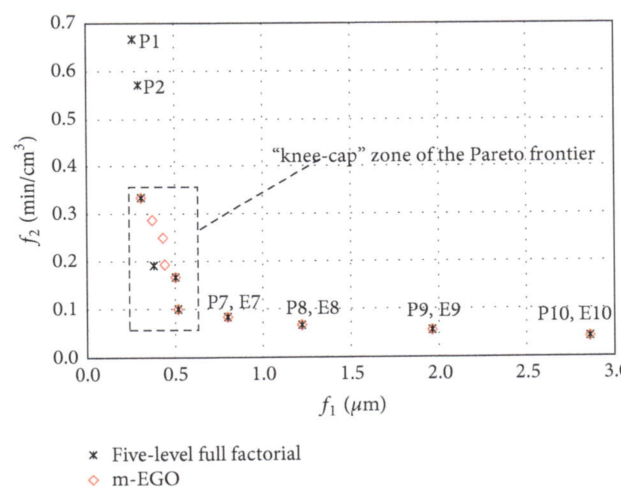

FIGURE 5: Comparison of obtained Pareto frontiers.

5. Discussion

J-Steels have relatively high hardness that correlates well with good surface quality in machining processes. This is mainly due to possessing low plastic flowability which results in better surface finish. This is attributed to the brittle nature of the interaction between the cutting tool and the workpiece surface, for the hard materials, that leads to material separation rather than plastic flow that would result in surface irregularities. Surface roughness was found to increase with increasing feed rate and depth of cut which both result in bigger cut areas that are consequently associated with higher cutting forces and higher friction which lead to poor surface finish. It was noticed from the surface roughness profile that high feed rates were associated with lager roughness markings horizontal spacing. Also, with higher depth of cuts, the vertical spacing between peaks and troughs of the surface irregularities was larger. Thus, higher feed rates and depth of cuts generally lead to higher surface roughness. However, higher feed rates and depth of cut are favorable in terms of increasing material removal rate. Thus, there is good motivation for finding the right combination of settings of the cutting parameters in order to obtain a good compromise between surface quality and machining time. Search for this good compromise may be done via "dense sampling" (such as the five-level full-factorial design of experiments) or through optimization techniques (such as m-EGO).

A comparison of obtained Pareto frontiers via the two approaches (five-level full factorial and m-EGO) is shown in Figure 5. Noted observations include the following:

(i) Undominated samples to the right-side of the knee-cap zone of the Pareto frontier (five-level full-factorial samples P7–P10 in Table 3) were exactly identified by the m-EGO approach (samples E7–E10 in Table 5).

(ii) Undominated samples to the top of the knee-cap zone of the Pareto frontier (five-level full-factorial samples P1 P2 in Table 3) were not identified by the m-EGO approach.

(iii) Sample P4 (from five-level full factorial) is dominant with respect to samples E3 and E4 (from m-EGO). However, sample E4 is fairly close in performance to P4.

(iv) m-EGO approach achieved a higher density of Pareto points (more alternative choices) within the knee-cap zone of the Pareto frontier, which is the primary area of interest for making trade-off decisions.

Overall, neither approach is decisively better in terms of quality of the generated Pareto frontiers. However, the m-EGO approach only required 80 sample experiments, while the five-level full factorial required 125 experiments. Thus, m-EGO achieved a comparable quality Pareto frontier for tuning the cutting parameters with 36% fewer experiments. It is also notable that this study presents the first demonstrated success of m-EGO with real machining experiments, as opposed to previous m-EGO studies [10, 11] that used numerical model/simulations of the machining processes.

6. Conclusion

This paper presented a multiobjective optimization study for tuning the cutting parameters in turning operation of J-Steel material. Exploration of the search space was conducted via five-level full factorial, as well as a metamodeling based optimization approach (m-EGO). Both approaches had qualitatively similar results that revealed the Pareto trade-off frontier as having a knee-shape, where the knee-cap zone becomes the primary area of interest for reasonable trade-off decisions. The knee-cap zone included trade-off settings of the machining parameters that range from Ra surface roughness of 0.32 μm to 0.52 μm at machining time of 0.25 min/cm^3 to 0.1 min/cm^3, respectively. Testing of the m-EGO approach was demonstrated as a success in generating the Pareto frontier with 36% fewer sample experiments, which is sizable saving in time and resources. The m-EGO approach is thus recommended for similar future studies that would seek to tune cutting parameters for different tools and/or materials.

Competing Interests

The authors declare that they have no competing interests.

TABLE 6: Listing of five-level full-factorial samples.

ID #	x_1 (m/min)	x_2 (mm)	x_3 (mm/rev)	f_1 (μm)	f_2 (min/cm^3)
1	150	0.10	0.050	0.370	1.3333
2	150	0.10	0.075	0.790	0.8889
3	150	0.10	0.100	0.868	0.6667
4	150	0.10	0.150	1.873	0.4444
5	150	0.10	0.200	3.200	0.3333
6	150	0.15	0.050	0.527	0.8889
7	150	0.15	0.075	0.578	0.5926
8	150	0.15	0.100	0.868	0.4444
9	150	0.15	0.150	1.618	0.2963
10	150	0.15	0.200	2.817	0.2222
11	150	0.20	0.050	0.339	0.6667
12	150	0.20	0.075	0.504	0.4444
13	150	0.20	0.100	0.856	0.3333
14	150	0.20	0.150	2.162	0.2222
15	150	0.20	0.200	3.229	0.1667
16	150	0.30	0.050	0.328	0.4444
17	150	0.30	0.075	0.576	0.2963
18	150	0.30	0.100	0.907	0.2222
19	150	0.30	0.150	2.389	0.1481
20	150	0.30	0.200	3.152	0.1111
21	150	0.40	0.050	0.348	0.3333
22	150	0.40	0.075	0.660	0.2222
23	150	0.40	0.100	1.261	0.1667
24	150	0.40	0.150	1.929	0.1111
25	150	0.40	0.200	2.798	0.0833
26	175	0.10	0.050	0.427	1.1429
27	175	0.10	0.075	0.606	0.7619
28	175	0.10	0.100	0.817	0.5714
29	175	0.10	0.150	1.716	0.3810
30	175	0.10	0.200	2.376	0.2857
31	175	0.15	0.050	0.364	0.7619
32	175	0.15	0.075	0.492	0.5079
33	175	0.15	0.100	0.770	0.3810
34	175	0.15	0.150	1.663	0.2540
35	175	0.15	0.200	2.436	0.1905
36	175	0.20	0.050	0.296	0.5714
37	175	0.20	0.075	0.553	0.3810
38	175	0.20	0.100	0.668	0.2857
39	175	0.20	0.150	1.566	0.1905
40	175	0.20	0.200	2.621	0.1429
41	175	0.30	0.050	0.604	0.3810
42	175	0.30	0.075	0.480	0.2540
43	175	0.30	0.100	0.769	0.1905
44	175	0.30	0.150	1.450	0.1270
45	175	0.30	0.200	2.253	0.0952
46	175	0.40	0.050	0.467	0.2857
47	175	0.40	0.075	0.381	0.1905
48	175	0.40	0.100	0.665	0.1429
49	175	0.40	0.150	1.270	0.0952

Table 6: Continued.

ID #	x_1 (m/min)	x_2 (mm)	x_3 (mm/rev)	f_1 (μm)	f_2 (min/cm^3)
50	175	0.40	0.200	2.221	0.0714
51	200	0.10	0.050	0.357	1.0000
52	200	0.10	0.075	0.330	0.6667
53	200	0.10	0.100	0.821	0.5000
54	200	0.10	0.150	1.859	0.3333
55	200	0.10	0.200	3.501	0.2500
56	200	0.15	0.050	0.264	0.6667
57	200	0.15	0.075	0.428	0.4444
58	200	0.15	0.100	0.675	0.3333
59	200	0.15	0.150	1.366	0.2222
60	200	0.15	0.200	3.098	0.1667
61	200	0.20	0.050	0.389	0.5000
62	200	0.20	0.075	0.465	0.3333
63	200	0.20	0.100	0.770	0.2500
64	200	0.20	0.150	1.843	0.1667
65	200	0.20	0.200	3.417	0.1250
66	200	0.30	0.050	0.309	0.3333
67	200	0.30	0.075	0.470	0.2222
68	200	0.30	0.100	0.908	0.1667
69	200	0.30	0.150	1.837	0.1111
70	200	0.30	0.200	3.593	0.0833
71	200	0.40	0.050	0.434	0.2500
72	200	0.40	0.075	0.532	0.1667
73	200	0.40	0.100	0.956	0.1250
74	200	0.40	0.150	2.046	0.0833
75	200	0.40	0.200	3.048	0.0625
76	250	0.10	0.050	0.371	0.8000
77	250	0.10	0.075	0.713	0.5333
78	250	0.10	0.100	1.013	0.4000
79	250	0.10	0.150	1.692	0.2667
80	250	0.10	0.200	3.480	0.2000
81	250	0.15	0.050	0.549	0.5333
82	250	0.15	0.075	0.705	0.3556
83	250	0.15	0.100	1.131	0.2667
84	250	0.15	0.150	2.036	0.1778
85	250	0.15	0.200	2.807	0.1333
86	250	0.20	0.050	0.329	0.4000
87	250	0.20	0.075	0.715	0.2667
88	250	0.20	0.100	0.884	0.2000
89	250	0.20	0.150	1.601	0.1333
90	250	0.20	0.200	3.337	0.1000
91	250	0.30	0.050	0.462	0.2667
92	250	0.30	0.075	0.771	0.1778
93	250	0.30	0.100	0.649	0.1333
94	250	0.30	0.150	1.695	0.0889
95	250	0.30	0.200	3.102	0.0667
96	250	0.40	0.050	0.557	0.2000
97	250	0.40	0.075	0.784	0.1333
98	250	0.40	0.100	0.518	0.1000
99	250	0.40	0.150	1.222	0.0667
100	250	0.40	0.200	2.928	0.0500
101	300	0.10	0.050	0.559	0.6667
102	300	0.10	0.075	0.671	0.4444
103	300	0.10	0.100	0.842	0.3333
104	300	0.10	0.150	1.852	0.2222
105	300	0.10	0.200	3.364	0.1667
106	300	0.15	0.050	0.519	0.4444
107	300	0.15	0.075	0.838	0.2963
108	300	0.15	0.100	1.151	0.2222
109	300	0.15	0.150	2.009	0.1481
110	300	0.15	0.200	3.113	0.1111
111	300	0.20	0.050	0.475	0.3333
112	300	0.20	0.075	0.611	0.2222
113	300	0.20	0.100	0.866	0.1667
114	300	0.20	0.150	1.633	0.1111
115	300	0.20	0.200	3.638	0.0833
116	300	0.30	0.050	0.762	0.2222
117	300	0.30	0.075	0.621	0.1481
118	300	0.30	0.100	0.788	0.1111
119	300	0.30	0.150	1.556	0.0741
120	300	0.30	0.200	2.803	0.0556
121	300	0.40	0.050	0.504	0.1667
122	300	0.40	0.075	0.861	0.1111
123	300	0.40	0.100	0.801	0.0833
124	300	0.40	0.150	1.963	0.0556
125	300	0.40	0.200	2.861	0.0417

Acknowledgments

This Project was supported by King Saud University, Deanship of Scientific Research, College of Engineering Research Center.

References

[1] V. N. Gaitonde, S. R. Karnik, L. Figueira, and J. Paulo-Davim, "Machinability investigations in hard turning of AISI D2 cold work tool steel with conventional and wiper ceramic inserts," *International Journal of Refractory Metals & Hard Materials*, vol. 27, no. 4, pp. 754–763, 2009.

[2] A. K. Sahoo, A. N. Baral, A. K. Rout, and B. C. Routra, "Multi-objective optimization and predictive modeling of surface roughness and material removal rate in turning using grey relational and regression analysis," *Procedia Engineering*, vol. 38, pp. 1606–1627, 2012.

[3] A. Srithar, K. Palanikumar, and B. Durgaprasad, "Experimental investigation and surface roughness analysis on hard turning of AISI D2 steel using coated carbide insert," *Procedia Engineering*, vol. 97, pp. 72–77, 2014.

[4] R. Suresh, S. Basavarajappa, V. N. Gaitonde, G. L. Samuel, and J. P. Davim, "State-of-the-art research in machinability of hardened steels," *Proceedings of the Institution of Mechanical Engineers, Part B: Journal of Engineering Manufacture*, vol. 227, no. 2, pp. 191–209, 2013.

[5] D. R. Jones, M. Schonlau, and W. J. Welch, "Efficient global optimization of expensive black-box functions," *Journal of Global Optimization*, vol. 13, no. 4, pp. 455–492, 1998.

[6] M. J. Sasena, P. Papalambros, and P. Goovaerts, "Exploration of metamodeling sampling criteria for constrained global optimization," *Engineering Optimization*, vol. 34, no. 3, pp. 263–278, 2002.

[7] M. J. Sasena, M. Parkinson, M. P. Reed, P. Y. Papalambros, and P. Goovaerts, "Improving an ergonomics testing procedure via approximation based adaptive experimental design," *Journal of Mechanical Design*, vol. 127, no. 5, pp. 1006–1013, 2005.

[8] Y. Ren and P. Y. Papalambros, "Design preference elicitation using efficient global optimization," in *Proceedings of the ASME 2011 International Design Engineering Technical Conferences and Computers and Information in Engineering Conference (IDETC/CIE '11)*, pp. 591–600, Washington, Wash, USA, August 2011.

[9] J. Knowles, "ParEGO: a hybrid algorithm with on-line landscape approximation for expensive multiobjective optimization problems," *IEEE Transactions on Evolutionary Computation*, vol. 10, no. 1, pp. 50–66, 2006.

[10] M. Aly, K. Hamza, M. Tauhiduzzaman et al., "Multi-objective selection of cutting conditions in advanced machining processes via an efficient global optimization approach," in *Proceedings of the International Design Engineering Technical Conferences and Computers and Information in Engineering Conference (ASME '14)*, Paper No. DETC2014-34624, 8 pages, Buffalo, NY, USA, August 2014.

[11] A. Sadek, M. Aly, K. Hamza, M. Meshreki, A. O. Nassef, and H. Attia, "Optimization of cutting conditions in vibration assisted drilling of composites via a multi-objective EGO implementation," in *Proceedings of the ASME International Design Engineering Technical Conferences (ASME IDETC '15)*, Paper no. DETC2015-47532, pp. 1–9, Boston, Mass, USA, 2015.

[12] S. Coromant, "Turning Tools," 2012, http://www.sandvik.coromant.com/sitecollectiondocuments/downloads/global/catalogues/en-gb/turning/turn_g.pdf.

[13] K. Deb, *Multi-Objective Optimization Using Evolutionary Algorithms*, John Wiley & Sons, Hoboken, NJ, USA, 2003.

[14] T. W. Simpson, A. J. Booker, D. Ghosh, A. A. Giunta, P. N. Koch, and R.-J. Yang, "Approximation methods in multidisciplinary analysis and optimization: a panel discussion," *Structural and Multidisciplinary Optimization*, vol. 27, no. 5, pp. 302–313, 2004.

[15] F. Viana, C. Gogu, and R. Haftka, "Making the most out of surrogate models: tricks of the trade," in *Proceedings of the ASME International Design Engineering Technical Conferences (IDETC '10)*, DETC2010-28064, Montreal, Canada, 2010.

[16] P. Goovaerts, *Geostatistics for Natural Evaluation*, Oxford University Press, New York, NY, USA, 1997.

[17] E. Zitzler, M. Laumanns, and L. Thiele, "SPEA2: improving the strength pareto evolutionary algorithm," TIK-Report 103, Swiss Federal Institute of Technology, Zurich, Switzerland, 2001.

7

Wear and Friction Evaluation of Different Tool Steels for Hot Stamping

Maider Muro,[1] Garikoitz Artola,[1] Anton Gorriño,[2] and Carlos Angulo [2]

[1]*Metallurgy Research Centre IK4 AZTERLAN, Aliendalde Auzunea 6, 48200 Durango, Spain*
[2]*Department of Mechanical Engineering, University of the Basque Country (UPV/EHU), Plaza Ingeniero Torres Quevedo 1, 48013 Bilbao, Spain*

Correspondence should be addressed to Carlos Angulo; carlos.angulo@ehu.es

Academic Editor: Akihiko Kimura

The aim of this work is to investigate the durability of tool steels for hot stamping by comparing the wear resistance of three hot work tool steels. Friction and wear behaviours of different tool steels sliding against a 22MnB5 uncoated steel at elevated temperatures were investigated using a high-temperature version of the Optimol SRV reciprocating friction and wear tester at temperatures of 40 and 200°C. Our results show that friction decreased with increasing temperature, whereas wear of the tool steel increased with temperature for the second and the third tested tool steels. The slightly better wear behaviour of steel specimen 1 comes from the hardness of the carbides in the martensitic microstructure, which are rich in vanadium.

1. Introduction

Over the past several years, the automotive industry has experienced a large growth in the manufacturing of ultrahigh-strength steel (UHSS) components, especially those who can be processed by means of hot stamping technology. This increase is related to improvements obtained with these steels in terms of crash resistance and fuel consumption reduction. The benefits of employing UHSS components are accompanied by important technological challenges though, since the particularities of the transformation of these steels have nothing to do with those of conventional steels. One of the UHSS transformation related knowledge areas which is not yet well understood is the tribological interaction between the forming tools and the UHSS parts at high temperatures during the hot stamping process. In sheet metal forming, the wear of tool steels continues to be a great concern to the automotive industry because of increasing die maintenance costs and scrap rates. Cold forming tools are subjected to severe tribological stresses due to high contact pressures arising via sliding contact between the die and the sheet materials. This results in high frictional heat generation, which affects both the bulk material and wear properties of the tool steel [1]. Since hot stamping tools are subjected to high temperatures, the wear of tool steels and the prevailing wear mechanisms have been studied in detail [2, 3]. Several authors have also studied wear behaviour in friction processes using specific tribological tests, in order to characterize different tool steel grades, with and without coatings, at elevated temperatures [3–5].

Understanding the factors that influence the wear mechanisms is necessary to minimize the rate of tool wear in hot stamping. This knowledge could be used to aid tool material selection and die design and hence increase the life of die materials used in hot stamping [6, 7].

To ameliorate the surface deterioration in hot stamping tools, it is necessary to achieve a more in-depth understanding of wear failure mechanisms [8–10]. The wear mechanisms involved in tool damage are determined by factors that are directly related to the mechanical properties of the materials. It is considered that the response to wear can be improved by means of structural martensitic changes [11].

The purpose of this article is to compare three specimen tool steels and assess which behave best against wear during

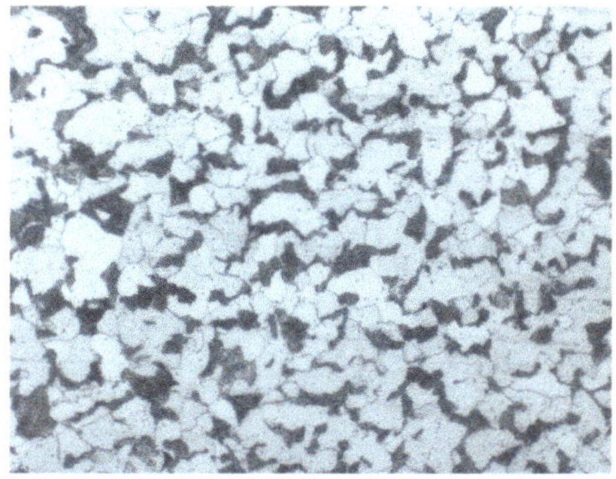

Figure 1: Microstructure of the pin is composed of ferrite and pearlite.

Table 1: Chemical composition of 22MnB5 (wt.%).

Material	C	Si	Mn	P	S	Al	Cr	B
22MnB5	0.24	0.11	0.97	0.009	0.002	0.034	0.25	0.0046

Table 2: Heat treatment of hot work tool steels.

Material	Hardening treatment			
	Austenitizing (°C)	Austenitizing (s)	Tempering (°C)	Tempering (s)
Steel 1	1030	1800	610–610	7200
Steel 2	1050	1800	510–510	7200
Steel 3	1040	1800	520–520	7200

Table 3: Chemical composition (wt.%) and hardness of hot work tool steels.

Material	Mn	Cr	Mo	V	C	Si	Hardness (HRC)	Hardness (HV)
Steel 1	0.75	2.6	2.25	0.9	0.38	0.3	51	527
Steel 2	0.25	4.5	3.0	0.55	0.50	0.2	57	632
Steel 3	0.4	6.5	1.3	0.8	0.42	0.5	56	612

press hardening. We interpret our results in the context on the microstructural features.

2. Experimental Procedure

2.1. Experimental Materials and Specimens. The study and the comparison of the wear behaviour of three tool steels were carried out with a tribopair composed of a boron steel without coating in annealed condition known as 22MnB5 (ferritic-pearlitic microstructure, Figure 1). The chemical composition of 22MnB5 is presented in Table 1.

The tool steels were tested in a hardened condition (quenched and tempered) with a tempered martensitic microstructure. The heat treatments applied to the tool steels are shown in Table 2. The austenitizing and tempering times are 30 minutes and 120 minutes, respectively. The nominal chemical composition of the tool steels (provided by material suppliers) and the final tool hardness are given in Table 3.

The tool steel specimens were flat disks (Ø24 mm and 7.9 mm thick) and were polished to a surface Ra roughness level below 0.09 microns in order to remove any marks and reduce the number of possible influencing variables during the tests. The counter specimens were cylindrical pins (Ø2 mm and 8 mm long) flat-end made from 22MnB5 boron steel. This geometry of the counter specimen was chosen with the objective of maintaining a constant contact pressure even if the pin specimen was subjected to high wear.

2.2. Test Equipment. The equipment used in this study was a reciprocating sliding friction and wear tester SRV model 8.110. The SRV machine, provided with an electromagnetic drive, allows the upper specimens (22MnB5 pins) to oscillate under normal load against a stationary lower test specimens (tool steel disks), as shown in Figure 2. The selected normal load was applied by a servo motor. The lower test specimen holder was provided with a heater

FIGURE 2: Upper and lower specimens in the SRV machine.

which permits to condition the test disks at the selected temperature. The SRV tribometer was equipped with a computerized data acquisition and control system, so that the applied load, temperature, stroke length, and frequency of the oscillatory movement were controlled and monitored.

The selection of the test parameters was based on typical hot stamping pressure used in industrial applications. The parameters were a load of 31 N, a nominal pressure of 10 MPa, a stroke length of 4 mm, temperatures of 40 and 200°C, a frequency of 25 Hz, and duration of 900 s. The total length of each test was 90 m.

2.3. Test Procedures. The tests were performed at temperatures of 40°C and 200°C. Before the tests started, all disks and pins were ultrasonically cleaned first in ether for 5 min and then for 5 more min in acetone. After that, all specimens were cleaned with paper and placed in a dryer to eliminate all possible moisture. The specimens were held in the dryer until the tests started, and before the beginning of each test, the specimens were weighted.

The tests started with the heating of the lower specimen (tool steel) to the desired temperature and held for 5 min at that set-point in order to ensure a homogenous temperature distribution along the disk. The upper pin specimen was kept separated from the disk during the heating sequence. After the disk reached the test temperature and the 5 min dwell time was over, the pin was brought into contact with the disk, the load was applied, and the test was performed.

Once the test was finalized, each specimen was cleaned again ultrasonically for 5 min in ether and 5 more min in acetone. After the cleaning, the wear of the disk and pin was measured. The measurements of the weight loss of the specimens were made with a precision weighting scale model Mettler Toledo XP205. The wear volume of the disks was also measured using a Nikon Eclipse ME600 confocal microscope.

The wear scars of the disk and the microstructure were examined with an Ultra Plus Zeiss Field Emission Gun-Scanning Electron Microscope (FEG-SEM), and the nature of the carbides in the tool steels was analysed via an Energy-dispersive X-ray Spectroscopy (EDS) analysis.

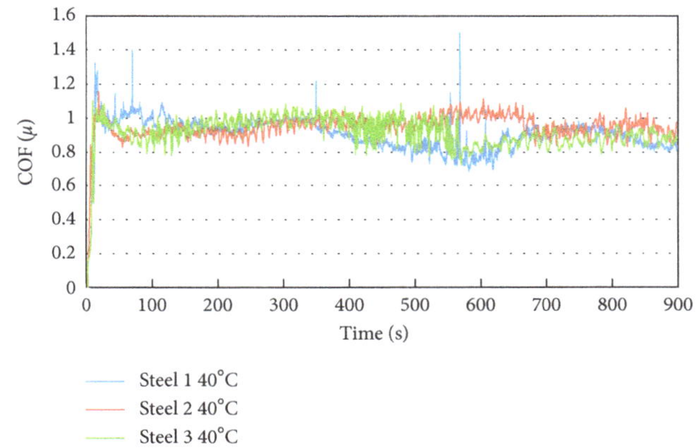

FIGURE 3: COF of the tests performed at 40°C.

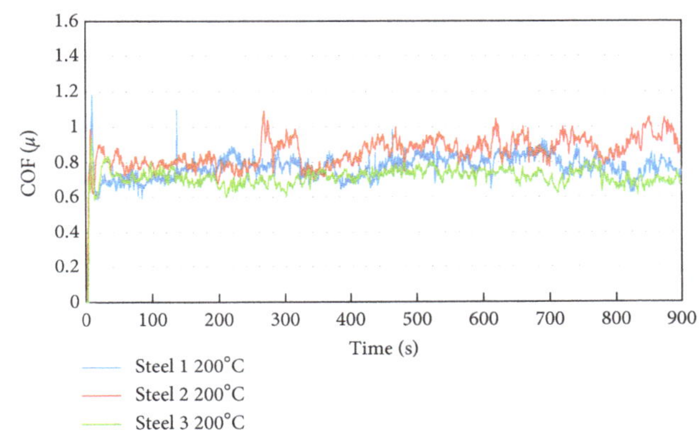

FIGURE 4: COF of the tests performed at 200°C.

TABLE 4: Average COF.

Material	Test temperature (°C)	Average COF
Steel 1	40	0.90 ± 0.005
	200	0.80 ± 0.005
Steel 2	40	1.00 ± 0.005
	200	0.90 ± 0.005
Steel 3	40	0.80 ± 0.005
	200	0.70 ± 0.005

3. Results and Discussion

3.1. Coefficient of Friction. During each test, the evolution of the coefficient of friction (COF) with time was recorded as shown in Figures 3 and 4.

The average values of the COF obtained in the tests are presented in Table 4.

When analysing the evolution of the COF with time during the test, it was observed that the three tool steels have a similar behaviour at both test temperatures, 40°C and 200°C. There is no relation between the hardness level and the COF for the three tool steels, and the COF decreases by around 15% as the test temperature increased

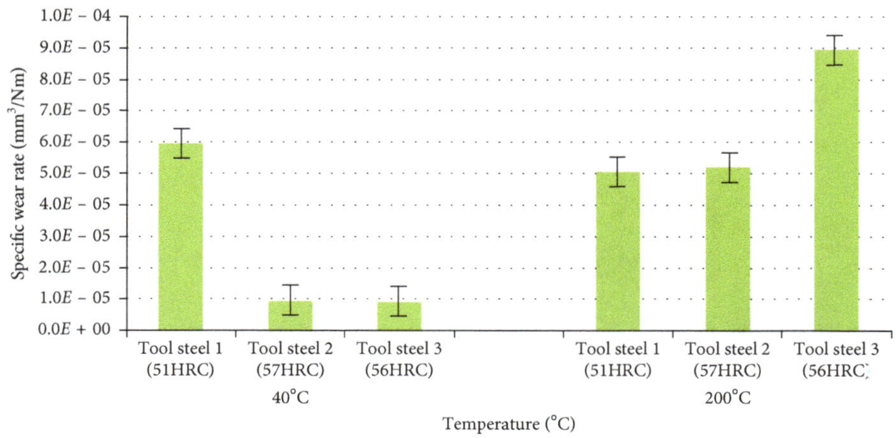

Figure 5: Specific wear rate of the disks at 40°C and 200°C.

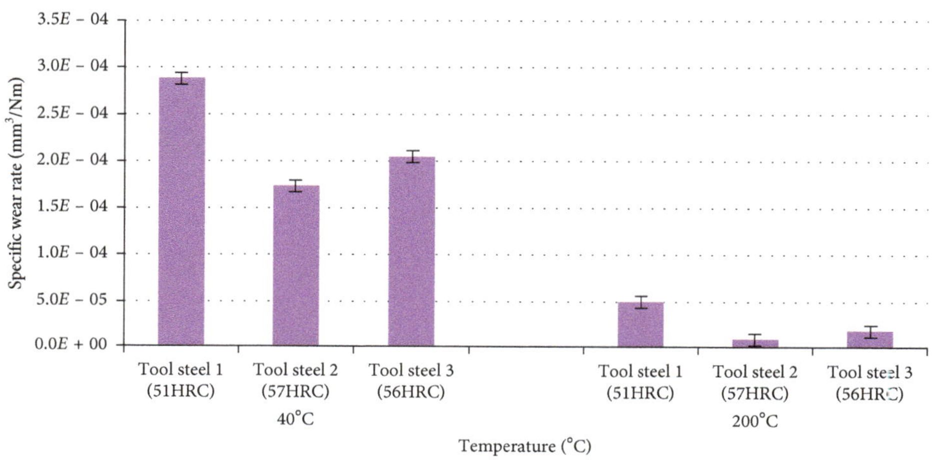

Figure 6: Specific wear rate of the pins at 40°C and 200°C.

to 200°C. This tendency was previously observed by others authors [3].

3.2. Wear. The evolution of the pin wear was measured in terms of specimen weight loss. For the disks, the wear characterization was made by quantifying the weight loss and the volume loss/increase with a confocal microscope. Figures 5 and 6 show the specific wear rates of the three tested group of materials (disks and pins).

3.2.1. Weight Loss. At a temperature of 40°C, the pin and disk weight losses for the tests with steel 2 and steel 3 were lower than the results obtained for steel 1. The tests with steel 3 stand out because they present the lowest wear in the disk and an intermediate wear on the pin. This behaviour is related with the hardness level of the steel 3 (56 HRC). Nevertheless, the differences in weight loss between the steel 2 and steel 3 disks were not large.

At a temperature of 200°C, the pin wear was lower than that at a temperature of 40°C. Regarding the disks, steel 1 and steel 2 present similar wear even though their hardness are different, that is, 51 HRC and 57 HRC, respectively. Steel 3 showed the highest wear despite its high hardness (56 HRC). Nevertheless, it is important to emphasise that the wear behaviour of these three steels at 200°C is similar despite their hardness differences.

A difference in behaviour, as a function of temperature, was clearly observed for the three tool steel samples. Higher differences were observed at 40°C, where steel 2 and steel 3 presented lower wear in comparison with steel 1. As the temperature increased to 200°C, oxide formation may occur, which decreases pin wear. In contrast, as the test temperature increased, disk wear increased, mainly for steel 2 and steel 3.

3.2.2. Wear Rate. In addition to the weight and volume loss, the wear rate of the pins and disks was calculated. The specific wear rate (k) is expressed by

$$k = \frac{V}{F_N \cdot s} = \frac{\Delta m/\rho}{F_N \cdot s} = \frac{\Delta m}{\rho \cdot F_N \cdot s}, \quad (1)$$

where V (mm^3) is the volume loss, Δm (kg) is the weight loss, ρ (kg/mm^3) is the density, F_N (N) is the applied normal load, and s (m) is the sliding distance. Figures 5 and 6 show the specific wear rates for the disks and pins, respectively.

Studying the results obtained from the specific wear calculation, it is concluded that the change in test temperature from 40°C to 200°C influences the tribological behaviour of the three analysed tool steels against 22MnB5, at least for steel 2 and steel 3. As mentioned by other authors [3], as the temperature increases, the COF decreases while the wear of the tool increases. In order to verify if the wear increase is related to the thermal softening (tempering) of the tool steels, the hardness of the disks tested at 200°C was checked on the wear tracks, and no variations from the initial values were found. Thus, the increase of the specific wear rate in the present work is not related with any softening, since the specimens tested at 200°C maintained their hardness after the test.

It was seen that steel 1 had similar wear rates at both 40°C and 200°C. This behaviour has been also reported by Deng et al. [6] who analysed the wear of hot work tool steel against 22MnB5 steel without coating and observed that the wear rate at 200°C was lower than that at 40°C. Similar results have been presented elsewhere [3] on studies made with 22MnB5 steel; as the test temperature increased, the wear of the disk of tool steel decreased. This behaviour is understood to be related to the formation of a compact oxide layer which protects the surface from wear.

It is worth remarking that, although the wear response of steel 1 was the poorest at 40°C, the three studied steels behaved nearly the same at 200°C. This effect of equalization at high temperature has also been reported for tool steels [2]. Regarding the wear rate of the pins, all specimens tested at 40°C showed a much higher rate than the tool steels.

By analysing the surface of the disks with electron microscopy (FEG-SEM/EDS), the disks of steel 2 and steel 3 show debris from an oxide layer on the surface, which is related with the low wear rate. The layer protects the disk surface against wear (Figures 7–10), which avoids metal-metal contact. The noncompacted wear debris particles caused an abrasive wear on the disk surfaces.

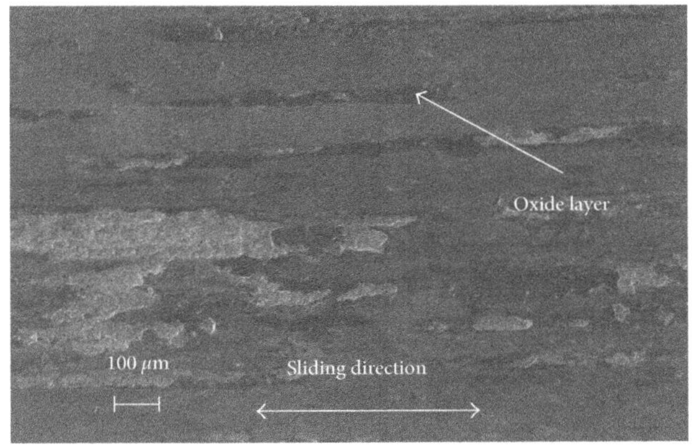

FIGURE 7: Scanning electron microscope micrograph of the steel 2 disk surface tested at 40°C.

3.3. *Contact Path.* After each test, the profile disk of the contact paths was characterised by measuring the maximum profile height and depth as shown in Figures 11–13, where the vertical axis shows the height (z) and the horizontal axis shows the width (x).

By analysing the contact path, the shallowest groove depths were measured for steel 2 and steel 3 at 40°C, with values of 7 and 10 μm, respectively (Figures 11 and 12). Additionally, from contact profilometry measurements, it was verified that the profile height increases due to the presence of an oxide layer in some zones on the surface.

In contrast, steel 1 showed a larger amount of wear. In this case, the disk surface presented numerous grooves, and zones covered by a compact oxide layer were hardly identified (Figure 14). The maximum groove depth reached 25 μm (Figure 13).

It is concluded that at 40°C test temperature, the harder the tool steel is, the higher the abrasive wear resistance, which explains the lower values of wear rate obtained for steel 2 and steel 3. In this case, the wear resistance is governed by the hardness of the martensitic matrix. Hence, the lower hardness of the steel 1 disk is not enough for it to withstand the large abrasive action occurring between the disk and the pin, which leads to a high groove formation and consequently high abrasive wear.

At a 200°C test temperature, the surface characteristics and the wear rates of three tool steels were very similar. Despite the large hardness of tool steels, a lower wear rate value is related to the presence of an oxide layer that partially covers the disk surface, thus protecting it against wear. It is assumed that the formation of the oxide layer is similar for the three tool steels. The surfaces were characterized by electron microscopy (Figures 15 and 16) and via contact profile measurements (Figures 17–19). It was observed that the wear of the three steel disks was abrasive, with maximum depths of the grooves between 29 and 38 μm.

In relation with the wear rate of the disks at the 200°C test temperature, the results are similar with those presented by other authors [3, 6], and the differences between results were probably caused by the different roughness of the specimens. In these works, the authors observed an agglomeration and compacting of oxidized wear debris in the tool steel surface that formed a protective layer against wear. This layer was able to stand the load and avoid the metal-metal contact during the test.

In this work, some zones with a protective oxide layer were also observed. In the case of steels 2 and 3, the wear rate increased by one order of magnitude from the test performed at 40°C to the test at 200°C (Figures 5 and 6). In the case of steel 1, an increase in the test temperature presented a slight decrease in the wear rate.

Regarding the analysis of the three steel surfaces tested at 200°C, an accumulation of wear debris due to the abrasive action suffered by the disks (Figures 15 and 16)

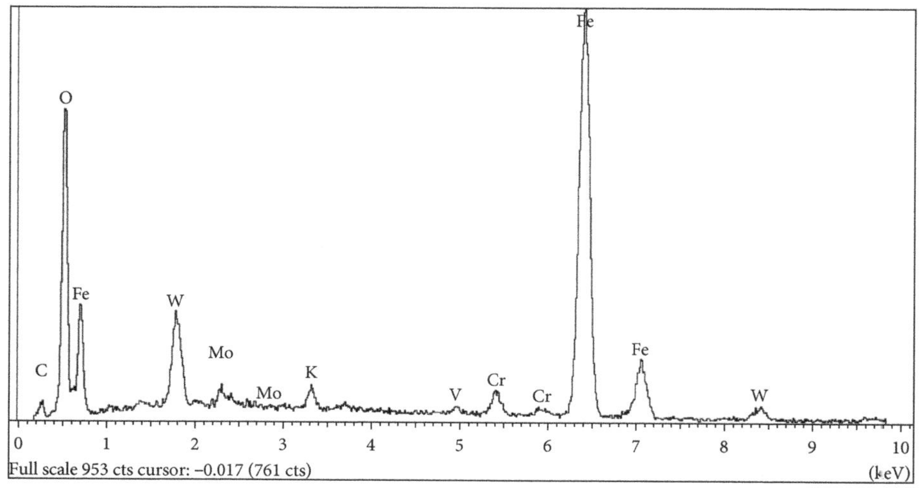

FIGURE 8: EDS spectrum of the oxides in Figure 7.

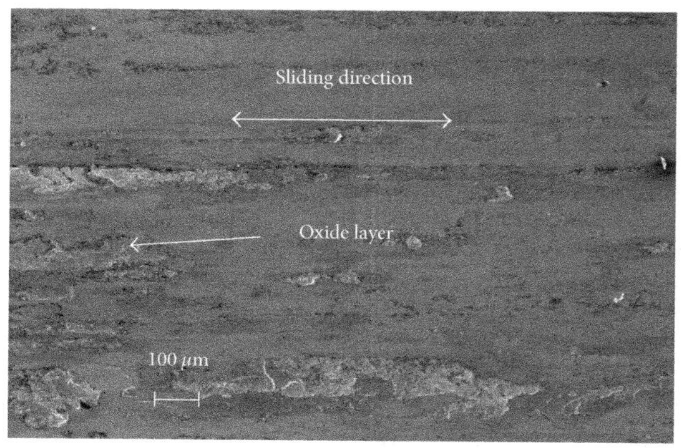

FIGURE 9: Scanning electron microscope micrograph of the steel 3 disk surface tested at 40°C.

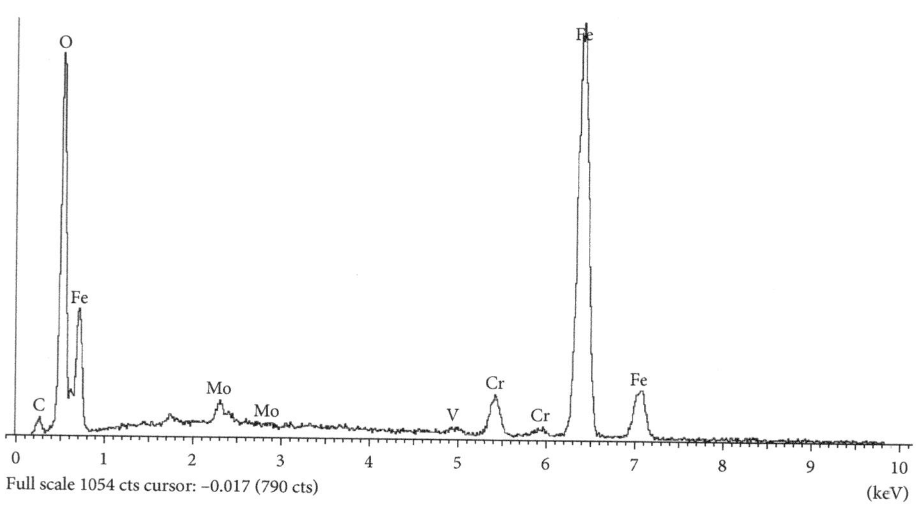

FIGURE 10: EDS spectrum of the oxides in Figure 9.

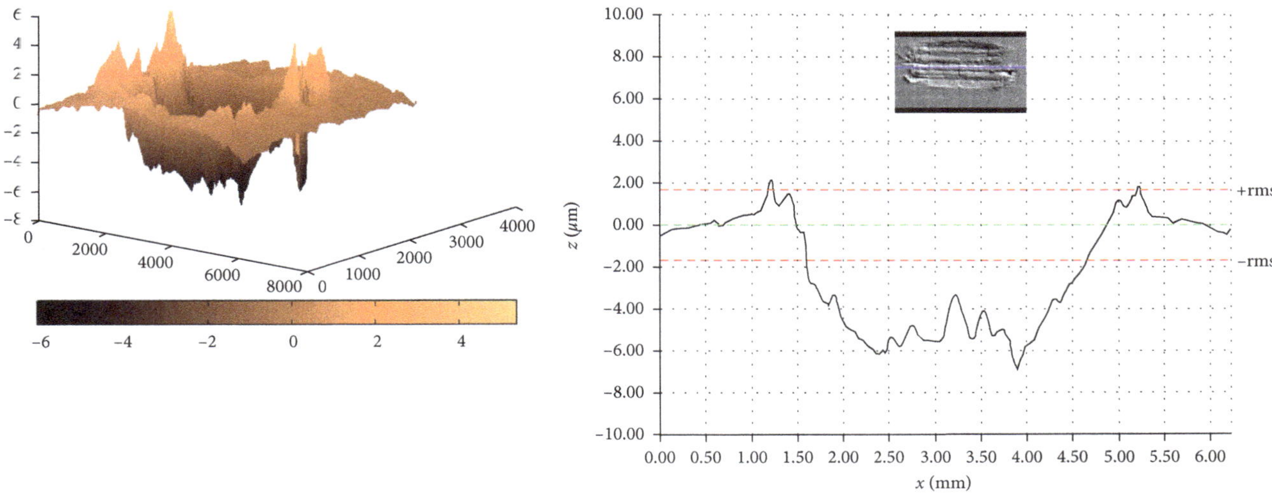

(a) (b)

Figure 11: Three-dimensional image of the disk track of steel 2 tested at 40°C (a). The maximum profile depth (7 μm) (b).

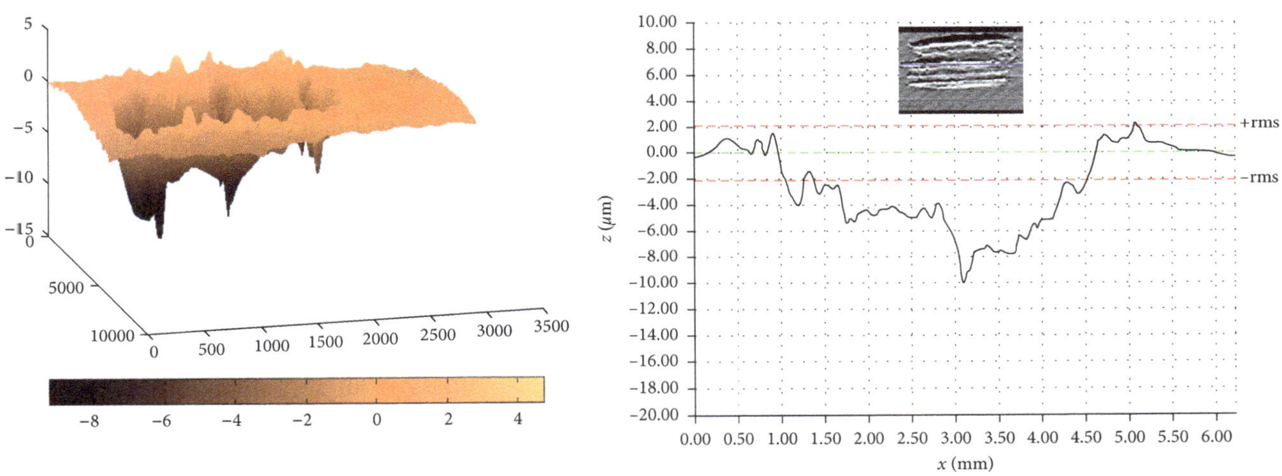

(a) (b)

Figure 12: Three-dimensional image of the disk track of steel 3 tested at 40°C (a). The maximum profile depth (10 μm) (b).

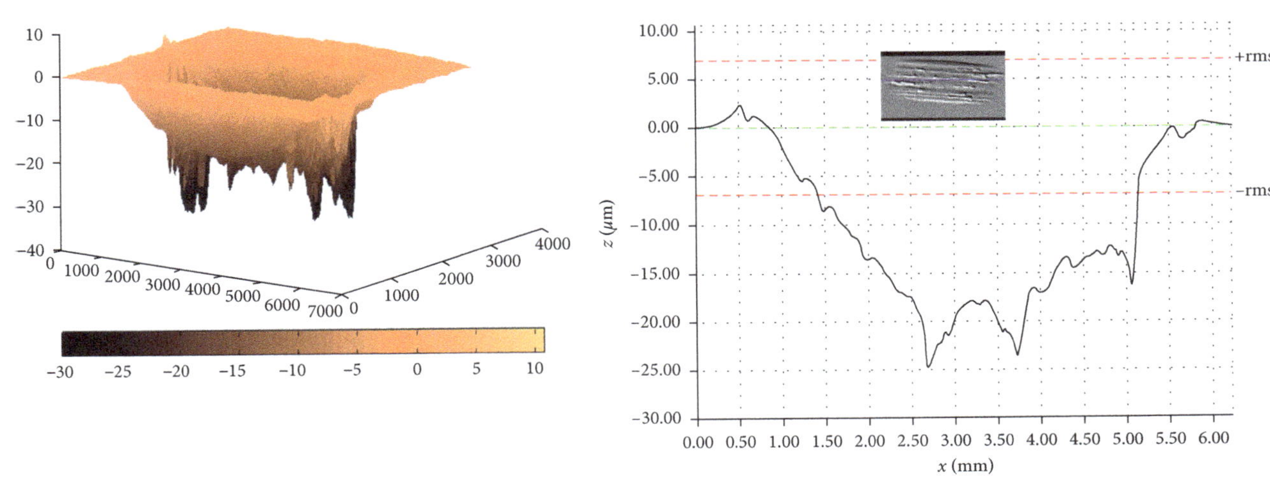

(a) (b)

Figure 13: Three-dimensional image of the disk track of steel 1 tested at 40°C (a). The maximum profile depth (25 μm) (b).

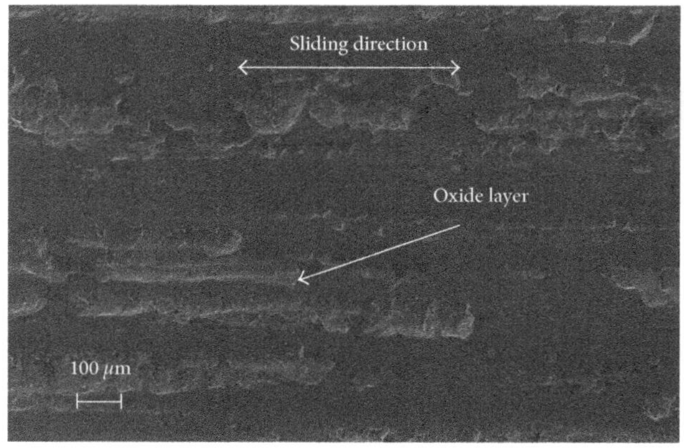

FIGURE 14: Scanning electron microscope micrograph of the steel 1 disk surface tested at 40°C.

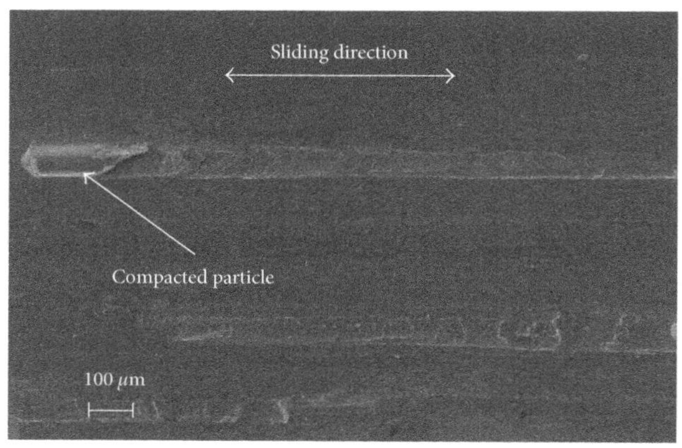

FIGURE 15: Scanning electron microscope micrograph of the steel 1 disk surface tested at 200°C.

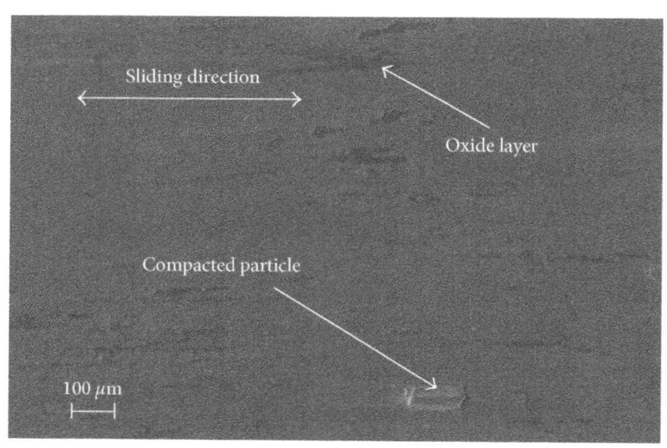

FIGURE 16: Scanning electron microscope micrograph of the steel 2 disk surface tested at 200°C.

was observed. Despite the wear debris accumulation, some zones of the surface were covered by an oxide layer (Figure 16). It seems that the release of the oxide layer particles causes a three-body wear that causes large abrasive wear in the disk surface and generates deep depth grooves.

When the test temperature increased from 40°C to 200°C, a high amount of oxide was observed on the disk surface, according to works in the literature [3, 6] the oxide layer should breaks into pieces easily below 300°C. The presence and release of oxides imply an increase in wear rate as the temperature increased. A larger amount of hard oxides

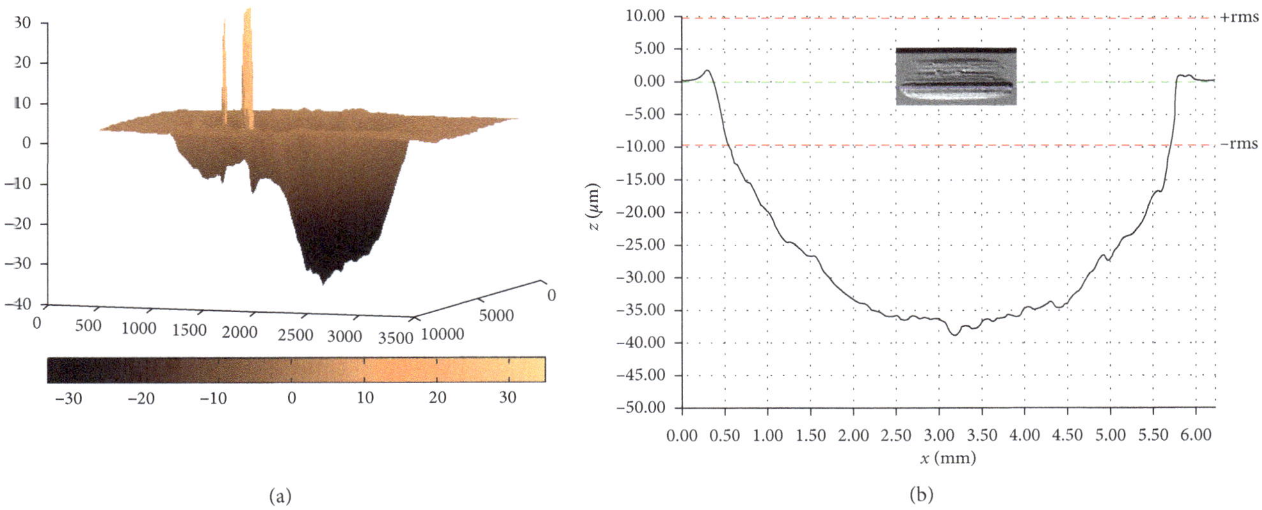

FIGURE 17: Three-dimensional image of the disk track of steel 1 tested at 200°C (a). The maximum profile depth (38 μm) (b).

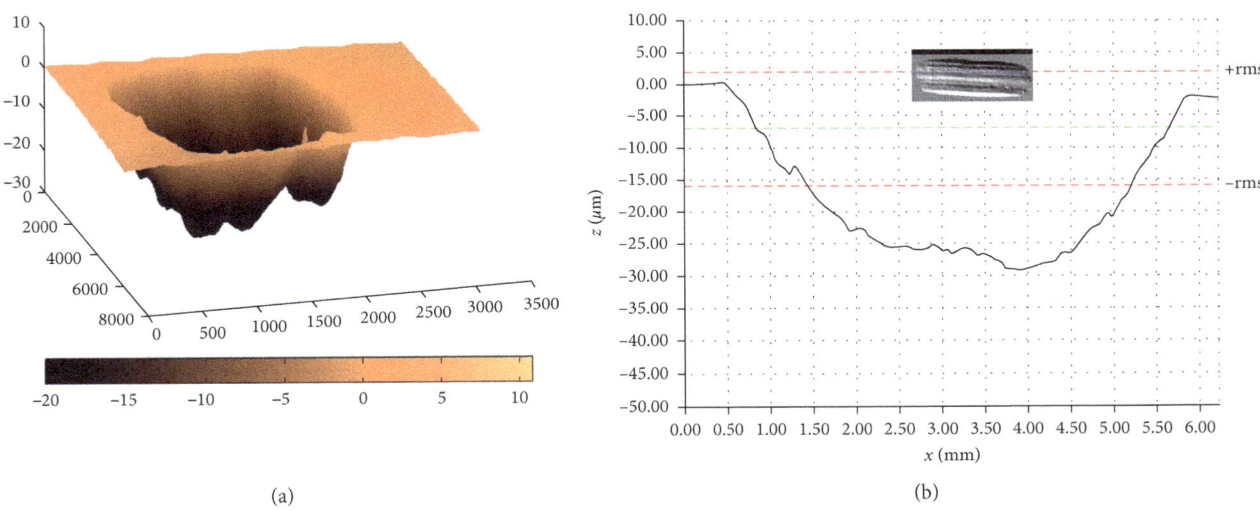

FIGURE 18: Three-dimensional image of the disk track of steel 2 tested at 200°C (a). The maximum profile depth (29 μm) (b).

directly affects the wear resistance of the tool steel when working against uncoated 22MnB5.

From the obtained results, it is concluded that at a test-temperature of 200°C, the release of oxide particles negatively affects the wear resistance of the tool steels. Also, the large hardness of tool steel 2 (57 HRC) was not enough for it to withstand the abrasive action of the oxide particles. Tool steel 1 presented the best wear behaviour, even though it had the smallest relative hardness. Analysing the chemical composition of the steels, steel 1 presented the largest vanadium content. Using a scanning electron microscope (SEM model Phillips), our EDS analysis identified the nature of the carbides in each steel sample. The carbides present in steel 2 were rich in molybdenum and the ones in steel 3 were rich in chromium. Hence, the better wear behaviour of steel 1 at 200°C is related with the nature of the carbides, which were rich in vanadium (Figure 20). In the test at 200°C, it seems that the wear behaviour of each steel was governed by the carbide hardness present in the steel sample, rather than the martensitic matrix hardness (Figure 21).

When analysing the three carbides, the vanadium carbides, due to their hardness and chemical nature, were the most efficient at improving the wear resistance. In contrast, the chromium carbides were the less efficient (Table 5).

4. Conclusions

The results obtained from pin-on-disk tests, using a SRV tribometer and temperatures of 40°C and 200°C, were presented as a method for determining the wear behaviour and the durability of the three selected hot work tool steels.

The slight decrease observed in the COF, as the temperatures increased to 200°C, was related to oxide layer formation. Despite the drop in the friction coefficient, the

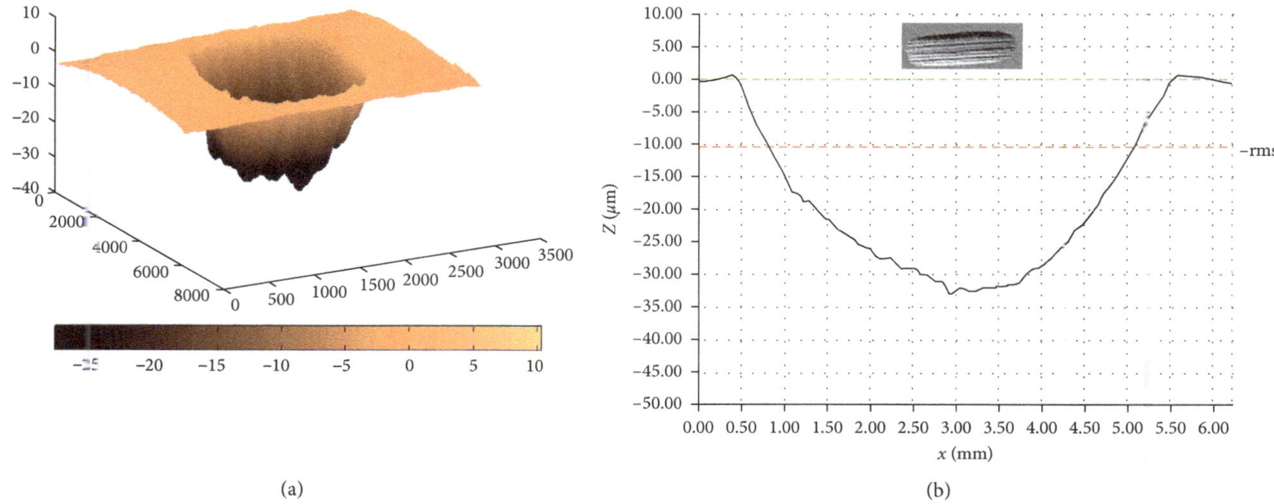

FIGURE 19: Three-dimensional image of the disk track of steel 3 tested at 200°C (a). The maximum profile depth (32 μm) (b).

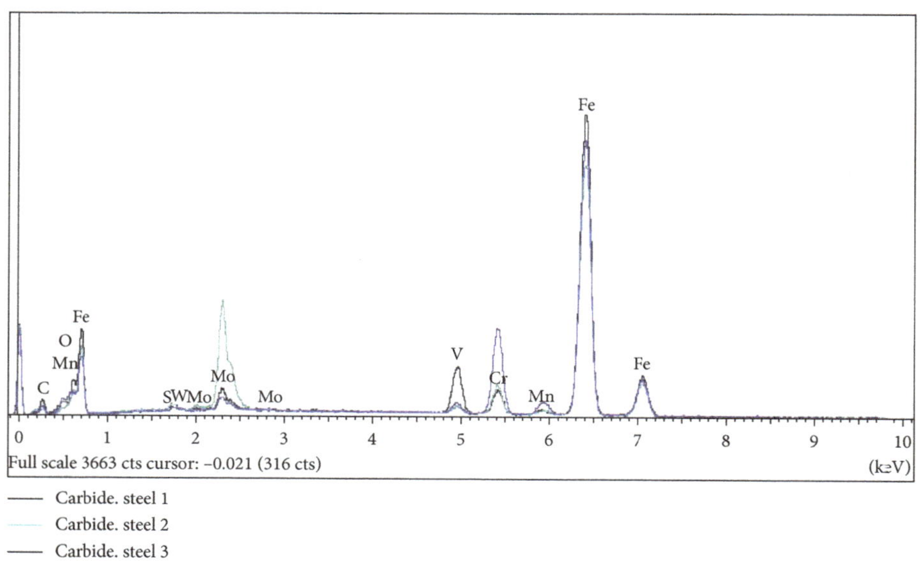

FIGURE 20: Comparison of the carbides' nature of the tool steels studied.

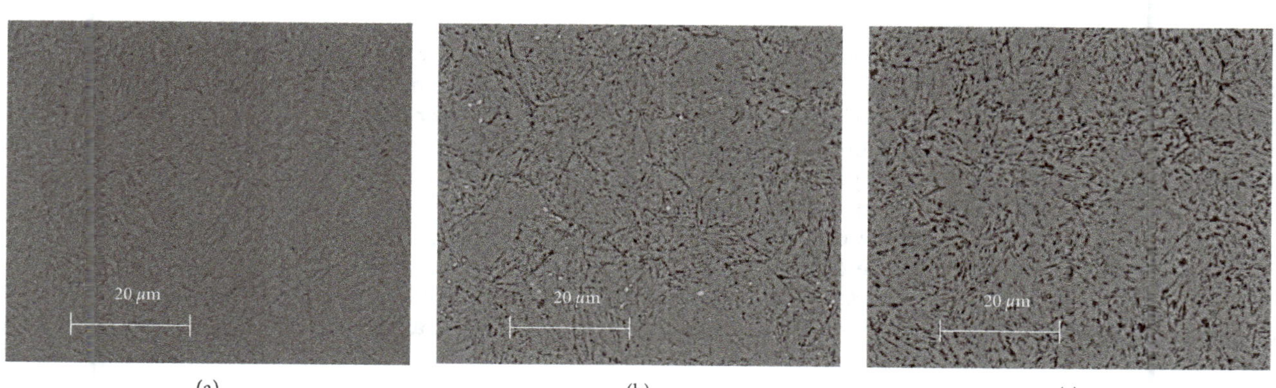

FIGURE 21: Detailed micrograph of steel 1 (a), steel 2 (b), and steel 3 (c).

Table 5: Different carbides hardness.

Nature of the carbide	Hardness (GPa)
Chromium carbides	10.2–20 [12]
Molybdenum carbides	13.39–28.87 [13]
Vanadium carbides	11.7–31.5 [14]

wear rate of the disks at 200°C was higher than that at 40°C for the steel 2 and steel 3 samples. At this temperature, steel 1 and steel 2 showed similar behaviour, while steel 3 performed worse.

SEM inspections confirmed that oxide layer debris, which is unstable at temperatures less than 300°C, is released from the steel surface during the SRV test. These released oxides are hard abrasive particles, leading to severe three-body wear and the formation of depth grooves. This wear mechanism affected each tool steel with different levels of severity, depending on the nature of the carbides in their microstructure. Steel 1 and steel 2, bearing vanadium and molybdenum carbides whose hardness is larger than those of the chromium carbides in steel 3, had greater wear resistance at 200°C.

It must be remarked that even though steel 1 outperformed steel 3 in terms of wear resistance at 200°C, it shows lower room temperature hardness. Thus the HRC hardness, which represents an average hardness of the martensitic matrix and the carbides of the tool steel, cannot be the only guidance when designing hot forming tool steels.

Conflicts of Interest

The authors declare that they have no conflicts of interest.

Acknowledgments

The authors gratefully acknowledge the funding provided by the Department of Research and Universities of the Basque Government under Grant no. IT947-16 and the University of the Basque Country UPV/EHU under Program no. UFI 11/29.

References

[1] Ö. N. Cora, K. Namiki, and M. Koc, "Wear performance assessment of alternative stamping die materials utilizing a novel test system," *Wear*, vol. 267, no. 5–8, pp. 1123–1129, 2009.

[2] J. Hardell and B. Prakash, "High-temperature friction and wear behaviour of different tool steels during sliding against Al–Si-coated high-strength steel," *Tribology International*, vol. 41, no. 7, pp. 663–671, 2008.

[3] J. Hardell, S. Hernandez, S. Mozgovoy, L. Pelcastre, C. Courbon, and B. Prakash, "Effect of oxide layers and near surface transformations on friction and wear during tool steel and boron steel interaction at high temperatures," *Wear*, vol. 330-331, pp. 223–229, 2015.

[4] C. Boher, S. Le Roux, L. Penazzi, and C. Dessain, "Experimental investigation of the tribological behavior and wear mechanisms of tool steel grades in hot stamping of a high-strength boron steel," *Wear*, vol. 294-295, pp. 286–295, 2012.

[5] A. Ghiotti, F. Sgarabotto, and S. Bruschi, "A novel approach to wear testing in hot stamping of high strength boron steel sheets," *Wear*, vol. 302, no. 1-2, pp. 1319–1326, 2013.

[6] L. Deng, S. Mozgovoy, J. Hardell, B. Prakash, and M. Oldenburg, "Press-hardening thermo- mechanical conditions in the contact between blank and tool," in *Proceedings of 4th International Conference on Hot Sheet Metal Forming of High-Performance Steel (CHS2)*, pp. 293–300, Luleå, Sweden, June 2013.

[7] A. Ghiotti, S. Bruschi, and F. Borsetto, "Tribological characteristics of high strength steel sheets under hot stamping conditions," *Journal of Materials Processing Technology*, vol. 211, no. 11, pp. 1694–1700, 2011.

[8] G. A. Fontalvo and C. Mitterer, "The effect of oxide-forming alloying elements on the high temperature wear of a hot work steel," *Wear*, vol. 258, no. 10, pp. 1491–1499, 2005.

[9] L. Pelcastre, J. Hardell, and B. Prakash, "Galling mechanisms during interaction of tool steel and Al–Si coated ultra-high strength steel at elevated temperature," *Tribology International*, vol. 67, pp. 263–271, 2013.

[10] K. Dohda, C. Boher, F. Rezai-Aria, and N. Mahayotsanun, "Tribology in metal forming at elevated temperatures," *Friction*, vol. 3, no. 1, pp. 1–27, 2015.

[11] G. A. Fontalvo, R. Humer, C. Mitterer, K. Sammt, and I. Schemmel, "Microstructural aspects determining the adhesive wear of tool steels," *Wear*, vol. 260, no. 9-10, pp. 1028–1034, 2006.

[12] I. Hussainova, E. Hamed, and I. Jasiuk, "Nanoindentation testing and modeling of chromium-carbide-based composites," *Mechanics of Composite Materials*, vol. 46, no. 6, pp. 667–678, 2011.

[13] Y. Z. Liu, Y. H. Jiang, J. Feng, and R. Zhou, "Elasticity, electronic properties and hardness of MoC investigated by first principles calculations," *Physica B: Condensed Matter*, vol. 419, pp. 45–50, 2013.

[14] L. Wu, T. Yao, Y. Wang, J. Zhang, F. Xiao, and B. Liao, "Understanding the mechanical properties of vanadium carbides: nano-indentation measurement and first-principles calculations," *Journal of Alloys and Compounds*, vol. 548, pp. 60–64, 2013.

8

Optimization of FCAW Parameters for Ferrite Content in 2205 DSS Welds Based on the Taguchi Design Method

Tianqi Li,[1] **Yingying Zhang**,[1] **Lei Gao**,[1] **and Yunhao Zhang**[2]

[1]*School of Mechanical Engineering, Liaoning Shihua University, Fushun 113001, Liaoning, China*
[2]*School of Aeronautical Manufacturing Engineering, Nanchang Hangkong University, Nanchang 330060, Jiangxi, China*

Correspondence should be addressed to Yingying Zhang; zyy_fs@163.com

Guest Editor: Mohamed Abdel-Hady Gepreel

This study presents the Taguchi design method with L9 orthogonal array which was carried out to optimize the flux-cored arc welding (FCAW) process parameters such as welding current, welding voltage, welding speed, and torch angle with reference to vertical for the ferrite content of duplex stainless steel (DSS, UNS S32205) welds. The analysis of variance (ANOVA) was applied, and a mathematical model was developed to predict the effect of process parameters on the responses. The results indicate that welding current, welding voltage, welding speed, torch angle with reference to vertical, and the interaction of welding voltage and welding speed are the significant model terms connected with the ferrite content. The ferrite content increases with the increase of welding speed and torch angle with reference to vertical, but decreases with the increase of welding current and welding voltage. Through the developed mathematical model, the target of 50% ferrite content in weld metal can be obtained when all the welding parameters are set at the optimum values. Finally, in order to validate experimental results, confirmation tests were implemented at optimum working conditions. Under these conditions, there was good accordance between the predicted and the experimental results for the ferrite content.

1. Introduction

Duplex stainless steel (DSS) typically contains the microstructures consisting of almost equal proportions of ferrite and austenite phases. The fine ferrite-austenite microstructure of these materials promotes a good combination of anticorrosion and mechanical properties [1–3]. Duplex stainless steel is a common engineering material used in oil refining, food processing, wastewater treatment, and pharmaceuticals as well.

The increased use of duplex stainless steels in industrial applications will demand a better understanding of those weld parameters that affect weldability. Most of the welding processes, such as laser beam welding (LBW), shielded metal arc welding (SMAW), tungsten inert gas (TIG) welding, and submerged arc welding (SAW), can be used for welding duplex stainless steels [4–6]. Flux-cored arc welding (FCAW) process is also one of the most popular technologies for duplex stainless in manufacturing industries because it produces better and more consistent mechanical properties and higher deposition rates. However, a major problem is that the thermal cycle can degrade the strength and corrosion resistance of weld metal by producing unbalanced ferrite/austenite content, and any attempt to increase heat input changes the welding parameters, probably leading to precipitation of the secondary phase such as σ phase which also reduces the corrosion resistance and strength of weld [1, 7]. Therefore, the studies that control the ferrite/austenite phase ratio of weld metal are necessary. Welding process parameters are important factors that establish the approximately equal proportion of the ferrite/austenite phase (1 : 1) in the weld metal area. Usually, the appropriate ferrite content is 40–60% for good performance of the dual-phase steel to satisfy the requirement for which it is expected in the weld joint.

The relationship between the welding process parameters and the ferrite content can be modeled by using mathematical means. Mathematical models can be used to understand the effect of welding process parameters on the ferrite content variables. The Taguchi method (TM) was a robust design method, originally developed to improve the quality of manufactured goods by Genichi Taguchi, and was effectively used in the optimization of multiple objectives by gray relational analysis [8–10]. The Taguchi method provides the research staff with a scientific and efficient way for ascertaining nearly optimal design parameters.

Though the study using Taguchi methods on weld geometries and mechanical properties has always been reported in literatures [11–14], it appears that the optimization of FCAW process parameters for the ferrite content in duplex stainless steel using the Taguchi method has hardly been studied. In view of the abovementioned facts, the Taguchi method with L9 orthogonal array and ANOVA is employed to analyze the effect of each weld processing parameter (welding current, welding voltage, welding speed, and torch angle with reference to vertical) for the ferrite content in flux-cored weld metal of 2205 DSS. The final intent of this research is to optimize the FCAW process parameters using the developed model to ensure that the ferrite content is close to 50% in 2205 DSS welds.

2. Experimental Work

The experiments were implemented by using a set of programmable welding equipment. The material chosen for the study was 12 mm DSS boards corresponding to UNS S32205, whose chemical composition is given in Table 1. Also, the chemical composition of filler metal (ER 2209, 1.2 mm diameter) is listed in Table 1. The welding joints were designed into a single V groove with an angle of 60° and a blunt edge of 1 mm. The welding parameters are given in Table 2. The interpass temperature was 150°C. The mean ferrite content in weld metal was measured by the FERITSCOPE MP30. Metallographic sample preparation has been carried out by grinding and polishing. To characterize the microstructures, the samples were etched by using a solution of 1 g $K_2S_2O_5$, 30 ml HCl, and 80 ml H_2O. Microstructures were observed by using an optical microscope and a field emission scanning electron microscopy. The scanning electron microscope was operated at an accelerating voltage of 15 kV. The chemical composition of different phases was analyzed by using an energy dispersive spectrometer connected to a scanning electron microscope.

3. Results and Discussion

3.1. Construction of Design Matrix. The experimental data required to relate the FCAW process parameters with response were arranged by using the Taguchi L9 (3^4) orthogonal array with nine experimental runs. The results of process parameters and their corresponding average ferrite content of weld metal are presented in Table 3. Figure 1 shows the optical micrographs of the weld metal microstructures. The dark regions are ferrite while the bright regions are austenite. Obviously, the weld ferrite content of sample 3 is higher than that of sample 9.

3.2. Effect of Parameters on Ferrite Content. The analysis of variance (ANOVA) was performed for the ferrite content with a confidence interval of 95 percent, and a p value less than 0.05 was considered to be statistically significant [15–17]. The test for significance of the developed model and the test for significance on individual coefficients were performed by using Design-Expert software. By using the stepwise regression method, which removes the insignificant model items automatically, the ANOVA (Table 4) for the reduced quadratic model summarizes the analysis of the ferrite content variance of the response and shows the significant model items. The F-test named after Fisher [18] can be used to find out which welding process parameters have a significant effect on the ferrite content. Usually, the change of the welding process parameter has a significant influence on the ferrite content when the F value is large [19]. In addition, the table also reveals the adequacy measures R^2, predicted R^2, and adjusted R^2. The adequacy measures are very close to 1, which is appropriate and indicates the adequate model. "Adeq Precision" measures the signal-to-noise ratio, and the value greater than 4 is considered to be desirable [20].

In this study, from the analysis of variance, it is clear that the ferrite content model is significant. The table indicates that welding current, welding voltage, welding speed, torch angle with reference to vertical, and the interaction of welding voltage and welding speed are the significant model terms affecting the ferrite content, and the result also shows that the highest F value was obtained for the welding speed equal to 5003.32 in the model. The F value for the interaction of welding voltage and welding speed was equal to 76.58, which indicates that the interaction term has a relatively less effect on the model. The Adeq Precision value of 146.135 indicates an adequate signal and implies that the model can navigate the design space.

The ANOVA shows that there is a connection between the main effects of the four parameters. The final mathematical model for the ferrite content is determined by the following procedures and represented as follows:

(a) In terms of coded factors:

$$\begin{aligned} \text{ferrite content} = 42.14 - 5.24I - 3.72U + 4.8S \\ + 1.82T + 0.92US. \end{aligned} \quad (1)$$

(b) In terms of actual factors:

$$\begin{aligned} \text{ferrite content} = 107.41556 - 0.131I - 2.92556U \\ - 0.56333S + 0.12111T + 0.076667US. \end{aligned} \quad (2)$$

Further, the validity of the developed regression model was confirmed by constructing a scatter diagram. A typical scatter diagram for the ferrite content is shown in Figure 2. The scatter diagram reveals that there is a good correlation between the experimental and predicted values.

Figure 3 is a perturbation plot, which helps to compare the effect of all the factors at the center place in the design

Table 1: Compositions of 2205 DSS and ER 2209 (wt.%).

Element	C	Si	Mn	P	S	N	Mo	Ni	Cr	Cu
2205 DSS	0.195	0.520	0.829	0.026	0.002	0.156	3.23	5.70	22.09	0.229
ER 2209	0.032	0.601	0.666	0.012	0.004	0.137	4.29	7.95	22.59	0.202

Table 2: Process parameters and design levels used in experiments.

Parameters	Symbol	Units	Level 1	Level 2	Level 3
Welding current	I	Ampere	160	200	240
Welding voltage	U	Volt	20	23	26
Welding speed	S	cm/min	18	22	26
Torch angle with reference to vertical	T	Degree	0	15	30

Table 3: Design matrix with measured responses.

Trial no.	Process parameters				Response
	I (ampere)	U (volt)	S (cm/min)	T (degree)	Average ferrite content (%)
1	160	20	18	0	45.5
2	160	23	22	15	47.2
3	160	26	26	30	51.3
4	200	20	22	30	47.6
5	200	23	26	0	45.1
6	200	26	18	15	32.8
7	240	20	26	15	44.6
8	240	23	18	30	33.9
9	240	26	22	0	31.3

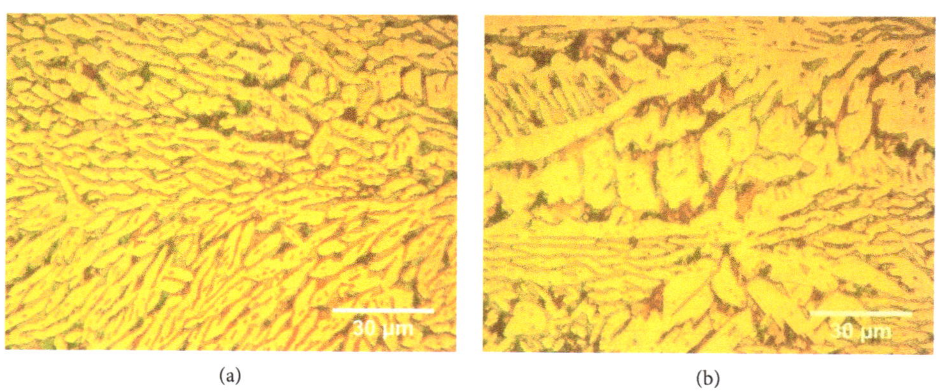

Figure 1: Optical micrographs of weld metal microstructures of (a) sample 3 with the highest ferrite content and (b) sample 9 with the lowest ferrite content.

Table 4: ANOVA and R-square table for ferrite content reduced quadratic model.

Source	Sum of squares	DF	Mean square	F value	Prob > F	
Model	437.98	5	87.6	3170.36	<0.0001	Significant
I	102.97	1	102.97	3726.65	<0.0001	Significant
U	82.88	1	82.88	2999.74	<0.0001	Significant
S	138.24	1	138.24	5003.32	<0.0001	Significant
T	19.8	1	19.8	716.68	0.0001	Significant
US	2.12	1	2.12	76.58	0.0031	Significant
Residual	0.083	3	0.028			
Cor. total	438.06	8				
	$R^2 = 0.9998$				Pred $R^2 = 0.9957$	
	Adj. $R^2 = 0.9995$				Adeq Precision = 146.135	

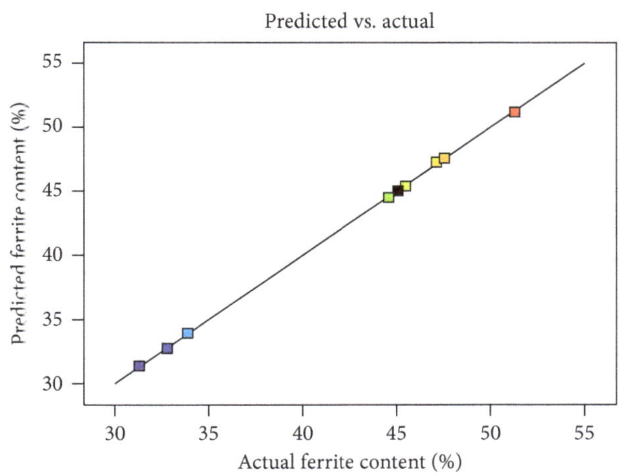

FIGURE 2: Predicted and experimental results of ferrite content.

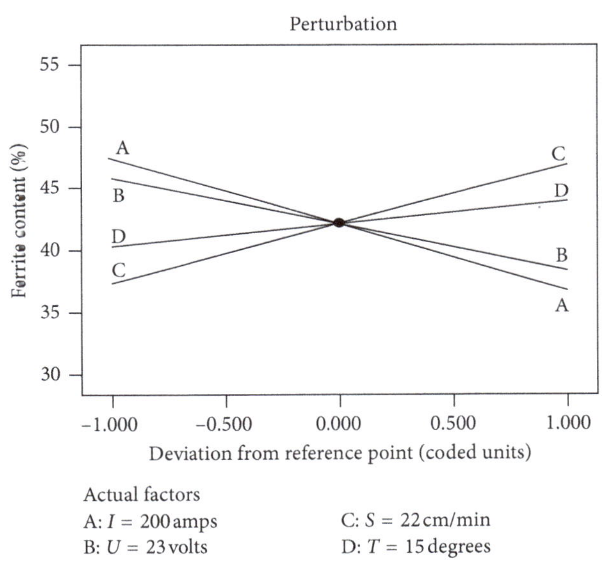

FIGURE 3: Perturbation plot showing the effect of all factors on the ferrite content.

space. It can be noticed that the ferrite content increases with the welding speed and torch angle with reference to vertical. This is because a higher welding speed reduces the heating time, and the electric arc becomes more decentralized from 0 degrees to 30 degrees for torch angle with reference to vertical. These two factors can cause lower heat input to the weld zone. Low heat input results in a fast cooling rate. There is no sufficient time for the diffusion of alloying elements (Cr, Ni, and Mo). The ferrite-to-austenite transformation ratio reduces during solidification [21]. From this figure, it is apparent that the welding current and the welding voltage have a negative effect on the ferrite content. It is due to the fact that heat input increases with welding current and welding voltage. There is sufficient time for the diffusion of alloying element under a slow cooling rate. Consequently, the ferrite-to-austenite transformation ratio increases under the effect of increased heat input, and a more austenite phase is generated during the process. In addition, σ phase precipitates in ferrite for a long time at high temperature, which further reduces the ferrite content. The SEM micrograph and EDS of the phases are shown in Figure 4. In contrast to the ferrite (α) and austenite (γ) phases, the light gray σ phase is richer in Cr.

4. Optimization of Results

In statistical software, the optimization part searches for a group of factor levels that simultaneously meet the goals placed on each of the factors and responses by using the developed mathematical model in the design space. The goals are combined into an overall desirability function. As displayed in Table 5, the optimization criterion has been executed for 2205 DSS welds. In the criterion, the process parameters are kept within the scope of the stationary design space. The goal response was set to the target of 50% ferrite content. However, the ferrite-austenite phase balance (1 : 1) imparts very desirable properties to the material [22–24]. The optimal results are listed in Table 6 with their desirability.

5. Confirmation of Model Validity

Through analysis of desirability function, optimal results have been validated by conducting three confirmatory tests. The process parameters and corresponding ferrite content have been selected randomly from Table 6. Table 7 displays the tested results at optimum condition. It is revealed from Table 7 that there is a small error ratio between predicted and the experimental values, which validates the applied optimization research. The error percentage in this table was calculated as follows:

$$\text{error\%} = \frac{\text{observed values} - \text{predicted values}}{\text{predicted values}}. \quad (3)$$

Figure 5 shows the SEM micrographs of the validation samples. The micrographs also prove that the ferrite content is very close to the ideal value (50% ferrite) and no detrimental intermetallic phases appear in it.

6. Conclusions

The influence of flux-cored arc welding parameters (welding current, welding voltage, welding speed, and torch angle with reference to vertical) on the ferrite content for 2205 DSS welds has been studied, and the following conclusions are drawn.

A relationship model was developed to predict the ferrite content of weld metal by using the Taguchi design method. Welding current, welding voltage, welding speed, torch angle with reference to vertical, and the interaction of welding voltage and welding speed are the significant terms for the model. Welding voltage and torch angle with reference to vertical have a negative effect on the ferrite content, while the welding speed and welding current have a positive effect on it. The developed model can be successfully used

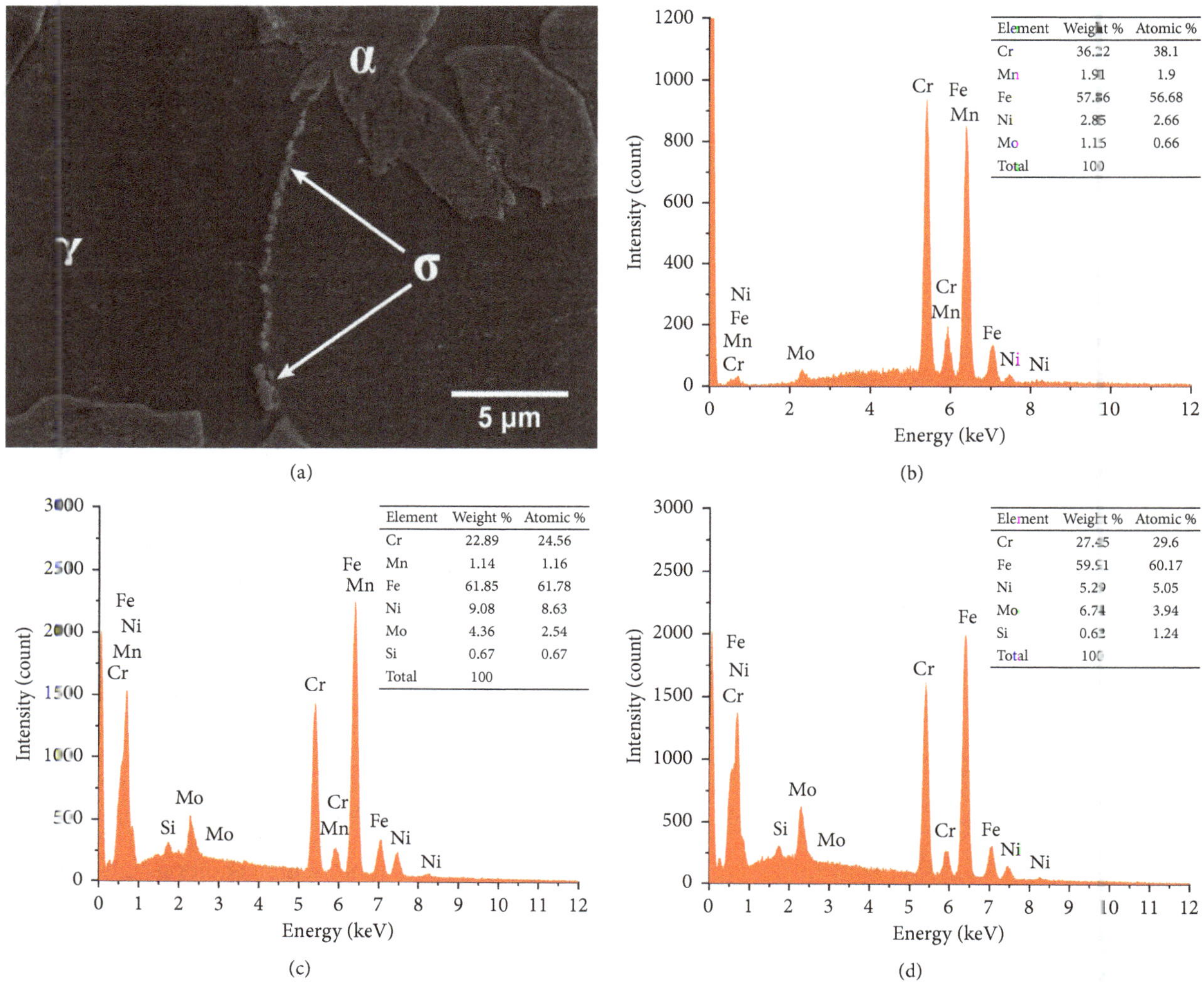

FIGURE 4: (a) SEM micrograph from 2205 DSS sample welded at trial no. 9 ($I = 240$ A, $U = 26$ V, $S = 22$ cm/min, and $T = 0$ degree); (b) EDS of σ phase corresponding to (a); (c) EDS of austenite phase corresponding to (a); and (d) EDS of ferrite phase corresponding to (a).

TABLE 5: Process parameter and response constraint.

Parameter or response	Limit		Goals	Importance
	Lower	Upper		3
Welding current	160	240	Is in range	3
Welding voltage	20	26	Is in range	3
Welding speed	18	26	Is in range	3
Welding torch angle with reference to vertical	0	30	Is in range	3
Ferrite content	31.3	51.3	Is target = 50	3

TABLE 6: Desirable solutions.

Number	I (A)	U (V)	S (cm/min)	T (degree)	Ferrite content	Desirability
1	161.97	21.81	22.94	17.96	49.9999	1.000
2	185.79	20.25	25.55	7.37	50.0001	1.000
3	196.95	20.63	25.70	21.17	50	1.000
4	160.82	20.06	20.47	19.71	50	1.000
5	171.10	20.71	21.84	26.51	49.9999	1.000
6	174.18	20.74	22.34	25.93	50.0001	1.000
7	185.80	20.07	22.68	29.02	50.0001	1.000
8	174.37	20.32	22.32	22.17	50	1.000
9	161.08	25.98	25.28	29.58	49.9999	1.000
10	163.85	20.09	19.71	29.49	49.9999	1.000

TABLE 7: Validation test results.

Number	Parameters				Ferrite content		Error (%)
	I (A)	U (V)	V (cm/min)	T (degree)	Predicted	Observed	
1	161.97	21.81	22.94	17.96	49.9999	49.6	−0.8
2	185.79	20.25	25.55	7.37	50.0001	49.4	−1.2
3	196.95	20.63	25.70	21.17	50	50.6	1.2

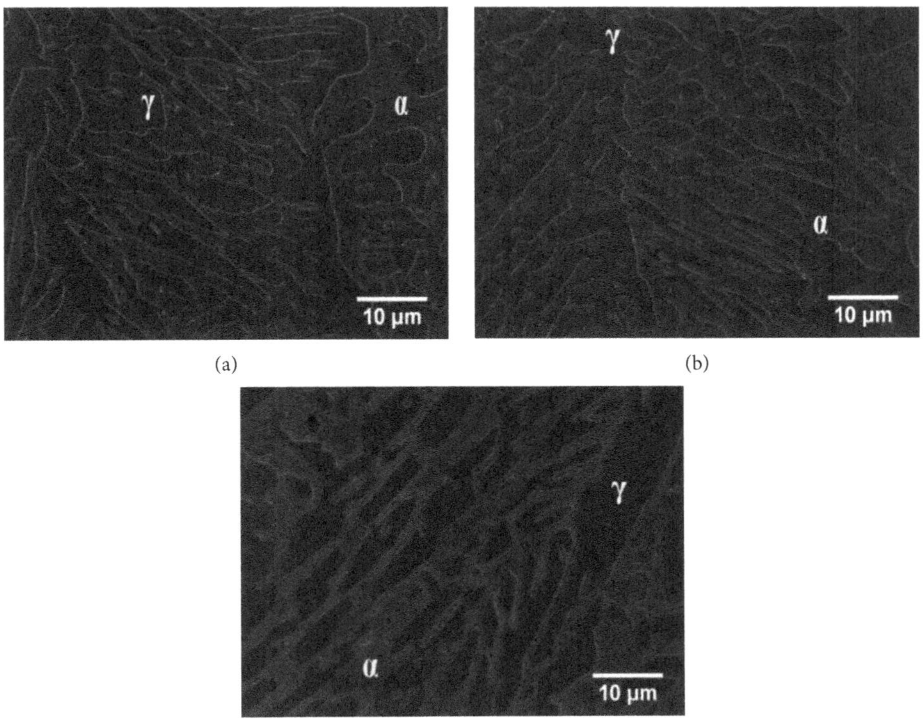

(a) (b) (c)

FIGURE 5: SEM micrographs of the validation samples.

to predict the ferrite content at 95% confidence level and verified by F-test by using the ANOVA table.

To verify the predicted results, three confirmation tests have been implemented, and in the results, errors have been found less than 2%. The results indicate that the developed methods can accurately determine welding parameters so that the desired ferrite content for 2205 DSS welds is achieved.

Conflicts of Interest

The authors declare that they have no conflicts of interest.

Acknowledgments

This work was financially supported by the Foundation of Liaoning Educational Committee (L2017LFW014).

References

[1] R. N. Gunn, *Duplex Stainless Steels: Microstructure, Properties and Applications*, Abington Publishing, Cambridge, England, 1997.

[2] D. H. Kang and H. W. Lee, "Study of the correlation between pitting corrosion and the component ratio of the dual phase in duplex stainless steel welds," *Corrosion Science*, vol. 74, pp. 396–407, 2013.

[3] H.-J. Kim, S.-H. Jeon, S.-T. Kim et al., "Investigation of the sensitization and intergranular corrosion of tube-to-tubesheet welds of hyper duplex stainless steel using an electrochemical reactivation method," *Corrosion Science*, vol. 87, pp. 60–70, 2014.

[4] A.-H. I. Mourad, A. Khourshid, and T. Sharef, "Gas tungsten arc and laser beam welding processes effects on duplex stainless steel 2205 properties," *Materials Science and Engineering: A*, vol. 549, pp. 105–113, 2012.

[5] J. Luo, Y. Yuan, X. Wang, and Z. Yao, "Double-sided single-pass Submerged Arc welding for 2205 duplex stainless steel," *Journal of Materials Engineering and Performance*, vol. 22, no. 9, pp. 2477–2486, 2013.

[6] S. Wang, Q. Ma, and Y. Li, "Characterization of microstructure, mechanical properties and corrosion resistance of dissimilar welded joint between 2205 duplex stainless steel and 16MnR," *Materials and Design*, vol. 32, no. 2, pp. 831–837, 2011.

[7] Y. S. Sato and H. Kokawa, "Preferential precipitation site of sigma phase in duplex stainless steel weld metal," *Scripta Materialia*, vol. 40, no. 6, pp. 659–663, 1999.

[8] E. Ozbay, A. Oztas, A. Baykasoglu, and H. Ozbebek, "Investigating mix proportions of high strength self compacting concrete by using Taguchi method," *Construction and Building Materials*, vol. 23, no. 2, pp. 694–702, 2009.

[9] M. Sarıkaya and A. Güllü, "Multi-response optimization of minimum quantity lubrication parameters using Taguchi-based grey relational analysis in turning of difficult-to-cut alloy Haynes 25," *Journal of Cleaner Production*, vol. 91, pp. 347–357, 2015.

[10] M. Pervez, F. Shafiq, Z. Sarwar, M. Jilani, and Y. Cai, "Multi-response optimization of resin finishing by using a Taguchi-based grey relational analysis," *Materials*, vol. 11, no. 3, p. 426, 2018.

[11] M. Yousefieh, M. Shamanian, and A. R. Arghavan, "Analysis of design of experiments methodology for optimization of pulsed current GTAW process parameters for ultimate tensile strength of UNS S32760 welds," *Metallography, Microstructure, and Analysis*, vol. 1, no. 2, pp. 85–91, 2012.

[12] S. Datta, A. Bandyopadhyay, and P. K. Pal, "Grey-based Taguchi method for optimization of bead geometry in submerged arc bead-on-plate welding," *International Journal of Advanced Manufacturing Technology*, vol. 39, pp. 1136–1143, 2008.

[13] N. Ghosh, P. K. Pal, and G. Nandi, "Parametric optimization of MIG welding on 316L austenitic stainless steel by grey-based Taguchi method," *Procedia Technology*, vol. 25, pp. 1038–1048, 2016.

[14] C. Schneider, C. Lisboa, R. Silva, and R. Lermen, "Optimizing the parameters of TIG-MIG/MAG hybrid welding on the geometry of bead welding using the Taguchi method," *Journal of Manufacturing and Materials Processing*, vol. 1, no. 2, p. 14, 2017.

[15] J. Joseph and S. Muthukumaran, "Optimization of pulsed current GTAW process parameters for sintered hot forged AISI 4135 P/M steel welds by simulated annealing and genetic algorithm," *Journal of Mechanical Science and Technology*, vol. 30, no. 1, pp. 145–155, 2016.

[16] N. Kumar, M. Mukherjee, and A. Bandyopadhyay, "Comparative study of pulsed Nd:YAG laser welding of AISI 304 and AISI 316 stainless steels," *Optics and Laser Technology*, vol. 88, pp. 24–39, 2017.

[17] G. Longlong, Z. Hualin, L. Shaohu, L. Yueqin, X. Xiaodong, and F. Chunyu, "Formation quality optimization and corrosion performance of inconel 625 weld overlay using hot wire pulsed TIG," *Rare Metal Materials and Engineering*, vol. 45, no. 9, pp. 2219–2226, 2016.

[18] R. Lokesh, V. S. Senthil Kumar, C. Rathinasuriyan, and R. Sankar, "Optimization of process parameters tool pin profile rotational speed and welding speed for submerged friction stir welding of AA6063 alloy," *International Journal of Technical and Research Applications*, vol. 12, pp. 35–38, 2015.

[19] S. C. Juang and Y. S. Tarng, "Process parameter selection for optimizing the weld pool geometry in the tungsten inert gas welding of stainless steel," *Journal of Materials Processing Technology*, vol. 160, pp. 128–137, 2005.

[20] H. Tebassi, M. A. Yallese, I. Meddour, F. Girardin, and T. Mabrouki, "On the modeling of surface roughness and cutting force when turning of Inconel 718 using artificial neural network and response surface methodology: accuracy and benefit," *Periodica Polytechnica Mechanical Engineering*, vol. 61, no. 1, pp. 1–11, 2017.

[21] J. D. Kordatos, G. Fourlaris, and G. Papadimitriou, "The effect of cooling rate on the mechanical and corrosion properties of SAF 2205 (UNS 31803) duplex stainless steel welds," *Scripta Materialia*, vol. 44, no. 3, pp. 401–408, 2001.

[22] C. T. Kwok, S. L. Fong, F. T. Cheng, and H. C. Man, "Pitting and galvanic corrosion behavior of laser-welded stainless steels," *Journal of Materials Processing Technology*, vol. 176, no. 1–3, pp. 168–178, 2006.

[23] J. H. Potgieter, P. A. Olubambi, L. Cornish, C. N. Machio, and El-S. M. Sherif, "Influence of nickel additions on the corrosion behaviour of low nitrogen 22% Cr series duplex stainless steels," *Corrosion Science*, vol. 50, no. 9, pp. 2572–2579, 2008.

[24] E. Taban and E. Kaluc, "Welding behaviour of duplex and superduplex stainless steels using laser and plasma ARC welding processes," *Welding in the World*, vol. 55, no. 7-8, pp. 48–57, 2013.

9

Experimental Analysis of the Feasibility of Shaving Process Applied for High-Strength Steel Sheets

Wiriyakorn Phanitwong, Arkarapon Sontamino, and Sutasn Thipprakmas

Department of Tool and Materials Engineering, King Mongkut's University of Technology Thonburi, Bangkok 10140, Thailand

Correspondence should be addressed to Wiriyakorn Phanitwong; wiriyakorn.wp@gmail.com

Academic Editor: Jingwei Zhao

In recent years, the engineered materials were developed to improve their mechanical properties. A high-strength steel sheet is one of them, developed to serve the requirement of reducing weight of vehicles. Therefore, as a new material, many researches have been carried out to examine the use of sheet metal forming process applied for high-strength steel sheet. However, the feasibility of shaving process applied for it has not been investigated yet. In the present study, this feasibility was revealed by using experiments on two types of high-strength steel sheets: SAPH 440 and SPFH 590Y (JIS). The relationship between shaved surface feature and shearing clearance of high-strength steel sheets corresponded well with those of their conventional metal sheets. However, due to the high ultimate strength of these materials, it was revealed in this present study that there were not any suitable conditions of shaving process that could be applied to achieve the requirements of smooth cut surface overall material thickness.

1. Introduction

In consideration of the fuel saving and thereby reducing of the CO_2 exhaustion, in recent years, the decrease in weight of vehicles is focused on in the automotive industry. One of the means applied to meet this requirement is the more use of high-strength steel sheets as the automotive parts instead of the conventional steel sheets. As for the merits of high-strength steel sheet, its mechanical property of ultimate tensile strength could be increased over 1000 MPa whereas the percentage of elongation could be maintained at the same level as that of conventional steel sheet. Nowadays, for this reason, it is usually applied more than 20 percentage of overall steel sheet used in the commercial car. Owing to these special mechanical properties, however, they prevented many conventional sheet metal forming processes from being suitably applied for high-strength steel sheet to fabricate the automotive part. For example, in terms of blanking process, many researches illustrated that the recommended blanking clearance commonly applied for conventional steel sheet is not suitable for applying in blanking process of high-strength steel sheet [1, 2]. It is also suggested that the suitable blanking clearance applied for high-strength steel sheet should be larger, and it was recommended of approximately 10–20% thickness [2]. Next, in terms of forming process, many past researches confirmed that the springback characteristics of high-strength steel sheet were different from that of conventional steel sheet, thereby the conventional bending process is not suitable for springback control of bent high-strength steel parts [3]. In addition, the development of improving of high-strength steel sheet properties is continuously carried out to serve the more requirements in automotive industry [4]. These have pushed a high corresponding development of advanced technologies for sheet metal forming process to serve the fabrication of high-strength steel automotive parts. However, especially for precision sheet metal forming process, the past researches were not enough to understand the interaction between materials of workpiece and the tool as well as the effect of process parameters when improving efficiency of process and product quality. Furthermore, most of them were performed on forming process such as bending, flanging, and deep drawing processes [3, 5, 6]. There are a few researches carried out on precision shearing and blanking processes [1, 5, 7]. For these reasons, in the present study, the shaving process which is one of the precision sheet metal die cutting processes and which could produce the smooth cut surface overall material thickness was investigated by the experiments to examine the feasibility of this process applied for high-strength steel sheets. The results showed that the

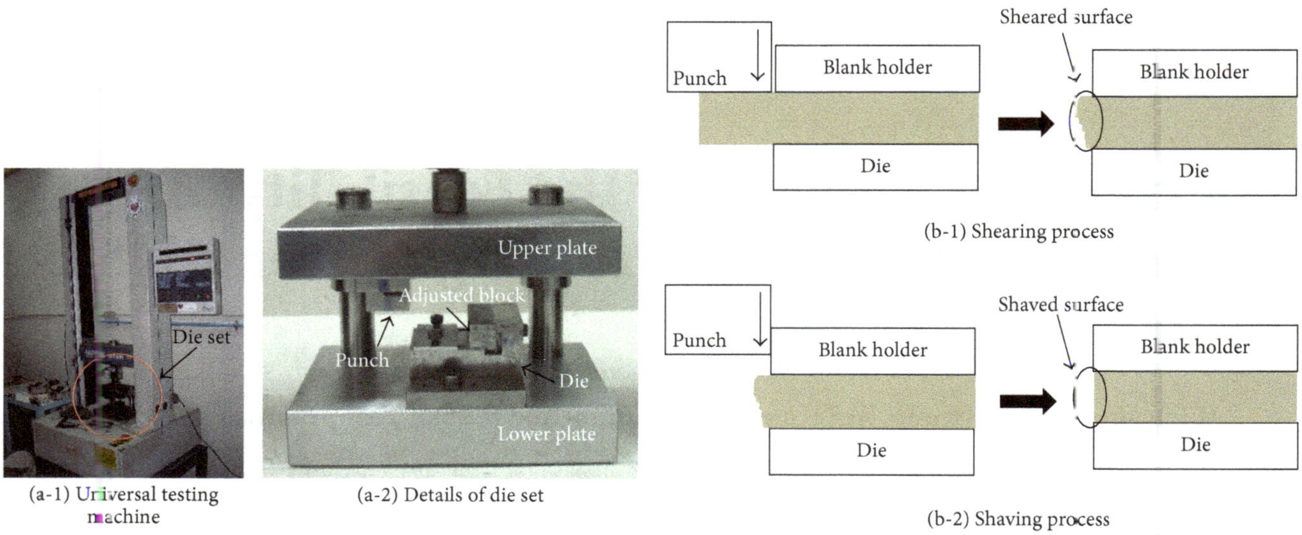

(a) Press machine and die set

(b) Principle of shaving process

FIGURE 1: Experimental equipment and principle of shaving process.

shaving process could not be applied to fabricate the smooth cut surface overall material thickness on high-strength steel sheets.

2. Experimental Procedures

The materials used, in the present study, were 2 types of high-strength steel sheets of SAPH 440 (JIS) and SPFH 590Y (JIS). In addition, the materials of aluminum sheets (A1100-O (JIS)) and carbon steel sheets (SPCC (JIS)) were also used as a conventional metal sheet to be compared with the high-strength steel sheets. The chemical compositions and mechanical properties of SAPH 440, SPFH 590Y, A1100-O, and SPCC are listed in Tables 1 and 2, respectively. The experimental equipment including the universal testing machine used as press machine and die set used for the experiments in the present study is shown in Figure 1(a). Next, as shown in Figure 1(b), the experimental model was illustrated. The shearing operation was first performed to predict cut edge geometries, as shown in Figure 1(b-1). Next, the sheared parts were again shaved by the shaving operation, as shown in Figure 1(b-2). As listed in Table 3, the shearing operation was performed with a 30 mm workpiece width and a 2 mm workpiece thickness at the clearance of 5, 10, and 20% of the workpiece thickness. The shaving operation was carried out using a shaving clearance of 0% of the workpiece thickness with four levels of 0.1, 0.3, 0.5, and 1.0 mm shaving allowance. The tool cutting edge radii were set at 0.01 mm on punch and die for both shearing and shaving processes. First, by using the optical microscopy, the workpiece was sheared and the sheared surface feature including die-roll, smooth cut surface, and cracks was observed, as shown in Figure 1(b-1). Next, as the second operation of shaving step, the shaving clearance of 0% of the workpiece thickness was adjusted as well as the shaving allowance which was also adjusted by setting the moving plate. The sheared workpiece was again shaved, and the shaved surface feature including die-roll, smooth cut surface, and cracks was observed by the optical microscopy as well, as shown in Figure 1(b-2). In the present study, five samples were investigated. The cut surface feature of die-roll, smooth cut, and crack was measured at five observation points in each sample.

3. Results and Discussion

3.1. Comparison of the Sheared Surface Feature in the Shearing Process. Figure 2 shows the comparison of the sheared surface feature in the shearing process regarding the shearing clearance. In terms of high-strength steel sheets, the results illustrated that as the shearing clearance increased, the amount of die-roll increased and the crack formations decreased, whereas the smooth cut surface was somewhat constant. These manners generally confirmed the agreement with the literature [6]. Comparing with the conventional metal sheet of A1100-O and SPCC, it was observed that the results of high-strength steel sheets were against those results of the conventional metal sheet of A1100-O and SPCC. Specifically, in terms of aluminum and carbon steel, the smooth cut surface decreased and again increased as the shearing clearance increased. The results illustrated that, in the shearing process, the relationship between sheared surface feature and shearing clearance of high-strength steel sheets was against the theory of the basic shearing process of conventional metal sheet from the literature. In the present study, for SAPH 440, as the shearing clearances were 5%t, 10%t, and 20%t, the amounts of die-roll were 0.24 mm, 0.27 mm, and 0.28 mm, respectively. Next, for SPFH 590Y, they were 0.15 mm, 0.19 mm, and 0.23 mm, respectively, Also, for SAPH 440, the smooth cut surface and crack were 0.35 mm and 1.41 mm, 0.35 mm and 1.38 mm, and 0.37 mm and 1.34 mm, respectively, as the shearing clearance were 5%t, 10%t, and 20%t. Again, for SPFH 590Y, the smooth

TABLE 1: Chemical composition of materials.

Material	Chemical compositions (%)												
	Mn	Fe	Si	P	C	S	Cr	Cu	Mo	Ni	Zn	Al	Mg
SAPH 440	1.026	98.712	0.050	0.011	0.126	0.003	0.022	0.020	0.012	0.018	—	—	—
SPFH 590Y	0.352	93.678	0.791	0.021	0.350	0.012	4.796	—	—	—	—	—	—
SPCC	0.440	99.285	—	0.101	0.131	0.043	—	—	—	—	—	—	—
Al1100-O	0.050	0.615	0.100	—	—	—	0.010	0.100	—	—	0.050	99.000	0.075

TABLE 2: Mechanical properties of workpiece material.

Material	Yield stress (MPa)	Tensile strength (MPa)	Elongation (%)
Al1100-O	88.00	105.00	30.00
SPCC	210.10	310.50	46.95
SAPH 440	340.05	461.18	33.83
SPFH 590Y	481.12	614.97	25.98

TABLE 3: Experimental conditions.

Punch, die	SKD 11 (60 HRC)
Workpiece	Al1100-O, SAPH 440, SPFH 590Y, and SPCC
Workpiece dimension: mm	Thickness: 2 Width: 30 Length: 70
Shearing clearance (Cl_s): %t	5, 10, and 20
Shaving clearance (Cl_{sh}): %t	0
Shaving allowance (D_{sh}): mm	0.1, 0.3, 0.5, and 1.0
Radius of cutting edge (R_p): mm	0.01

cut surface and crack were 0.43 mm and 1.42 mm, 0.46 mm and 1.35, and 0.43 mm and 1.34 mm, respectively, as the shearing clearance were 5%t, 10%t, and 20%t. Through these results, it was observed that the amounts of die-roll and smooth cut surface were, respectively, smaller and larger in the case of SPFH 590Y than those in the case of SAPH 440. In addition, it was also observed that, comparing with the conventional metal sheet of A1100-O and SPCC, the smooth cut surface was smaller in the case of high-strength steel sheets than those in the case of aluminum and carbon steel sheets. In addition, poor orthogonal cut surface was generated as the shearing clearance increased. Figure 3 shows an example of comparison of the sheared surface feature and poor orthogonal cut surface in the shearing process regarding the shearing clearance. As per the theory of the basic shearing process [8–10], the poor orthogonal cut surface was generated by shearing clearance. In the present study, the poor orthogonal cut surface were 0.10 mm, 0.20 mm, and 0.40 mm in both cases of SAPH 440 and SPFH 590Y as the shearing clearance were 5%t, 10%t, and 20%t, respectively, which corresponded well with the basic shearing process from the literature [8–10].

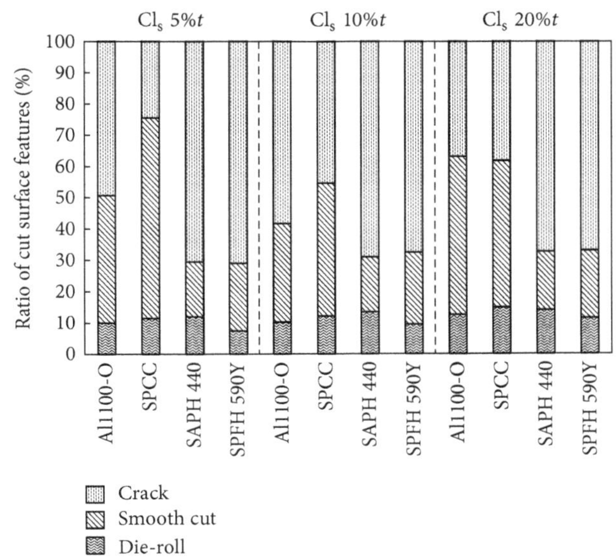

FIGURE 2: Comparison of cut surface after shearing process with respect to the shearing clearances.

3.2. Comparison of the Shaved Surface Feature in the Shaving Process. As the abovementioned sheared surface feature of high-strength steel sheets, the sheared surface feature with respect to shearing clearances was against those of aluminum sheets, as well as the basic shearing process from the literature [8–10]. As in the past researches [8], in terms of aluminum and carbon steel sheets, the shaving process could be applied to fabricate the smooth cut surface overall material thickness. Therefore, in the present study, the feasibility of this process applied for high-strength steel sheets was investigated. Figure 4 shows an example of comparison of the shaved surface feature in the shaving process regarding the shearing clearance and shaving allowance. The results showed that as the small shearing clearance and small shaving allowance were applied, the tearing was generated as shown in Figure 4(a). These manners generally confirmed the agreement with the theory of the shaving process of conventional metal sheet from the literature [8]. Next, as the shearing clearance increased, the tearing decreased as shown in Figure 4(b). However, it was also observed that the smooth cut surface decreased as well. Specifically, the crack formations were easily generated. These manners generally confirmed the agreement with the theory of the shaving process of conventional metal sheet from the literature [8]. As the shaving allowance increased as shown in Figure 4(c), the tearing could not be generated. However, with the large

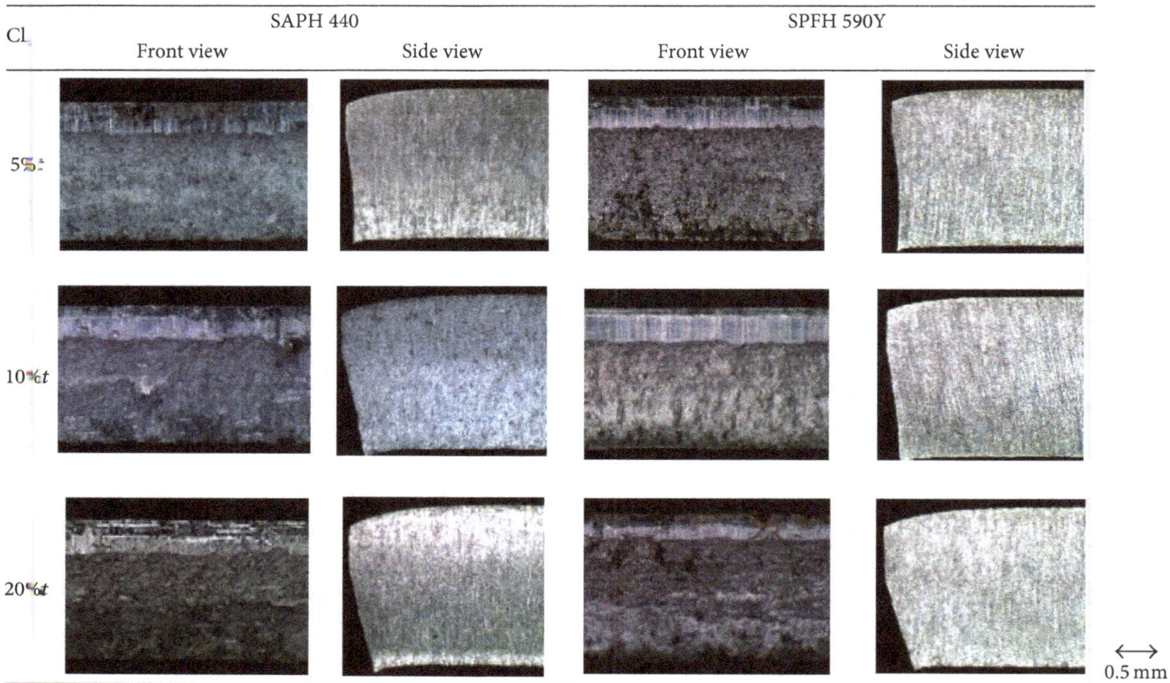

FIGURE 3: An example of cut surface after shearing process of high-strength steel sheet.

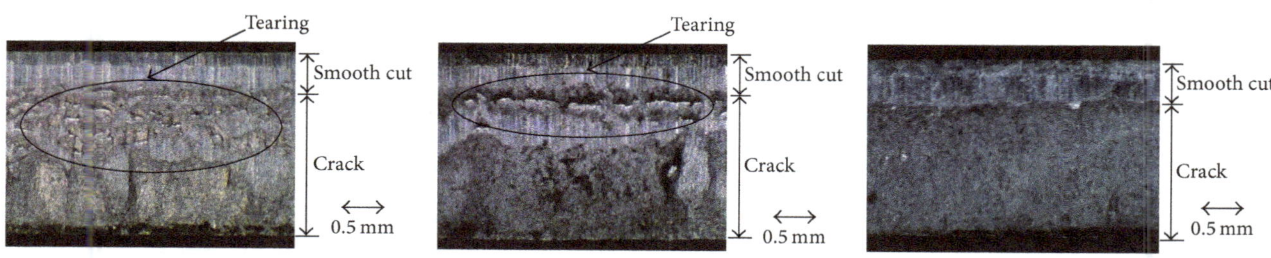

(a) Cl_s 5%t, D_{sh} 0.1 mm (b) Cl_s 10%t, D_{sh} 0.1 mm (c) Cl_s 5%t, D_{sh} 1.0 mm

FIGURE 4: An example of shaved surface of SAPH 440.

shaving allowance, the crack formations were easily generated and then the smooth cut surface decreased. Again, these manners generally confirmed the agreement with the theory of the shaving process of conventional metal sheet from the literature [8]. Figure 5 illustrates the relationship between shaved surface feature and shearing clearance. As the results show, the tendency of shaved surface feature of high-strength steel sheets corresponded well with that of aluminum and carbon steel sheets. Specifically, as the shearing clearance increased, the amount of die-roll increased but the smooth cut surface decreased. However, as aforementioned, it was observed that the tearing was generated as the shearing clearance of 5%t was applied. It was also noted that as comparing with aluminum and carbon steel sheets, the smooth cut surface of high-strength steel sheets were smaller than that of aluminum and carbon steel sheets. These results could be explained by the mechanical properties of material. Specifically, the mechanical property of ultimate tensile strength of high-strength steel sheets was larger than those of aluminum and carbon steel sheets. However, the results elucidated that the relationship between shaved surface feature and shearing clearance of high-strength steel sheets was well corresponded with those of conventional metal sheets from the literature [8–10]. Next, to obtain the smooth cut surface overall material thickness, the successful area of shaving conditions was investigated. As shown in Figure 6, the successful area for aluminum and carbon steel sheets could be found which corresponded well with the basic shaving process from the literature [8]. Specifically, as per the past research [8], the successful area could be found by applying shearing clearance smaller than 5%t with shaving allowances of 0.1 mm. In the present study, the results showed the successful area of using shearing clearance of 5%t with shaving allowances of 0.1 mm. However, it was interesting that the successful area could not be found for high-strength steel sheets. Therefore, through these results, this research revealed that, in terms of high-strength steel sheets, the shaving process could not be applied to fabricate the smooth cut surface overall material thickness.

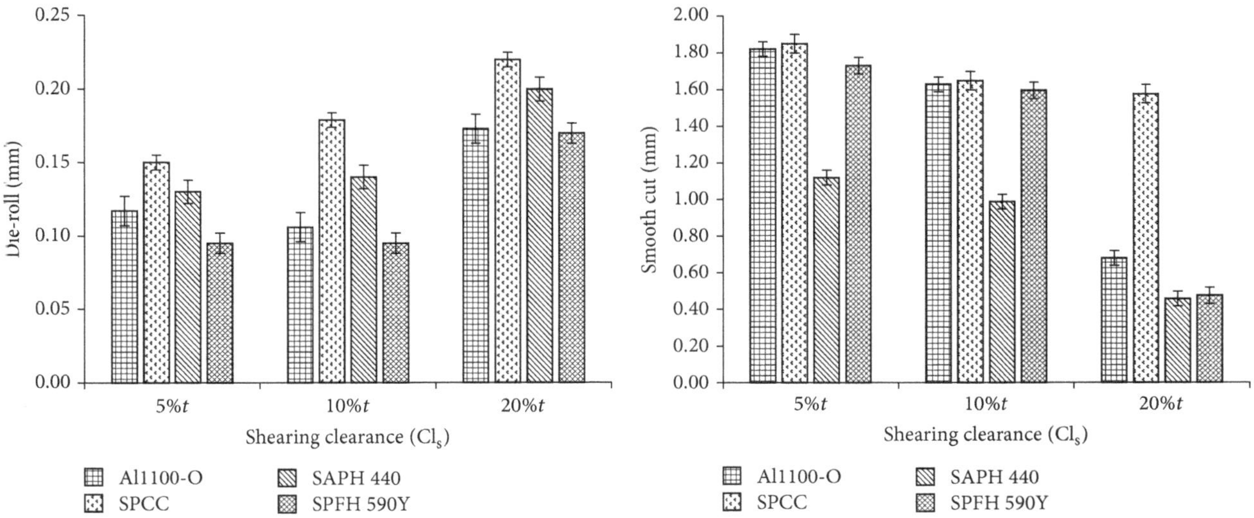

(a) Die-roll (b) Smooth cut

FIGURE 5: Relationship between shaved surface features and shearing clearances (D_{sh}: 0.1 mm).

Cl_s	Al1100-O				SPCC				SAPH 440				SPFH 590Y			
D_{sh} (mm)	0.1	0.3	0.5	1.0	0.1	0.3	0.5	1.0	0.1	0.3	0.5	1.0	0.1	0.3	0.5	1.0
5%t	○	✕	✕	✕	○	✕	✕	✕	✕ (T)	✕	✕	✕	✕(T)	✕	✕	✕
10%t	✕	✕	✕	✕	✕	✕	✕	✕	✕ (T)	✕	✕	✕	✕(T)	✕	✕	✕
20%t	✕	✕	✕	✕	✕	✕	✕	✕	✕	✕	✕	✕	✕	✕	✕	✕

○ Good
✕ Not good
T Tearing

FIGURE 6: Successful ranges of shaving conditions.

4. Conclusions

In the present study, the feasibility of shaving process applied for high-strength steel sheets was investigated by using the experiments. The shearing process of high-strength steel sheets was performed as the first operation of shaving process. The sheared surface feature in this process was examined and compared with that of conventional metal sheet of aluminum and carbon steel. The results illustrated that, in the shearing process, the relationship between sheared surface feature and shearing clearance was well against the theory of the basic shearing process of conventional metal sheet from the literature. As the shearing clearance increased, the amount of die-roll increased, but the crack formation decreased, whereas the smoot cut surface was somewhat constant. In addition, poor orthogonal cut surface increasingly was generated as the shearing clearance increased. Next, in the shaving process, the relationship between shaved surface feature and shearing clearance confirmed the agreement with the theory of the shaving process of conventional metal sheet from the literature. As the shearing clearance increased, the amount of die-roll increased but the smooth cut surface decreased. In addition, the tearing usually was generated as the small shearing clearance and shaving allowance were applied. Although the relationship between shaved surface feature and shearing clearance of high-strength steel sheets confirmed the agreement with the theory of those of conventional metal sheet from the literature, the successful area of shaving process application on high-strength steel sheets could not be found. Through these results, it was revealed in the present study that, owing to the special mechanical properties of high-strength steel sheets, the shaving process could not be applied to fabricate the smooth cut surface overall material thickness. To fabricate the smooth cut surface overall material thickness, in the next future work, the new process should be invented.

Competing Interests

The authors declare that they have no competing interests.

Acknowledgments

This research was partially supported by the Higher Education Research Promotion and National Research University

Project of Thailand, Office of the Higher Education Commission, under Grant no. 56000519 and Grant no. 57000618.

References

[1] M. Feistle, R. Golle, and W. Volk, "Determining the influence of shear cutting parameters on the edge cracking susceptibility of high-strength-steels using the edge-fracture-tensile-test," *Procedia CIRP*, vol. 41, pp. 1078–1083, 2016.

[2] K. Mori, Y. Abe, Y. Kidoma, and P. Kadarno, "Slight clearance punching of ultra-high strength steel sheets using punch having small round edge," *International Journal of Machine Tools and Manufacture*, vol. 65, pp. 41–46, 2013.

[3] X. Yang, C. Choi, N. K. Sever, and T. Altan, "Prediction of springback in air-bending of Advanced High Strength steel (DP780) considering Young's modulus variation and with a piecewise hardening function," *International Journal of Mechanical Sciences*, vol. 105, pp. 266–272, 2016.

[4] A. G. Kalashami, A. Kermanpur, A. Najafizadeh, and Y. Mazaheri, "Development of a high strength and ductile Nb-bearing dual phase steel by cold-rolling and intercritical annealing of the ferrite-martensite microstructures," *Materials Science and Engineering A*, vol. 658, pp. 355–366, 2016.

[5] K. Mori, S. Maki, and Y. Tanaka, "Warm and hot stamping of ultra high tensile strength steel sheets using resistance heating," *CIRP Annals—Manufacturing Technology*, vol. 54, no. 1, pp. 209–212, 2005.

[6] K. Mori, Y. Abe, and Y. Suzui, "Improvement of stretch flangeability of ultra high strength steel sheet by smoothing of sheared edge," *Journal of Materials Processing Technology*, vol. 210, no. 4, pp. 653–659, 2010.

[7] M. Murakawa, K. Nakamura, T. Shionome, F. Komuro, G. Muro, and N. Koga, "Press-shaving characteristics of ultrahigh-strength steel sheets," *Key Engineering Materials*, vol. 554–557, pp. 1879–1886, 2013.

[8] M. Murakawa, S. Thipprakmas, and M. Jin, "Investigate of relationship between shaved surface and shaving allowance," *Journal of Japan Society for Technology of Plasticity*, vol. 44, no. 513, pp. 53–57, 2003 (Japanese).

[9] S. Thipprakmas, W. Phanitwong, M. Chinwithee, and T. Morkprom, "Reciprocating shaving approach to eliminate crack and burr formation in pressed parts," *Key Engineering Materials*, vol. 443, pp. 217–224, 2010.

[10] S. Thipprakmas and W. Phanitwong, "Finite element analysis of shaving direction effects in reciprocating shaving process," *Steel Research International*, vol. 81, no. 9, pp. 1058–1061, 2010.

10

Influence of Vanadium and Boron Additions on the Microstructure, Fracture Toughness, and Abrasion Resistance of Martensite-Carbide Composite Cast Steel

Waleed Elghazaly,[1] R. Rashad,[2] Sayed Elmohr,[3] and Saied Elghazaly[1]

[1]*Steel Technology Lab, Central Metallurgical R&D Institute, Cairo, Egypt*
[2]*Mechanical Design, Production Engineering, Cairo University, Cairo, Egypt*
[3]*Siegwart Co., Cairo, Egypt*

Correspondence should be addressed to Saied Elghazaly; saiedelghazaly@yahoo.com

Academic Editor: Patrice Berthod

High chromium cast steel alloys are being used extensively in many industrial services where dry or wet abrasion resistance is required. Such steel castings are demanded for cement, stoneware pipes, and earth moving industries. In this research, five steel heats were prepared in 100 kg and one-ton medium frequency induction furnaces and then sand cast in both Y-block and final impact arm spare parts, respectively. Vanadium (0.5–2.5%) and boron (120–150 ppm) were added to the 18Cr-1.9C-0.5Mo steel heats to examine their effects on the steel microstructure, mechanical properties especially impact, fracture toughness and abrasion resistance. Changes in the phase transformation after heat treatment were examined using inverted, SEM-EDX microscopy; however, the abrasion resistance was measured in dry basis using the real tonnage of crushed and milled stoneware clay to less than 0.1 mm size distribution.

1. Introduction

Steel designers and researchers seek always to develop alloys that can give a balanced strength and toughness, for example, high wear resistant together with high fracture toughness [1, 2]. One of the well-known abrasion resistant ferrous materials is high chromium white cast iron (20–30% Cr, 3.5–4% C, and 1–5% Mo) which has excellent wear resistance, but unfortunately poor toughness or ductility due to the bigger matrix grain size and the massive brittle leduburitic eutectic carbide precipitations that occur in a continuous, three-dimensional network surrounding the dendrite matrix grains [3–5]. The fracture toughness of the cast iron shows minimal values (10–25 $Mpa{\cdot}m^{1/2}$) as a result of brittle carbide precipitations [6]. Many researches were trying to increase the abrasion resistance of cast iron by alloying with wolfram, vanadium, or niobium but the product toughness was always deteriorated [7–9]. In this research and aiming to keep high wear resistance and more tough matrices, trials to design cast steel alloys with lower carbon content, well-distributed harder carbides, and a matrix of tempered martensite and some retained austenite were studied.

The idea of producing steel alloys with lower carbon content together with vanadium and boron additions can be a promising solution to increase the toughness and wear resistance of such alloys.

2. Experimental Procedure

In this investigation, five steel heats were prepared using coreless, medium frequency, induction furnace. The alloys under investigation are the products of melting and casting special tool steel scraps collected from used tools and dies (DIN X210Cr12) scrap offered by Universal Engineering Group. Ferroalloys, low sulfur graphite, and clean steel scrap were used to adjust the final compositions designed in this investigation. Ferroalloys used comprised low-carbon ferrochromium (0.1% C, 80% Cr), ferrovanadium (80% V),

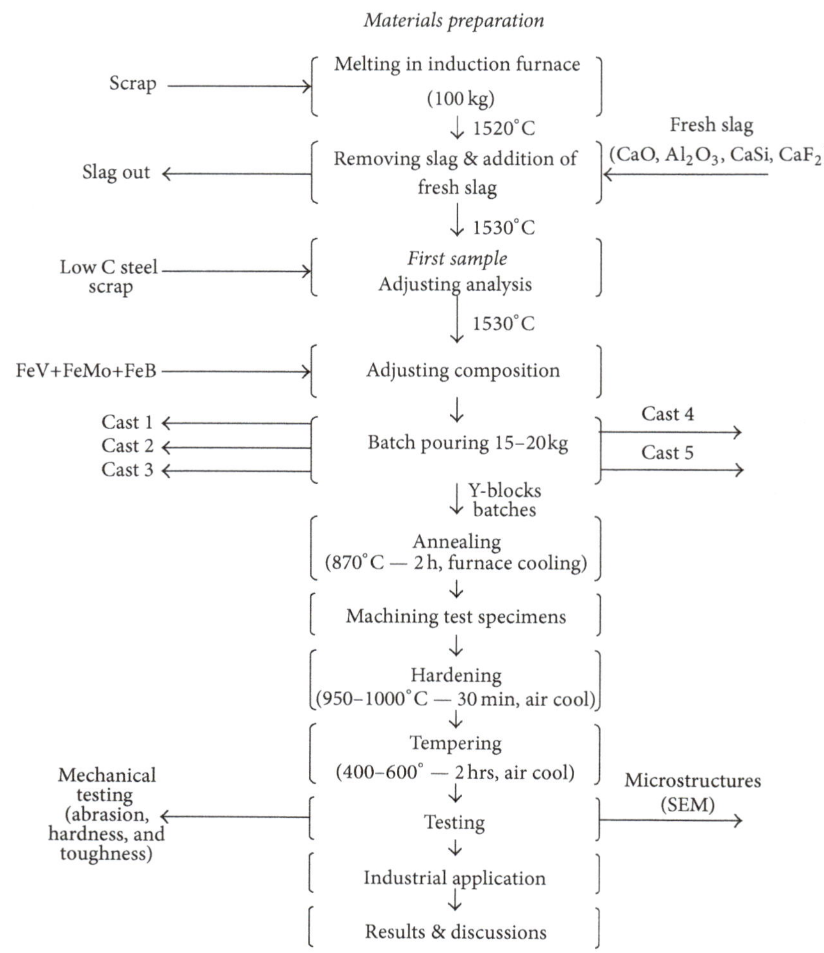

FIGURE 1: Overall plan of melting, casting, and testing of experimental alloys.

ferroboron (35% B), ferromolybdenum (80% Mo), and pure aluminum (99.9% Al). Fresh slag materials were used in composition of low melting calcium aluminates together with Al-CaSi powder. The region of low melting Ca-aluminates of compositions $12CaO{\cdot}7Al_2O_3$ in binary CaO-Al_2O_3 system, which is liquid at about 1415°C, was chosen to form the slag constituents. Thermite mixture of aluminum and iron oxide powders was also added to the top of slag under the furnace cover to give exothermic heat to keep continuous metal/slag active mass transfer. The lining was chosen to be high alumina one not only to withstand melting temperature but also to not interact chemically with both slag and the alloy melt. The furnace cover was also rammed from high alumina bonded powder, to preserve the heat of the top slag to ensure good metal protection and better mass transfer through metal/slag interface. After complete melting (1520°C), first oxidizing slag (negligible) was removed and the other 3–5 kg fresh slag materials were added to refine melt from any impurities. Ferroalloys were then added to correct the required final analysis of every alloy. The overall plan of work is projected as shown in Figure 1, where addition of correcting ferroalloys, carbon and aluminum, was performed to adjust the final alloy compositions. The final casting temperature was adjusted at 1530°C using dipping thermocouple or optical pyrometer. Every melt was tapped and cast in designed cavity and experimental Y-block as well, preheated (250°C) sand moulds sprayed with refractory ZrO_2 according to DIN EN 1536 type 3. Figure 2 represents a schematic drawing of the experimental cast Y-block and casting pattern. Before casting of the first heat, addition of scrap and/or ferroalloys was performed to adjust the composition of the final alloy. The cover lid of the furnace was always kept closed, to ensure no further oxidation of the melt by atmospheric air; moreover, inert atmosphere under the furnace lid was achieved using Argon lancing. This is to ensure effective removal of shrinkage cavity from the main cast blocks.

3. Results and Discussions

The final chemical analysis of the five experimental steel patches is shown in Table 1.

In this investigation, the melting procedure was designed to get final steel casts different in vanadium and boron content, but with nearly the same basic composition. It is evident that the content of dissolved aluminum in all steel ranged

TABLE 1: Chemical analysis of the steel patches.

Material	C	Mn	Si	Cr	Mo	V	B	Al
GS-1	1.98	0.63	0.80	18.56	0.57	—	—	0.031
GS-2	1.99	0.73	0.79	18.91	0.61	0.59	—	0.027
GS-3	1.98	0.76	0.83	18.56	0.59	1.51	—	0.035
GS-4	1.94	0.69	0.76	18.32	0.63	1.56	0.012	0.032
GS-5	1.97	0.80	0.81	18.69	0.59	2.56	0.015	0.029

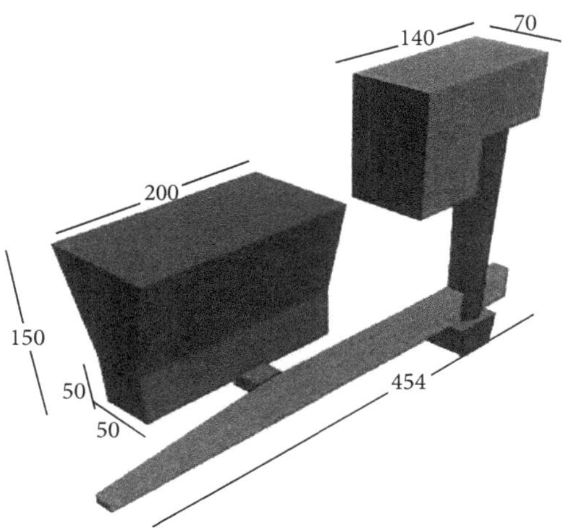

FIGURE 2: Geometry of casting Y-block mould.

from 0.027 to 0.039 as an indication that the experimental melts were killed to have 3–5 ppm maximum dissolved oxygen; this ensures minimized casting cavities.

3.1. Microstructure Investigation

3.1.1. As-Cast Conditions. Vanadium, chromium, molybdenum, and boron are reported to affect the transformation behavior of austenite [10]. The as-cast microstructures of hypereutectoid 18% Cr-1.9% C-0.5% Mo experimental GS.1–5 alloys consist of mainly metastable dendrites of primary austenite (dark phase) + mixed primary and eutectic carbides (white phase) and martensite as shown in Figures 3, 4, 5(a), and 5(b). Emission-diffraction X-ray (EDX) analyses revealed that the primary carbides consists mainly of chromium carbides of type Cr_7C_3 colonies which have a hexagonal structure with lattice parameters $a = 13.982$ Å and $c = 4.507$ Å; this type of carbide has many structural defaults and forms during solidification as austenite rejects some of its carbon. The morphology of such complex carbides depends to a great extent on the solidification and casting technology; however, rod or blade (multiple rods) and radiating shaped carbides of type Cr_7C_3 are normally formed during and after solidification as shown in Figures 6(a) and 6(b). The austenite phase remains in metastable condition at room temperature as in alloy GS-1 due to suppression of martensite start temperature; however, by alloying with vanadium and boron more carbides are formed by reacting with adjacent carbon, this creates in turn areas of depleted carbon that transforms in turn to martensite during further cooling depending on the content of vanadium added.

The grain refinement of dendrites as well as type and morphology of carbides changed to some extent on adding vanadium as in alloys GS-2 and GS-3 having 0.59% and 1.51% vanadium, respectively, where vanadium-rich carbide is identified by EDX analysis as V_6C_5 or complex Fe-Cr-V depending on the content of vanadium in the alloy.

During solidification, austenite dendrite forms first and rejects some of its saturation carbon which reacts in sequence with vanadium, chromium, and iron to form primary single or complex carbide depending on their affinity, diffusivity, temperature, and concentration. The carbide phases nucleate side by side with/within the austenite dendrites forming rods or blades. In alloy GS-3 where vanadium content is about 1.5%, the volume fraction of primary austenite (as detected by X-ray diffraction) is decreased to some extent and a massive precipitation of V_6C_5 and Cr_7C_3 (as detected by EDX analyzer) resulting in narrowing dendrite arm spacing; this can be noticed by comparing the microstructure of GS-3 with GS-1 and GS-2.

The as-cast microstructures of alloy GS-5 having 2.5% vanadium together with 0.012 and 0.015% boron showed massive precipitation of both primary and secondary carbides hindering grain growth of dendrite arms. Volume fraction of austenite and carbides for the cast alloys, as detected using X-ray diffraction and EDX analyzer, is shown in Table 2.

3.1.2. Heat Treated Metal Matrix. The experimental heats GS-1 to GS-5 were heat treated using the published data on cooling curves for near compositions [11] where the start of transformation during cooling is retarded for about 3–5 h depending on the contents of carbon and alloying elements of the steel. For the overall heat treatment cycle used in this research, the annealing temperature, suitable to machine the test samples, was chosen at 870°C for two hours, while hardening temperatures were chosen in the range 950–1000°C for 30 minutes, quenching in still air or oil and then tempering in the range 400–600°C for 2 h. The overall microstructure of the alloys consists of tempered martensite, metastable or retained austenite, complex primary carbides, and secondary carbides as shown in Figures 7 and 8(a)–8(d). Alloying of the steels with variable amounts of vanadium and boron retards the grain growth of primary carbides as in (a) and (d) microstructures as compared with alloy GS-1 having (c) microstructure.

The presence of more alloying elements of vanadium and boron as in alloy GS-5 leads to dispersive carbide colonies

TABLE 2: Volume fraction of austenite and carbides in cast alloys.

Heat	Condition	Vol.% $\gamma + \alpha$	Vol.% M_7C_3	Vol.% $V_6C_5 + BC$	Dendrite arm spacing, μm
GS-1	As cast	42.3	28.7	—	18.4
GS-2	As cast	37.9	21.2	8.5	15.1
GS-3	As cast	30.7	19.5	10.4	14.9
GS-4	As cast	30.3	19.1	11.9	10.6
GS-5	As cast	27.8	21.6	17.5	8.3

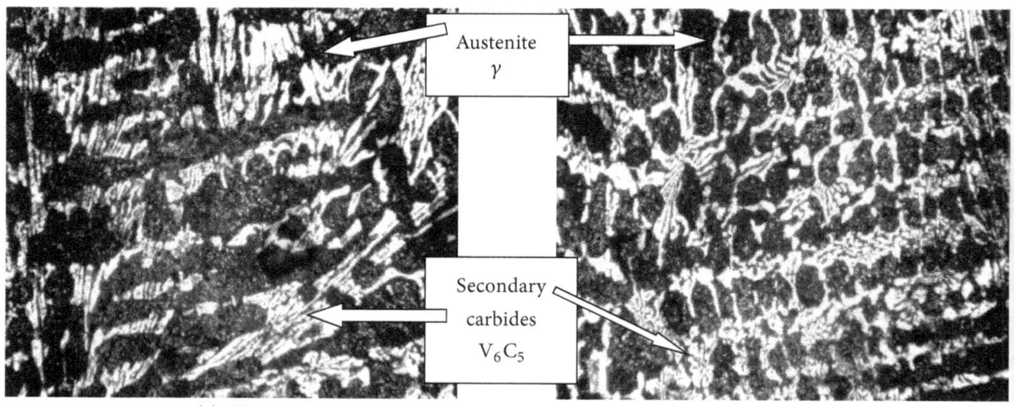

(a) GS-1 (b) GS-2, 0.5% V, ×100

FIGURE 3: As-cast microstructure of GS-1 and GS-2 experimental steel.

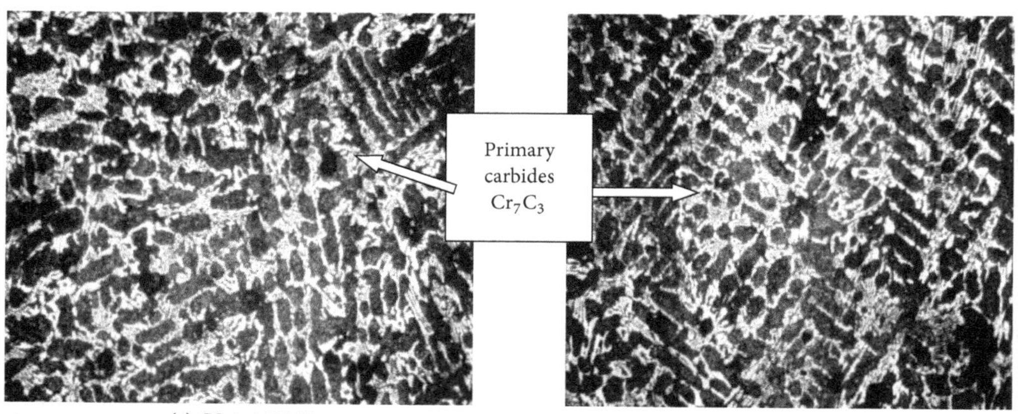

(a) GS-3, 1.5% V (b) GS-4, 1.5% V + 0.012% B, ×100

FIGURE 4: As-cast microstructure of GS-3 and GS-4 experimental steel.

and the absence of dendrite structure and ensures more fine distribution of carbides as in (b) microstructure. X-ray diffraction techniques together with emission-diffraction EDX scanning analyses were used to determine the types and compositions of phases and carbides transformed after heat treatment; however, it was not precise to determine the volume fractions of phases due to the interference of dominated martensite phase. More measurements for the heat treated microstructures for such steels especially GS-3, GS-4, and GS-5 need further investigation in the coming future researches.

3.1.3. Mechanical Properties

(1) Hardness. The effect of obtained microstructures on the hardness values of GS alloys is summarized as in Figure 9; the bulk or microhardness of the GS improved indeed by increasing the vanadium content in the as-cast, quenched, and quenched tempered as compared with 25% Cr cast iron. This can be attributed to the increased volume fraction of the primary carbides Cr_7C_3 and the secondary V_6C_5 carbides; however, BC intermetallic precipitates harden the matrix as well. After quenching and tempering at 400–600°C,

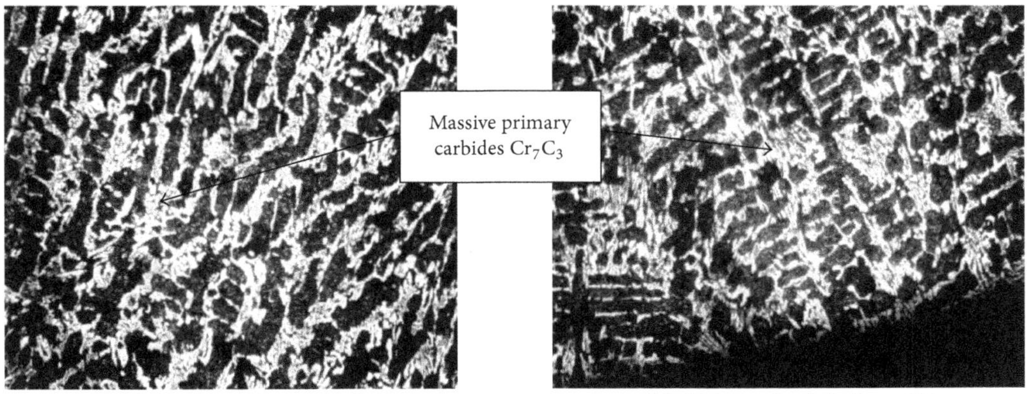

(a) GS-5, 2.5% V + 0.015% B (b) GS-5 near the chill edge, ×100

FIGURE 5: As-cast microstructure of GS-5 (2.5 V + 0.015 B) experimental steel.

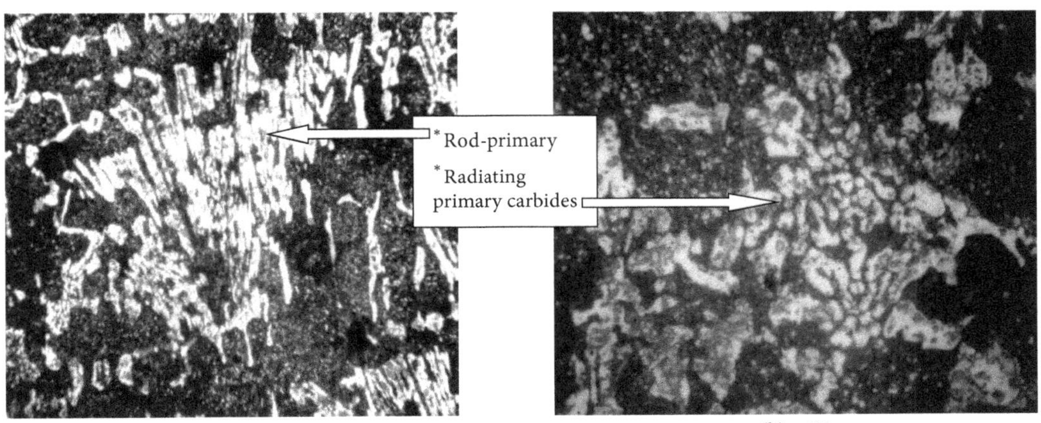

(a) (b) ×600

FIGURE 6: Morphology of rod-like primary carbides and radiating and secondary carbides.

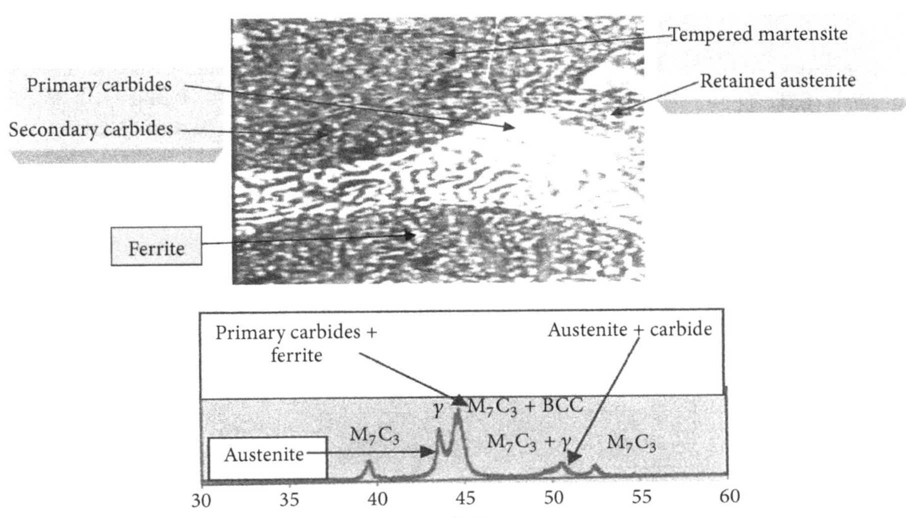

FIGURE 7: Typical microstructure of experimental steel after heat treatment, as revealed by X-ray diffraction [hardening 980°C-AC, tempering 60C°C, 2 h].

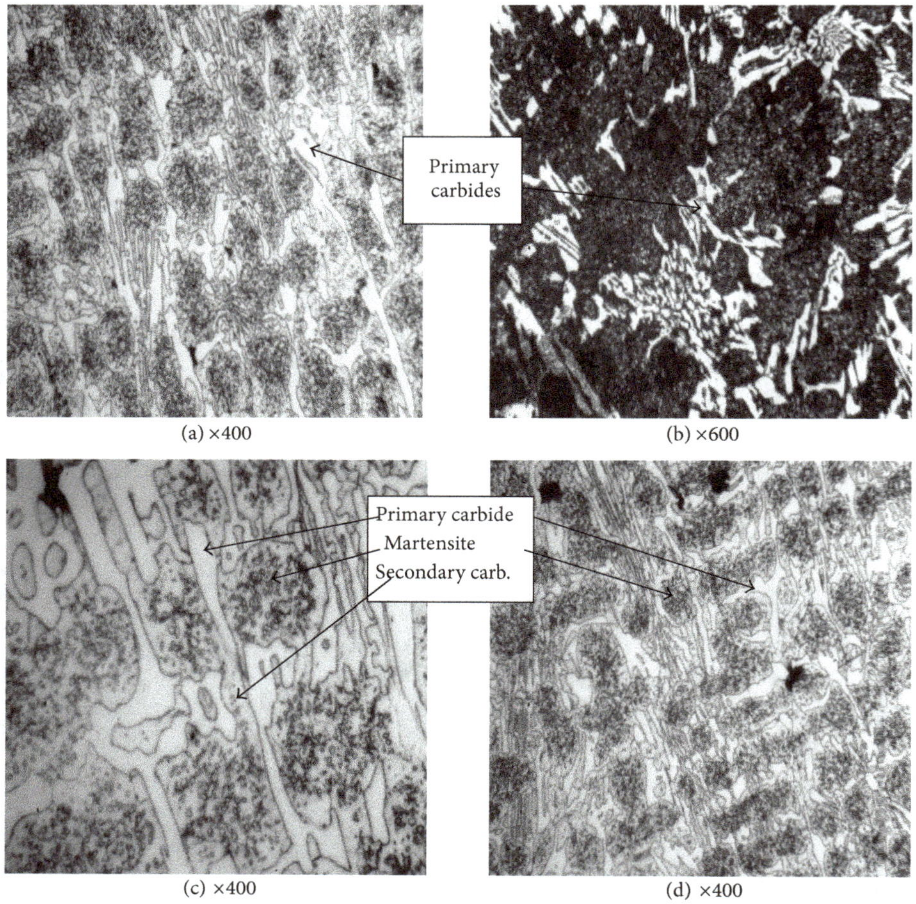

FIGURE 8: (a-b) Morphology of primary carbides in heat treated microstructures of experimental steel GS-4 and steel GS-5. (c-d) Appearance of quenched and tempered microstructures.

the precipitation of carbides and the transformation of primary and secondary austenite to martensite improve the hardness of GS alloys.

3.2. Influence of V and B on Fracture Toughness. Due to nonhomogenous cast structures and their high volume fraction of brittle carbides, the alloys under investigation showed low ductility values (5–15 J/cm^2); for those reasons the Charpy V or U notch cannot give real measurements. Therefore and from fracture mechanics point of view, the most straightforward parameter to characterize fracture toughness is the critical stress intensity factor (K) or dynamic fracture parameter (K_{id}). Fracture toughness K_c (Pa$\cdot\sqrt{\mathrm{m}}$) measurements were made on primary carbide matrix at room temperature by the indentation factor method [12] which can be calculated from the width of crack induced on applying the Vickers Pyramid hardness test as follows (Figure 12):

$$K_c = \frac{\alpha\sqrt{EP}\,(d/2)}{\sqrt[3]{a2}}, \qquad (1)$$

where, for those cast alloys, there is total crack edge distance (m), E Young's modulus (310–350 GPa), P load (about 30 kg), and a half length of crack (m), d is the diagonal length of indentation, and α is calibration factor (about 0.025 depending on the surface finish).

A comparison of calculated fracture toughness of produced alloys GS1–5 in as-cast, quenched tempered conditions and 25% Cr cast iron, using bulk Vickers hardness testing machine, is shown in Figure 10. It is evident that alloys having vanadium content higher than 1.5% suffer from deteriorated fracture toughness in as-cast or even in tempered conditions; however, the toughness for all alloys in as-cast or heat treated conditions is higher than that for chromium cast iron. This can be attributed to the grain refinement of the alloy matrix due to the pinning action caused by secondary carbides precipitated and solid solution hardening caused by vanadium and boron and the lower carbon contents as well.

3.3. Industrial Abrasion Monitoring. Impact arms and grinding segments were produced from these experimental alloys as spare parts for Hazemag and Eirich stoneware (lean or fired) crushing and milling systems. The results were collected as actual product of crushed and milled powders (0.1 mm max). It is evident from Figure 11 that the mechanism of reinforcing the produced alloys with vanadium and boron

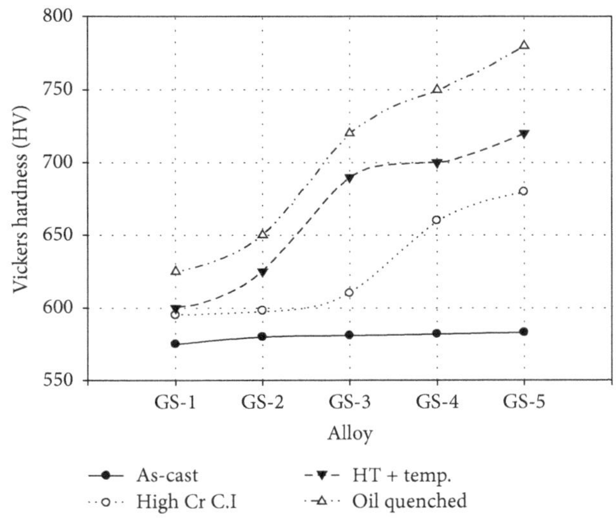

FIGURE 9: Hardness values of GS alloys as compared with 25% Cr cast iron.

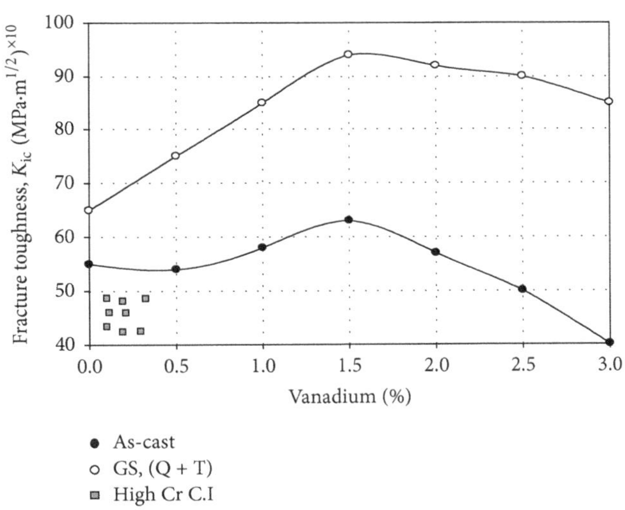

FIGURE 10: Influence of vanadium on fracture toughness coefficient as measured for GS alloys in comparison with 25% Cr cast iron.

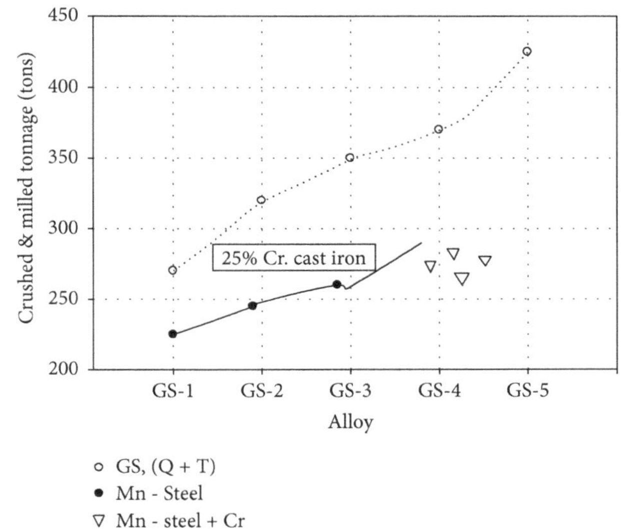

FIGURE 11: Industrial performance of GS alloys compared with Mn-modified steel and Cr cast iron.

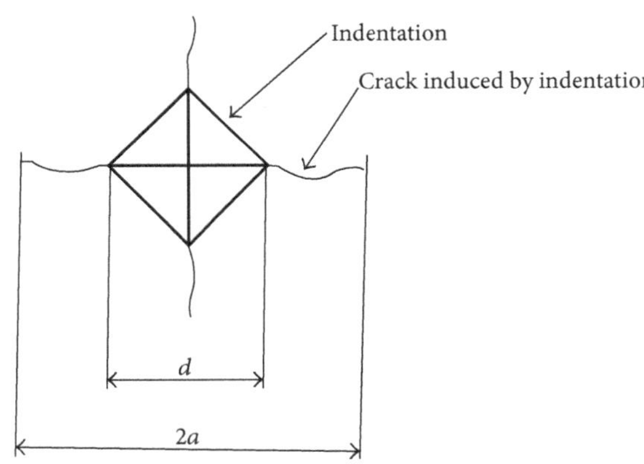

FIGURE 12

carbides at that lower carbon level resulted in much higher wear resistance than that for high chromium cast iron.

4. Conclusions

From the above measurements and discussions, the following conclusions can be put as follows:

(1) The microstructures of the experimental steel GS-1 to GS-5 in both as-cast and heat treated conditions are more fine and homogeneous if compared with those for white chromium cast iron; this is due to the lower carbon content and the effects of adding vanadium and boron. This leads to harder matrices than those for cast iron.

(2) The fracture toughness of the experimental steels is higher than that for cast iron due to refinement of the overall matrices and the well distributed primary and secondary carbides; this leads to high abrasion resistance and higher productivity.

(3) The wear resistance of the experimental quenched and tempered steel is 1.5 times higher than that for 25% Cr white cast iron.

Competing Interests

The authors declare that they have no competing interests.

References

[1] A. Kootsookos and D. Gates, "Development of a white cast iron," *Cast Metals*, vol. 4, no. 7, pp. 239–247, 1995.

[2] C. P. Tabrett and I. R. Sare, "Fracture toughness of high-chromium white irons: influence of cast structure," *Journal of Materials Science*, vol. 35, no. 8, pp. 2069–2077, 2000.

[3] M. Filipovic and Z. Kamberovic, "Correlation of microstructure with wear resistance and fracture toughness of the white cast iron," *Met. Mater. Rev*, vol. 41, pp. 59–82, 2013.

[4] Ö. N. Doğan, "Columnar to equiaxed transition in high Cr white iron castings," *Scripta Materialia*, vol. 35, no. 2, pp. 163–168, 1996.

[5] C. P. Tabrett, I. R. Sare, and M. R. Ghomashchi, "Microstructure-property relationships in high chromium white iron alloys," *International Materials Reviews*, vol. 41, no. 2, pp. 59–82, 1996.

[6] K.-H. Z. Gahr and D. V. Doane, "Optimizing fracture toughness and abrasion resistance in white cast irons," *Metallurgical Transactions A*, vol. 11, no. 4, pp. 613–620, 1980.

[7] A. Bedolla-Jacuinde, "Microstructure of vanadium-, niobium- and titanium-alloyed high-chromium white cast irons," *International Journal of Cast Metals Research*, vol. 13, no. 6, pp. 343–361, 2001.

[8] M. Fiset, K. Peev, and M. Radulovic, "The influence of niobium on fracture toughness and abrasion resistance in high-chromium white cast irons," *Journal of Materials Science Letters*, vol. 12, no. 9, pp. 615–617, 1993.

[9] P. Dupin, "Influence of addition of silicon, molybdenum, vanadium and tungsten on the structural evolution of the as cast state of a high Cr-white cast iron," *AFS Transactions*, vol. 92, pp. 355–360, 1984.

[10] G. Changqing and W. Caidong, "Effect of variable heat treatment modes on microstructure of Fe-Cr-B cast irons," *Journal of Research and Development*, vol. 5, no. 1, 2008.

[11] ASM International metals handbook V.1, 2000.

[12] K. Yamamoto, S. Inthidech, N. Sasaguri, and Y. Matsubara, "Influence of mo and w on high temperature hardness of M7C 3 carbide in high chromium white cast iron," *Materials Transactions*, vol. 55, no. 4, pp. 684–689, 2014.

Numerical Investigations on Residual Stress in Laser Penetration Welding Process of Ultrafine-Grained Steel

Dezheng Liu,[1] Yan Li,[1] Haisheng Liu,[1] Zhongren Wang,[1] and Yu Wang[2]

[1]*Department of Mechanical Engineering, Hubei University of Arts and Science, Xiangyang, Hubei 441053, China*
[2]*Shanghai Institute of Applied Physics, Chinese Academy of Sciences, Shanghai 201800, China*

Correspondence should be addressed to Haisheng Liu; 185669911@qq.com

Academic Editor: Antonio Boccaccio

Weld solidification crack prevention in the laser penetration welding process is essential for the strength of the welded component. The formation of solidification cracks can ultimately be attributed to welding residual stresses, and preventive measures should be taken during welding. In this study, the effects of residual stresses on the laser penetration welding quality of ultrafine-grained steels were investigated. A heat source model was established through the analysis of the metallography of the cross section of the heat-affected zone (HAZ) of ultrafine-grained AN420s-grade steel, and the chemical composition of the weld bead was obtained using an FLS980-stm Edinburgh fluorescence spectrometer. Furthermore, the constitutive coupling relation between the temperature and material flow stress was established based on the Gibbs function, and the welding residual stress was obtained by setting trace points in a finite element analysis (FEA) model based on experimental data of the weld bead cross section under different welding conditions. The results show that weld solidification cracks will form when the residual stresses exceed the material flow stresses in the weld bead, and the residual stresses can be decreased through a reasonable increase of the welding speed. The results indicate that the proposed criterion has high accuracy and can be used to predict the formation of weld solidification cracks in the laser penetration welding process.

1. Introduction

Laser penetration welding, which is a stable and noncontact joining process, provides high welding speeds, good flexibility (narrow weld width and large weld depth), and high accuracy in process control [1]. The laser penetration welding technology has potential uses in additive manufacturing of ultrafine-grained steels [2, 3]. However, solidification crack formation in the fusion welding process continues to be one of the major challenges in the use of laser penetration welding technology to join ultrafine-grained steels [4]. For example, the use of laser penetration welding for new structural components still requires significant amounts of engineering data to determine its effects on the mechanical properties, such as tensile strength, fatigue strength, and the generation of residual stresses [5]. During the laser penetration welding process, the heat transfer and liquid metal flow in the molten pool affect the grain growth direction, which may affect the weld seam structure and the properties of the weld [6]. If thermomechanical strains caused by the time-dependent temperature distribution during welding and material-specific shrinkage during the liquid- or solid-phase transition cannot be compensated for by the ductility of the weld metal or by reflow of the melt, solidification cracks will appear [7]. Furthermore, the residual stress caused by the changes in the chemical composition and the heating and cooling rates during the welding process can aggravate the initiation and growth of solidification cracks [8]. Measured residual stress data can be used for fracture analysis, safety design, and strength evaluations. However, it is difficult to experimentally measure the evolution of residual stress and solidification cracks due to the strong interference from the arc light, the rapid change in temperature, and the phase transition [5].

Thus, numerical simulations can provide valuable information to understand the laser penetration welding process of ultrafine-grained steels.

Numerical simulation methods can use multiple physics models to describe the complex physical phenomena in the molten pool [9, 10]. Many of the studies on the susceptibility to solidification cracking during the welding process using thermomechanical modelling focused on the resulting deformation and residual stresses. Frewin and Scott [11] presented a three-dimensional (3D) finite element (FE) model of the heat flow during pulsed laser beam welding. They also investigated the transient temperature profiles, dimensions of fusion, and heat-affected zone (HAZ) by incorporating temperature-dependent thermophysical properties and experimentally measured beam profiles. Shibahara et al. [12] developed a temperature-dependent interface element technique to analyze solidification cracking for mild steels. They suggested that this technique could be used to predict solidification cracks with a high accuracy. Volpp and Vollertsen [13, 14] presented an analytical method to quickly calculate the keyhole dynamics for different spatial laser intensity distributions during laser deep penetration welding. They concluded that a top hat intensity distribution led to higher maximum frequencies of the keyhole oscillations compared to a Gaussian beam profile. Dong et al. [15] indicated that the prediction of solidification cracking could be obtained by comparing the crack driving force and resistance curves. Fadaei and Mokhtari [16] presented an FE simulation and experimental measurements of the residual stresses in repair butt welds using the Goldak double-ellipsoid heat source model.

However, most recent studies on residual stresses in the laser penetration welding process were limited to numerical simulations of the evolution of temperature and residual stress fields and rarely involved the accurate modelling and analysis of the effects of transient residual and material flow stresses for weld solidification cracking.

The objective of this study was threefold. First, a heat source model was established through the analysis of the metallography of the cross section of the weld bead of ultrafine-grained AN420s-grade steel, and the chemical composition of the weld bead was measured using an FLS980-stm Edinburgh fluorescence spectrometer. Second, the constitutive coupling relation between temperature and material flow stress in the molten pool was derived. The Gibbs interface thermodynamic method and the Sente Software JMatPro were used in the derivation of the constitutive coupling relation. Third, an FE model of the molten pool was established based on the experimental data of the weld bead cross section under two different welding conditions, and four trace points were used to calculate the residual stresses in different positions of the molten pool. Emphasis was placed on the accuracy of predicting solidification cracking using the proposed criterion. Finally, a comparison between the predicted results and experimental data is given.

2. Materials and Methods

2.1. Heat Source Model and Chemical Composition in Weld Bead. The material used in the laser penetration welding was ultrafine-grained AN420s-grade steel in the form of a thick plate formed by pressure welding. Due to the dramatic changes in temperature and stress fields, the choice of a heat source model for laser welding should consider the influence of the stability of keyholes and molten pool formation [17]. Sun et al. [18] studied a new type of heat source model for laser deep welding. They suggested the use of a combined heat source model comprising a 3D cone heat source model and Gaussian surface heat source. This model exhibits Gaussian decay of the heat source radius on the surface and an exponential increase in the power density of the heat source with depth, which captures the penetration laser welding screw shape structure observed in actual welding experiments. Hence, a compound [18] cone and Gaussian surface heat source model was adopted in this work. The Gaussian surface heat source controls the surface of the molten pool and the head of the weld bead, and it can capture the effect of fluid agitation at the opening in the molten pool. The cone heat source controls the heat transfer of the liquid in deep layers and the cone body of the weld bead, and it can capture the behavior of a keyhole caused by the beam flow mining. In this compound heat source model, the power of the laser welding machine [19] can be expressed as follows:

$$\begin{aligned} P_t &= W_s + W_v + P_l, \\ W_s &= \iint \rho_m \cdot e^{-ar^2/\sigma^2} ds, \\ W_v &= \iiint R_m e^{-br^2/r_0^2} \cdot \frac{mz + r_0}{r_0} dv, \end{aligned} \tag{1}$$

where P_t is the power of the laser welding machine, W_s is the power of the surface heat source, W_v is the power of the volume heat source, P_l is the power loss caused by the reflection of the workpiece surface, ρ_m is the density of the heat flux of the surface heat source, R_m is the density of the heat flux of the volume heat source, σ is the heat flux distribution parameter, a is the heat flux concentration coefficient, b is the endogenous heat concentration coefficient, r is the diameter of the cone, r_0 is the effective diameter of the cone, m is the coefficient of the cone shape, and z is the depth of the cone. The schematic diagram of the compound heat source model is illustrated in Figure 1.

In this heat source model, the governing equation of thermal conduction [20] can be expressed as follows:

$$\rho c \frac{\partial T}{\partial t} = k \frac{\partial^2 T}{\partial \Delta^2} + q, \tag{2}$$

where q is the intensity of the laser heating source, T is the temperature, t is the time, Δ is the point spatial vector (x, y, z), and ρ, c, and k are the density, specific heat, and thermal conductivity, respectively.

The HAZ in the ultrafine-grained steel was a major concern in many studies [21, 22]. In the HAZ, the microstructure of the ultrafine-grained steel undergoes severe

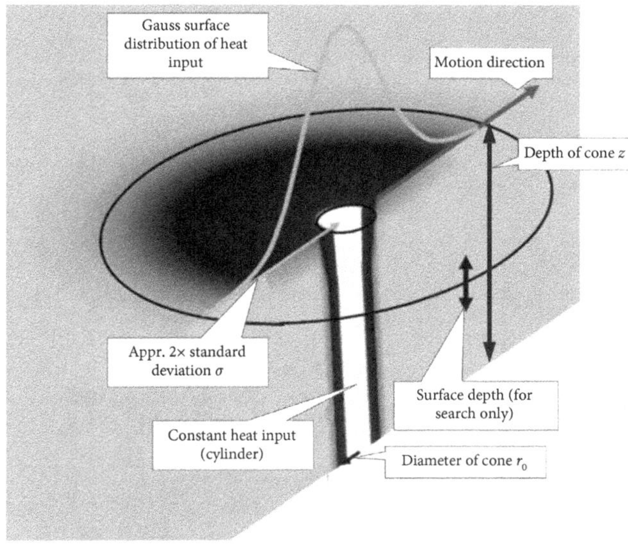

FIGURE 1: The compound heat source model.

transformations due to the influence of the welding thermal cycle, and the structure of the overheated zone changes significantly under the influence of the welding heat [21]. To investigate the microstructure and mechanical properties of the ultrafine-grained steel under laser penetration welding, the metallography of the cross section of the weld bead should be studied. The samples of ultrafine-grained steel were provided by Angang Steel Company Limited. The mechanical properties of the plate of AN420s-grade steel with dimensions of 60 mm × 30 mm × 8 mm were obtained through an experimental tension test using a SUNS UTM5000 electronic tensile testing machine under an ambient temperature of about 0°C. The mechanical properties of the AN420s-grade steel plate are shown in Table 1, where σ_s is the yield strength, σ_b is the tensile strength, and δ_s is the extensibility.

In this study, an IPG-YSL-10000 fiber laser welder with a minimum spot diameter of 0.72 mm and an argon gas flow rate of 15 L/min were used to weld the samples. The chemical composition of the weld bead was determined using an FLS980-stm Edinburgh fluorescence spectrometer and an Olympus GX71 metallographic microscope (Figure 2). The instruments used in this study are shown in Figure 2.

2.2 Constitutive Coupling Relation. Residual stresses introduced in the weld bead are a consequence of incompatible thermal strains caused by heating and cooling cycles during the laser penetration welding process, and the solidification cracks can ultimately be attributed to the welding residual stresses [5]. There is a stage between the start and end of solidification where the material exhibits brittle behavior, even though the material is ductile. The end of solidification in the molten pool occurs at the temperature where all the material is solidified, which may be below the temperature where the body exhibits solid behavior macroscopically [9]. Thus, the investigation of the effects of the stress and temperature fields on the mechanical properties of the weld bead was significant in this study. The Gibbs interface thermodynamic method [23] was adopted to establish the constitutive coupling relation between the temperature and material flow stress. The basic equation for the Gibbs energy of a multicomponent solution phase can be expressed as follows:

$$G_{\mathrm{m}} = \sum_i x_i G_i^0 + \mathrm{RT}\sum_i x_i \ln x_i + \sum_i \sum_{j>i} x_i x_j \sum_V \Omega_V \left(x_i - x_j\right)^V, \tag{3}$$

where $\sum_i x_i G_i^0$ is the Gibbs energy of the pure components, $\mathrm{RT}\sum_i x_i \ln x_i$ is the ideal entropy, and $\sum_i \sum_{j>i} x_i x_j \sum_V \Omega_V (x_i - x_j)^V$ accounts for pairwise interactions of the species. According to the Gibbs interface thermodynamic method [23], the equilibrium pressure of the system is achieved when the Gibbs free energy attains a minimum. Solute mass fraction in the solid can be expressed as follows:

$$\tau(x, T) = \frac{1}{\beta 2^{N/8} D \Delta T^q} \int_0^x \frac{dx}{x^{(2(1-x))/3}(1-x)^{(2x)/3}}, \tag{4}$$

where τ is the solute mass fraction in the solid for different amounts of undercooling, β is an empirical coefficient, N is the size of a grain, D is an effective diffusion coefficient, ΔT is the undercooling, q is an exponent that depends on the effective diffusion mechanism, and x is the fraction transformed. Based on equations (2)–(4) and the data in Table 2, the phase transition diagram of the weld bead can be calculated using JMatPro using $N = 9$ scale by ASTM (American Society of Testing Materials).

The relationship between the stress and strain can be determined by the yield strength, hardening exponent, and reference strain [23, 24] as follows:

$$\sigma = \frac{\sigma_{0.2}}{\varepsilon_0^n} \cdot \varepsilon^n, \tag{5}$$

where $\sigma_{0.2}$ is the yield strength, n is the hardening exponent, and ε_0 is a reference strain. Using equation (5) and Figure 3,

Table 1: Mechanical properties of AN420s-grade steel plate.

Steel grade	Thickness (mm)	σ_s (MPa)	σ_b (MPa)	δ_s (%)	Remarks
AN420s	8 mm	950	1080	12	Transverse
		940	1050	12	Longitudinal
		945	1025	13	45°

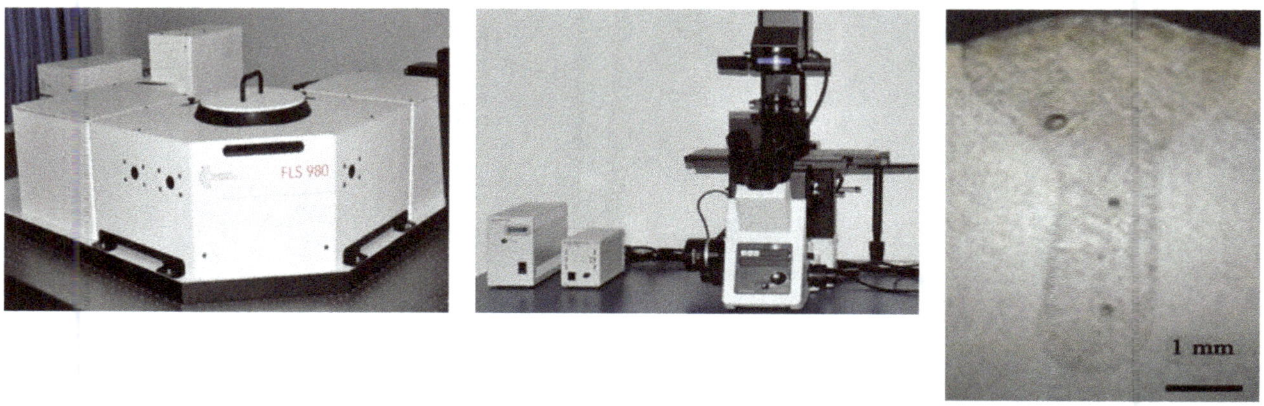

(a) (b) (c)

Figure 2: The instruments used for the analysis of the chemical composition of the weld bead. (a) FLS980-stm Edinburgh fluorescence spectrometer. (b) Metallographic microscope. (c) The cross section of the weld bead.

Table 2: Chemical compositions of the weld bead (wt.%).

Al	C	Fe	Mn	Nb	P	S	Si	Ti	V
0.015	0.11	97.062	1.71	0.087	0.22	0.016	0.46	0.15	0.17

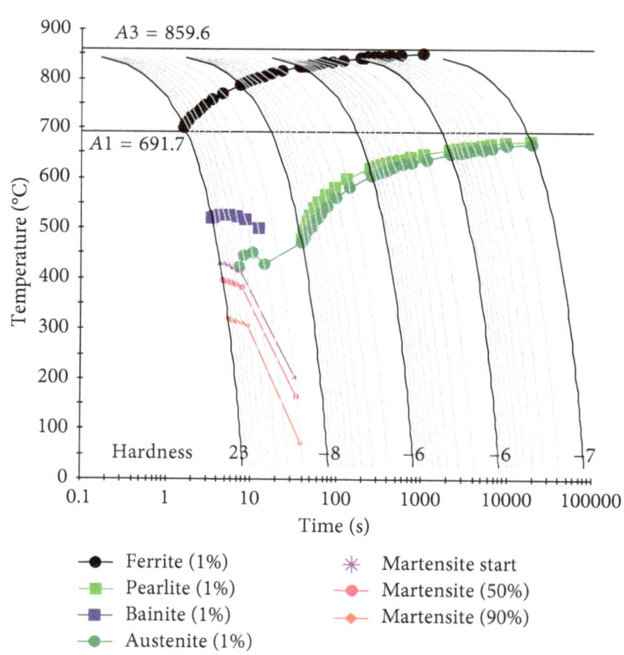

Figure 3: Phase transition diagram of the weld bead.

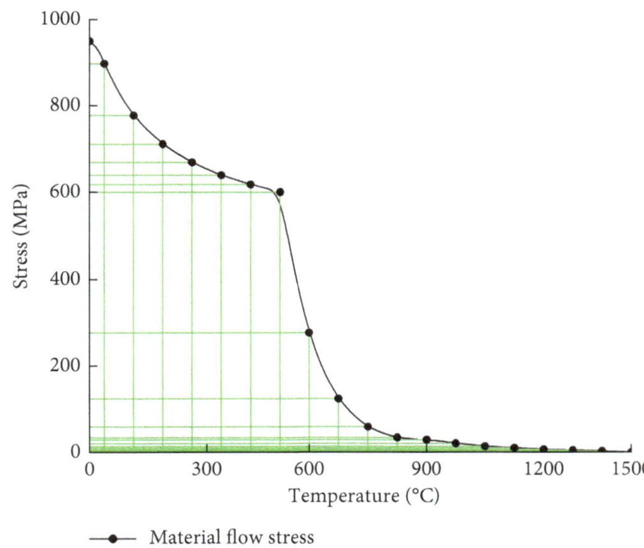

FIGURE 4: Material flow stress vs. temperature.

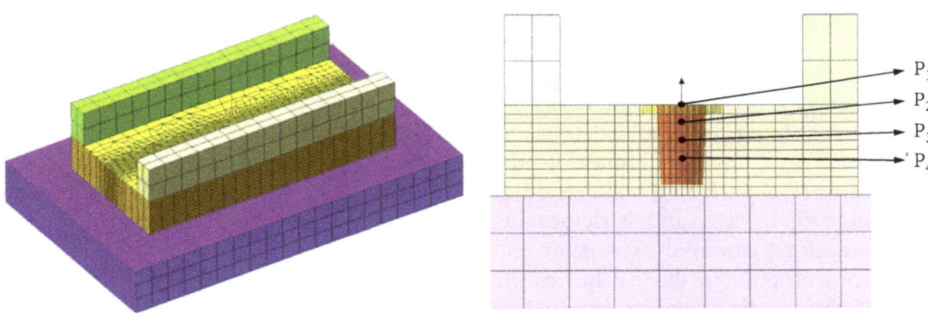

(a) (b)

FIGURE 5: The FEA model: (a) the welding model (from top to bottom: two clamps, the welding plate, and the support plate) and (b) trace points.

TABLE 3: Input data for the FEA model.

Input data	Welding plate	Clamp	Support plate
Element size	Weld bead: 0.5 mm; base metal: 1.5 mm	2.5 mm	3 mm
Clamping force (kN)	0.3	0.3	0.3
Heat exchange coefficient	Air heat exchange coefficient: 20 W/(m^2·K); weldment heat exchange coefficient: 1000W/(m^2·K)		

a constitutive relation between the temperature and material flow stress can be obtained using JMatPro.

Figure 4 shows the material flow stress of ultrafine-grained AN420s-grade steel at different temperatures between the solidus and liquidus temperatures. A softening phenomenon occurs at a temperature above 500°C, and liquefaction of the material occurs at a temperature of about 1455°C, at which point the metallic state in the molten pool can be regarded as liquid. The material flow stress dramatically decreases from 610 MPa to about 50 MPa at a temperature above 500°C. The region beneath the material flow stress curve in Figure 4 can be considered a safety zone in which cracks will not form.

2.3. FEA Model. The finite element analysis (FEA) model of the ultrafine-grained AN420s-grade steel in the form of a vehicle component with a joint fabricated by pressure welding was created using the HyperWorks software. Subsequently, the FEA model was imported into the Simufact Welding software to analyze the laser penetration welding process. Two different welding conditions were used in this investigation. The FEA model is shown in Figure 5, the input data for the FEA model are shown in Table 3, and the welding conditions are shown in Table 4.

As shown in Figure 5(a), the total number of elements was 20040, and the ambient temperature was 20°C. To investigate the changes in the temperatures and stresses in the molten pool,

TABLE 4: The welding conditions.

Welding parameters	Power (kW)	Speed (mm/s)	Holding force (kN)	Defocusing distance (mm)
1st condition	5.0	10.0	0.3	3
2nd condition	5.0	20.0	0.3	3

four trace points were assigned. As shown in Figure 5(b), the trace points P_1, P_2, P_3, and P_4 were set to collect the residual stress and temperature information in the molten pool.

The characteristics of the laser welding heat source were the local concentration and the instantaneous and heterogeneous temperature field, and an accurate heat source model can improve the accuracy of the FE simulation results [25]. A compound heat source model comprising a cone heat source and Gaussian surface heat source was adopted in this study, and the accurate heat source model was obtained through careful collation from the surface to the root of the weld between the FEA model and the cross section of the weld bead. The heat source model collation is shown in Figure 6, and the error between the simulated and actual molten pools is shown in Table 5.

3. Results and Discussion

3.1. Simulated Results under the 1st Weld Operating Condition. The effects of the welding process parameters, including the laser power, welding speed, holding force, and defocusing distance, were studied. For the 1st welding condition, the parameters were a laser power of 5.0 kW, a weld speed of 10.0 mm/s, a holding force of 0.3 kN, and a defocusing distance of 3.0 mm. Nonuniform thermal expansion and contraction due to different temperatures during the heating and cooling sequence of the welding process resulted in tensile and compressive residual stress fields in and near the weld region [26]. Figure 7 shows the temperature evolution at the trace points in the molten pool. Figures 8(a)–8(c) show the residual stresses at the trace points at different times.

As shown in Figure 7, the temperature varied with time at the four trace points. The temperature in the molten pool increased dramatically from 20°C to a maximum temperature above 1570°C in a short period of 0.5 s, after which the temperature deceased during the cooling process. As shown in Figure 8, the residual stresses varied with time at the four trace points, and the residual stress curves can be divided into four stages: (1) the metal melted, and the stress value was below 3 MPa; (2) the metal solidified, and the stress value was below 200 MPa; (3) the molten pool began to cool, and the maximum stress occurred at about 7 s; and (4) the stress fluctuated about a certain value and finally reached a steady state. The residual stresses at the four trace points eventually stabilized to constant values, and the steady-state values are shown in Table 6.

According to Table 4, the steady-state residual stress values at the trace points were in the following order: $P_2 > P_3 > P_1 > P_4$. Under the 1st welding condition, the highest residual stress occurred in the middle of the welding plate.

3.2. Simulated Results under the 2nd Weld Operating Condition. For the 2nd welding condition, the parameters were a laser power of 5.0 kW, a weld speed of 20.0 mm/s, a holding force of 0.3 kN, and a defocusing distance of 3.0 mm. Figure 9 shows the temperature evolution at the trace points in the molten pool. Figures 10(a)–10(c) show the transient residual stresses at the trace points at different times.

As shown in Figure 9, the temperature varied with time at the four trace points. Compared to the 1st welding condition, the highest temperature in the molten pool under the 2nd welding condition increased by about 5% from 1530 to 1600°C. As shown in Figure 10, the residual stresses varied with time at the four trace points, and the residual stress curve can be divided into four stages: (1) the metal melted, and the stress value was below 3 MPa; (2) the metal solidified, and the stress value was below 200 MPa; (3) the molten pool began to cool, and the maximum stress occurred at about 6 s; and (4) the stress fluctuated about a certain value and finally reached a steady state. The residual stresses at the four trace points eventually stabilized to constant values, and the steady-state values are shown in Table 7.

According to Table 4, the steady-state residual stress values at the trace points were in the following order: $P_2 > P_1 > P_4 > P_3$. Under the 2nd welding condition, the highest residual stress occurred in the middle of the welding plate.

3.3. Discussion. The results from the two different sets of welding parameters indicated that the highest residual stresses occurred in the middle of the welding plate, which corresponds to the trace point P_2. However, the highest value of residual stress under the 2nd welding condition was significantly less than that under the 1st welding condition. When the welding speed increased and the weld pool geometry became tapered, the solidification cracks will be more concentrated and in positions near the centerline. Under this condition, the liquid flow rate was high and the rate of volume change was low. To investigate the effects of the residual stress on the solidification cracking behavior, a comparison between the experimental results and the predicted stress distributions under the two operating conditions is shown in Figure 10. The metallography of the cross sections of the weld beads under the two welding conditions was obtained using the Olympus GX71 metallographic microscope.

Figure 11 shows the cracks observed in the molten pool under the 1st welding condition. Upon increasing the weld speed from 10.0 to 20.0 mm/s for the 2nd welding condition, no cracks formed. Sheikhi et al. [27] reported that a low solidification rate and a small vulnerable zone were ideal for producing a weld without any cracks. Figures 11(c) and 11(d) show that the stresses in the molten pool under the 1st welding condition were higher than those under the 2nd welding condition, and high stresses in the molten pool increased the liquid flow. The constitutive coupling relation

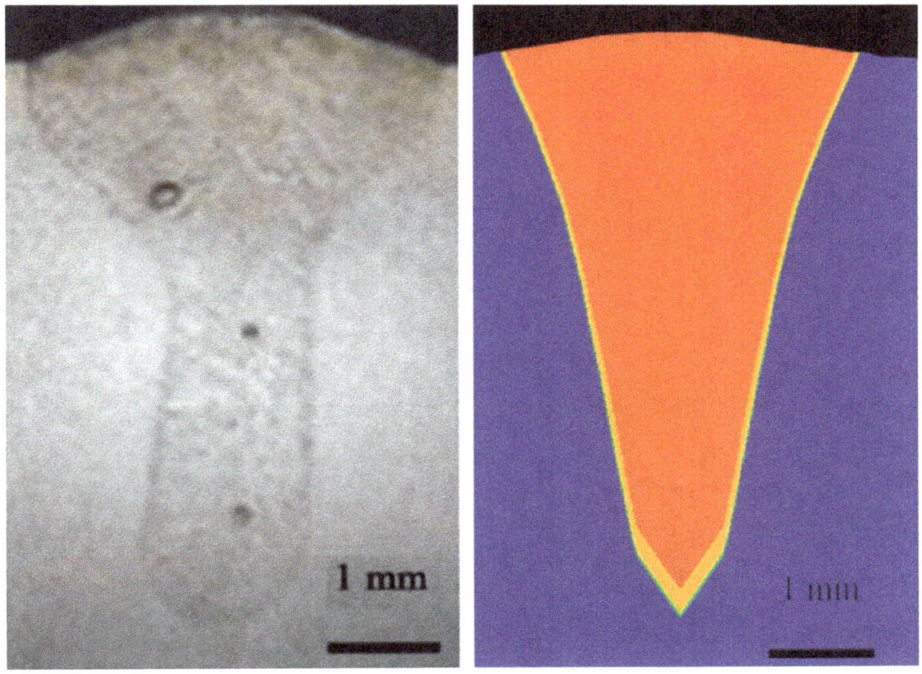

(a)

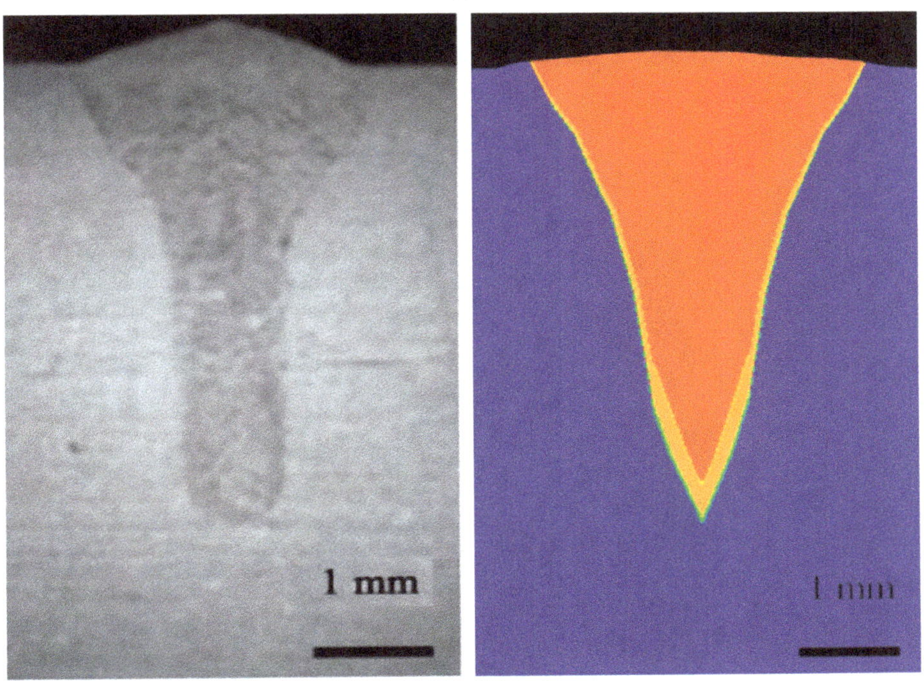

(b)

FIGURE 6: The heat source model collation: the collation under the (a) 1st and (b) 2nd welding conditions.

TABLE 5: The percentage error between the simulated and actual molten pools.

	Weld width (mm)		Weld depth (mm)		Maximum percentage error (%)
	Experimental data	FEA model	Experimental data	FEA model	
1st condition	4.0	4.068	5.4	5.387	0.2
2nd condition	3.5	3.523	4.2	4.186	0.3

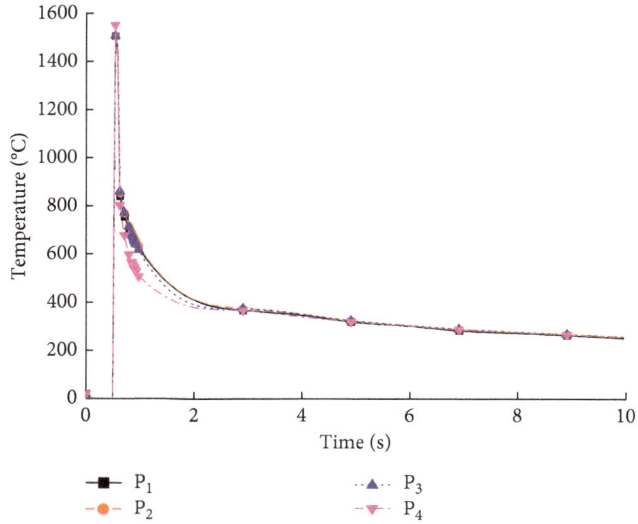

FIGURE 7: The temperature evolution at the trace points (1st welding condition).

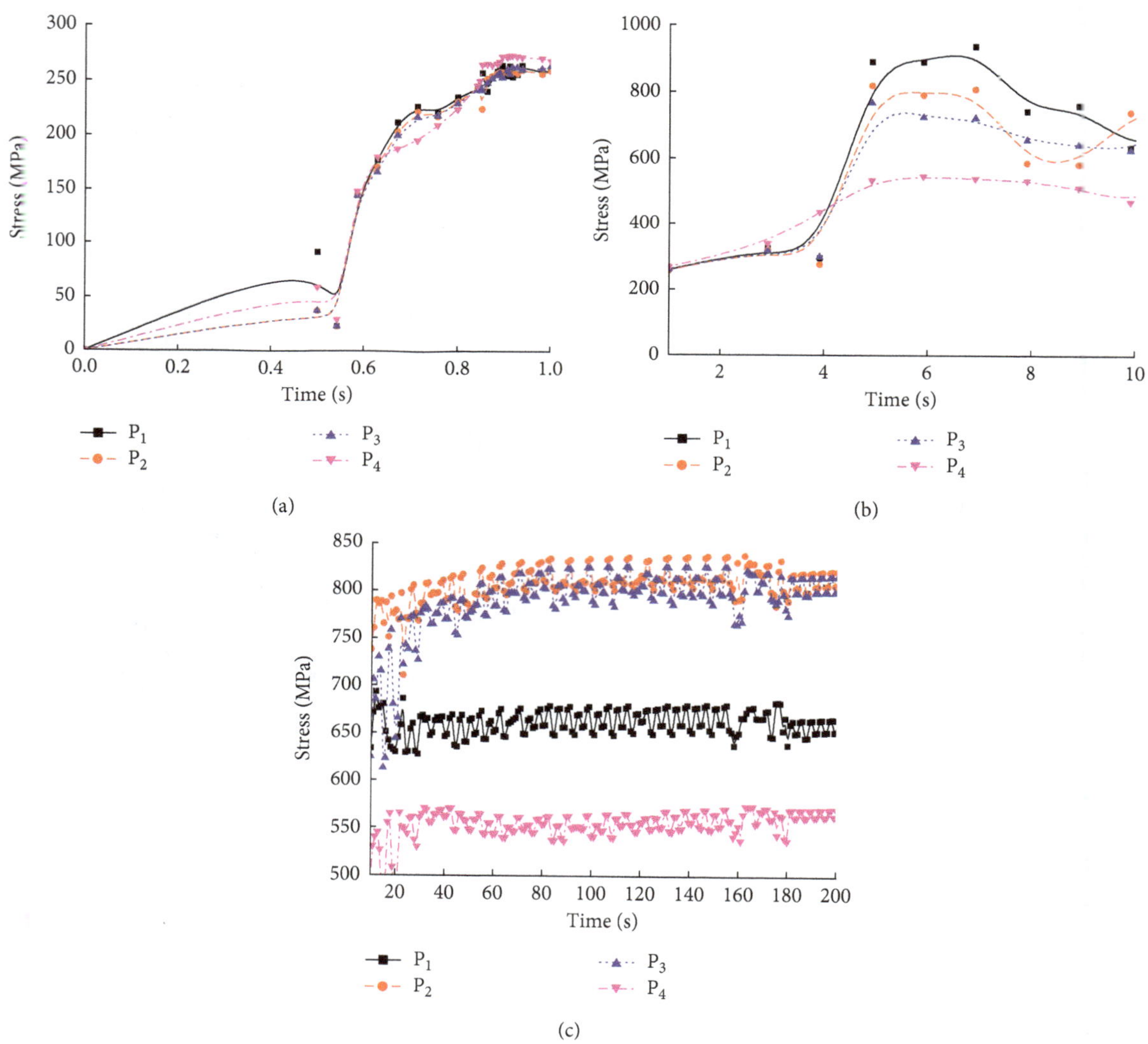

FIGURE 8: The residual stresses at the trace points at different times (1st welding condition): (a) 0–1 s; (b) 1–10 s; (c) 10–200 s.

Table 6: Steady-state residual stresses at the trace points.

Trace point	P_1	P_2	P_3	P_4
Steady-state value of stress (MPa)	650	820	800	570

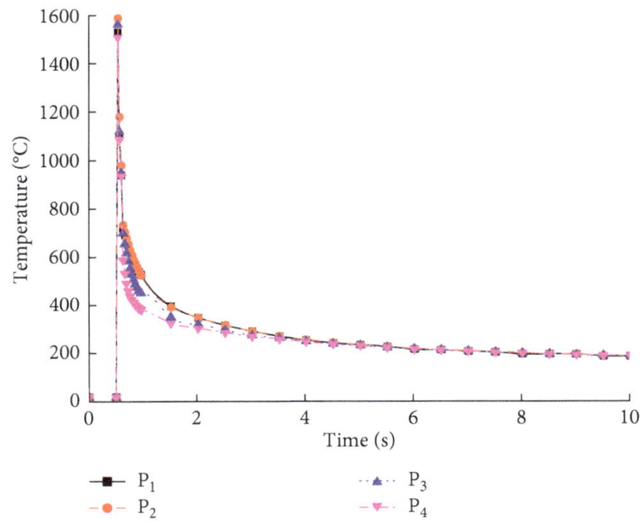

Figure 9: The temperature evolution at the trace points (2nd welding condition).

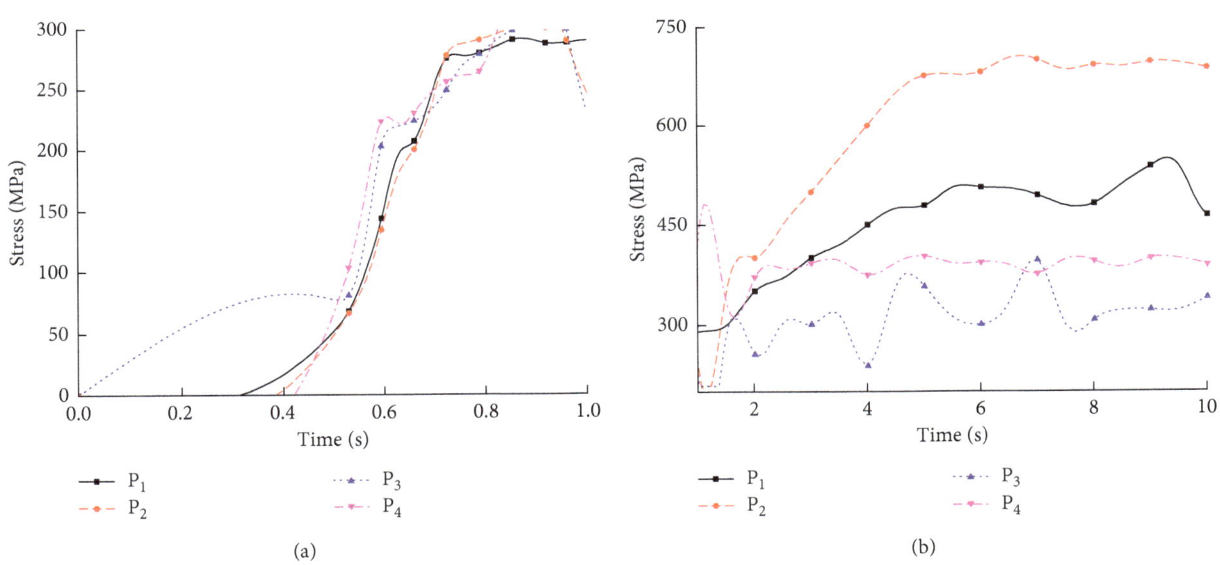

Figure 10: Continued.

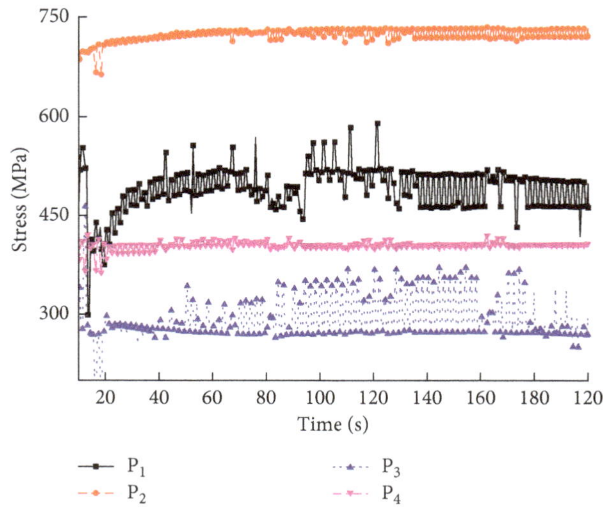

(c)

FIGURE 10: The residual stresses at the trace points at different times (2nd welding condition): (a) 0–1 s; (b) 1–10 s; (c) 10–200 s.

TABLE 7: Steady-state residual stresses at the trace points.

Trace point	P_1	P_2	P_3	P_4
Steady-state value of stress (MPa)	475	720	280	410

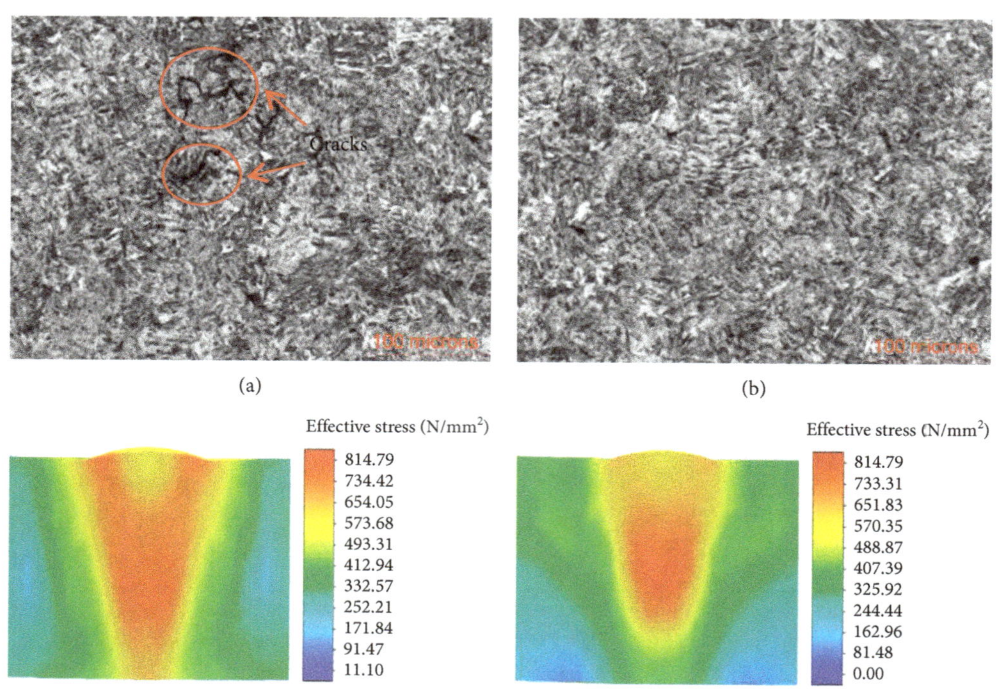

(a) (b) (c) (d)

FIGURE 11: Comparison between the measured and the simulated results: the cross section of the weld bead under the (a) 1st and (b) 2nd welding conditions; the stress distributions under the (c) 1st and (d) 2nd welding conditions.

between the temperature and material flow stress is shown in Figure 4. The residual stresses at the trace point P_2 at different temperatures were extracted and compared to the material flow stress curve.

Figure 12 shows a comparison of the highest residual stress curve under the two welding conditions and the material flow stress curve. Based on Figures 11 and 12, the effects of residual stress on the laser penetration welding

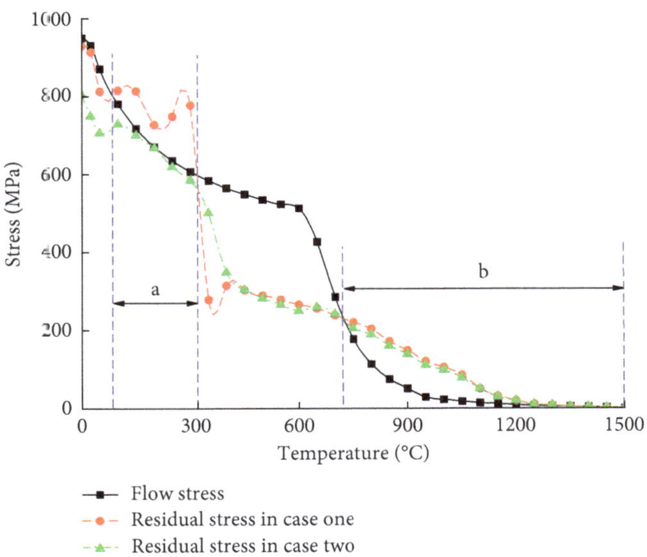

Figure 12: Comparison between the residual stress and the material flow stress.

process of ultrafine-grained steel can summarized as follows:

(1) In the temperature range of interval *b*, the residual stresses under the two different welding conditions were higher than the material flow stresses. The weld bead was not completely solidified, and the state of the molten pool was a mixture of solid and liquid. The liquid still available between the solid cells was continuous and could move easily, since the solid cells have not yet coalesced, and compensates for the deformation induced by shrinkage and thermal stresses. Thus, for temperatures in the interval *b*, cracks will occur.

(2) In the temperature range of interval *a*, the weld bead was completely solidified, and the solid cells coalesced. The residual stresses under the 1st welding condition were higher than the material flow stresses, and cracks will form in this case. The residual stresses under the 2nd welding condition were lower than the material flow stresses, and cracks will not form in this case. These conclusions are in good agreement with the experimental results.

(3) The residual stresses under the two different welding conditions were nearly equal when the temperature was above 400°C. The residual stresses under the 1st welding condition quickly exceeded the residual stresses under the 2nd welding condition, indicating that cracks will form during the cooling process of the molten pool. This also explains why a reasonable increase of the welding speed can prevent solidification cracks in the laser penetration welding process of ultrafine-grained steel.

4. Conclusions

In this study, the residual stresses in the laser penetration welding process of ultrafine-grained steel were investigated. A constitutive coupling relation between the temperature and material flow stress in the molten pool was developed based on the Gibbs interface thermodynamic method. An FE model of the molten pool was established based on experimental data of the weld bead cross section under two different welding conditions. Lastly, a criterion for assessing the formation of solidification cracks in the laser penetration welding process was proposed. From the experimental work and modelling, the following conclusions were drawn:

(1) At the same temperature and position in the molten pool, when the residual stress is higher than the material flow stress of ultrafine-grained steel, cracks will form in the weld bead.

(2) A reasonable increase of the welding speed can decrease the residual stress in the molten pool and enhance the laser penetration welding quality for ultrafine-grained steel.

(3) The proposed criterion obtained through the comparison of the residual stress and material flow stress curves of the ultrafine-grained steel can provide a new method to evaluate the formation of solidification cracks in weld beads.

Conflicts of Interest

The authors declare that they have no conflicts of interest.

Acknowledgments

The authors gratefully acknowledge the use of the services and facilities of AVIC Hubei Aviation Precision Machinery Technology Co, Ltd. and Hubei Key Laboratory of Power System Design and Test for Electrical Vehicle. This research was funded by the Youth Project of Hubei Provincial Department of Education (No. Q20172603), Hubei Superior and Distinctive Discipline Group of Mechatronics and Automobiles (No. XKQ2018065), and Hubei Superior and Distinctive Discipline Group of Mechatronics and Automobiles (No. XKQ2018009).

References

[1] S. Chen, D. Yang, M. Li et al., "Laser penetration welding of an overlap titanium-on-aluminum configuration," *International Journal of Advanced Manufacturing Technology*, vol. 87, no. 9–12, pp. 3069–3079, 2016.

[2] Y. Zhang, F. Li, Z. Liang, Y. Ying, Q. Lin, and H. Wei, "Correlation analysis of penetration based on keyhole and plasma plume in laser welding," *Journal of Materials Processing Technology*, vol. 256, pp. 1–12, 2018.

[3] C. C. Zhao, W. N. Li, and Z. B. Liu, "Laser welding of ultrafine grained steel ss400," *Journal of Iron and Steel Research (International)*, vol. 10, no. 3, pp. 32–36, 2003.

[4] H. L. Wei, J. J. Blecher, T. A. Palmer, and T. Debroy, "Fusion zone microstructure and geometry in complete-joint-

penetration laser-arc hybrid welding of low-alloy steel," *Welding Journal*, vol. 94, no. 4, pp. 135–144, 2015.

[5] R. A. Sindhu, M. K. Park, S. J. Lee, and K. D. Lee, "Effects of residual stresses on the static and fatigue strength of laser-welded lap joints with different welding speeds," *International Journal of Automotive Technology*, vol. 11, no. 6, pp. 857–863, 2010.

[6] L.-J. Zhang, G.-F. Zhang, X.-Y. Bai, J. Ning, and X.-J. Zhang, "Effect of the process parameters on the three-dimensional shape of molten pool during full-penetration laser welding process," *The International Journal of Advanced Manufacturing Technology*, vol. 86, no. 5–8, pp. 1273–1286, 2016.

[7] E.-J. Chun, H. Baba, K. Nishimoto, and K. Saida, "Development of laser beam welding transverse-varestraint test for assessment of solidification cracking susceptibility in laser welds," *Metals and Materials International*, vol. 21, no. 3, pp. 543–553, 2015.

[8] C. Lee, H. Park, J. Yoo, C. Lee, W. Woo, and S. Park, "Residual stress and crack initiation in laser clad composite layer with co-based alloy and wc + nicr," *Applied Surface Science*, vol. 345, pp. 286–294, 2015.

[9] X. Li, F. Lu, H. Cui, X. Tang, and Y. Wu, "Numerical modeling on the formation process of keyhole-induced porosity for laser welding steel with t-joint," *The International Journal of Advanced Manufacturing Technology*, vol. 72, no. 1–4, pp. 241–254, 2014.

[10] G. Pastras, A. Fysikopoulos, C. Giannoulis, and G. Chryssolouris, "A numerical approach to modeling keyhole laser welding," *The International Journal of Advanced Manufacturing Technology*, vol. 78, no. 5–8, pp. 723–736, 2014.

[11] M. R. Frewin and D. A. Scott, "Finite element model of pulsed laser welding," *Welding Research Supplment*, vol. 78, pp. 15–22, 1999.

[12] M. Shibara, H. Serizawa, and H. Murakawa, "Finite element method for hot cracking using temperature dependent interface element (Report II)," *Trans JWRI*, vol. 29, no. 1, pp. 59–64, 2000.

[13] J. Volpp, "Investigation on the influence of different laser beam intensity distributions on keyhole geometry during laser welding," *Physics Procedia*, vol. 39, no. 9, pp. 17–26, 2012.

[14] J. Volpp and F. Vollertsen, "Modeling keyhole oscillations during laser deep penetration welding at different spatial laser intensity distributions," *Production Engineering*, vol. 9, no. 2, pp. 167–178, 2014.

[15] Z. Dong, M. A. Rui, W. Yong, X. Zhan, and Y. Wei, "Post-data treatment design of software package for three dimensional simulation and prediction of weld solidification cracks," *Transactions of the China Welding Institution*, vol. 32, no. 3, pp. 17–20, 2011.

[16] A. Fadaei and H. Mokhtari, "Finite element modeling and experimental study of residual stresses in repair butt weld of ST-37 plates," *Iranian Journal of Science & Technology Transactions of Mechanical Engineering*, vol. 39, pp. 291–307, 2015.

[17] L. Zhang, J. Zhang, G. Zhang, W. Bo, and S. Gong, "An investigation on the effects of side assisting gas flow and metallic vapour jet on the stability of keyhole and molten pool during laser full-penetration welding," *Journal of Physics D: Applied Physics*, vol. 44, no. 13, pp. 135–201, 2011.

[18] Q. D. Sun, L. I. Meng-Sheng, and R. W. Song, "Research on a new type of heat source model on laser deep welding," *Welding Technology*, vol. 39, no. 12, pp. 11–14, 2010.

[19] G. L. Qin, X. B. Qi, Y. B. Yang, X. Y. Wang, and S. Y. Lin, "Acquisition and processing of coaxial image of molten pool and keyhole in nd:yag laser welding with high power," *China Welding*, vol. 13, no. 1, pp. 51–55, 2004.

[20] C. Lampa, A. F. H. Kaplan, J. Powell, and C. Magnusson, "An analytical thermodynamic model of laser welding," *Journal of Physics D: Applied Physics*, vol. 30, no. 9, pp. 1293–1299, 1999.

[21] H. Halfa, "Recent trends in producing ultrafine grained steels," *Journal of Minerals and Materials Characterization and Engineering*, vol. 2, no. 5, pp. 428–469, 2014.

[22] P. Layus, P. Kah, A. Zisman, M. Pirinen, and S. Golosienko, "Recrystallization-based formation of uniform fine-grained austenite structure before polymorphic transition in high-strength steels for arctic applications," *International Journal of Mechanical & Materials Engineering*, vol. 11, no. 1, pp. 1–7, 2016.

[23] J. Gernert, A. Jäger, and R. Span, "Calculation of phase equilibria for multi-component mixtures using highly accurate helmholtz energy equations of state," *Fluid Phase Equilibria*, vol. 375, no. 31, pp. 209–218, 2014.

[24] S. A. English and N. K. Arakere, "Effects of the strain-hardening exponent on two-parameter characterizations of surface-cracks under large-scale yielding," *International Journal of Plasticity*, vol. 27, no. 6, pp. 920–939, 2011.

[25] A. Belitzki and M. F. Zaeh, "Accuracy of calculated component distortions using the weld pool length to calibrate the heat source," *Journal of Laser Applications*, vol. 28, no. 2, article 022424, 2016.

[26] J. Ahn, E. He, L. Chen et al., "Fem prediction of welding residual stresses in fibre laser-welded aa 2024-t3 and comparison with experimental measurement," *International Journal of Advanced Manufacturing Technology*, vol. 95, no. 5, pp. 1–21, 2018.

[27] M. Sheikhi, F. Malek Ghaini, and H. Assadi, "Prediction of solidification cracking in pulsed laser welding of 2024 aluminum alloy," *Acta Materialia*, vol. 82, pp. 491–502, 2015.

12

Prediction Model of Cutting Parameters for Turning High Strength Steel Grade-H: Comparative Study of Regression Model versus ANFIS

Adel T. Abbas,[1] Mohanad Alata,[1] Adham E. Ragab,[2] Magdy M. El Rayes,[1] and Ehab A. El Danaf[1]

[1]Department of Mechanical Engineering, College of Engineering, King Saud University, P.O. Box 800, Riyadh 11421, Saudi Arabia
[2]Department of Industrial Engineering, College of Engineering, King Saud University, P.O. Box 800, Riyadh 11421, Saudi Arabia

Correspondence should be addressed to Adel T. Abbas; atabbas1954@yahoo.com

Academic Editor: Fernando Lusquiños

The Grade-H high strength steel is used in the manufacturing of many civilian and military products. The procedures of manufacturing these parts have several turning operations. The key factors for the manufacturing of these parts are the accuracy, surface roughness (R_a), and material removal rate (MRR). The production line of these parts contains many CNC turning machines to get good accuracy and repeatability. The manufacturing engineer should fulfill the required surface roughness value according to the design drawing from first trail (otherwise these parts will be rejected) as well as keeping his eye on maximum metal removal rate. The rejection of these parts at any processing stage will represent huge problems to any factory because the processing and raw material of these parts are very expensive. In this paper the artificial neural network was used for predicting the surface roughness for different cutting parameters in CNC turning operations. These parameters were investigated to get the minimum surface roughness. In addition, a mathematical model for surface roughness was obtained from the experimental data using a regression analysis method. The experimental data are then compared with both the regression analysis results and ANFIS (Adaptive Network-based Fuzzy Inference System) estimations.

1. Introduction

High strength steel is widely used in industrial civilian and military products. Many grades of high strength steel are available based on its chemical composition and heat treatment procedures [1]. One of the famous military products that uses grade (H) is the heavy cannons such as the Russian howitzer 122 mm and the 105 British cannon.

Using tungsten-carbide tools and employing multiobjective optimization, Abbas et al. [2] investigated the effect of various cutting parameters in turning of heat-treated alloy steel using tungsten-carbide tools. The experiments covered a practical range of studied parameters that allows maximizing material removal rate while keeping good surface quality within reasonable industrial limits.

In order to establish a model predicting surface roughness for AISI 1019 steel machined by turning, in which cutting speed, feed rate depth of cut, and nose radius were selected as the cutting variables, Bhardwaj et al. [3] employed response surface methodology (RSM) based on center composite rotatable design. The application of Box–Cox transformation improved the prediction of the model. In this work it was proven that the feed rate was the main affecting parameter on surface roughness, whereas the effect of cutting speed and feed rate was negligible.

The work conducted by Mokhtari Homami et al. [4] applied full factorial design analysis approach in which the flank wear and surface roughness were response variables whereas feed rate, cutting speed, tool nose radius, and approach angle were process variables. The authors concluded that approach angle, feed rate, and tool nose radius had a major influence on the surface roughness and tool flank wear. However, cutting speed individually affected the tool flank wear. The results showed that as the cutting speed increased the surface quality was improved; however, the tool life was shortened due to excessive flank wear. Also, when the feed rate or approach angle increased, the surface roughness increased and the tool life decreased. They, also, concluded that the surface roughness increased with increasing nose radius and related this to the increased contact surface between the chip and the tool.

Svalina et al. [5] suggested evolutionary neurofuzzy system using multiobjective optimization of cutting parameters aiming for the minimum machining time as well as maximum metal removal rate while controlling the obtained surface roughness. Through their experimental investigation, they justified the application and further development of the suggested evolutionary neurofuzzy system for evaluation of surface roughness within the given constraints.

The influence of machining parameters on the chip shape classification was studied [6] using straight turning of mild steel (A500/A500M-13) and AISI 304 stainless steel, where the resulting chip shapes were evaluated. Adaptive neurofuzzy inference system (ANFIS) was used in order to determine the inputs influence on the chip shape classification and also to estimate the most dominant factors which affect it. The inputs were, namely, cutting speed, feed rate, depth of cut, and surface roughness of the material. From results it was found that surface roughness has the highest influence on the chip shape classification. The obtained model could be used as optimal parameter settings for the best chip shape classification.

A fuzzy model was developed [7] for predicting the surface roughness at a given set of inputs, cutting speed, feed rate, and depth of cut representing realistic industrial environment. This model is capable of predicting the surface roughness for a given set of inputs helping the operator to select a set of working parameters that enables the setting of machining parameters to achieve a certain surface quality. For model verification, comparison between fuzzy output with the experimental data and an empirical model was built consequently. It was found that the experimental results were in good agreement with those predicted using the fuzzy model, where the system accuracy was 95%, as applied experimentally under real operating conditions in milling of aluminum blocks.

For achieving minimum surface roughness through the prediction and optimization of machining parameters, Kant and Sangwan [8] presented an artificial neural network (ANN) technique coupled with genetic algorithm (GA). The predicted results are found to be close to the experimental values. They showed that the developed model outperforms the regression and fuzzy logic models. From modeling point of view, the ANN was validated as a prediction model which has statistically satisfactory prediction performance.

Raja et al. [9] applied Taguchi's L27 orthogonal array to investigate the surface roughness of AISI316 and AISI410 machined by CNC turning using different coated cutting tools in dry conditions. Analysis of variance (ANOVA) showed that linear feed rate and then quadratic feed rate are the most significant factors affecting surface quality. It was reported that TiAlN composition coated insets exhibited the best cutting performance compared to other inserts.

Chandrasekaran et al. [10] employed response surface methodology (RSM) to analyze the CNC turning of AISI 316 with multilayer coated Ti (C, N, B) and single layer coated with TiAlN cutting tool for different cutting conditions. The model predicted surface roughness and tool wear values fairly close to the experimental values. Feed rate, followed by the depth of cut, and quadratic depth of cut, followed by linear feed rate, were the most significant factors for surface roughness and tool wear, respectively, for the Ti (C, N, B) cutting tool. Feed rate, followed by interaction between feed rate and depth of cut, and quadratic depth of cut, followed by interaction of feed rate and depth of cut, were the most significant factors for surface roughness and tool wear, respectively, for TiAlN cutting tool.

Jafarian et al. [11] predicted surface roughness, resultant cutting forces, and tool wear by three artificial neural networks (ANN). They proposed a novel method for training the ANNs by using evolutionary algorithms rather than conventional methods that use backpropagation. The optimization was carried to minimize surface roughness and cutting forces and maximize tool life in turning process. Genetic algorithm (GA) and particle swarm optimization (PSO) were utilized to optimize each of the outputs, while the other outputs were kept in the suitable range. The obtained results showed that trained neural networks with genetic GA as optimization objective functions presented a powerful tool to analyze the effect of each parameter on the output model with high accuracy level.

D'Addona and Raykar [12] proposed using wiper inserts to achieve significantly better machined surface in hard turning process. The performance of wiper inserts in hard turning of oil hardened nonshrinking steel was investigated compared to conventional inserts. Analysis of variance (ANOVA) was used to examine the effect of four process parameters, namely, cutting speed, feed rate, depth of cut, and nose radius on surface roughness. The analysis showed the positive influence of wiper inserts on surface finish.

Acayaba et al. [13] applied Multiple Linear Regression and artificial neural network integrated with an optimization algorithm, known as simulated annealing (SA) methodologies to predict the minimum surface roughness for a set of cutting parameters in turning stainless steel at low cutting speeds. Neural network was reported to give better prediction compared to linear model by a fair margin. It was concluded that low speed results in a surface with higher roughness values due to built-up edge formation. In contrary, as speed increases the surface roughness improves, due to the low contact time between the chip and the tool.

Sahu and Choudhury [14] investigated the performance of multilayer TiN coated tool in machining of hardened steel (AISI 4340 steel) under high speed turning, which was also compared with that of uncoated tool. The influences of cutting parameters (speed, feed, and depth of cut) on surface roughness were analyzed using Taguchi methodology. It was reported that the machining of hard materials at higher speeds and lower feeds improved the surface roughness by using coated tools.

Agrawal et al. [15] developed three regression models, multiple regression, random forest, and quantile regression, to model the relation between surface roughness and several cutting parameters for dry turning of AISI 4340. It was concluded that the random forest regression model is a superior choice over multiple regression models for the prediction of expected surface roughness.

Azam et al. [16] investigated that the surface roughness is very important indicator to determine the surface quality of machined parts. Accurate predictive models for surface roughness will support the selecting of best machining parameters, to maximize the productivity without any deviation on quality. In this paper, an average surface roughness (R_a) model has been developed for turning operation of high strength low-alloy steel using multilayer coated carbide tools. Multiple of tests was carried out to determine surface quality. The surface methodology (RSM) has been employed to develop a relationship between R_a and machining parameters (feed, speed, and depth of cut). The feed rate has been noticed as the main parameter that affect the value of surface roughness. Contour plots of "feed versus speed" and "feed versus depth of cut" have a significant effect on R_a value and can be achieved through optimal combination of cutting parameters. The accuracy of proposed model has been validated through the experimental data with average prediction of error 3.38%.

In the paper of Mia and Dhar [17], the artificial neural network based on the predictive model of average surface roughness in turning operation of the hardened EN 24T steel has been predicted. The prediction was performed by using Neural Network Tool Box 7 of MATLAB R2015a for different levels of cutting conditions: cutting speed, feed rate, and different material hardness. The dry and high pressure cutting fluid jet environments were explored as cutting conditions. The experimental tests were carried out using full factorial design of experiment. Afterward the 3-n-1, 3-n-2, and 4-n-1 ANN architectures were trained by utilizing the Levenberg–Marquardt, Bayesian regularization, and scaled conjugate gradient algorithms and evaluated based on the lowest root mean square error. The 3-10-1 and 3-4-2 ANN models, trained by Bayesian regularization, revealed the lowest root mean square error. A good fitting of the models was established by the regression coefficients higher than 0.997. Finally, the effect of cutting speed, feed, and hardness surface roughness at dry and high pressure cutting fluid conditions were studied. The high pressure cutting fluid leading to effective cooling and lubrication had the effect of reducing the surface roughness, whereas it was significantly increased when cutting hard materials because of the higher cutting forces arriving on the cutting tool.

Theoretical and empirical investigation were proposed by He et al. [18] to predict the surface finish obtained in a turning process using a diamond single-point tool in which the surface finish is considered to be composed of the defined and undefined parts. Theoretical formulation of the defined components such as the effects of the kinematics and minimum undeformed chip thickness were established. The undefined components in relation to the material spring back, plastic side flow, microdefects on the cutting edge of diamond tool, and others are empirically predicted by a RBF (radial basis function) neural network, which is established by referring to the experimental data. Finally, the particle swarm optimization algorithm is employed to find the optimum cutting conditions for the best surface finish. It was found that the optimization was satisfied through the experimental validation and the satisfactory level of prediction with an error of 0.59–10.11%. This proves the effectiveness of the novel method used in surface roughness prediction.

In this paper a mathematical model for surface roughness was obtained from the experimental data using two techniques: regression analysis method and adaptive neurofuzzy inference system of MATLAB (ANFIS). These models have been validated using fifteen experimental runs that were not included in the regression analysis or in the ANFIS and a comparison between the two models is presented.

TABLE 1: Chemical composition for H-steel material according to DEFSTAN 10-13/2005.

C	Si	Mn	Ni	Cr	Mo	V	S	P
0.30	0.17	0.46	2.86	0.76	0.47	0.01	0.002	0.006
0.36	0.32	0.53	3.10	0.94	0.58	0.03	0.01	0.011

2. Materials and Methods

The material used in this work is high strength steel Grade-H, commercially known as "gun steel," which is normally used in the manufacturing of gun barrel and muzzle brake. The material is initially forged into cylindrical form and then annealed to remove residual stresses induced by forging. The cylinders are then deeply drilled and rough-machined to form the primary shape of gun barrel. Table 1 shows the chemical composition of high strength steel Grade-H according to UK ministry defense standard DEFSTAN 10-13/2005 [1]. The heat treatment was carried out by heating the gun barrel to about 870°C for 4 hours before quenching in oil. This was immediately followed by a tempering procedure in which the sample was heated to 650°C over a period of 1.5 hours and maintained at this temperature for further 10 hours and then removed from the tempering furnace and cooled in air. The gun barrel is then ready for final machining. Table 2 shows the mechanical properties of Grade-H steel after heat treatment. The test samples were cut from the actual gun barrel from breech side (thick part) and machined to cylindrical shape in which they will be divided into five equally separated lengths, as shown in Figure 1, for conducting the turning runs.

EMCO Concept Turn 45 CNC lathe equipped with Sinumeric 840-D was used to conduct experimental work. An uncoated tungsten-carbide insert was used in the turning process. The insert and tool holder are SVJCL2020K16 and VCMT160404. The clearance angle, cutting edge angle, and nose radius were maintained at 7°, 75°, and 0.4 mm, respectively. Figure 2 shows the test rig for machining specimens. All cutting parameters were controlled via CNC part program. The test plan was implemented through 125 turning runs, in which the cutting speed, depth of cut, and feed rate had five levels each. Details of all testing conditions are provided in Table 3. TESA-Rugosurf-90 was used for

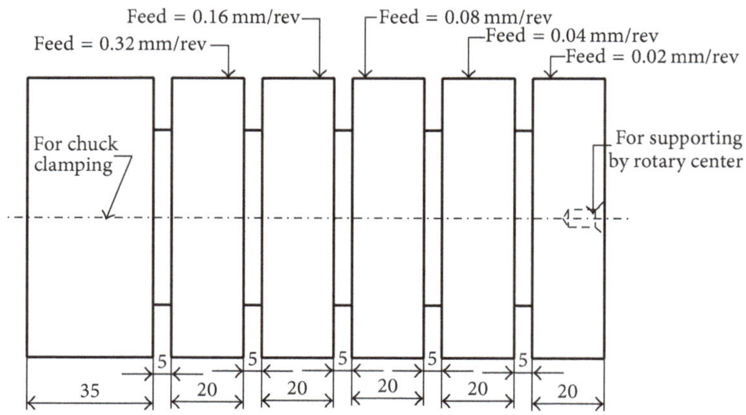

FIGURE 1: Test specimen drawing.

TABLE 2: Mechanical properties for H-steel material according to DEFSTAN 10-13/2005.

Mechanical properties	Value
Ultimate tensile strength	>1200 N/mm^2
0.2% proof strength	850–1000 N/mm^2
Elongation%	≥10%
Reduction of area	>27%
Charpy at (−40°C)	≥24.4 J
Hardness	32–38 HRc

FIGURE 2: Test rig for samples machining.

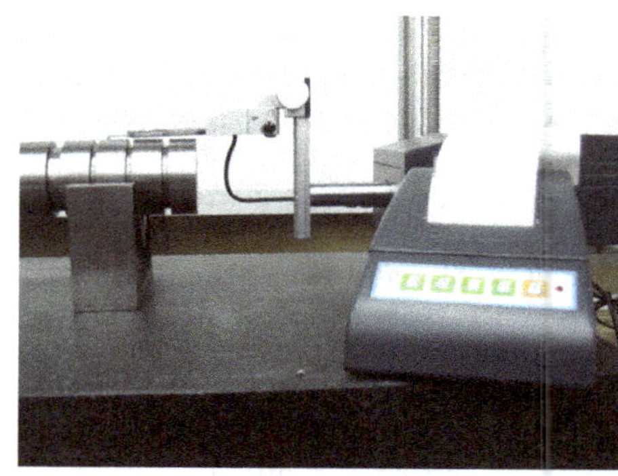

FIGURE 3: Test rig for measuring surface roughness.

surface roughness testing, where the test rig is shown in Figure 3. Full listing of all samples and the resulting measured surface roughness is provided in Appendix.

Full factorial experimental design was used to investigate the effect of three numerical factors (cutting speed, depth of cut, and feed rate) on the produced surface roughness. The studied range of each factor was divided into five levels creating 125 experiments. Regression was used to build the relation between surface roughness and studied factors. However, only 110 experimental readings were used for the regression, as 15 readings were excluded to be used for the sake of testing the predictability of the regression model. The 15 testing cases were selected such as to represent low, medium, and high levels of machining conditions in the experimental design. Material removal rate (MRR) was calculated using (1) for each run. Desirability function approach was used to maximize MRR maintaining R_a below 0.8 μm as a maximum limit for the surface roughness value.

$$\text{MRR} = 1000V * f_r * d, \quad (1)$$

where MRR is volume removed per unit time (mm^3/min.), V is the cutting speed (m/min.), f_r is the feed rate (mm/rev.), and d is depth of cut (mm).

In fuzzy inference systems, all expert knowledge about the process is modeled by a set of rules. The set of rules has two parts: the antecedent part which covers all states of the system and the other part, the consequent one, which shows the resulting activities. Fuzzy systems may use neural networks as learning techniques. The numerical training data should be used in the hybrid network to learn the parameters and create the fuzzy rules to establish a linguistic model of the problem in hand. The Adaptive Neurofuzzy Inference System (ANFIS) is an algorithm to be used to create a fuzzy model from input/output data. In this paper, fuzzy inference system and ANFIS of MATLAB in Fuzzy Logic Toolbox are used to build a prediction model for surface roughness.

3. Results and Discussions

3.1. Regression Model. Minitab 17 multilinear regression analysis was used to build a mathematical model relating the

TABLE 3: The cutting parameters and their levels used in experiments.

Designation	Process parameter	Level 1	Level 2	Level 3	Level 4	Level 5
A	Cutting speed (m/min)	75	100	125	150	175
B	Depth of cut (mm)	0.15	0.30	0.45	0.60	0.75
C	Feed rate (mm/rev)	0.02	0.04	0.08	0.16	0.32

independent machining parameters (cutting speed, depth of cut, and feed rate) with the surface roughness represented by R_a. Stepwise technique was used to eliminate the insignificant terms from the model. The model was fitted in the form given by (2) [19]:

$$R_a = \beta_o + \sum_{i=1}^{k} \beta_i X_i + \sum_{i=1}^{k} \beta_{ii} X_i^2 + \sum\sum_{i<j} \beta_{ij} X_i X_j + \sum\sum_{i<j<k} \beta_{ijk} X_i X_j X_k + \sum\sum_{i \neq j} \beta_{iij} X_i^2 X_j + \epsilon_i, \quad (2)$$

where β_o is the constant term, β_i represents the linear effects, β_{ii} represents the pure quadratic effects, β_{ij} represents the second-level interaction effects, β_{ijk} represents the third-level interaction effects, β_{iij} represents the effect of interaction between linear and quadratic terms, and ϵ_i represents the error in predicting experimental surface roughness.

Equation (3) represents the regression analysis fitted model. The adequacy of the model fit, measured by the coefficient of determination (adjusted R-squared), was calculated as 0.999 showing that 99% of the variability in R_a readings is explained by the fitted model. The predicted R-square was calculated as 0.999, proving that the model is not overfitted and has a good predictability. Figure 4 shows a scatter plot for the predicted R_a versus the measured R_a. It is clear from the figure that the relation between them is linear. Anderson-Darling test was conducted to check the normality of residuals with a result of p value = 0.163 > 0.05 assuming normally distributed residuals.

$$R_a = 0.0614 - 0.00301A + 0.012B + 6.331C + 0.000008A^2 + 0.299B^2 + 1.878C^{*2} + 0.001924AB + 0.030293AC + 5.195BC - 3.995B^2C. \quad (3)$$

Optimization process was run using the desirability function approach to maximize the material removing rate (MRR) while maintaining R_a below 0.8 μm. The results, illustrated in Figure 5, show that the maximum expected MRR equals 5668 mm^3/min. and happens at these machining conditions: cutting speed of 175 m/min., depth of cut of 0.75 mm, and feed rate of 0.043 mm/rev while maintaining R_a at about 0.79 μm.

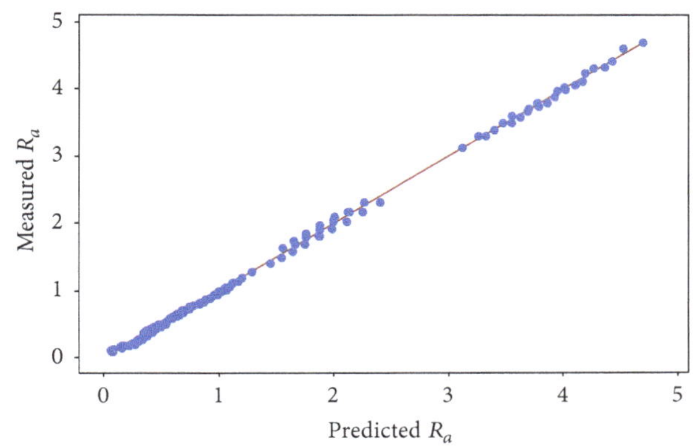

FIGURE 4: Scatter plot of measured R_a versus predicted R_a.

3.2. Fuzzy Inference System. A fuzzy inference system is basically described in Figure 6.

Fuzzy inference system can be classified into three types: Tsukamoto type (also called type 1), Mamdani type (also called type 2), and Sugeno type (also called Takagi-Sugeno type, Takagi-Sugeno-Kang type, and type 3). Most of the differences among these types come from the specification of the consequent part. Sugeno type fuzzy inference system is extremely suitable for smoothly interpolating linear gains applied across an input space; it is a natural and efficient gain scheduler. Similarly, it is also suited for modeling nonlinear systems by interpolating between multiple linear models.

Fuzzy inference process consists of the following steps: (1) fuzzification, (2) application of fuzzy operation in the antecedent, (3) implication to the consequent, (4) aggregation, and (5) defuzzification. Figure 7 describes how these steps are performed for the three different types of fuzzy inference system.

Firstly, crisp data input is converted into fuzzy form through fuzzification. The fuzziness is determined by the membership function used. If the antecedent part (also called premise part) has more than one input, fuzzy operation using AND (intersection) and/or OR (union) is applied. AND operation can be conducted using T-norm operators such as minimum or product, whereas OR operation can be conducted using T-conorm (also called S-norm) operators such as maximum or probabilistic OR. Implication method is then applied to the results of the fuzzy operator. In Sugeno type fuzzy system, the fuzzy operation and the implication method are directly governed by its right hand side (RHS) equations which are mathematical functions of the inputs. The fuzzy operation followed by the implication method creates the consequent part of the fuzzy system.

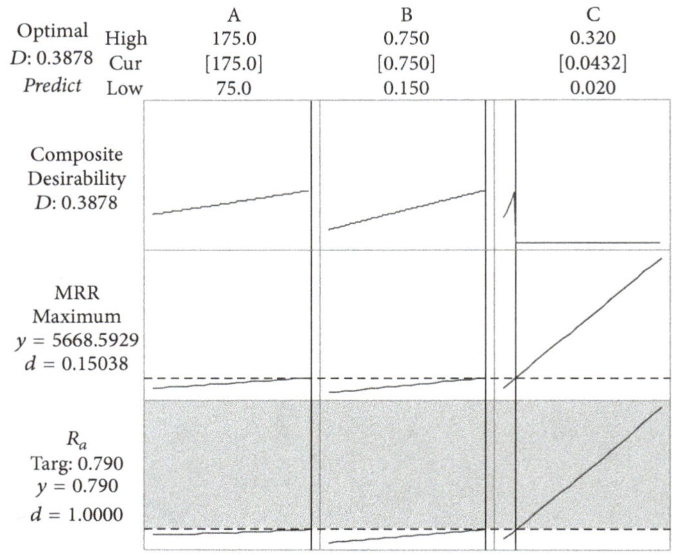

FIGURE 5: Optimization plot for R_a and MRR.

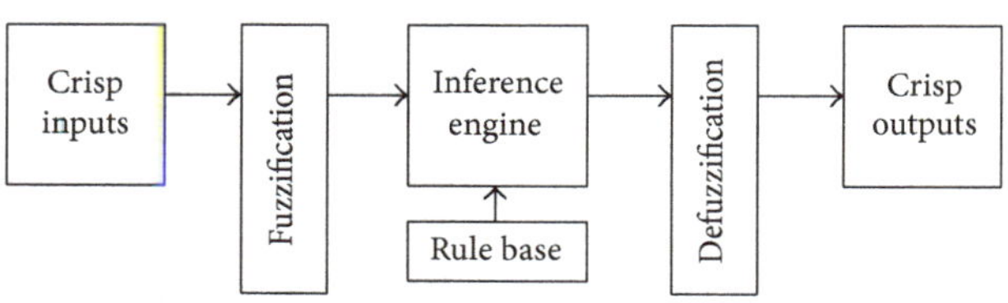

FIGURE 6: Fuzzy inference system.

After the implication step is accomplished for all of the rules, all of the consequents are combined into one value. This is called aggregation. In Mamdani type fuzzy system, the aggregation is conducted by using either maximum method, probabilistic OR method, or sum method, whereas the defuzzification is applied by using either centroid, bisector, middle of maximum, largest of maximum, or smallest of maximum method. In Sugeno type and Tsukamoto type fuzzy system, the aggregation-defuzzification is usually defined by the weighted average $\overline{z}$ given as follows:

$$\overline{z} = \frac{\sum_{i=1}^{N} w_i z_i}{\sum_{i=1}^{N} w_i}, \tag{4}$$

where z_i is the output of rule i, w_i is firing strength of rule i, and N is the number of rules.

3.3. Adaptive Network. Adaptive network is a special case of neural network with supervised-learning capability. After a training data (input-output data) is provided, the adaptive network will adjust its parameters (i.e., weights) in order to satisfy a certain termination criteria, usually number of iterations or error tolerance. The error is defined as the difference between the desired output and the output obtained from the learning process. The basic learning algorithm of the adaptive network to train the input-output data is steepest descent (backpropagation). In this learning algorithm, error is fed back to the network. The error signal originates from the output neuron of the network and it propagates backward in the network. The error signal eventually adjusts the weights. One iteration through all of the training data is called epoch. Number of epoch and the required error tolerance should be specified prior to the learning process.

To make the learning process faster, a hybrid learning algorithm which combines least square estimates (LSE) to the back propagation is usually used. Furthermore, the learning process can be done online or offline. In the online learning, the weights are updated based on the error signal of each input-output pair in the training data. Consequently, at each epoch, the number of adjustments is equal to the number of input-output pairs. In the offline learning, the weight adjustment is conducted based on the error signal of the entire training set. In other words, the weights are adjusted once only after all of the training data have been processed. The weights are only adjusted once at each epoch.

A typical architecture of adaptive network based on multilayer perceptron network is shown in Figure 8. It is a feedforward network with circle and square nodes. The circle nodes have fixed weight. On the other hand, the square nodes are adjustable weights which are updated during the learning process.

ANFIS combines the advantages of both the learning capability of an adaptive network and the modeling and approximation capabilities of a fuzzy inference system (FIS). Applying fuzzy inference system on a neural network frame, ANFIS has smoothness and learning capability at the same time. Given an input-output data, fuzzy rules are created through the learning process, and the FIS will interpolate (FIS is a universal approximate) among the rules to give smooth output.

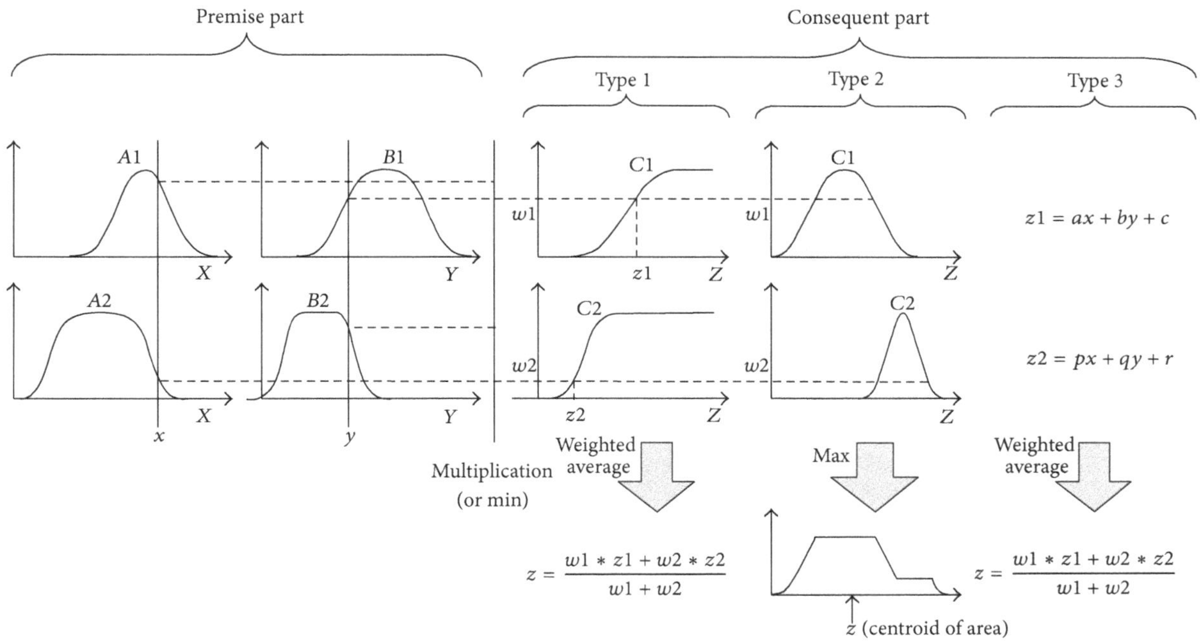

FIGURE 7: Fuzzy inference process for different types of fuzzy system.

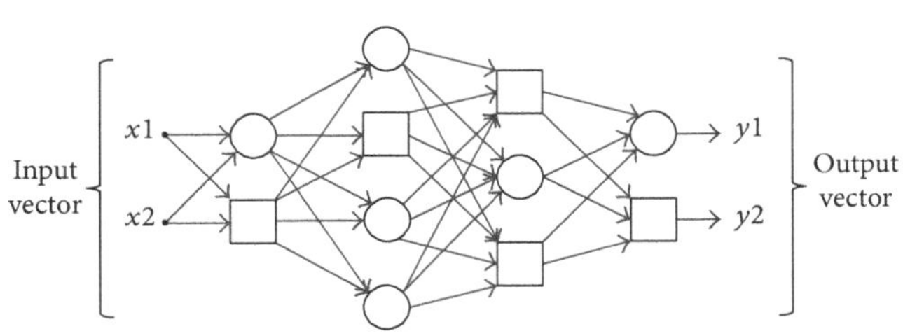

FIGURE 8: A typical adaptive network.

Figure 9 shows a two-input Sugeno type fuzzy model with two rules and its equivalent ANFIS model. The two rules states are as follows.

Rule 1. If x is A_1 and y is B_1 then $f_1 = p_1x + q_1y + r_1$.

Rule 2. If x is A_2 and y is B_2 then $f_2 = p_2x + q_2y + r_2$.

In the ANFIS model [20], nodes in the same layers are of the same function family. Subscript i represent a rule. As in this example there are 2 rules; i are positive integers from 1 to 2. Every node i in layer 1 has a node function which represents its membership function. Each node contains parameters called premise parameters. These parameters define the shape of the membership function. Circle nodes in layer 2 serve as fuzzy operators which can be T-norm or T-conorm operators, depending on how the inputs in the antecedent are connected. As the inputs in this example are connected by AND operation, any T-norm operator such as minimum or product can be applied in the circle nodes. Each circle node in layer 2 represents the firing strength of a rule. In layer 3, each circle node calculates the ratio of the ith rule's firing strength w_i to the sum of all rules' firing strengths. The ratio, therefore, is called normalized firing strength $\overline{w}_i$. As in this example there are two rules; the normalized firing strength is given by

$$\overline{w}_i = \frac{w_i}{w_1 + w_2}. \tag{5}$$

In layer 4, each square node multiplies the normalized firing strength by its corresponding consequent function. Hence, the output of the square node is given by

$$O_i^4 = \overline{w}_i f_i = \overline{w}_i \left(p_i x + q_i y + r_i\right), \tag{6}$$

where p_i, q_i, and r_i are called consequent parameters.

Layer 5 represents the aggregation. In this layer, a circle node labeled Σ computes the weighted average as given in (1).

According to Jang [20], learning in ANFIS may use backpropagation algorithm. In MATLAB, the default option is the hybrid learning of backpropagation and least squares estimation (LSE). The hybrid learning converges faster compared to only backpropagation. In the hybrid learning, the set of parameters consists of nonlinear antecedent parameters and linear consequent parameters. Each epoch in the learning process is composed of forward pass and backward pass. In the forward pass, the node outputs are calculated layer by layer, until layer 4. In this pass, the nonlinear antecedent parameters are fixed while the linear consequent parameters are computed using LSE. In the backward pass, the error is sent back through the network. In this pass, the linear consequent parameters are fixed while the nonlinear antecedent parameters are computed using backpropagation. Table 4 summarizes these learning procedures. Table 5 summarizes the information of ANFIS for the surface roughness model.

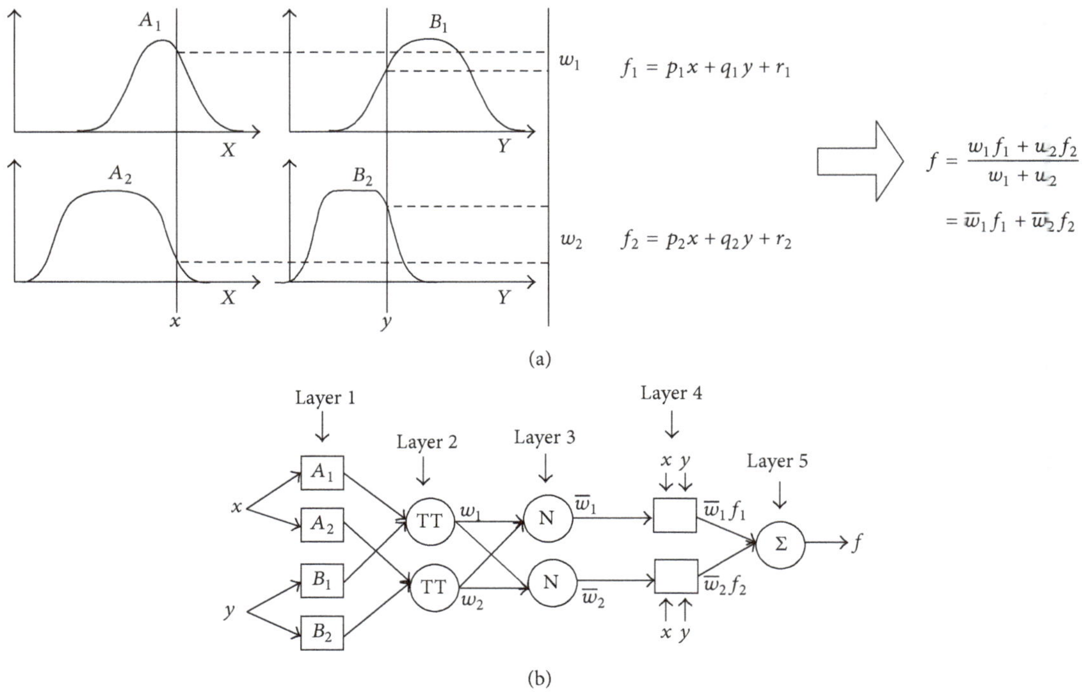

FIGURE 9: (a) A Sugeno type fuzzy model and (b) its equivalent ANFIS architecture.

TABLE 4: Two passes in the hybrid learning of ANFIS.

	Forward pass	Backward pass
Nonlinear premise parameters	Fixed	Back propagation
Linear consequent parameters	LSE	Fixed
Signals	Node outputs	Error signals

TABLE 5: ANFIS information.

Number of inputs	3 (cutting speed, depth of cut, feed rate)
Input MF	"gbellmf"
Number of outputs	1 (roughness)
Output type	Linear (first-order Sugeno type)
Number of fuzzy rules	81
Number of input-output pairs (training data)	125
Learning algorithm	Hybrid
Number of epochs	1000
Error tolerance	0
Evaluation method	Training data versus ANFIS output

3.4. Comparison. Fifteen experimental validation tests, not included in the regression analysis or in the ANFIS, were used to compare the predictability of both models. Table 6 illustrates the results of this comparison. Figure 10 shows a column chart illustrating the differences between measured R_a and the predictions of regression and ANFIS in the fifteen runs. The table and the graph illustrate that there is a slight difference between the ANFIS and the regression predictions.

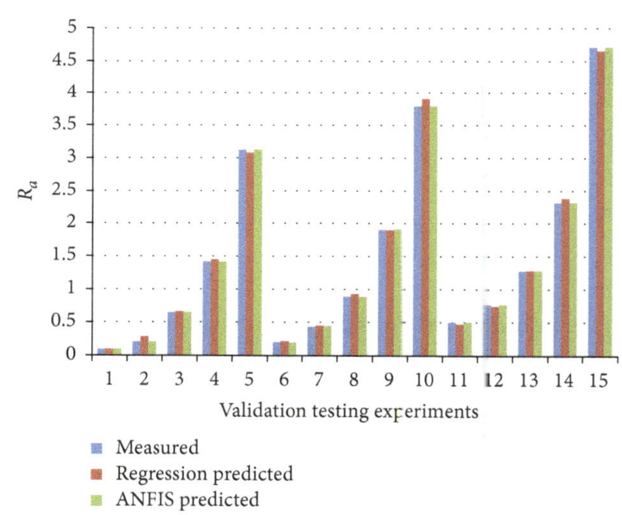

FIGURE 10: Column chart for measured R_a versus predicted R_a.

However, the mean absolute error (MAE) of the regression analysis is one order of magnitude higher than the MAE of ANFIS analysis with values of 0.035 μm and 0.003 μm, respectively.

4. Conclusions

Full factorial experimental design was used to examine the effect of three machining parameters (cutting speed, depth of cut, and feed rate) on the produced surface roughness. The studied range of each factor was divided into five levels

TABLE 6: Comparison between regression model and ANFIS predictions.

Machining parameters			Measured R_a (μm)	Regression model		ANFIS	
A	B	C		Predicted R_a (μm)	Residual	Predicted R_a (μm)	Residual
75	0.15	0.02	0.1	0.097	0.003	0.0996	0.0004
75	0.15	0.04	0.212	0.286	−0.074	0.212	0
75	0.15	0.08	0.65	0.666	−0.016	0.650	0
75	0.15	0.16	1.415	1.446	−0.031	1.410	0.005
75	0.15	0.32	3.121	3.077	0.044	3.120	0.001
125	0.45	0.02	0.203	0.218	−0.015	0.197	0.006
125	0.45	0.04	0.435	0.453	−0.018	0.443	−0.008
125	0.45	0.08	0.893	0.928	−0.035	0.889	0.004
125	0.45	0.16	1.902	1.896	0.006	1.91	−0.008
125	0.45	0.32	3.792	3.903	−0.111	3.79	0.002
175	0.75	0.02	0.504	0.476	0.028	0.502	0.002
175	0.75	0.04	0.767	0.744	0.023	0.769	−0.002
175	0.75	0.08	1.278	1.284	−0.006	1.28	−0.002
175	0.75	0.16	2.318	2.382	−0.064	2.32	−0.002
175	0.75	0.32	4.707	4.652	0.055	4.71	−0.003

TABLE 7: Full listing of experimental samples.

Sample ID	Group	Cutting speed (m/min)	Depth of cut (mm)	Feed rate (mm/rev)	Surface R_a (μm)	Material removal rate (MRR) mm^3/min
1		75	0.15	0.020	0.100	225
2		75	0.15	0.040	0.212	450
3	1	75	0.15	0.080	0.650	900
4		75	0.15	0.160	1.415	1800
5		75	0.15	0.320	3.121	3600
6		75	0.30	0.020	0.152	450
7		75	0.30	0.040	0.368	900
8	2	75	0.30	0.080	0.738	1800
9		75	0.30	0.160	1.645	3600
10		75	0.30	0.320	3.312	7200
11		75	0.45	0.020	0.182	675
12		75	0.45	0.040	0.389	1350
13	3	75	0.45	0.080	0.798	2700
14		75	0.45	0.160	1.699	5400
15		75	0.45	0.320	3.386	10800
16		75	0.60	0.020	0.261	900
17		75	0.60	0.040	0.499	1800
18	4	75	0.60	0.080	0.898	3600
19		75	0.60	0.160	1.811	7200
20		75	0.60	0.320	3.599	14400
21		75	0.75	0.020	0.394	1125
22		75	0.75	0.040	0.599	2250
23	5	75	0.75	0.080	0.999	4500
24		75	0.75	0.160	1.811	9000
25		75	0.75	0.320	3.678	18000
26		100	0.15	0.020	0.106	300
27		100	0.15	0.040	0.225	600
28	6	100	0.15	0.080	0.688	1200
29		100	0.15	0.160	1.497	2400
30		100	0.15	0.320	3.311	4800
31		100	0.30	0.020	0.162	600
32		100	0.30	0.040	0.389	1200
33	7	100	0.30	0.080	0.782	2400
34		100	0.30	0.160	1.743	4800
35		100	0.30	0.320	3.504	9600

TABLE 7: Continued.

Sample ID	Group	Cutting speed (m/min)	Depth of cut (mm)	Feed rate (mm/rev)	Surface R_a (μm)	Material removal rate (MRR) mm^3/min
36		100	0.45	0.020	0.193	900
37		100	0.45	0.040	0.412	1800
38	8	100	0.45	0.080	0.845	3600
39		100	0.45	0.160	1.798	7200
40		100	0.45	0.320	3.585	14400
41		100	0.60	0.020	0.276	1200
42		100	0.60	0.040	0.527	2400
43	9	100	0.60	0.080	0.950	4800
44		100	0.60	0.160	1.916	9600
45		100	0.60	0.320	3.807	19200
46		100	0.75	0.020	0.416	1500
47		100	0.75	0.040	0.633	3000
48	10	100	0.75	0.080	1.057	6000
49		100	0.75	0.160	1.916	12000
50		100	0.75	0.320	3.891	24000
51		125	0.15	0.020	0.112	375
52		125	0.15	0.040	0.237	750
53	11	125	0.15	0.080	0.728	1500
54		125	0.15	0.160	1.585	3000
55		125	0.15	0.320	3.495	6000
56		125	0.30	0.020	0.170	750
57		125	0.30	0.040	0.412	1500
58	12	125	0.30	0.080	0.826	3000
59		125	0.30	0.160	1.843	6000
60		125	0.30	0.320	3.709	12000
61		125	0.45	0.020	0.203	1125
62		125	0.45	0.040	0.435	2250
63	13	125	0.45	0.080	0.893	4500
64		125	0.45	0.160	1.902	9000
65		125	0.45	0.320	3.792	18000
66		125	0.60	0.020	0.292	1500
67		125	0.60	0.040	0.558	3000
68	14	125	0.60	0.080	1.005	6000
69		125	0.60	0.160	2.028	12000
70		125	0.60	0.320	4.030	24000
71		125	0.75	0.020	0.441	1375
72		125	0.75	0.040	0.670	3750
73	15	125	0.75	0.080	1.118	7500
74		125	0.75	0.160	2.028	15000
75		125	0.75	0.320	4.119	30000
76		150	0.15	0.020	0.123	450
77		150	0.15	0.040	0.254	900
78	16	150	0.15	0.080	0.783	1800
79		150	0.15	0.160	1.698	3600
80		150	0.15	0.320	3.745	7200
81		150	0.30	0.020	0.184	900
82		150	0.30	0.040	0.442	1800
83	17	150	0.30	0.080	0.885	3600
84		150	0.30	0.160	1.974	7200
85		150	0.30	0.320	3.974	14400
86		150	0.45	0.020	0.218	1350
87		150	0.45	0.040	0.467	2700
88	18	150	0.45	0.080	0.957	5400
89		150	0.45	0.160	2.038	10800
90		150	0.45	0.320	4.063	21600

TABLE 7: Continued.

Sample ID	Group	Cutting speed (m/min)	Depth of cut (mm)	Feed rate (mm/rev)	Surface R_a (μm)	Material removal rate (MRR) mm^3/min
91		150	0.60	0.020	0.313	1800
92		150	0.60	0.040	0.598	3600
93	19	150	0.60	0.080	1.077	7200
94		150	0.60	0.160	2.173	14400
95		150	0.60	0.320	4.318	28800
96		150	0.75	0.020	0.472	2250
97		150	0.75	0.040	0.718	4500
98	20	150	0.75	0.080	1.198	9000
99		150	0.75	0.160	2.173	18000
100		150	0.75	0.320	4.413	36000
101		175	0.15	0.020	0.128	525
102		175	0.15	0.040	0.272	1050
103	21	175	0.15	0.080	0.832	2100
104		175	0.15	0.160	1.812	4200
105		175	0.15	0.320	3.994	8400
106		175	0.30	0.020	0.195	1050
107		175	0.30	0.040	0.471	2100
108	22	175	0.30	0.080	0.945	4200
109		175	0.30	0.160	2.106	8400
110		175	0.30	0.320	4.239	16800
111		175	0.45	0.020	0.232	1575
112		175	0.45	0.040	0.497	3150
113	23	175	0.45	0.080	1.021	6300
114		175	0.45	0.160	2.174	12600
115		175	0.45	0.320	4.334	25200
116		175	0.60	0.020	0.334	2100
117		175	0.60	0.040	0.638	4200
118	24	175	0.60	0.080	1.149	8400
119		175	0.60	0.160	2.318	16800
120		175	0.60	0.320	4.606	33600
121		175	0.75	0.020	0.504	2625
122		175	0.75	0.040	0.767	5250
123	25	175	0.75	0.080	1.278	10500
124		175	0.75	0.160	2.318	21000
125		175	0.75	0.320	4.707	42000

creating 125 experiments, among which 110 runs were used in the modeling phase. Prediction of surface roughness was conducted using both a mathematical model built through regression analysis and ANFIS. The predictions from the regression model were then compared to the ANFIS predictions. Fifteen experimental test runs, not included in the regression analysis or in the ANFIS, were used in the validation and comparison of the models. The results illustrate that the ANFIS technique has slightly better accuracy than the regression analysis. The mean absolute error (MAE) of the regression analysis is one order of magnitude higher than the MAE of ANFIS analysis with values of 0.035 μm and 0.003 μm, respectively. Optimization process was run using the desirability function approach to maximize the material removing rate (MRR) while maintaining R_a below 0.8 μm.

Conflicts of Interest

The authors declare that they have no conflicts of interest.

Acknowledgments

Authors would like to thank Twasol Research Excellence Program (TRE Program), King Saud University, Riyadh, Saudi Arabia, for support.

References

[1] UK ministry of defense report no: DEFSTAN 10-13/2005.

[2] A. T. Abbas, K. Hamza, M. F. Aly, and E. A. Al-Bahkali, "Multiobjective optimization of turning cutting parameters for J-steel material," *Advances in Materials Science and Engineering*, vol. 2016, Article ID 6429160, 8 pages, 2016.

[3] B. Bhardwaj, R. Kumar, and P. K. Singh, "Surface roughness (R_a) prediction model for turning of AISI 1019 steel using response surface methodology and Box–Cox transformation," *Proceedings of the Institution of Mechanical Engineers, Part B: Journal of Engineering Manufacture*, vol. 228, no. 2, pp. 223–232, 2014.

[4] R. Mokhtari Homami, A. Fadaei Tehrani, H. Mirzadeh, B. Movahedi, and F. Azimifar, "Optimization of turning process using artificial intelligence technology," *International Journal of Advanced Manufacturing Technology*, vol. 70, no. 5–8, pp. 1205–1217, 2014.

[5] I. Svalina, G. Šimunović, T. Šarić, and R. Lujić, "Evolutionary neuro-fuzzy system for surface roughness evaluation," *Applied Soft Computing*, vol. 52, pp. 593–604, 2017.

[6] S. Jović, N. Arsić, V. Vukojević, O. Anicic, and S. Vujičić, "Determination of the important machining parameters on the chip shape classification by adaptive neuro-fuzzy technique," *Precision Engineering*, vol. 48, pp. 18–23, 2017.

[7] T.-L. (Bill) Tsenga, U. Konadaa, and Y. (James) Kwonb, "A novel approach to predict surface roughness in machining operations using fuzzy set theory," *Journal of Computational Design and Engineering*, vol. 3, no. 1, pp. 1–13, 2016.

[8] G. Kant and K. S. Sangwan, "Predictive modelling and optimization of machining parameters to minimize surface roughness using artificial neural network coupled with genetic algorithm," in *Proceedings of the 15th CIRP Conference on Modelling of Machining Operations (CMMO '15)*, vol. 31, pp. 453–458, June 2015.

[9] K. Raja, P. Marimuthu, and K. Chandrasekaran, "Performance study on AISI316 and AISI410 using different layered coated cutting tools in CNC turning," *Journal of Engineering Science and Technology*, vol. 10, no. 1, pp. 25–34, 2015.

[10] K. Chandrasekarana, P. Marimuthu, and K. Raja, "Prediction model for CNC turning on AISI316 with single and multilayered cutting tool using Box Behnken Design," *International Journal of Engineering, Transactions A: Basics*, vol. 26, no. 4, pp. 401–410, 2013.

[11] F. Jafarian, M. Taghipour, and H. Amirabadi, "Application of artificial neural network and optimization algorithms for optimizing surface roughness, tool life and cutting forces in turning operation," *Journal of Mechanical Science and Technology*, vol. 27, no. 5, pp. 1469–1477, 2013.

[12] D. M. D'Addona and S. J. Raykar, "Analysis of surface roughness in hard turning using wiper insert geometry," *Procedia CIRP*, vol. 41, pp. 841–846, 2016.

[13] G. M. A. Acayaba and P. M. D. Escalona, "Prediction of surface roughness in low speed turning of AISI316 austenitic stainless steel," *CIRP Journal of Manufacturing Science and Technology*, vol. 11, pp. 62–67, 2015.

[14] S. Sahu and B. B. Choudhury, "Optimization of surface roughness using taguchi methodology & prediction of tool wear in hard turning tools," *Materials Today: Proceedings*, vol. 2, no. 4-5, pp. 2615–2623, 2015.

[15] A. Agrawal, S. Goel, W. B. Rashid, and M. Price, "Prediction of surface roughness during hard turning of AISI 4340 steel (69 HRC)," *Applied Soft Computing*, vol. 30, pp. 279–286, 2015.

[16] M. Azam, M. Jahanzaib, A. Wasim, and S. Hussain, "Surface roughness modeling using RSM for HSLA steel by coated carbide tools," *International Journal of Advanced Manufacturing Technology*, vol. 78, no. 5–8, pp. 1031–1041, 2015.

[17] M. Mia and N. R. Dhar, "Prediction of surface roughness in hard turning under high pressure coolant using Artificial Neural Network," *Measurement: Journal of the International Measurement Confederation*, vol. 92, pp. 464–474, 2016.

[18] C. L. He, W. J. Zong, Z. M. Cao, and T. Sun, "Theoretical and empirical coupled modeling on the surface roughness in diamond turning," *Materials and Design*, vol. 82, pp. 216–222, 2015.

[19] E. A. Al Bahkali, A. E. Ragab, E. A. El Danaf, and A. T. Abbas, "An investigation of optimum cutting conditions in turning nodular cast iron using carbide inserts with different nose radius," *Proceedings of the Institution of Mechanical Engineers, Part B: Journal of Engineering Manufacture*, vol. 230, no. 9, pp. 1584–1591, 2016.

[20] J.-S. R. Jang, "ANFIS: adaptive-network-based fuzzy inference system," *IEEE Transactions on Systems, Man and Cybernetics*, vol. 23, no. 3, pp. 665–685, 1993.

13

Numerical Study on the Local Buckling Behaviour of Bolted Steel Plates in Steel Jacketing

Xiao-liang Xu, Zhou-dao Lu, Ling-zhi Li, and Chang-jiu Jiang

Research Institute of Structural Engineering and Disaster Reduction, College of Civil Engineering, Tongji University, 1239 Siping Road, Shanghai 200092, China

Correspondence should be addressed to Ling-zhi Li; tjlilingzhi@gmail.com

Academic Editor: Jun Liu

A numerical simulation was conducted to investigate the local buckling behaviour of the bolted steel plates in steel jacketing technique. The numerical model was firstly validated by the results of a previous experimental study. Then a parametric study was conducted to investigate the influence of different restraint measures on the local buckling behaviour and the sensitivity of the buckling behaviour to the initial imperfection. Fitted formulae were developed to calculate the structural field capacity of the bolted steel plates, and recommended values of stiffener size were also provided to facilitate the strengthening design of steel jacketing.

1. Introduction

Reinforced concrete (RC) members in existing buildings may need to be retrofitted due to defective construction, accidental damage, material degradation, and being imposed higher loads than initial design. Among all of the available retrofitting techniques, external jacketing, which includes jacketing with concrete, steel, and composite such as fibre reinforced polymers (FRPs) [1], is the most convenient method. To bond steel plates or FRPs using adhesive mortar is the most popular technique, but it would lead to premature debonding and peeling failure at the ends of the bonded plates or FRPs [2, 3]; thus considerable efforts have been made to suppress and evaluate these kinds of premature failures [4, 5]. On the other hand, steel plates attached mechanically using anchor bolts are immune to these adverse effects [6–8]. Although attaching steel plates to beam soffits can effectively increase their flexural strength and stiffness, it may lead to overreinforcement and a subsequent serious decrease in ductility [6–9], while attaching steel plates to the side faces of beams by anchor bolts has been proven to possess enhanced flexural strength without a significant reduction in ductility [10, 11]. Nevertheless, another problem arises; that is, local buckling would occur in the upper compressive region of the bolted steel plates in steel jacketing [11, 12], as shown in Figure 1(a). In the aspect of columns, the same problem occurs when the bolted steel plates are under axial or eccentric compression [13–15], as shown in Figure 1(b).

The local buckling mode of the bolted steel plates in steel jacketing can be seen as a unilateral buckle, since the inward displacement of plates is restrained by the RC beams or columns. Comprehensive experimental studies have been conducted by researchers worldwide to investigate this kind of unilateral buckle. In the field of RC columns strengthened by steel jacketing, Chai et al. [16] encased the critical regions of circular columns with steel jacketing that was bonded to the column using grout, and high load capacity and ductility was obtained because the jacketing was effective in passive confinement. Wang [1] found that local buckling would occur in the plastic hinge area of rectangular RC columns with steel jacketing. Xiao and Wu [14] applied stiffeners to avoid local buckling of steel jacketing on five rectangular RC columns with different buckling restraint measures, while, in the field of bolted-side-plated (BSP) RC beams, Smith et al. [17] investigated the local buckling behaviour of bolted steel plates, by introducing the so-called local buckling push test that was proposed by Uy and Bradford in their experimental study on composite profiled beams [18]. Su and Zhu [7] investigated the local buckling behaviour in BSP coupling beams and pointed out that the problem could hardly be suppressed by increasing the number of anchor bolts. Li et al. [11] investigated the variation of the overall performance

(a) (b)

FIGURE 1: Local buckling of the bolted steel plates in (a) RC beams and (b) columns strengthened using the steel jacketing technique.

of BSP beams after local buckling occurring in the bolted steel plates, and the capacity and ductility decreased by 6% and 56%, respectively, directly owing to this problem. They [19] also studied the buckling suppressing effect of different restraint measures, which showed that the local buckling deformation of the bolted steel plates can be restricted by employing longitudinal stiffeners and narrower anchor bolts spacing, and the structural field capacity could also be improved.

Since experimental studies of RC structures are expensive thus the number of specimens is usually very limited, researchers conducted finite element analysis (FEA) to extend specimen geometries and kinds of loading conditions. Lubell et al. [20] simulated the postbuckling behaviour of unstiffened steel plate shear wall under cyclic loading with simplified tension field analytical models and provided good predictions of the postyield strength of the specimens. Wright [21] researched the buckling of a long thin plate in contact with a rigid medium unilaterally. Ge and Usami [22] studied the local buckling of short columns with concrete infilled and thin-walled steel and developed an ultimate strength formula for steel plates under uniform compression. For the numerical efforts into the local buckling behaviour of bolted steel plates, Smith et al. [23, 24] developed the buckling formula of bolted steel plates by using Ritz method and by simplifying the boundary conditions of the plates as being simply supported, clamped, and free and carried out a series of simply supported BSP beams to verify the proposed analytical model [25].

Although comprehensive studies have been conducted, it is still lack of a convenient way to estimate the structural field capacity of the bolted steel plates in steel jacketing that can account for the influence of the local buckling effect, especially a simple formula that can be used in the strengthening design practice. Since the number of specimens in experimental study is relatively limited due to expenditure consideration, a parametric study with FEA models built the ABAQUS programme and validated by experimental results was conducted as a feasible tool to investigate the local buckling behaviour of bolted steel plates with different plate thickness, bolt spacing, and different configuration of longitudinal and transverse stiffeners. Fitted formulae and recommended stiffener size were also proposed to facilitate the strengthening design of the steel jacketing technique.

TABLE 1: Constitutive relation of steel.

Steel 1		Steel 2	
Stress/MPa	Strain	Stress/MPa	Strain
302.0	0.00000	602.0	0.00000
302.0	0.01856	703.0	0.00075
370.1	0.10928	730.0	0.00247
404.2	0.15464	736.0	0.00393
438.2	0.20000	739.0	0.00597
438.2	1.00000	739.0	1.00000

2. Numerical Modelling

The three-dimensional nonlinear finite element analysis (NLFEA) program ABAQUS was employed to simulate the local buckling behaviour of the bolted steel plates in steel jacketing. The instance assembly in ABAQUS is shown in Figure 2. The size of the concrete block is 360 mm × 400 mm × 150 mm, with six or seven holes used for the installation of anchor bolts. The two investigated steel plates of 300 mm × 560 mm were bolted to its two opposite sides by the anchor bolts 12 mm in diameter. The bolt spacing is 300 mm and 100 mm in transverse and longitudinal directions, respectively. With the restraint of the anchor bolts, the steel plates had no deflection at the connected region of bolts, while the region between bolts can deflect in the transverse direction. And the loading condition can be changed by adjusting the location of the steel hinge on the bearing plates that were fixed to both ends of the bolted steel plates.

Table 1 shows the material constitutive relation of steels that were used in this numerical model, and Table 2 shows the choice of element type and material for each component. The modelling of the main components and the analysis steps will be described in details hereafter in the subsequent subsections.

2.1. Modelling of the Concrete Block. Considering the main purpose of this study is to investigate the local buckling behaviour of the bolted steel plates, so, in this numerical model, the concrete was idealized as an elastic material whose density, Young's modulus, and Poisson's ratio is 2500 Kg/m^3, 2.95 × 10^{10} N/m^2, and 0.2, respectively. The comparison showed that the model with elastic material for concrete

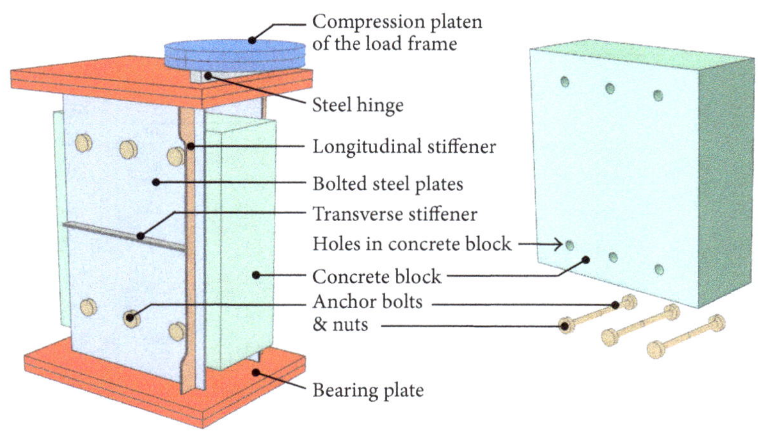

FIGURE 2: Instance assembly of the local buckling test modeling.

TABLE 2: The modelling of main components.

Components	Element type	Material
Concrete block	C3D8R	Elastic material
Steel plates	S4R	Steel 1
Stiffener (4 × 20)	S4R	Steel 1
Stiffener (6 × 25 &) 8 × 30	S4R	Steel 2
Steel hinges/bolts/nuts	C3D8R	Steel 1
Bearing plates	S4R	Steel 1

had a better convergence, thus, less time-consuming without declining in accuracy, so it is appropriate to employ this elastic simplification. An element size of 0.05 and element type of C3D8R was chosen for the concrete block.

2.2. Modelling of the Steel Plates and Stiffeners. In order to get the best simulating results, all the material used in this model was defined based on the material testing. Table 1 showed material property of steel. And the steel plates and the stiffeners in 4 mm × 20 mm were assigned the section with material of steel 1, while the rest of the stiffeners in the other sizes were assigned the section with material of steel 2. The steel hinge and the anchor bolts and nuts were assigned elements in the type of C3D8R with an element size of 0.05, while the bolted steel plates, the stiffeners, and the bearing plates were assigned S4R with an element size of 0.05.

2.3. Modelling of the Bolt Connection. In most of existing numerical studies with ABAQUS, the connection between different parts was simulated by defining an interaction type of "all with self." This simplification could simulate the friction action effectively in most cases but not for the connection in this study, where the precompression was loaded on by tightening the nuts. Therefore, a solution was proposed to simulate the precompression of the anchor bolts, by preshortening the anchor bolts for 0.05 mm in longitudinal direction after defining the "all with self" between different parts with a friction coefficient 0.15, and then a longitudinal displacement of 3.5 mm was imposed on the compression platen to apply concentric or eccentric compression to the bolted steel plates. The measure was proved effective by the subsequent investigation.

2.4. Initial Imperfection. The initial imperfection should be considered because it may have a significant influence in investigation of buckling. It is common to introduce the first buckling mode as the initial imperfection. To get the first buckling mode and use it as the initial imperfection of steel plates, the following progress was conducted in sequence:

(1) Clamp the concrete block and nuts; then conduct a "BUCKLIE Step" analysis for the model, which produces an input file "job-1.inp".

(2) Write the statement of "*Nodefile//U" into the input file "job-1.inp" after "Restart, write, frequency=0", then re-executed the file "job-1.inp" and obtain the file "job-2.fil", which includes the buckling information and modes of the bolted steel plates, as shown in Figure 3.

(3) Redefine the friction connection by preshortening the anchor bolts for 0.05 mm, and impose a longitudinal displacement of 3.5 mm on the compression platen; then conduct a "Static General Step" analysis by executing the input file "job-1.inp".

(4) Write the statement of "*imperfection, file=job-a,sep=1//N, A" into the file "job-1.inp" between "*Boundary" and "*Step"; then the nonlinear analysis can be realized by reexecuting the input file "job-1.inp". The letter "N" in the statement means introducing the Nth Buckling mode as the initial imperfection. And the letter "A" is the magnitude of the imperfection (usually set to be 10% of the plate thickness), which will be multiplied with the selected buckling mode and applied to the bolted steel plates automatically in the nonlinear analysis step.

TABLE 3: The information of specimens.

Number	Name	Loading condition	Thickness of plate t (mm)	Size and category of stiffeners (mm × mm)	
				Longitudinal	Transverse
1	C4NN	Axial-compressive (C)	4.3 (4)	No (N)	No (N)
2	E4NN	Eccentric-compressive (E)	4.3 (4)	No (N)	No (N)
3	E6NN	Eccentric-compressive (E)	5.7 (6)	No (N)	No (N)
4	E4AN	Eccentric-compressive (E)	4.3 (4)	Bolt fixing (A)	No (N)
5	E4SN	Eccentric-compressive (E)	4.3 (4)	4 × 20 (small, S)	No (N)
6	E4MN	Eccentric-compressive (E)	4.3 (4)	6 × 25 (medium, M)	No (N)
7	E4LN	Eccentric-compressive (E)	4.3 (4)	8 × 30 (large, L)	No (N)
8	E4SS	Eccentric-compressive (E)	4.3 (4)	4 × 20 (small, S)	4 × 20 (small, S)

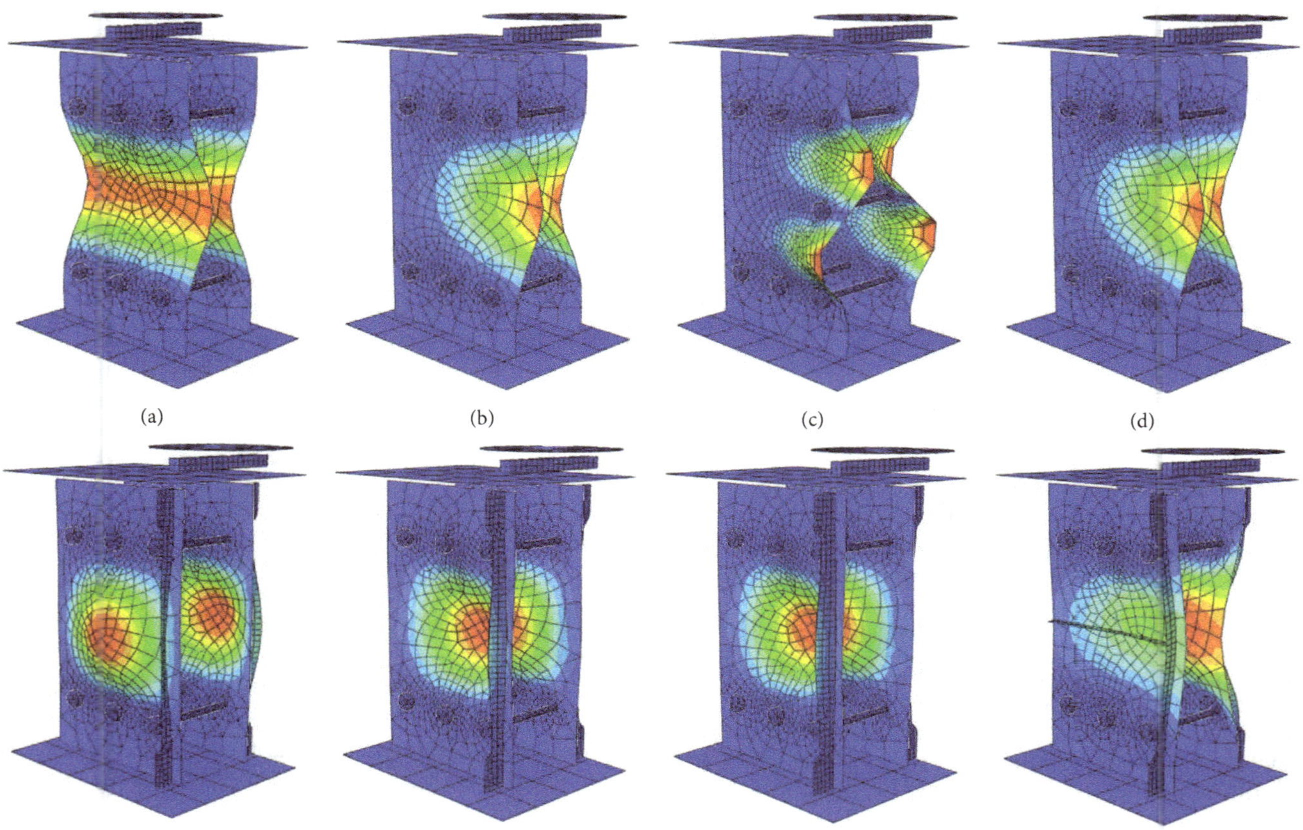

(a) (b) (c) (d) (e) (f) (g) (h)

FIGURE 3: Buckling modes of each of the specimens (a) C4NN, (b) E4NN, (c) E4AN, (d) E6NN, (e) E4SN, (f) E4MN, (g) E4LN, and (h) E4SS.

3. Validation of the Numerical Model

3.1. A Brief Introduction to the Experimental Study. In a previous experimental study [19], eight specimens with different buckling restraint measures were investigated to study how the loading condition, plate thickness, arrangement of longitudinal and transverse stiffeners, and bolt spacing affect the local buckling behaviour of the bolted steel plates. The design drawings and the general information of the specimens are shown in Figure 4 and Table 3, and more details can be found in [19].

3.2. Comparison of the Test and Numerical Results. In order to verify the efficiency and accuracy of the numerical models, comparison between the experimental and numerical results is shown in Figures 5 and 6 and Table 4.

It is shown in Figure 5 that the failure modes derived from the numerical study agree with those observed in the experimental study. There is no significant buckling at the beginning of loading; then the steel plates buckles outwards in lateral direction as the increase of compressive loading; the steel plates finally fails due to lateral buckling as a result of the rapid degradation of stiffness. When the bolted steel plates are

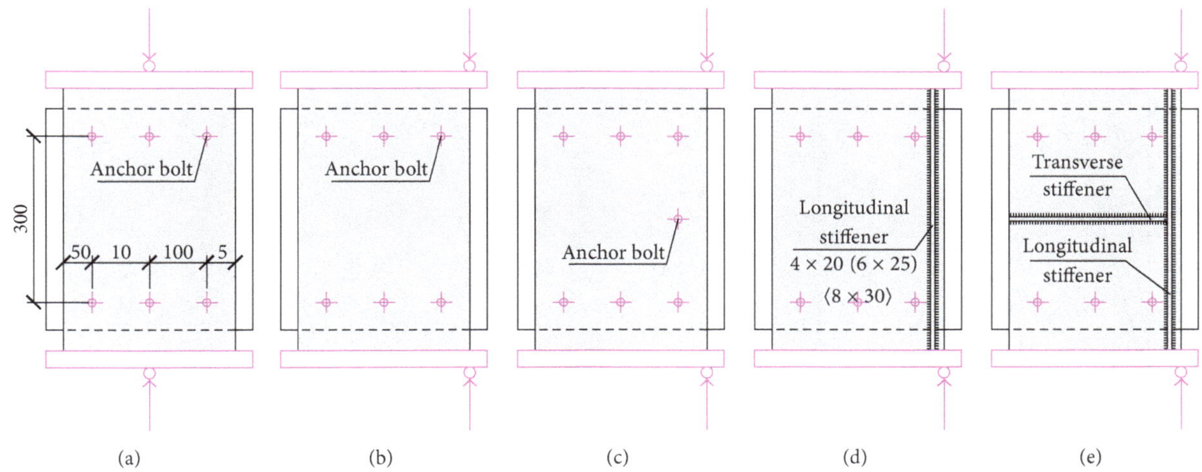

FIGURE 4: Restrain configuration of specimens (a) C4NN, (b) E4NN (E6NN), (c) E4AN, (d) E4SN (E4MN) ⟨E4LN⟩, and (e) E4SS.

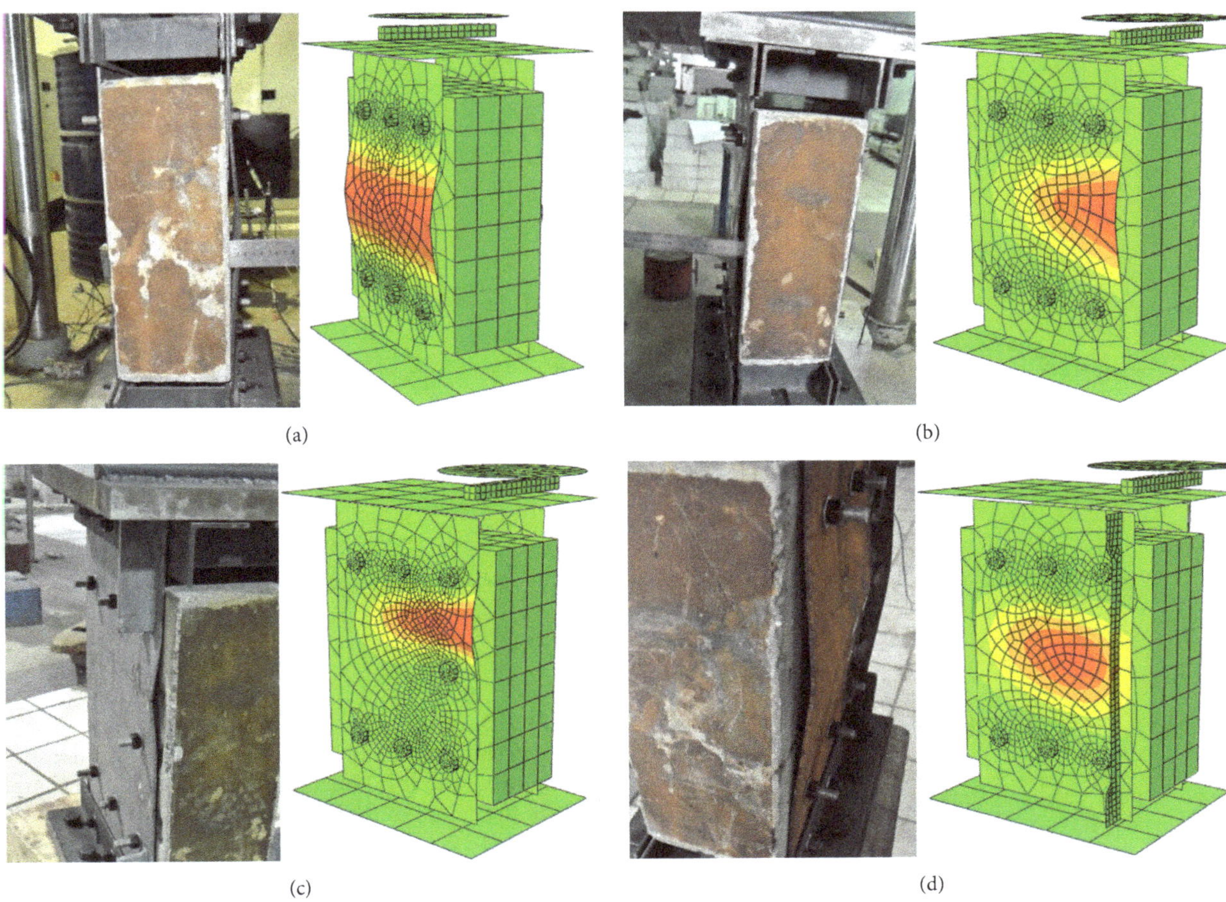

FIGURE 5: Experimental and simulative failure modes: (a) C4NN, (b) E6NN, (c) E4AN, and (d) E4MN.

under concentric compression, the whole horizontal middle section buckles outwards with a uniform buckling deformation in both the experimental study and the numerical simulation, as shown in Figure 5(a). On the other hand, when the bolted steel plates are under eccentric compression, local buckling only occurs on the compressive edge of the horizontal middle section, as shown in Figure 5(b). When the bolt spacing on the compressive edge is halved, the local

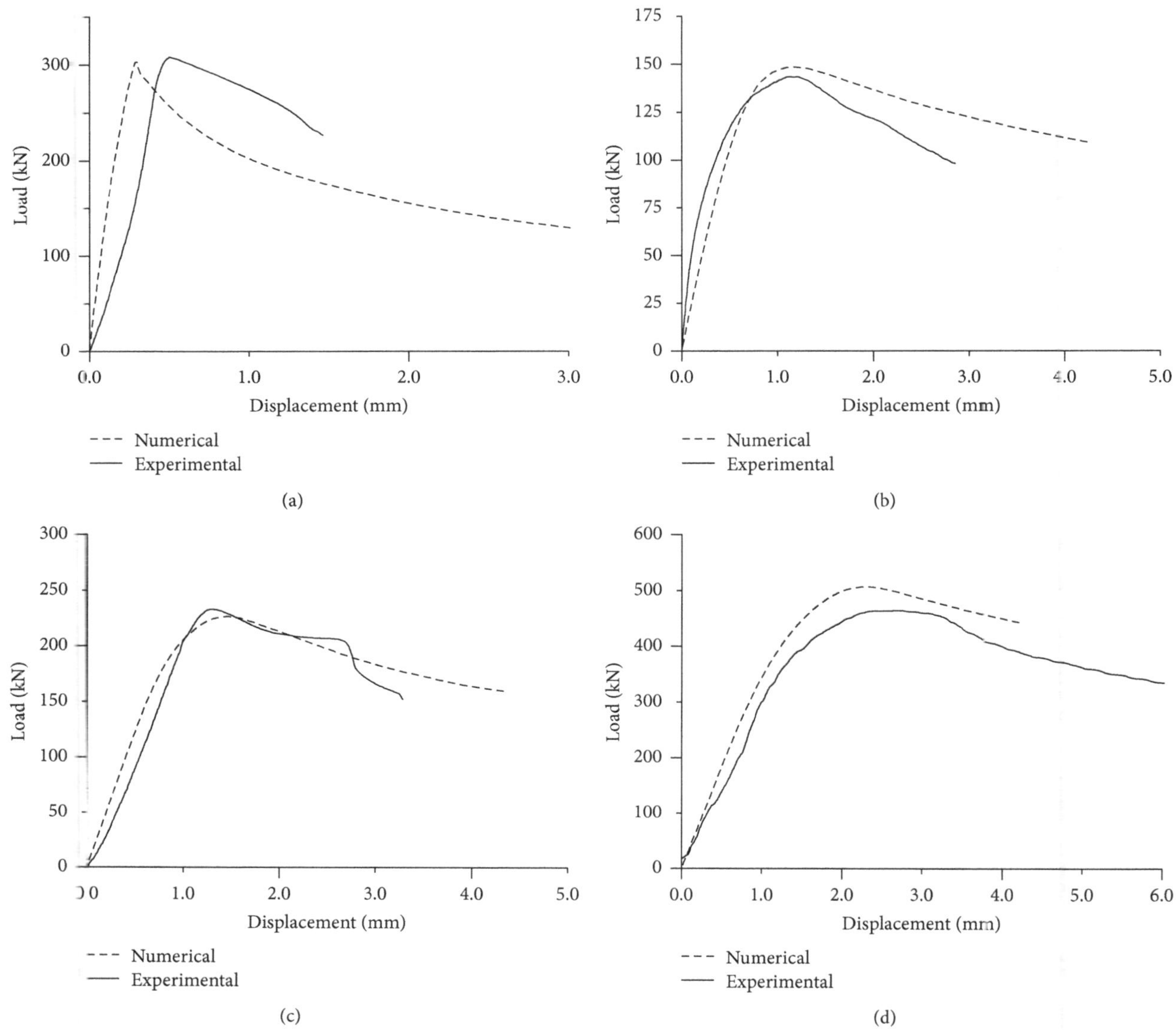

FIGURE 6: Experimental and numerical load-displacement curves for (a) C4NN, (b) E4NN, (c) E4AN, and (d) E4MN.

TABLE 4: Comparison of the experimental and numerical field capacity.

Specimen	Experimental value (kN)	Simulative value (kN)	Error
C4NN	309	306	1.0%
E4NN	144	149	3.5%
E4AN	233	228	2.1%
E6NN	266	269	1.1%
E4SN	314	347	10.5%
E4MN	464	508	9.5%
E4LN	603	670	11.1%
E4SS	318	350	10.1%
			Average error: 6.1%

buckling shifts upwards (or downwards) to one of the two spans, as shown in Figure 5(c). And if a longitudinal stiffener is employed, the local buckling shifts from the compressive edge to the centre of the horizontal middle section, as shown in Figure 5(d).

Figure 6 shows the comparison of the overall load-displacement curves derived from the experimental and numerical studies. And it can be found that the simulation results generally agree with the behaviour of the specimens in test, except for a slight error of both stiffness and peak load; this may be caused by the experimental discrepancy, such as the difference in the roughness of plate and concrete surfaces, the precompression provided by the torque of anchor bolts, the gaps between the bolt shafts, and the holes in the bolted steel plates. The comparison in structural field capacity was listed in Table 4, and the maximum margin of error was

TABLE 5: Information of models.

Compression	Restraint measure		Thickness of plates (mm)		
			2.0	4.3	5.7
Concentric	No restraint measure		C2NN	C4NN	C6NN
eccentric	No restraint measure		E2NN	E4NN	E6NN
	Bolt densification		E2AN	E4AN	E6AN
	Longitudinal stiffener (mm × mm)	4 × 20	E2SN	E4SN	E6SN
		6 × 25	E2MN	E4MN	E6MN
		8 × 30	E2LN	E4LN	E6LN
	Longitudinal and transverse stiffeners (mm × mm)	4 × 20	E2SS	E4SS	E6SS
		6 × 25	E2MM	E4MM	E6MM
		8 × 30	E2LL	E4LL	E6LL

11.1% and the average was 6.1%. Therefore, the error in the load-displacement curves and the structural field capacity was acceptable, and the FEA model can simulate the local buckling behaviour of the bolted steel plates satisfactorily.

4. Parametric Study

Since the accuracy of the numerical model was validated, it is feasible to conduct a more detailed parametric study on the local buckling behaviour of the bolted steel plates, by analysing a series of models with different thickness of steel plates, with or without stiffeners of different sizes.

4.1. A Brief Introduction to the Numerical Models. In this parametric study, twenty-seven models were conducted with different buckling restraint measures. To investigate the effect of the thickness of the bolted steel plates, it was varied from 2.0 mm to 4.3 mm and 5.7 mm. Similarly, to calibrate the sensitivity of the buckling behaviour to the initial imperfection, another series of models with the initial imperfection of 1.08 mm, 2.15 mm, 3.23 mm, and 4.3 mm were conducted. The detailed information of each model can be found in Table 5, of which the parameters have the same meanings as those in Table 3.

4.2. Effect of Plates Thickness and Bolt Spacing. Several particular models (as shown in Figure 7, whose initial imperfection were all 1.08 mm) were investigated to study the influence of plate thickness and bolt spacing on the local buckling behaviour of the bolted steel plates.

It is obvious from Figure 7 that the plate thickness has no significant effect on the failure modes, while narrower bolt spacing due to an additional anchor bolt on the compressive edge can make a very different failure mode by changing one span into two spans; thus the buckling moved into either of the two spans. In the other aspect, the load-displacement curves in Figure 8(a) show that a thicker plate distinctly increases the stiffness, the field capacity, and the longitudinal displacement (which means an improved ductility). However, although narrower bolt spacing can improve the field capacity and the longitudinal displacement (thus the ductility) significantly, it does not affect the stiffness; for instance, the load-displacement curves of E6NN and E6AN in Figure 8(a) generally overlapped with each other at the beginning of loading.

The lateral buckling deformation of three different positions on the bolted steel plates was shown in Figures 8(b)–8(d): Point A represents the midpoint of the compressive edge, Point B represents the centre of plate, and Point C represents the midpoint of the tensile edge. It is shown that a thicker plate has no significant influence on the growing trend of the lateral buckling deformation except in large size, but it can decrease the buckling deformation especially when it was a large one. Additionally, in the same plate thickness, narrower bolt spacing can decrease the magnitude and the increase rate of the lateral buckling deformation to a certain extent.

4.3. Effect of Longitudinal and Transverse Stiffeners. Similarly, the influence of different arrangement of stiffeners on the buckling behaviour of the bolted steel plates was investigated by comparing the models with stiffeners in variable sizes and different categories. Only the models with plate thickness of 4.3 mm and initial imperfection of 1.08 mm were illustrated for simplicity.

Figure 9 shows the deformation distribution diagrams, which demonstrate that the longitudinal stiffener can visibly decrease the lateral buckling deformation of the plates and transfer the region of buckling from compressive edge of the plates to the centre even to the tensile edge (as shown in Figures 9(a)–9(c)), while the transverse stiffener can only decrease the buckling deformation of the central region of the plates and transfer the region of buckling to the upward region of the transverse stiffener (as shown in Figures 9(c) and 9(d)). And it is also evident that the bigger stiffener has a stronger buckling restraint.

The load-displacement curves shown in Figure 10(a) indicate that the longitudinal stiffeners can distinctly improve the stiffness and the field capacity, and an increase in stiffener size led to a more obvious improvement. Besides, it is also shown that the longitudinal displacement at the peak load increases. The transverse stiffeners, by contrast, do not improve the stiffness obviously, although they can improve

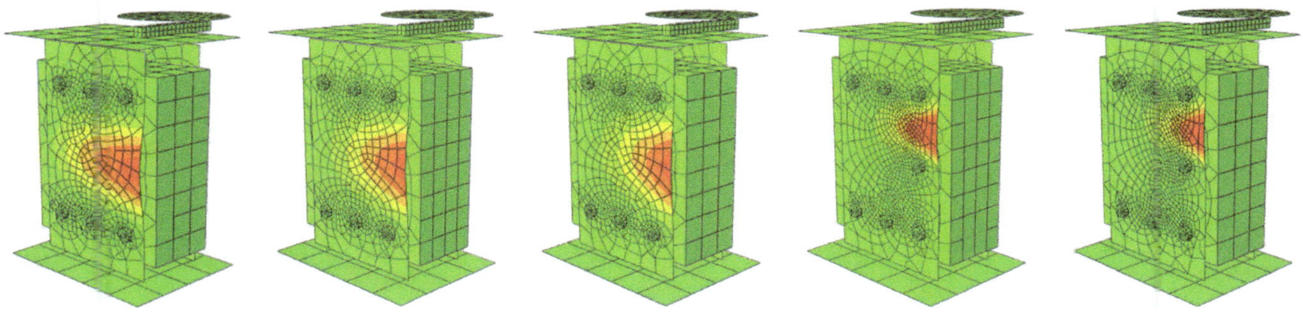

(a) E2NN (b) E4NN (c) E6NN (d) E4AN (e) E6AN

FIGURE 7: Failure modes of models with different plate thickness and narrower bolt densification: (a) E2NN, (b) E4NN, (c) E6NN, (d) E4AN, and (e) E6AN.

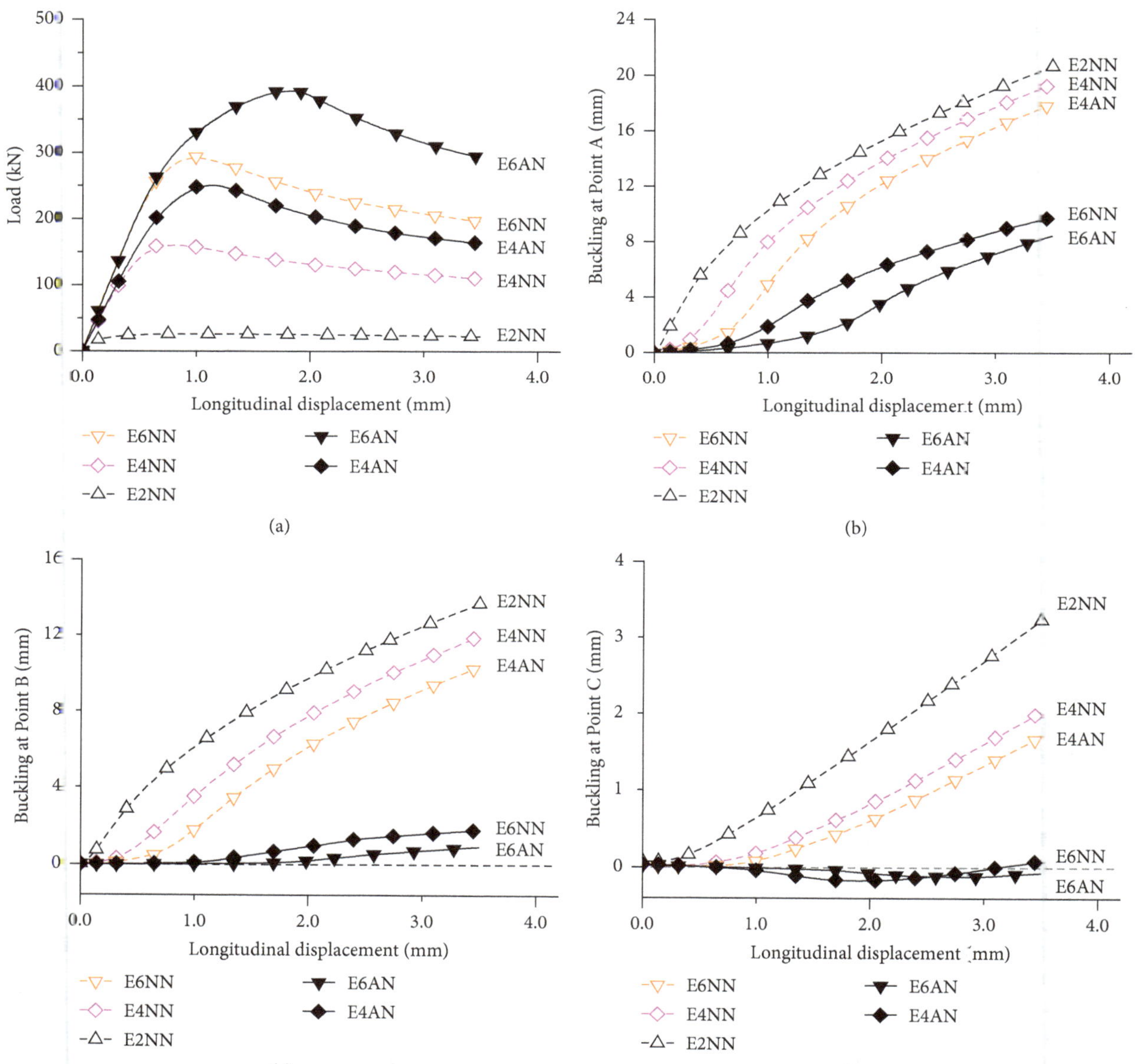

FIGURE 8: Load-displacement and buckling-displacement curves 1: (a) load-displacement curves; buckling-displacement curves at (b) Point A, (c) Point B, and (d) Point C.

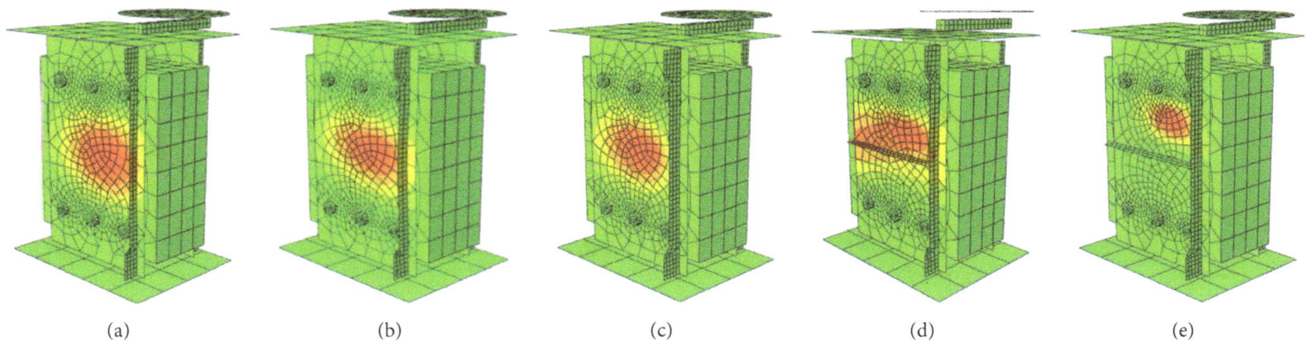

FIGURE 9: Failure modes of models with arranging different stiffeners: (a) E4SN, (b) E4MN, (c) E6LN, (d) E4MM, and (e) E4LL.

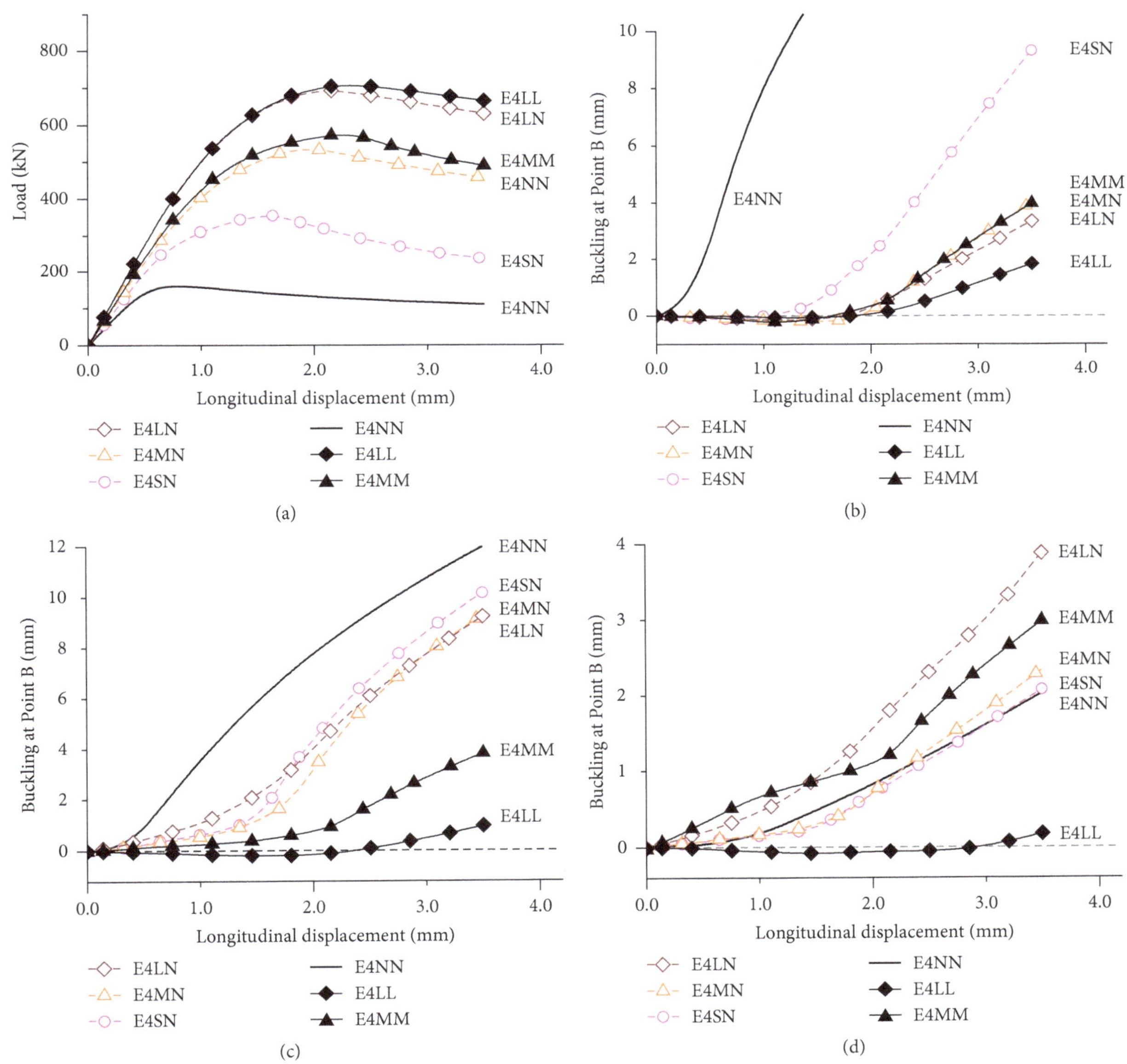

FIGURE 10: Load-displacement and buckling-displacement curves 2: (a) load-displacement curves; buckling-displacement curves at (b) Point A, (c) Point B, and (d) Point C.

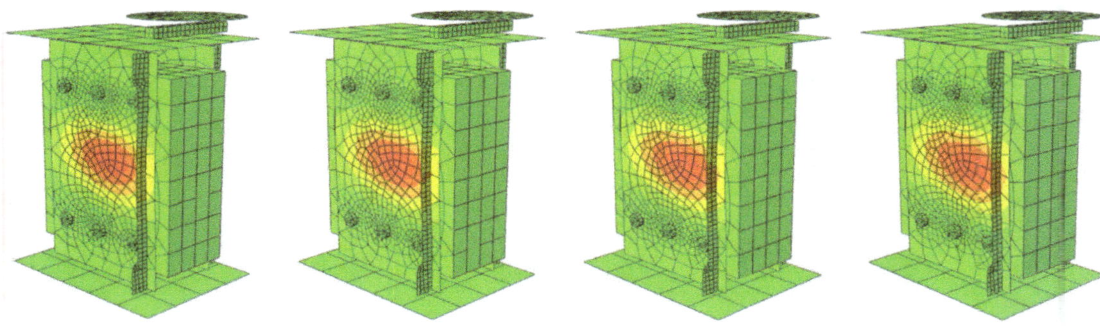

FIGURE 11: Failure modes of models with different initial imperfection.

the field capacity in a certain degree; for instance, the load-displacement curves of E4LN and E4LL in Figure 10(a) generally overlapped with each other at the beginning of loading.

The development of the lateral buckling deformation in Figures 10(b)–10(d) demonstrates that the longitudinal stiffeners effectively decrease buckling deformation at Point A (the midpoint on the compressive edge) but cause no obvious decrease at Point B (the centre of the plate) and even increase the buckling deformation at Point C (the midpoint on the tensile edge) when it is a large one. By contrast, arranging transverse stiffeners do not decrease the buckling deformation at Point A in a visible way when it have already arranged longitudinal stiffeners, and only a big transverse stiffener can lead to an obvious decrease in the buckling deformation at Point C, although it causes a substantial decrease in Point B at the centre of the plate.

4.4. Sensitivity to Initial Imperfection. To investigate the sensitivity of the buckling behaviour of the bolted steel plates to the initial imperfection, the buckling behaviour of models with a series of initial imperfections, that is, 1.08 mm, 2.15 mm, 3.23 mm, and 4.3 mm, was compared.

It is shown from Figure 11 that the failure modes were almost the same for the models with different imperfection; thus it is insensitive to the magnitude of initial imperfection.

In the aspect of the load-displacement curves, a high coherence between curves with different initial imperfection is shown in Figure 12(a), which nearly overlapped with each other at the beginning of loading. It is shown that the field capacity is insensitive to the initial imperfection, although a smaller initial imperfection leads to a little higher field capacity. And the development of the lateral buckling deformation in Figures 12(b)–12(d) indicates that a bigger initial imperfection has no substantial influence on the growing trend though it can increase the lateral buckling deformation in a certain degree. So the buckling behaviour of the bolted steel plates is insensitive to initial imperfection.

5. Design Recommendation

5.1. Formula Fitting

5.1.1. Components of the Structural Field Capacity. Based on the numerical models, a formula can be built to fit the structural field capacity of the bolted steel plates that considers the difference in restraint measures. Since the arrangement of longitudinal stiffeners can substantially improve the structural field capacity, for simplicity the improvement can be divided into two parts: the bearing capacity of the stiffeners and the enhancement to the steel plates beneficial from the buckling restraint of the stiffeners. Therefore, the structural field capacity F can be divided into three parts: the capacity of the bolted steel plates (F_0), the yield strength of the longitudinal stiffeners (F_1), and the enhancement due to the buckling restraint of the longitudinal stiffeners (F_2), as follows:

$$F = F_0 + F_1 + F_2. \quad (1)$$

(1) According to the buckling theory, the capacity of the bolted steel plates (F_0) is equal to the value of structural field capacity without any restraint measure and is related to the value of Young's modulus (E), the breadth and thickness (B & t) of the bolted steel plates, and the bolt spacing (L). It can be assumed in the form of (2) by taking E4NN (whose field capacity is 149 kN) as reference. And solving (3) by introducing the related parameters (i.e., E = 210 GPa, B = 0.3 m, L = 0.3 m, and t = 4.3 mm) yields $\gamma_0 = 0.2713$.

$$F_0 = \gamma_0 \times \frac{\pi^2 E B t^3}{L^2} \times \left(\frac{B}{0.3}\right)^x \left(\frac{L}{0.3}\right)^y \left(\frac{t}{0.0043}\right)^z, \quad (2)$$

$$F_0 = \gamma_0 \times \frac{\pi^2 E B t^3}{L^2} = 149\,\text{kN}. \quad (3)$$

(2) According to the theory in material mechanics, the capacity of the longitudinal stiffeners (F_1) is related to the sectional size of the stiffeners and the field strength of steel, thus in the form of

$$F_1 = 2abf_y. \quad (4)$$

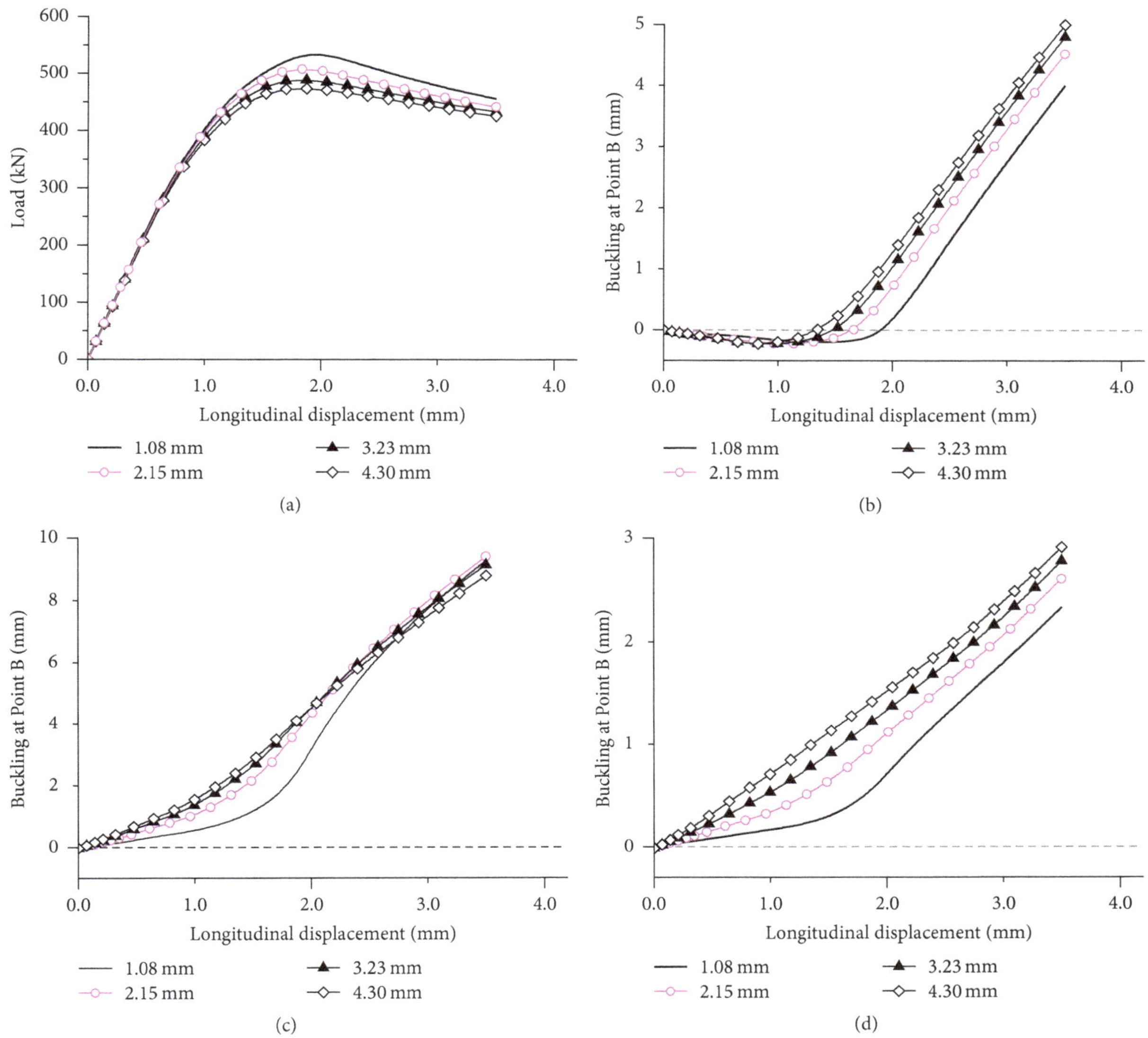

FIGURE 12: Load-displacement and buckling-displacement curves 3: (a) load-displacement curves; buckling-displacement curves at (b) Point A, (c) Point B, and (d) Point C.

(3) The enhancement benefitting from the buckling restraint of the longitudinal stiffeners (F_2) may be related to the width and thickness (a and b) of the stiffener and the thickness of the steel plate (t). Besides, the breadth of the bolted steel plate (B) and the bolt spacing (L) may possibly affect the value of (F_2). So the formula can be defined as follows. And similar to F_0, taking E4SN (F_2 = 127 kN) as reference and solving (6) yield $\gamma_2 = 0.2312$.

$$F_2 = \gamma_2 \times \frac{\pi^2 E B t^3}{L^2} \times \left(\frac{a}{0.004}\right)^j \left(\frac{b}{0.02}\right)^k \left(\frac{B}{0.3}\right)^l \left(\frac{L}{0.3}\right)^m \left(\frac{t}{0.0043}\right)^n, \quad (5)$$

$$F_2 = \gamma_2 \times \frac{\pi^2 E B t^3}{L^2} = 127\,\text{kN}. \quad (6)$$

5.1.2. Solution of the Undetermined Coefficients. To solve the undetermined coefficient x in (2), a series of models are investigated by adjusting the breadth of the bolted steel plates (B = 0.25, 0.275, 0.3, 0.325, and 0.35 m) and keeping the other parameters the same as those of E4NN (i.e., E = 210 GPa, B = 0.3 m, L = 0.3 m, and t = 4.3 mm). Substituting E, B, L, and t into (2) gives an inconsistent equation series as follows:

$$124 \times \left(\frac{0.25}{0.3}\right)^x = 141,$$

$$137 \times \left(\frac{0.275}{0.3}\right)^x = 146,$$

Table 6: Comparison of the field capacities derived from the theoretical and numerical studies.

Number	t (mm)	B (mm)	L (mm)	Longitudinal stiffener	Transverse stiffener	F_{num} (kN)	F_{the} (kN)	Errors
1	4.3	300	300	4 × 20	—	324.3	324	0.1%
2	4.3	300	300	4 × 20	4 × 20	338.3	350	3.3%
3	4.3	300	300	6 × 25	—	513.3	508	1.0%
4	4.3	300	300	6 × 25	6 × 25	530.2	530	0.0%
5	4.3	300	300	8 × 30	—	663.9	670	0.9%
6	4.3	300	300	8 × 30	8 × 30	683.5	682	0.2%
7	5.7	300	300	4 × 20	—	446.1	448	0.4%
8	5.7	300	300	4 × 20	4 × 20	460.3	462	0.4%
9	5.7	300	300	6 × 25	—	635.4	658	3.4%
10	5.7	300	300	6 × 25	6 × 25	652.6	679	3.9%
11	5.7	300	300	8 × 30	—	786.5	823	4.4%
12	5.7	300	300	8 × 30	8 × 30	806.4	840	4.0%
13	5.6	310	340	5.6 × 24	—	437.1	456	4.1%
								Average error: 2.0%

$$149 \times \left(\frac{0.3}{0.3}\right)^{x} = 149,$$
$$161 \times \left(\frac{0.325}{0.3}\right)^{x} = 156,$$
$$174 \times \left(\frac{0.35}{0.3}\right)^{x} = 162. \tag{7}$$

The equation series can be solved by a least square fitting (LSF); thus the undetermined coefficient x can be solved as $x = -0.60$.

Similarly, other series of models can also be investigated by adjusting the bolt spacing (L) and the thickness of steel plates (t) to solve the undetermined coefficients y and z, respectively, as $y = 1.02$ and $z = -0.91$. Therefore, (2) can be determined as follows:

$$F_0 = 0.2713 \times \frac{\pi^2 EBt^3}{L^2} \times \left(\frac{B}{0.3}\right)^{-0.60} \left(\frac{L}{0.3}\right)^{1.02} \left(\frac{t}{0.0043}\right)^{-0.91}. \tag{8}$$

In the same way, the undetermined parameters of (5) can also be solved as $j = 0.20$, $k = 0.49$, $l = -0.75$, $m = 2.70$, and $n = -2.94$, by sequentially changing the thickness and breadth of the longitudinal stiffeners (a and b), the breadth and the thickness of the bolted steel plates (B & t), and the bolt spacing (L). After that, the formula of F_2 can be expressed as follows:

$$F_2 = 0.2312 \times \frac{\pi^2 EBt^3}{L^2} \times \left(\frac{a}{0.004}\right)^{0.20} \left(\frac{b}{0.02}\right)^{0.49} \cdot \left(\frac{B}{0.3}\right)^{-0.75} \left(\frac{L}{0.3}\right)^{2.70} \left(\frac{t}{0.0043}\right)^{-2.94}. \tag{9}$$

In addition to the longitudinal stiffeners, the transverse stiffeners can also improve the structural field capacity to a certain extent. Considering the fact that, under normal conditions, the transverse stiffeners are always arranged together with the longitudinal ones with the same sectional size, so it is reasonable to consider the improvement of the transverse stiffeners by introducing an amplification coefficient α to the enhancement of the longitudinal stiffeners (F_2). And thereby the total field capacity (F) can be expressed more accurately in a modified form as

$$F = F_0 + F_1 + \alpha F_2. \tag{10}$$

By investigating a series of models in different combinations of stiffeners, the undetermined coefficient α can also be solved as $\alpha = 1.11$. And, for those without transverse stiffeners, $\alpha = 1.0$.

In order to test the degree fitting of (10), a series of models with different buckling restraint measures were built and their field capacities derived from the numerical FEA simulation (F_{num}) and those computed using the aforementioned formulae (F_{the}) are compared in Table 6. It is shown that the errors between the calculated and the numerical results agree with each other very well. The maximum error was 4.4% and the average error is 2.0%. So the fitted formulae have a high fitting degree thus being reasonable and authentic.

5.1.3. Verification of the Fitted Formulae by Experimental Results. To verify the applicability of the fitted formulae in evaluating the structural field capacity of the bolted steel plates with different buckling restraint measures, the experimental results (F_{exp}) of the previous experimental study [19] were also extracted and compared with the calculated results (F_{the}). The comparison in Table 7 indicates that the maximum error is 10.6% and the average is 5.8% So the accuracy and effectivity of the fitted formulae are validated.

5.2. Recommended Size of Stiffeners. It is evident from the aforementioned study that the arrangement of stiffeners, especially the longitudinal stiffeners, can substantially improve the structural field capacity and restrict the lateral

TABLE 7: Comparison of the field capacities derived from the theoretical and experimental studies.

Specimens	F_{the} (kN)	F_{exp} (kN)	Errors
E4NN	149.0	144	3.5%
E6NN	268.6	266	1.0%
E4SN	324.3	314	3.3%
E4MN	513.3	464	10.6%
E4LN	663.9	603	10.1%
E4SS	338.3	318	6.4%
			Average error: 5.8%

local buckling of the bolted steel plates. The efficiency of such improvement is highly related to the size of the stiffeners, so providing a recommended stiffener size to obtain a maximum enhancement with an economic consumption of stiffener steel is not only necessary, but also practical and useful.

In the Chinese steel structural specifications [26], when a transverse stiffener is employed in a double T-steel, its size is recommended as follows:

$$b \geq \frac{h}{30} + 40, \qquad a = \frac{b}{15}, \tag{11}$$

where h is the clear height of the web plate of the double T-steel and b and a, respectively, represent the width and thickness of the stiffener. Nevertheless, it only specifies the minimum geometric dimension of the stiffeners. So such recommended value is not convenient for utilization. And, furthermore, there is even no such recommended value for the stiffeners used for the bolted steel plates in steel jacketing.

In order to solve this problem, a parametric study of more models restricted by stiffeners in variable sizes was conducted. It is reasonable to use the value of F_2/F to evaluate the efficiency of stiffener utilization, in which F represents the total structural field capacity and F_2 represents the capacity enhancement beneficial from the buckling restraint of the stiffeners. Then, the relation between the value of F_2/F and the dimensional parameters of stiffeners can be studied to optimize the stiffener size.

Figure 13(a) shows the relationship between F_2/F and ab/Bt, which represents the area ratio. It can be found that the efficiency of stiffener utilization (F_2/F) is stable and relatively large when $ab/Bt = 0.1$. Figures 13(b) and 13(c) respectively, show the relationship between F_2/F and a/t or b/t, which represents the ratio of the thickness or the breadth of the stiffeners to the thickness of the bolted steel plates, respectively. Similarly, b/t and a/t can be, respectively, recommended as 1.2 and 6. And Figure 13(d) shows F_2/F-b/a curve, which indicates that b/a is suitable to be chosen as 5. By comparing the recommended values in this study (ab/Bt, a/t, b/t, and b/a) and that in the Chinses steel structural specifications, the former are more specific and can provide guideline to the strengthening design practice of the bolted steel plates in steel jacketing.

6. Conclusion

By a comprehensive parametric study on the local buckling behaviour of the bolted steel plates with different buckling restraint measures, the following conclusions can be drawn:

(1) The local buckling behaviour of steel plates can be accurately simulated by the NLFEA program ABAQUS, if the suggested simulation procedure is followed.

(2) The arrangement of less bolt spacing, longitudinal stiffeners, and thicker steel plates can substantially improve the structural field capacity of the bolted steel plates. The first two restraint measures can also considerably decrease the lateral buckling deformation.

(3) The arrangement of transverse stiffeners can limit the lateral buckling deformation of the central region of the bolted steel plates, while it is less obvious in improving the structural field capacity.

(4) The local buckling of the bolted steel plates is insensitive to initial imperfection. And the field capacity and the growing trend of the lateral buckling deformation of the bolted steel plates with different initial imperfection are generally the same.

(5) The improvement of the structural field capacity due to the arrangement of longitudinal stiffeners can be divided into two parts: the yield strength capacity of the stiffeners and the capacity enhancement of the bolted steel plates beneficial from the buckling restraint of the stiffeners. And the proposed fitted formulae can evaluate the structural field capacity with satisfactory accuracy.

(6) To get a high efficiency of stiffener utilization, the area ratio (ab/Bt), the ratio of the breadth of stiffeners to the plate thickness (b/t), and the breadth to thickness ratio of the stiffeners (b/a) can be recommended as 0.1, 6, and 5, respectively.

Notations

- B: The width of steel plate
- E: The elastic modulus of steel plate
- F_0: The capacity of the bolted steel plates
- F_1: The yield strength of the longitudinal stiffeners
- F_2: The improvement of yield strength due to the restraint of the longitudinal stiffeners
- F_{exp}: The experimental results
- F_{num}: The numerical results
- F_{the}: The calculated results
- L: The spacing of bolts
- a: The thickness of longitudinal stiffener
- b: The width of longitudinal stiffener
- f_y: The field capacity of stiffener
- h: The clear height of the web plate of a double T-steel
- t: The thickness of steel plate

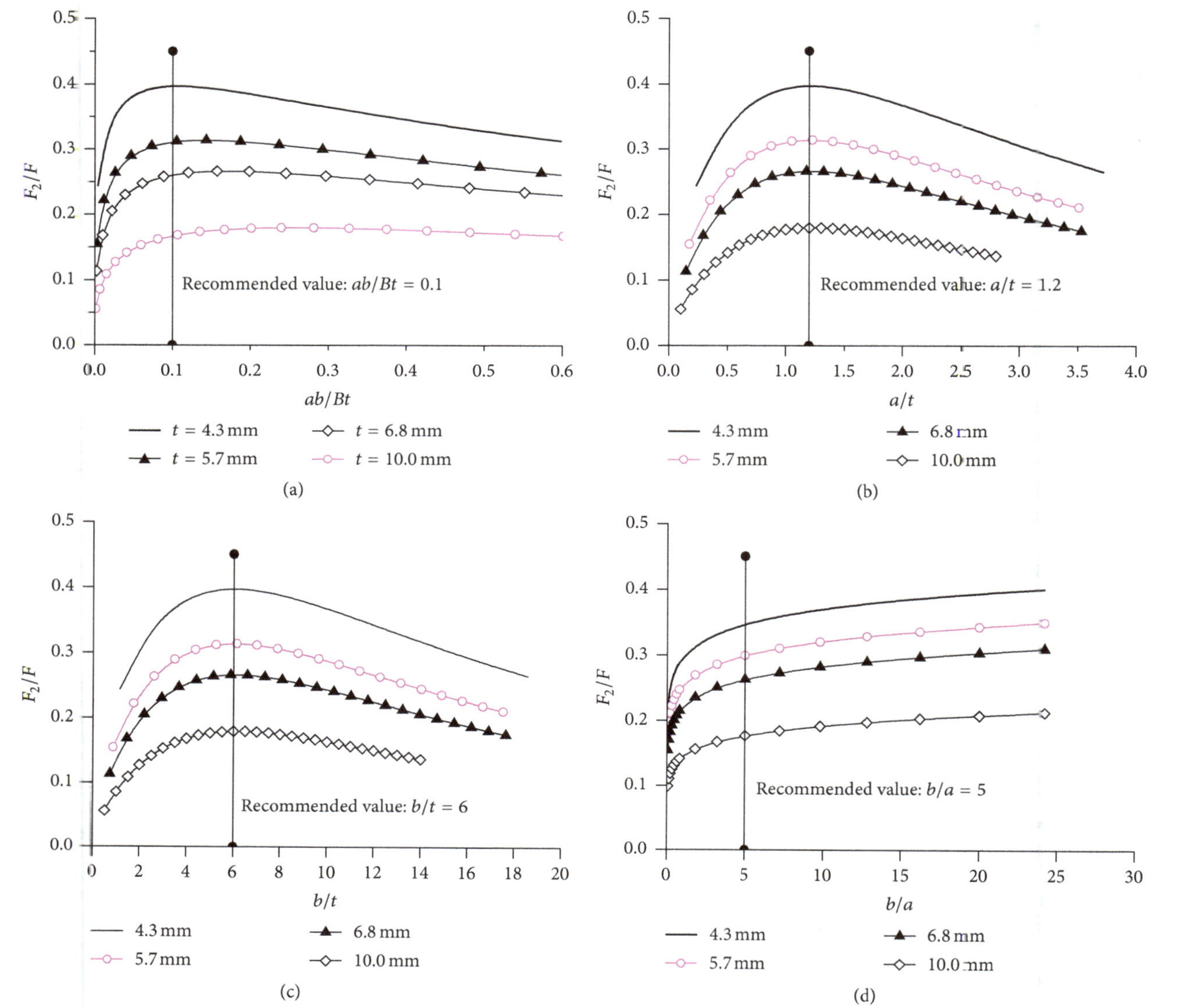

FIGURE 13: The relation between the value of F_2/F and the geometric dimensional parameters of stiffeners: (a) F_2/F-ab/Bt, (b) F_2/F-a/t, (c) F_2/F-b/t, and (d) F_2/F-b/a.

α: The factor for the improvement due to transverse stiffeners

γ_0, x, y, and z: Parameters in fitting F_0

γ_2, j, k, l, m, and n: Parameters in fitting F_2.

Competing Interests

The authors declare that there is no conflict of interests regarding the publication of this paper.

Acknowledgments

The research described here received financial support from the National Natural Science Foundation of China (Project no. 51408436).

References

[1] L. Wang, *Post-compressed plates for strengthening preloaded reinforced concrete columns [Ph.D. dissertation]*, The University of Hong Kong, Hong Kong, 2013.

[2] S. A. Hamoush and S. H. Ahmad, "Debonding of steel plate-strengthened concrete beams," *Journal of Structural Engineering*, vol. 116, no. 2, pp. 356–371, 1990.

[3] O. Buyukozturk, O. Gunes, and E. Karaca, "Progress on understanding debonding problems in reinforced concrete and steel members strengthened using FRP composites," *Construction and Building Materials*, vol. 18, no. 1, pp. 9–19, 2004.

[4] S. J. E. Dias and J. A. O. Barros, "Performance of reinforced concrete T beams strengthened in shear with NSM CFRP laminates," *Engineering Structures*, vol. 32, no. 2, pp. 373–384, 2010.

[5] U. Ebead and H. Saeed, "Hybrid shear strengthening system for reinforced concrete beams: an experimental study," *Engineering Structures*, vol. 49, pp. 421–433, 2013.

[6] T. M. Roberts and H. Haji-Kazemi, "Strengthening of under-reinforced concrete beams with mechanically attached steel plates," *International Journal of Cement Composites and Lightweight Concrete*, vol. 11, no. 1, pp. 21–27, 1989.

[7] R. K. L. Su and Y. Zhu, "Experimental and numerical studies of external steel plate strengthened reinforced concrete coupling beams," *Engineering Structures*, vol. 27, no. 10, pp. 1537–1550, 2005.

[8] A. Souici, J. F. Berthet, A. Li, and N. Rahal, "Behaviour of both mechanically connected and bonded steel—concrete composite beams," *Engineering Structures*, vol. 49, pp. 11–23, 2013.

[9] C. M. Foley and E. R. Buckhouse, "Method to increase capacity and stiffness of reinforced concrete beams," *Practice Periodical on Structural Design and Construction*, vol. 4, no. 1, pp. 36–42, 1999.

[10] R. K. L. Su, W. H. Siu, and S. T. Smith, "Effects of bolt-plate arrangements on steel plate strengthened reinforced concrete beams," *Engineering Structures*, vol. 32, no. 6, pp. 1769–1778, 2010.

[11] L. Z. Li, S. H. Lo, and R. K. L. Su, "Experimental study of moderately reinforced concrete beams strengthened with bolted-side steel plates," *Advances in Structural Engineering*, vol. 16, no. 3, pp. 499–516, 2013.

[12] D. J. Oehlers, N. T. Nguyen, M. Ahmed, and M. A. Bradford, "Transverse and longitudinal partial interaction in composite bolted side-plated reinforced-concrete beams," *Structural Engineering and Mechanics*, vol. 5, no. 5, pp. 553–563, 1997.

[13] K. Fukuyama, Y. Higashibata, and Y. Miyauchi, "Studies on repair and strengthening methods of damaged reinforced concrete columns," *Cement and Concrete Composites*, vol. 22, no. 1, pp. 81–88, 2000.

[14] Y. Xiao and H. Wu, "Retrofit of reinforced concrete columns using partially stiffened steel jackets," *Journal of Structural Engineering*, vol. 129, no. 6, pp. 725–732, 2003.

[15] J. Garzón-Roca, J. M. Adam, and P. A. Calderón, "Behaviour of RC columns strengthened by steel caging under combined bending and axial loads," *Construction & Building Materials*, vol. 25, no. 5, pp. 2402–2412, 2011.

[16] Y. H. Chai, M. J. N. Priestley, and F. Seible, "Analytical model for steel-jacketed RC circular bridge columns," *Journal of Structural Engineering*, vol. 120, no. 8, pp. 2358–2376, 1994.

[17] S. T. Smith, M. A. Bradford, and D. J. Oehler, "Buckling tests on steel plates restrained at discrete points in the retrofit of reinforced concrete beams," *Proceedings of the Institution of Civil Engineers: Structures and Buildings*, vol. 146, no. 2, pp. 115–127, 2001.

[18] B. Uy and M. A. Bradford, "Local buckling of thin steel plates in composite construction: experimental and theoretical study," *Proceedings of the ICE-Structures and Buildings*, vol. 110, no. 4, pp. 426–440, 1995.

[19] L. Z. Li, C. J. Jiang, L. J. Jia, and Z. D. Lu, "Local buckling of bolted steel plates with different stiffener configuration," *Engineering Structures*, vol. 119, no. 7, pp. 186–197, 2016.

[20] A. S. Lubell, H. G. L. Prion, C. E. Ventura, and M. Rezai, "Unstiffened steel plate shear wall performance under cyclic loading," *Journal of structural engineering New York, N.Y.*, vol. 126, no. 4, pp. 453–460, 2000.

[21] H. D. Wright, "Buckling of plates in contact with a rigid medium," *Structural Engineer*, vol. 71, no. 12, pp. 209–215, 1993.

[22] H. B. Ge and T. Usami, "Strength analysis of concrete-filled thin-walled steel box columns," *Journal of Constructional Steel Research*, vol. 30, no. 3, pp. 259–281, 1994.

[23] S. T. Smith, M. A. Bradford, and D. J. Oehlers, "Local buckling of side-plated reinforced-concrete beams. I: Theoretical Study," *Journal of Structural Engineering*, vol. 125, no. 6, article 622, 1999.

[24] S. T. Smith, M. A. Bradford, and D. J. Oehlers, "Unilateral buckling of elastically restrained rectangular mild steel plates," *Computational Mechanics*, vol. 26, no. 4, pp. 317–324, 2000.

[25] S. T. Smith, M. A. Bradford, and D. J. Oehlers, "Local buckling of side-plated reinforced-concrete beams. II: experimental study," *Journal of Structural Engineering*, vol. 125, no. 6, pp. 635–642, 1999.

[26] GB50017-2003, *Code for Design of Steel Structures*, China Architecture & Building Press, Beijing, China, 2003.

14

The Adhesion of CrN Thin Films Deposited on Modified 42CrMo4 Steel

O. Lupicka and B. Warcholinski

Faculty of Technology and Education, Koszalin University of Technology, Sniadeckich 2, 75-453 Koszalin, Poland

Correspondence should be addressed to B. Warcholinski; bogdan.warcholinski@tu.koszalin.pl

Academic Editor: Hiroshi Noguchi

Here, the effect of adhesion of CrN hard coatings on modified 42CrMo4 steel substrate is presented. Modifications of the substrate are shot peening, nitriding, shot peening, and nitriding joined process. In the shot peening process, two variable process parameters were used: exposure time (t) and air pressure (p). The nitriding process was conducted in the following parameters: nitriding potential $N_p = 4.86$, nitriding time $t_n = 3$ h, and temperature process $T = 530°C$. Modified substrates were characterized by surface hardness HV5 and hardness profiles on the cross section of samples and by surface roughness parameters. On such prepared substrates, the CrN coating was deposited. The adhesion of CrN coating on modified substrates was defined by the scratch test. Chemical and phase composition of the films was determined using EDS method and X-ray diffraction, respectively. The surface hardness of deposited films was also defined. The substrate of 42CrMo4 steel without mechanical and heat treatment coated by hard CrN film was used as a reference.

1. Introduction

Adhesion and hardness of thin coatings are the most important quantities that determine the various applications. Qualitative assessment of both of these parameters allows understanding the properties of the coatings, as well as their contribution to the development of others. Among numerous techniques used to evaluate the adhesion of the coating to the substrate, the scratch method is the most commonly used which allows to get quantitative image of adhesion. Critical load is defined as the force at which the coating is removed from the substrate, and it depends on many factors, including the hardness of the coating and the substrate, film thickness, roughness of the coating and the substrate, and the type of the substrate material [1, 2]. On the one hand, it depends on the deposition technology of thin coatings and, on the other hand, on scratch test parameters (diamond indenter radius, the intenders travelling speed and load increase rate). Some of these factors also affect the results of another test—qualitative assessment of the durability of the coating, the Daimler–Benz test. Both of these methods are complementary and provide a deeper assessment of the phenomena affecting the adhesion of the coating.

Among the plastic treatments modifying the surface layer, the shot peening plays an important role. During the process of shot peening on the treated surface dynamically, peening elements work in the form of balls (steel, glass, and ceramic). This method is used only as a strain hardening treatment. It is irreplaceable for machining limp and/or thin-walled objects, where the use of large static forces is impossible due to the deformation of the shape of the workpiece. Furthermore, after machining, the layer shows an increase in defects in the structure—dislocations, texture, shape, and size of the grains [3]. Shot peening is a method to improve the resistance of metal pieces to fatigue by creating regions of residual stress.

The results of thermochemical treatments are the layers formed by diffusion introducing a foreign element, which forms compounds (nitrides, borides, etc.) with very varied properties. Gas nitriding, one of the most common surface treatments used, consists in introduction of a nitrogen atom into the surface layer of the steel object usually thermally enhanced. The source of atomic nitrogen is ammonia, which dissociates at a temperature of nitriding on atomic hydrogen and nitrogen. The most important feature of the nitrided layers, especially on alloy steels, is their high hardness. The maximum hardness is achieved at low-temperature nitriding (500–520°C), since at these temperatures, finely nitrides of alloying elements are formed.

The nitriding at elevated temperatures, in which nitrides of larger dimensions and lower coherence with the substrate are formed, leads to a reduction in hardness. Another advantage of nitriding is the possibility of protecting the surface against corrosion; to this purpose, so-called anticorrosive nitriding is used. In order to meet the above requirements, it is necessary to select the parameters of the nitriding process which enable the formation of a nonporous surface carbonitrides zone with a thickness of several micrometers protecting steel against corrosion. Nitrided layer also has a sufficiently high resistance to fatigue [3–5], abrasion [3], and cavitation erosion.

Each of the analyzed surface treatments is widely used in various industries (automotive, machinery, and aerospace). For several years, attempts have been made to their sequential use, in particular to the connection of nitriding and PVD processes or shot peening and nitriding. Nitriding of steel and PVD coating deposition have been the subject of many publications [6–9] as shot peening and PVD coating deposition [10, 11]. To ensure good adhesion of coatings to the substrate nitrided, the formation of so-called "white layer" of iron nitrides ε-$Fe_{2\text{-}3}N$ i γ'-Fe_4N [12] on the surface should be reduced or completely eliminated. The improvement of the performance of tools and machine parts was obtained after the deposition of hard thin coatings on them, which significantly alter their properties. Chekour et al. [6] emphasized that the nitriding has a significant effect on the increase in durability. Tools with only CrN coating show 1.3 times higher durability compared with the uncoated tools. Nitrided substrates with CrN coating increase the tool's durability fourfold.

Chromium nitride has a relatively low coefficient of friction, good wear resistance, and high corrosion resistance. Due to the excellent resistance to abrasion, chromium nitride coating has been used as the coating on tools for cutting, milling, and threading of components made of titanium and its alloys, brass, copper, and other nonferrous metals, covering molds, stamps, and machine parts. Chromium nitride is characterized by high chemical resistance and has a very low affinity for nonferrous workpieces. These coatings can be prepared by various PVD including arc and magnetron techniques [13].

One of the critical parameters in the selection of the type of coating on the tool is the adhesion of the coating to the substrate. The hardness of the coating affecting the wear resistance, fracture toughness [7], structure, or the coefficient of friction will have little significance when the adhesion of the coating to the substrate is insufficient. Low adhesion results in an almost immediate flaking off of the coating on contact of the tool with the workpiece. It is the adhesion that characterizes the ability of the coating-substrate system to load transfer. The energy of chemical bonds, the processes at the substrate-coating interface such as mutual diffusion of atoms, and the substrate surface development are responsible for the adhesion. Reducing the surface roughness promotes the growth of adhesion [14].

Liu et al. [10] found that the hard coating or surface hardening increases wear resistance. It should be remembered that the brittle surface layer accelerates the initiation of cracks and reduces the fatigue life. In this case, it is helpful; for example, shot peening treatment increases considerably the material fatigue resistance.

One of the most important trends in surface engineering includes activities aimed at improving the mechanical properties of the coating-substrate systems. Additive synergistic effect can be achieved by combining existing surface treatments, such as mechanical with thermochemical or searching for new solutions. In particular, the objective of this study was to find an answer to the question of whether favorable changes in the properties of the substrate material resulting from the combined process of burnishing (shot peening) and gas nitriding are achieved.

In the present work, the adhesion of CrN thin coatings deposited on the substrate, previously modified by cold working (shot peening) or thermochemical treatment (gas nitriding) and by the combined process of shot peening and nitriding, was tested. As a reference, the ground substrate was used.

2. Experimental

The tests were conducted on samples made of 42CrMo4 structural steel with the following chemical composition (wt.%): C (0.22), Si (0.35), Mn (0.70), P (0.01), S (0.001), Cr (1.12), Mo (0.26), and Fe balanced. All samples were hardened and tempered. The hardening was carried out for 8 min at a temperature of 845°C and tempering at a temperature of 620°C for 120 min. After that, the hardness of the samples increased from 93 HRB (19 HRC) to 30 ± 1 HRC.

2.1. Surface Modification. Quenched and tempered samples were subjected to shot peening, nitriding, and combined shot peening and nitriding process.

In the shot peening process, the shot marked as SW170 of a diameter of 0.43 mm and a hardness of 470 HV was used. The head of a diameter of 5.3 mm, positioned 200 mm from the shot-peened substrate, was also used. Shot peening was performed at 120 s, 360 s, or 600 s, under the air pressure of 0.45 MPa.

The nitriding was conducted in a laboratory furnace with a vertical quartz retort of a diameter of 50 mm using the following process parameters: temperature $T = 530$°C, nitriding time $t_n = 3$ h, nitrogen potential $N_p = 4.86$, and nitriding atmosphere 70% NH_3 + 30% H_2. After the nitriding process, the samples were cooled down in the furnace.

CrN coatings were deposited using cathodic arc evaporation on the 42CrMo4 steel substrate. The substrate surface was previously modified by shot peening or nitriding or by the combined process of shot peening and nitriding.

Before PVD process, the substrates were cleaned in an ultrasonic alkaline bath and then placed on the turntable in the working chamber at a distance of 18 cm from the arc source. The vacuum chamber was evacuated to the base pressure of 1×10^{-3} Pa. The substrates were heated up to a temperature of 300°C. Before the deposition of the coating, the substrates were cleaned by ion etching of metal ions at a voltage of 600 V for 10 minutes, and the argon pressure in the chamber was maintained at 0.5 Pa to remove weakly bonded particles on the surface. The improvement in the coating adhesion to the substrate was obtained by deposition of the thin chromium layer with a thickness of about 0.1 μm directly on the substrate. The coating deposition was carried out with a substrate bias voltage bias of −150 V, arc current of 80 A, and the nitrogen pressure of 1.8 Pa. Rotation speed of the table was 2 min^{-1}.

2.2. Sample Characterization. The chemical composition of the CrN coating was determined using EDX (Oxford Link ISIS 300). The X-ray diffraction (XRD) with a glancing angle ($\omega = 3°$) and Bragg–Brentano geometry was recorded on an X'Pert PANalytical device using Cu-Kα radiation in the range of 2 theta angles between 30 and 120°. The grain size was calculated using Scherrer's formula [15]. Due to instrumental peak broadening, 0.2° for silicon standard, the Warren–Biscoe correction method was used [15].

The phase composition of the nitrided layer was determined by X-ray phase analysis. One aim of this analysis was to determine the volume fraction of the μ and γ' phases in the layer. For this purpose [16], the adjusted intensities of the main diffraction lines of the phases and iron ({101}-ε, {200}-γ', and {110}-α-Fe) were used. Layer thickness was calculated from the formula resulting from the absorption of Beer's law:

$$g = \frac{\ln(I_0/I_\alpha)\sin\theta}{2\mu}, \tag{1}$$

where I_0 is the diffraction line for initial state sample, I_α the diffraction line for nitrided sample, μ the linear absorption coefficient for the nitride layer, and θ the Bragg angle.

The studies of the sample structure on the cross sections were conducted on a Neophot 2 optical microscope, and the morphology studies on a Jeol JSM 5500LV scanning electron microscope were conducted. For the surface quality evaluation, the Nikon Eclipse Imaging MK200 optical microscope, equipped with Imaging Software NIS-Elements software, was applied. The data were automatically collected for quantitative and dimensional analyses of the macroparticles. All images were taken at the same magnification (400x), the same contrast, sharpness, and threshold of sensitivity. The measurements were taken for each sample at five points, located on a line at an equal distance from each other.

Hardness measurements on cross sections of the samples were determined by the Vickers method under the load of 0.1 kgf (0.98 N). For measuring the surface hardness, the loads of 0.5 kgf (4.903 N) and 5 kgf (49.03 N) were applied. The measurement was performed 15 times for each load, at different locations on the surface of each sample. The hardness of CrN coatings was determined using a FISCHERSCOPE® HM2000 nanoindenter with the indenter penetration depth of about 0.2 μm. The results were averaged based on 10 measurements.

The coating stresses were determined using Stoney's formula [17]:

$$\sigma = \frac{E}{1-\nu} \cdot \frac{t_s^2}{6t_f}\left(\frac{1}{R} - \frac{1}{R_s}\right), \tag{2}$$

where E is the Young modulus of the substrate, ν the Poisson coefficient of the substrate, t_f the thickness of a layer, R the radius of the substrate curvature together with the coating, and R_s the radius of the substrate curvature. The curvature radius was calculated from the data obtained from the surface roughness analyzer (Hommel Werke Profilometer T8000). For this test, a silicon substrate of 0.3 mm thick, 30 mm long, and 4 mm width covered by CrN coatings was applied.

Surface roughness measurements were performed using a Hommel Tester T8000 profilometer.

The adhesion of the coatings was determined using the scratch test (CSEM Revetest® Scratch-Tester) with a Rockwell type C diamond indenter. It was moved at a speed of 10 mm/min by linearly changing the load from 0 to 100 N at 100 N/min. Critical load Lc_1 was defined as the force at which the first damage of the coating was observed, and Lc_2 was determined as the load at which the total detachment of the coating from the substrate was observed. This load was determined by observing using an optical microscope with an average of at least three measurements. To check the adhesion of the coatings, the Daimler–Benz test [4] was also used. It depends on the assessment of the coating damage caused by the Rockwell intender pressed with constant load (ca. 1470 N = 150 kgf) in the substrate-coating system. Depending on the type and quantity of the coating damage, it is assigned to one HF1 to HF6 groups, where HF1 to HF4 show good adhesion to the substrate and the marks HF5 to HF6 show poor adhesion.

3. The Test Results

3.1. Properties of the Surface Layer after the Shot Peening Process. As a result of the shot peening process, surface layers were obtained, whose properties were characterized, among others, through the study of surface roughness, surface hardness, and the hardness distribution on the cross sections of the samples.

In Figures 1 and 2, obtained by SEM, the characteristic change in the geometry of the surface after shot peening process can be seen. From the analysis presented in pictures, it ensures that the increase in the value of the exposure time t increases the number of traces in the form of pits on the surface of the workpiece (Figure 1). It is shown explicitly in Figure 2, which presents the results of the roughness measurements.

The value of surface roughness parameter, R_a, of the samples before the shot peening had been 0.29 μm, and for the samples after shot peening, the value was in the range from 5.24 to 5.91 μm. The presented data show that the

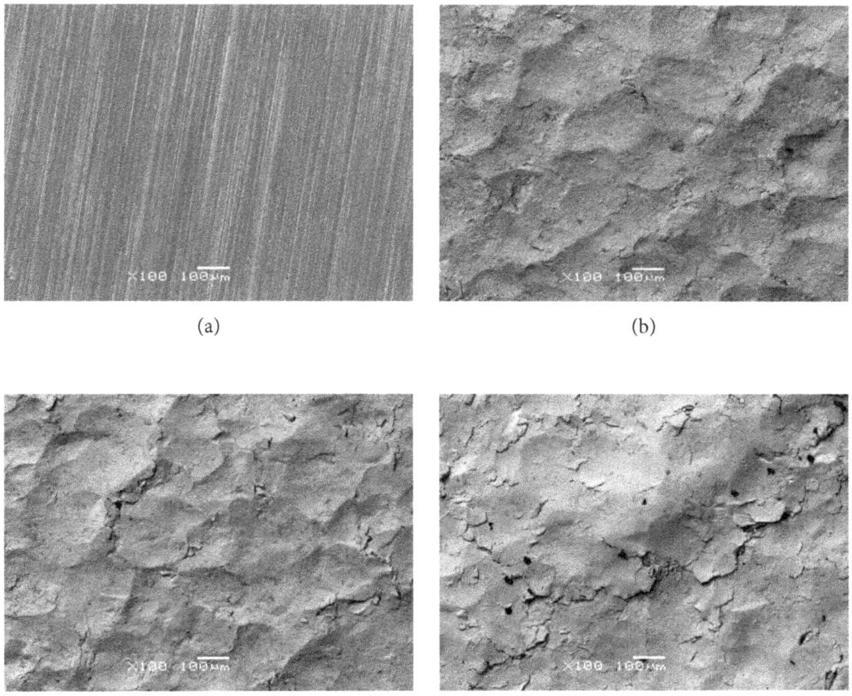

FIGURE 1: Metallographic photos taken on SEM: (a) initial substrates; (b) shot-peened substrate ($t = 120$ s, $p = 0.45$ MPa); (c) shot-peened substrate ($t = 360$ s, $p = 0.45$ MPa); (d) shot-peened substrate ($t = 600$ s, $p = 0.45$ MPa).

surface roughness parameter, R_a, increased by about 20 times after the shot peening process, as compared to the surface roughness in the initial state. The lowest value of the parameter ($R_a = 5.24\,\mu$m) was obtained in the case of shot peening surface with the shortest exposure time of 120 s, while the highest $R_a = 5.91\,\mu$m was obtained for the longest exposure time 600 s. It follows that, with increasing exposure time, the surface roughness of shot-peened steel increases. In the same shot peening process, parameter ratio of roughness R_t and R_z is 17 times. Because of the roughness of the surface, hindering the performance of subsequent testing adhesion of CrN, all samples were polished to R_a about $0.029\,\mu$m.

The degree of deformation of the layer obtained by shot peening is different from that of the material in the initial state. The analysis of metallographic pictures shows that, as in the case of changes in surface roughness (Figure 2), there is an increase of crumple zones (Figure 3) with the increase of exposure time.

As a result of shot peening, the structure was obtained (Figure 3) with visible irregular deformations of layers, which has resulted from the repeated hitting of balls on the surface of the material being processed. The peening process also caused essential, from the strength properties point of view, microstructural changes in the surface layer (Figure 3). As a result of the applied heat treatment of steel (Section 2.1), a sorbitic structure was obtained. For burnished steels with sorbitic structure the formation of fine-grained amorphous surface layer containing separated high-dispersive carbides is characteristic. This layer is difficult to etch. Both the thickness of this layer and the amount of the carbide present

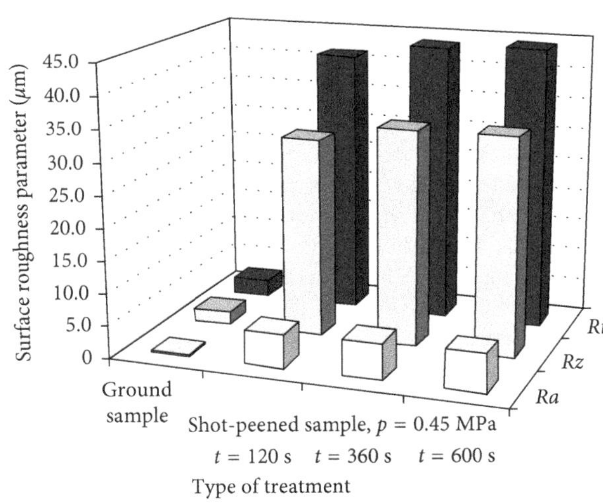

FIGURE 2: The parameters of surface roughness after grinding process (initial state) and after shot peening process.

in it are dependent on the degree of deformation and increase with the increase of degree of deformation [18].

In surface layer on account of its characteristic structure zone, smaller or larger differences (gradients) of virtually all elastic and plastic properties occur, represented by the increase of hardness. The surface hardness was measured on all samples subjected to shot peening process with the parameters set in Section 2.1. Comparing the results of changes of surface hardness (Figure 4) and structural changes (Figure 3), it can be concluded that the shot peening process causing larger changes

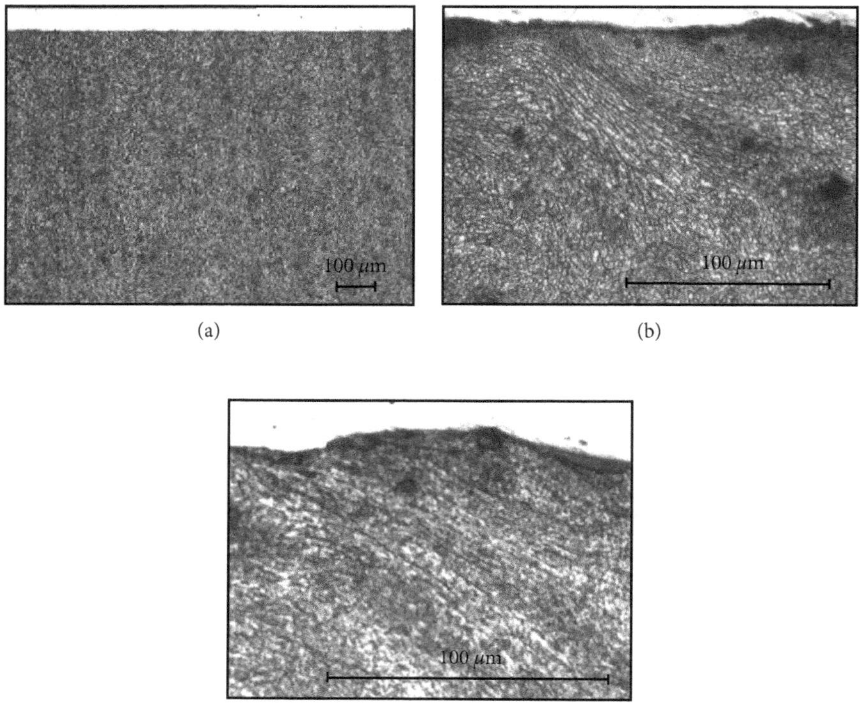

Figure 3: Photos taken on a cross section of an optical microscope: (a) sample ground, larger 50.4x; (b) shot-peened sample (the process parameters: $t = 120$ s, $p = 0.45$ MPa); (c) shot-peened sample (the process parameters: $t = 600$ s, $p = 0.45$ MPa).

in the structure of the surface layer results in simultaneous larger increase of surface hardness (Figure 4). Moreover, the analysis of the obtained results indicated that the process parameters for assumed maximum increment of surface hardness amounted to 216 HV5, representing an increase of 72% relative to the hardness of the samples in the initial state. The surface hardness of the shot-peened sample in time 120 s amounted to 427 HV5 (relative to hardness of the sample in the initial state 299 HV5), extra time for shot peening causes an increase of hardness to 515 HV5.

The essence of the shot peening is a direct impact of peening elements on the treated surface. The surface layer is therefore characterized by high hardness, diminishing with the distance from the surface (Figure 5). To determine the hardness of the shot-peened zone, additionally, the measurements of hardness at a load of 0.5 kg (4.9 N) were taken (Figure 4). In this case, the analogous character of changes in hardness was obtained, compared with the larger load (5 kg). The determined values are higher by about 150 HV0.5. It should be noted that the depth of penetration of the indenter into the surface hardened zone was about 8 μm. The higher load, where depth of penetration was around 20 μm, includes a zone of varying hardness values, which results in a lower hardness value.

On the basis of measurements on the hardness of the cross sections of samples (HV0.1) before and after shot peening, hardness profiles were also determined. Figure 5 shows a hardness profile characteristic for this type of treatment, on which relatively small changes in hardness were observed. Hardness distribution curve indicates the area of enhanced zone extending about 180 μm from the surface. Shorter times of shot peening cause the reduction of the hardness but the hardness profiles are similar. Simultaneously with time reduction of shot peening, the depth of the enhanced zone is smaller. The research [18] concludes that sorbitic steels have the least capacity to strengthening by deformation of superficial (15% strengthening). This is related, among others, with lower susceptibility of fine-grained steel structures (sorbitic steel) to cold plastic deformation. Despite the small ability to strain hardening, sorbitic steels are treated by burnishing.

The surface layer formed in the shot peening process is characterized by a high density of structural defects, such as dislocations. In the surface layer, fragmentation into smaller grains also occurs. From the viewpoint of forming a nitrided layer, defecting the structure, which makes easy diffusion paths, affects the growth of kinetics. The defected surface in shot peening process determines the phase of nucleation of iron (carbo)nitrides zone. Plastic deformation and gradients forming stress have also influenced the process of diffusion of nitrogen atoms to the center of the material and carbon toward the surface.

3.2. The Properties of the Surface Layer after Shot Peening and Nitriding Process. In the next stage, shot-peened samples were treated by nitriding in time $t_n = 3$ h (2.2). The properties of the layers were determined, as in the previous case, by measuring the surface hardness and the hardness distribution on the cross sections of the samples. In addition,

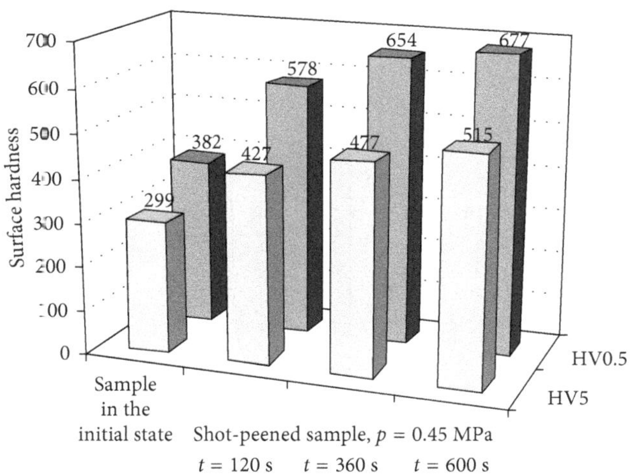

FIGURE 4: The surface hardness of the sample in the initial state and after the shot peening process.

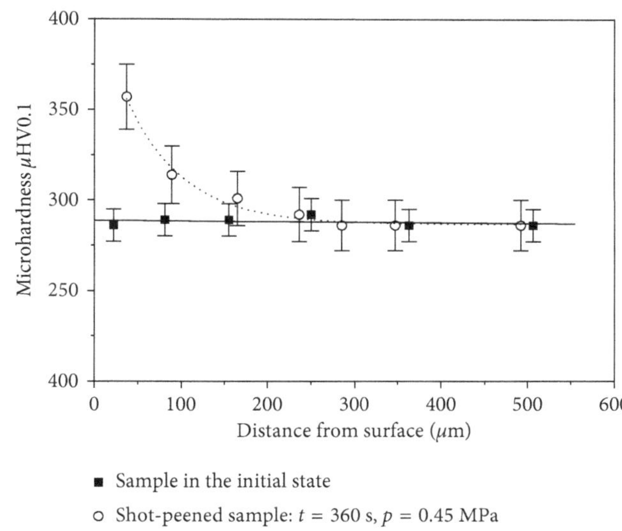

FIGURE 5: Distribution of microhardness profiles μHV0.1 in samples before and after shot peening.

structural studies and phase changes of the nitrided layer by optical microscopy and X-ray diffraction were conducted.

The nitrided layer produced in the process consisted of the external zone of the phase composition $\varepsilon(Fe_{2-3}N)$ and $\gamma'(Fe_4N)$ and the zone of internal nitriding, so-called a diffusion zone. From the analysis of X-ray diffractions presented in Figure 6(a), it comes out that the use of shot peening process as an additional treatment prior to nitriding causes changes in the volume fraction of phases ε and γ' in iron (carbo)nitrides zone. For shot peened samples, compared to surface unmodified by plastic treatment samples, one can observe increased ε phase content in iron (carbo) nitrides zone. The ε phase content is 67 and 54%, respectively (Figure 6(b)). Consequently, the thickness of iron (carbo) nitride zone grows faster on samples in which the surface layer was modified by shot peening process (Figure 6(c)).

Samples subjected to nitriding process retained variation of hardness caused by various times of shot peening process. These results are illustrated in the bar graph shown in Figure 7. As shown, in the case of samples subjected only to nitriding process, there was an increase in hardness of 402 HV5 (299 HV5 initial state), an increase of 134%. After the shot peening process and then nitriding, the variation in hardness values was, as already mentioned, 216 HV5 after shot peening, and increased to the amount of 248 HV5. Maximal increase of 471 HV5 in hardness was recorded for the sample shot peened at time $t = 360$ s and then nitrided, compared with the sample that was only shot peened in which the hardness amounted to 477 HV5 (Figure 4).

Hardness profiles (Figure 8) after the combined process are characterized by a shape similar to that obtained after the shot peening process. If a process is combined, the area of enhanced zone reaches about 250 μm from the surface. An increase in hardness in the surface zone of approximately 85 HV0.1 on samples of the combined shot peening and nitriding process was observed. Moreover, hardness profiles are characterized by greater gradient of hardness after combined process compared to the hardness profiles designated after the nitriding process (Figure 8).

3.3. Adhesion of the Coatings on Modified Substrates

3.3.1. General Remarks. CrN coatings deposited on 42CrMo4 steel substrates: pure and after thermochemical treatment and/or shot peening are characterized by a uniform thickness throughout the area of its occurrence, good adhesion to the substrate material, and a compact structure without visible delamination. There are no areas of weak adhesion of the coating to the substrate. Basic properties of the coatings summarized in Table 1 are comparable to data presented by other authors [19].

Mechanical properties (H, E) of CrN coatings (Table 1) are similar to those presented in [20, 21]. Compressive stress amounted to 1.4 GPa is relatively small compared to 8 GPa [21] or 5 GPa [22]. Similar value of stress −1.2 GPa is shown in [23].

For single-phase monolayer coatings, macrostress is connected with microstructural damage caused by high-energy ion bombardment during film growth (internal stresses) and mismatch of the thermal expansion coefficients of the coating and the substrate (thermal stress) occurring during cooling from a deposition to room temperature [24]. The value of thermal stress σ_{th} can be calculated using the formula [24]

$$\sigma_{th} = \frac{E_{CrN}}{1 - \nu_{CrN}} \times (\alpha_{CrN} - \alpha_{sub}) \times (T_{dep} - T_0), \quad (3)$$

where E_{CrN} is the Young modulus of CrN coating, ν_{CrN} the Poisson ratio of CrN coating, α_{CrN} the thermal expansion coefficient of CrN, α_{sub} the thermal expansion coefficient of substrate, T_{dep} the deposition temperature, T_0 the room temperature.

The thermal stress in CrN coating-Si substrate system amounts to about −0.1 GPa and for CrN coating-steel substrate system about −0.7 GPa. It indicates that the internal compressive stresses for CrN-steel substrate system are higher of about 0.6 GPa and reaches about 2 GPa.

On the surface of the coating, macroparticles and craters can be observed (Figure 9). The macroparticles are

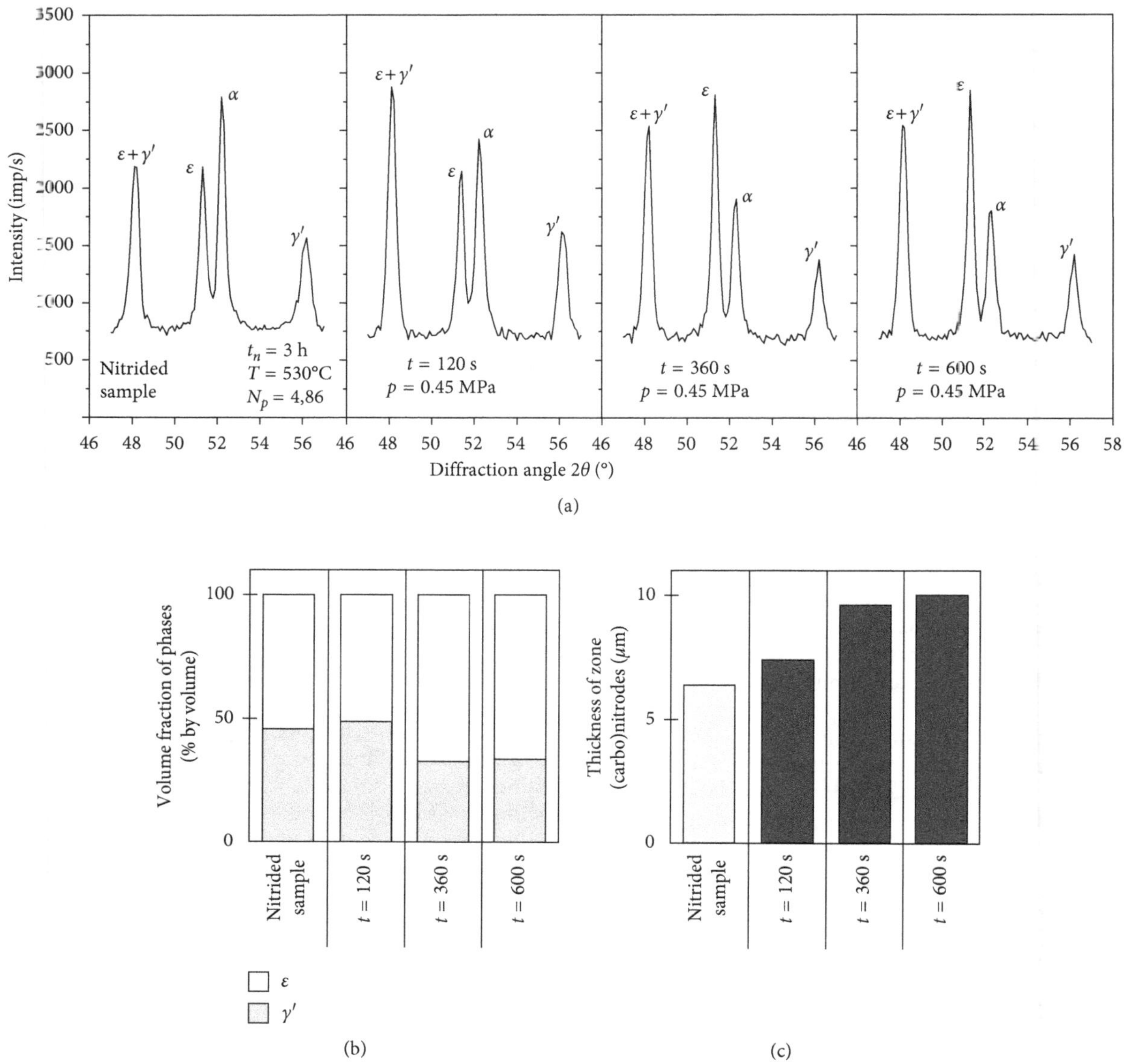

Figure 6: The phase composition of (carbo)nitride zone after nitriding process and after combined shot peening and nitriding process presented by means of X-ray diffraction pattern (a), distribution of volume contents of ε and γ′ phases in a compound layer (b), and the (carbo)nitride zone thickness (c).

characterized by different sizes, from fractions to some micrometers (Figure 10).

It was found that the macroparticle density decreases with the increase in size above a maximum of around 0.25–0.5 μm. Only a small number of particles are found below 0.25 mm. The particles with the size of 1 μm constitute about 72% of all particles. The macroparticles are the typical effect of the deposition technique applied, that is, cathodic arc evaporation, and their quality depends on the deposition parameters [25, 26].

The composition of the coatings was determined by means of the EDX method. These measurements confirm the presence in the coating, obtained at a nitrogen pressure of 1.8 Pa, of chromium and nitrogen in an amount of about 48 at. % and 51 at. %, respectively (Figure 11). This proportion is typical for cubic CrN phase [27]. Creation of chromium nitride was then confirmed by X-ray studies. It must also be noted that, in all samples, about 1 at. % of oxygen was recorded.

3.3.2. X-Ray Phase Structure Analysis. Figure 12 shows the diffraction pattern of the obtained CrN coating. All identified diffraction lines, that is, (111), (200), (220), (311), (222), etc., are characteristic in cubic phase of chromium nitride. Diffraction lines from the substrate are not visible due to the identification method—grazing incidence X-ray diffraction.

The observed diffraction lines are shifted about 0.3–0.5° toward the lower angles with respect to the ICDD data. This is related to the compressive stress in the layer caused by ion

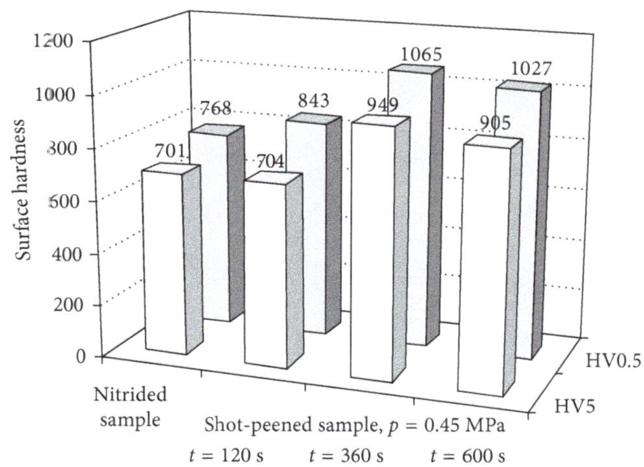

FIGURE 7: The surface hardness after nitriding process (nitrided sample) and after combined process of shot peening and nitriding.

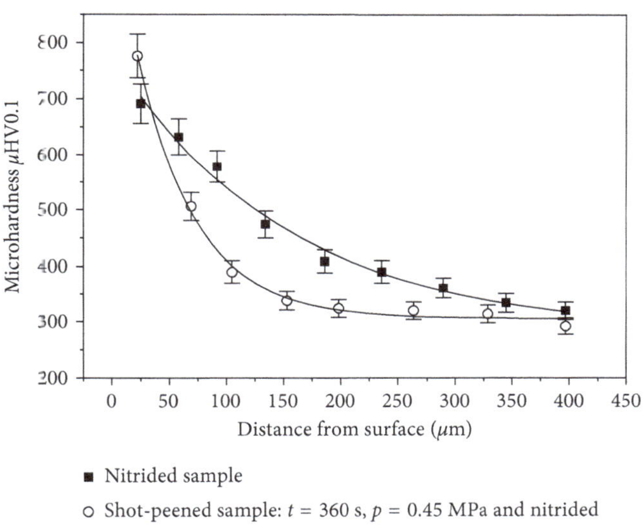

FIGURE 8: Microhardness (μHV0.1) profiles for the samples after nitriding and after combined shot peening and nitriding process.

bombardment during the deposition process. A similar diffraction pattern is shown in [28–31]. The size of the crystallites, calculated using Scherrer's formula, in the coating is about 10 nm.

The diffraction pattern (Figure 12) indicates the difference in the preferred orientation compared to the standard (ICDD 11-0065). It can be noted that the highest intensity is for (111) and then for (200) diffraction line. The intensities of the remaining diffraction lines are close to one another, but more than 10 times less than the intensity of the (111) line. It should be noted that according to the ICDD, the intensity of the diffraction lines is as follows: (111)-80%, (200)-100%, (220)-80%, (311)-60%, (222)-30%, (400)-30%, and (420)-50%. It means that the coating is textured. The values of the texture coefficient, Tc, for the planes are proportional to the number of grains oriented with the corresponding crystallographic plane parallel to the sample surface. The texture coefficient Tc(hkl) indicates the relative degree of preferred

TABLE 1: The properties of CrN coatings on 42CrMo4 steel.

Thickness (μm)	1.8 ± 0.2
Hardness, H (GPa)	23.6 ± 0.8
Young's modulus, E (GPa)	285 ± 8
H/E index	0.082 ± 0.005
Stress, σ (GPa)	-1.4 ± 0.2

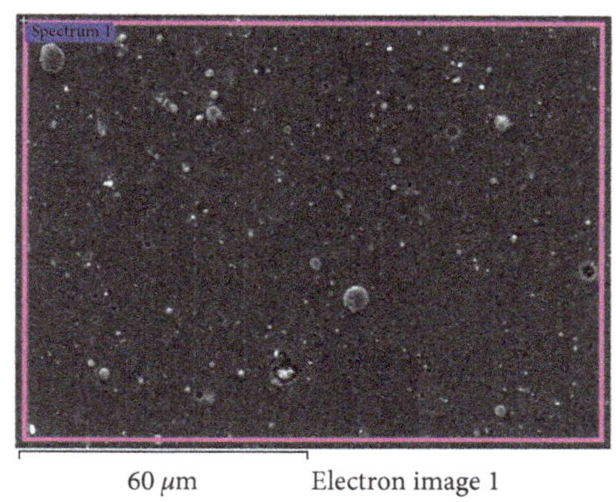

FIGURE 9: SEM image of CrN coating surface in EDS analysis.

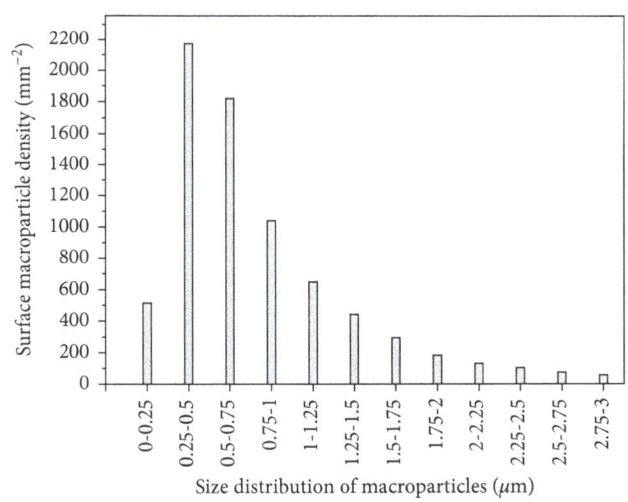

FIGURE 10: Size distribution and macroparticle density of the CrN coating on 42CrMo4 steel substrate.

orientation among crystal planes. It was calculated using the following equation [32]

$$\mathrm{Tc}(\mathrm{hkl}) = \frac{I(\mathrm{hkl})/I_0(\mathrm{hkl})}{1/n \cdot \sum_1^n I(\mathrm{hkl})/I_0(\mathrm{hkl})}, \quad (4)$$

where I(hkl) is the diffraction line intensity from the (hkl) plane for the textured sample, I_0(hkl) the respective intensity corresponding to the bulk CrN data from JCPDS File number 11-0065, and n the number of diffraction lines analyzed.

The Tc(hkl) data are gathered in Table 2. The value of the Tc is proportional to the number of preferentially oriented (hkl) planes. Tc(hkl) close to unity indicates a randomly distributed powder sample, and Tc(hkl) > 1

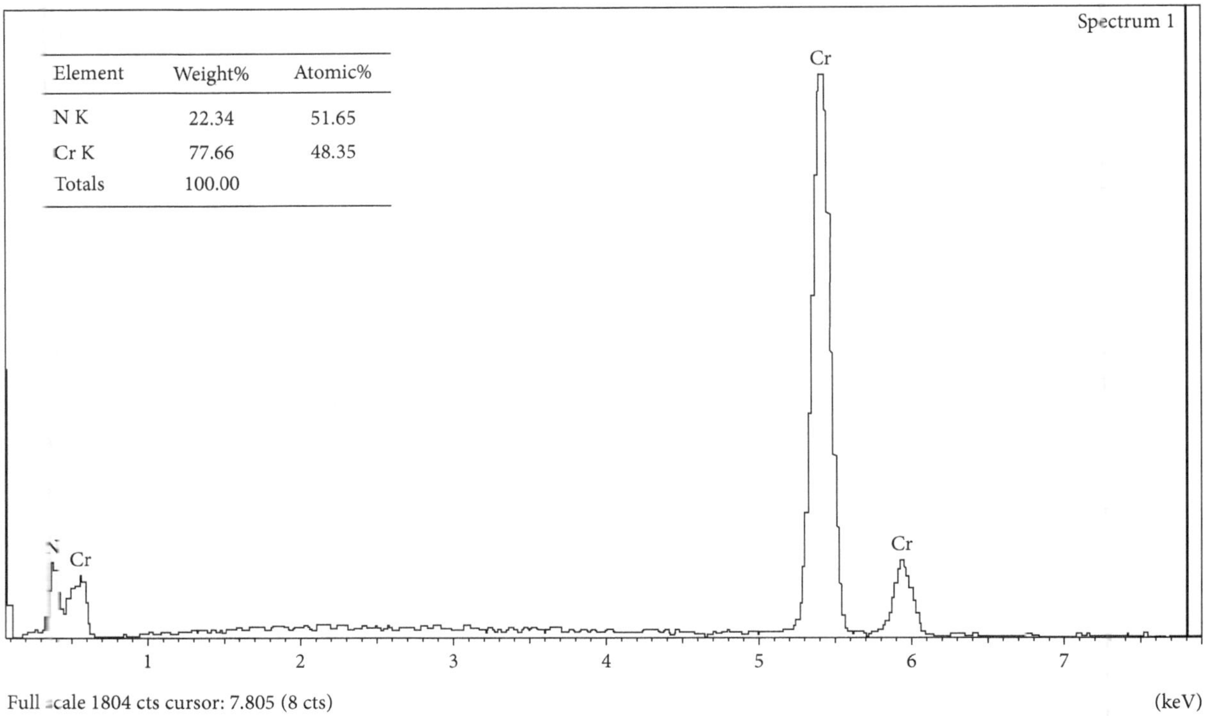

FIGURE 11: The EDS analysis of CrN coating from area shown in Figure 9.

indicates preferential orientation. High preferred orientation is attributed to many grains in this orientation.

3.3.3. Adhesion of the Coatings. Of many methods for determining the adhesion, one of the simplest and widely used is a scratch test. This method is characterized by a relatively short measurement time and high repeatability of easily interpretable results, and it simulates the load (stress) occurring during the use of the product with the coating. Applying a linearly increasing normal force to the coating-substrate system tested generates a stress, which may have both a tensile and compressive character. These stresses commonly affect the stresses arising in the coating occurring as a result of difference in thermal expansion coefficients of the substrate material and the coating. After exceeding the critical value of stress, the coating is removed—this corresponds to the critical load Lc. Such damage is an adhesive. Damage to the cohesion of the layer-substrate system occurring outside the surface of the layer-substrate contact shows good adhesion of the coating to the substrate. The sliding of a loaded diamond indenter on the coating is accompanied by plastic deformation of the substrate and pressing the indenter into the coating. There is also a shear force acting on the indentation-coating boundary.

Damage to the coating can take several forms such as cohesive, adhesive, or conformal crack, coating delamination from the substrate, chipping, flaking off, etc. [33]. The variety of possible defects of the coating in the scratch test makes consistent interpretation of results difficult.

Figure 13 shows the results obtained from the scratch tests carried out on CrN coatings deposited on the substrate

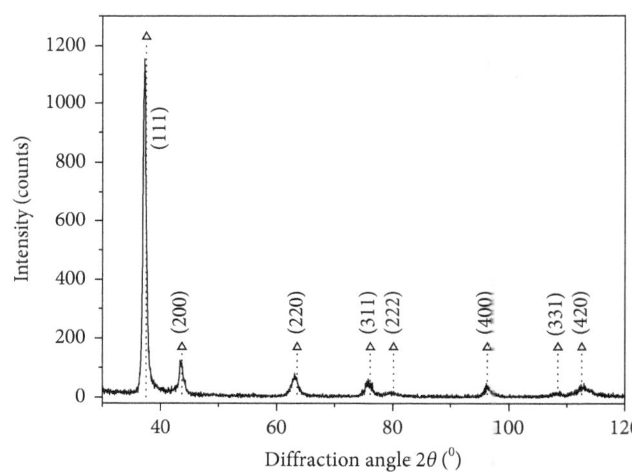

FIGURE 12: Diffraction pattern of the CrN coating on the steel 42CrMo4 substrate.

TABLE 2: The Tc(hkl) texture coefficient for CrN coating.

(hkl) plane	(111)	(200)	(220)	(311)	(222)	(400)	(331)	(420)
Tc(hkl)	5.76	0.44	0.35	0.36	0.13	0.56	0.14	0.26

without shot peening and on the modified shot peening substrates. All samples were characterized by the critical force Lc_2 in the range of 35–40 N, comparable to the critical force of the coating on the substrate in the initial state. As shown in Figures 1(b)–1(d), after shot peening, the surface layer was not compact, which resulted in a high surface roughness (Figure 2). In this case, prior to deposition of the

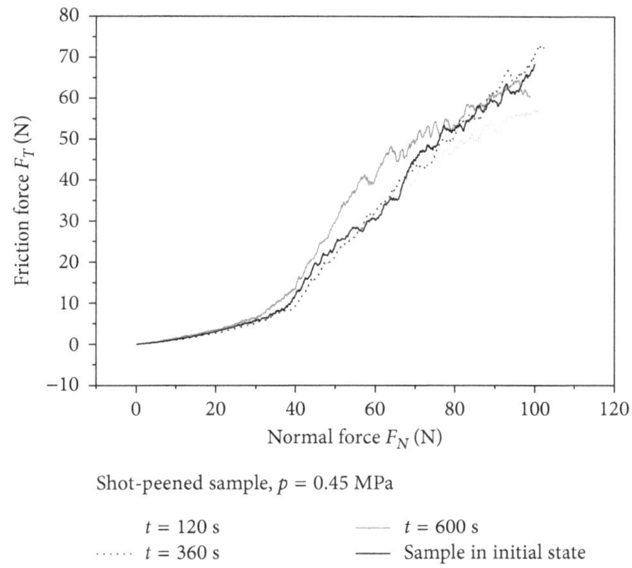

FIGURE 13: The friction force for a progressive load scratch test on CrN coating deposited on substrate without shot peening and after shot peening with different treatment parameters plotted versus the applied load.

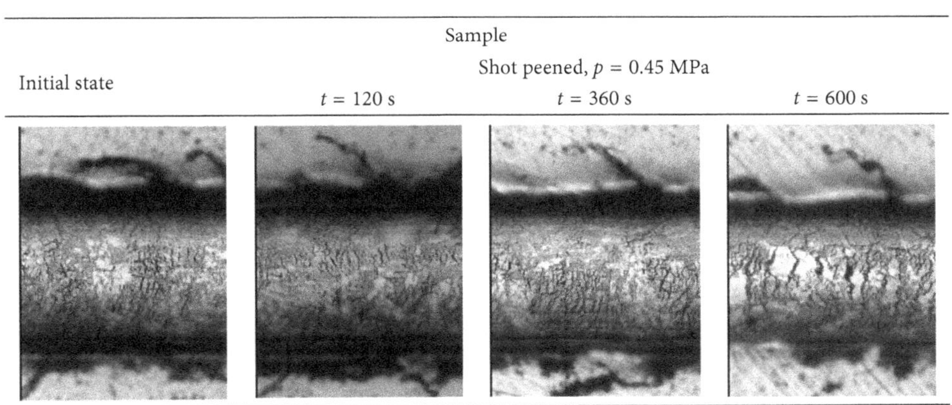

FIGURE 14: A fragment of the scratches corresponding to the load from the range of 37–42 N for CrN coatings deposited on shot peening-treated substrates and the untreated substrate.

coating, surface grinding was performed in order to improve the quality of the surface and reduce the roughness. Hardness measurements performed showed that the hardness of the substrate was reduced by about 50 HV0.5. Despite the increase in the hardness of the shot peening processed substrates, there was no difference in the critical load Lc of the samples. Detailed analysis indicates even a slight decrease in the critical load for the sample with the longest time of shot peening, 600 s. This may result from greater number of defects in the structure, increasing the surface hardness.

Comparison of coating damage in scratch test is rather difficult because of the different properties of the coating-substrate system tested. Film thickness, hardness of the substrate, and the stresses arising at the diamond indenter-coating interface are the main factors that have an impact on the critical force Lc [2]. Therefore, the measurements were performed for systems in which the coatings show the uniform thickness of $(1.8 \pm 0.2)\,\mu$m. The strong effect on the adhesion of the coating to the substrate, determining the critical load Lc, is adequate for substrate surface preparation. Particularly important is the ionic cleaning, conducted directly before the coating process, which removes the surface oxide layer and the weakly bonded impurities. The procedure was used for the deposition of the coatings studied.

Figure 14 shows the photos of scratches made with a load of 37–42 N and the damages of the coatings. It was found that in all the coatings, there is brittle fracture, particularly evident in the bottom and the edge of the scratch. In the case of a coating deposited on the substrate after shot peening treatment (600 s), it can be observed as definitely getting the most spallings in the track of friction.

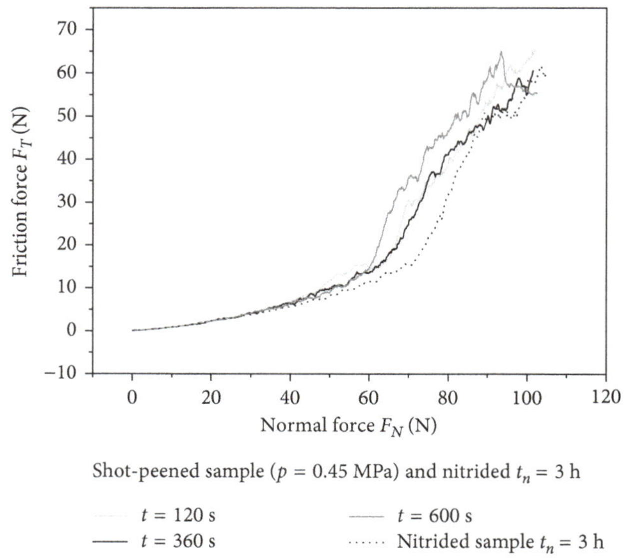

FIGURE 15: The results of the scratch test of the CrN coating on nitrided substrate and substrates subjected by combined shot peening and nitriding processes, nitriding time $t_n = 3$ h.

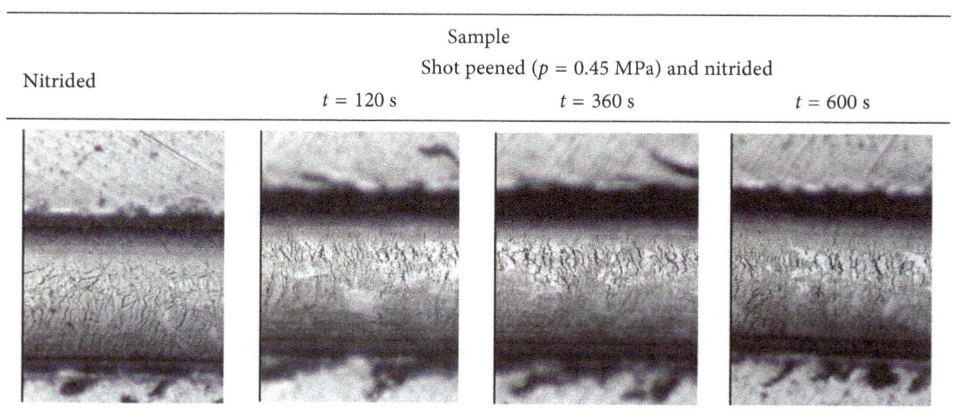

FIGURE 16: A fragment of the scratches corresponding to the load in the range of 60–65 N for CrN coatings deposited on nitrided substrate and combined shot peening and nitriding process-treated substrates.

Substrates after successive shot peening and nitriding show much higher hardness (about 50%) than after the shot peening process. This certainly contributed to the observed (Figure 15) increase in the critical load, that is 60–67 N, and for a substrate only nitrided, even 74 N. The slightly lower value for the substrate with the largest shot peening time is probably connected with the lower surface hardness (Figure 7). The nature of these changes, that is, a relationship of substrate hardness and critical force was previously presented by Ichimura and Rodrigo [1]. Slightly lower critical load for the CrN coating deposited on the substrate subjected to shot peening and nitriding treatment compared to the nitrided substrate results from the nature of the surface layer after these treatments. After shot peening, there is a large development of the substrate surface (higher roughness, Figure 2), and nitriding impairs the condition. The parts of the machined surface are weakly associated with the core result in lower adhesion of the coating to the substrate.

The photos of the scratches performed for loads in the range of 60–65 N (Figure 16), at the same magnification, show the slightly different nature of the coating damage than those shown in Figure 16. CrN coating cracks are smaller due to the cohesive coating cracking and/or bending of the coating in the direction of the motion of a diamond indenter.

The flaking off of the coating which may lead to partial exposure of the substrate or even force in coating into the substrate can be seen. Gas nitriding of the substrate (first stage of treatment causes an increase in its hardness) limits the fragmentation of coating by the restriction of the substrate plastic deformation. The total detachment of coating from the substrate was not observed.

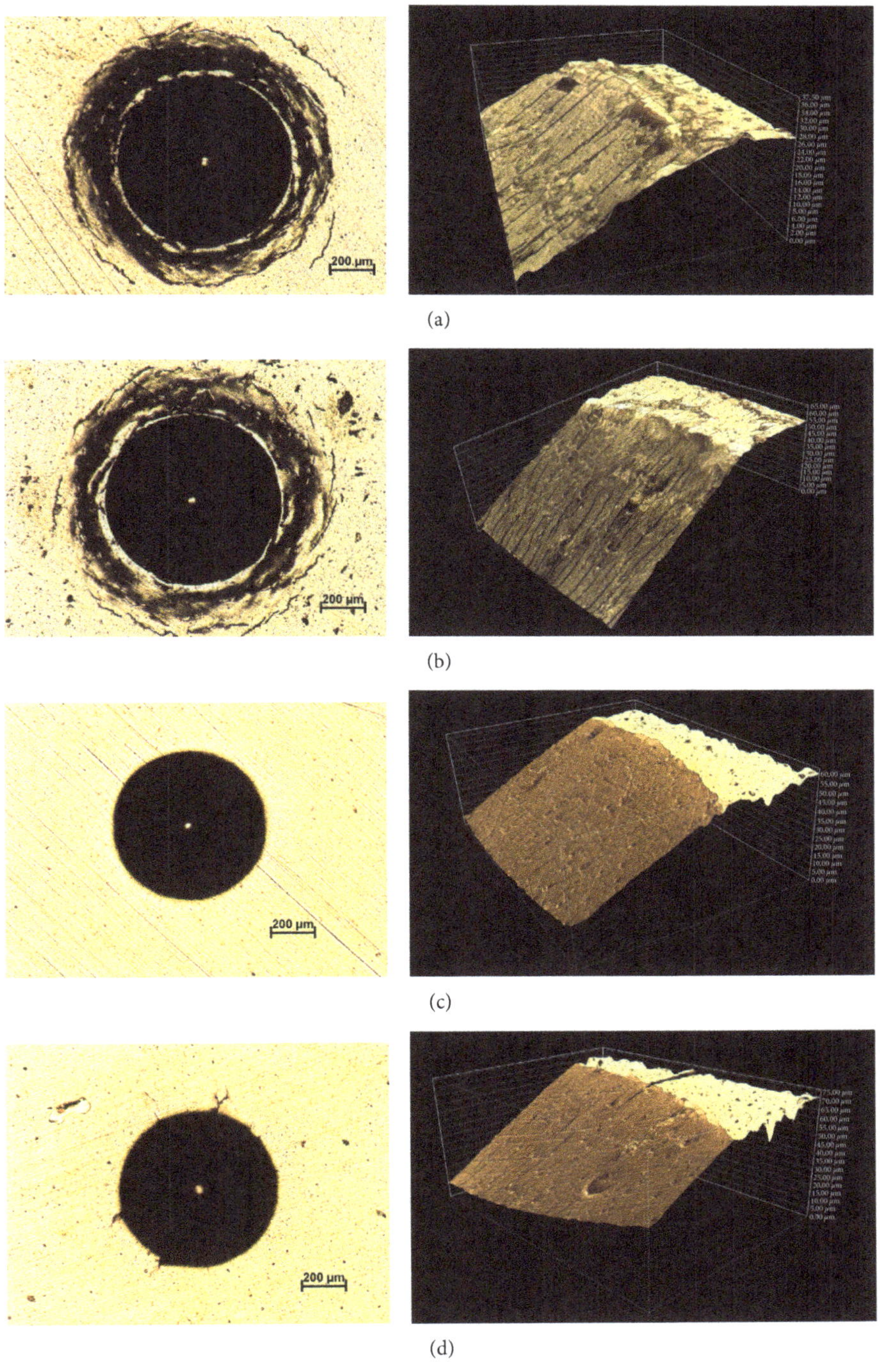

FIGURE 17: Failure modes after Daimler–Benz test (left column) and 3D visualization of the selected indent morphology (right column) for CrN coatings deposited on (a) initial substrates, (b) shot-peened substrate ($t = 600$ s, $p = 0.45$ MPa), (c) nitrided substrate ($t_n = 3$ h), and (d) shot-peened ($t = 600$ s, p = 0.45 MPa) and nitrided substrate ($t_n = 3$ h).

Rockwell diamond indenter movement may cause plastic deformation of the coating and the substrate. Despite this, there is no delamination of the coatings. Similar observations in the complex of nitriding and PVD thin coating deposition processes were reported by Chang et al. [34]. This indicates the excellent toughness and adhesion of coatings.

An easy-to-use test for the evaluation of hard coating adhesion to the substrate is the Daimler–Benz test. The examination shall cover an area both inside and outside the indentation in terms of the type of coating damage: radial cracks, angular intersecting cracks, coating delamination, or the combination of its defects. In the hard coatings with good coherence arise only small radial cracks if any. These coatings with good adhesion do not exhibit delamination neither in indentation nor in its direct surrounding. The test results for CrN coatings deposited on different substrates without any treatment, after the shot peening treatment, nitriding, and the combination of the nitriding and shot peening treatments are shown in Figures 17(a)–17(d), respectively. For a proper evaluation the images of coating-substrate systems are presented in the same magnification. One can see the different "responses" of the substrate-coating system on the load. In Figures 17(a) and 17(b), there is a so-called pile up of substrate materials on the border of the coating-indentation resulting in peripheral cracks. These effects can be seen in the accompanying 3D images, which also reveal a number of radial cracks and the chipping of the coating inside the indentation. In Figure 17(c) illustrating the test result for a coating deposited onto the nitrided only substrate, previously described damages to the coating are not present. 3D image of the sample shows a smooth surface within the indentation. CrN coating deposited onto multistage- (shot peening and nitriding-) treated substrate (Figure 17(d)) shows a few radial cracks which are also revealed in the 3D image.

The test results confirm a very good adhesion of the coating to the nitrided substrates. The substrate subjected also to shot peening treatment, despite higher hardness, shows a greater amount of damage. Thus, a process of nitriding in the multistage substrate treatment is crucial to improve the adhesion of the coating to the substrate. The test results confirm those obtained with the scratch test.

4. Conclusions

The use of modern technology significantly improves the tool materials. These modified substrate-coating systems provide a significant increase in service performance, both for dry cutting and at high cutting speeds.

In this study, the substrates were modified by machining (shot peening), by thermochemical treatment (nitriding), and by the combination of these two processes. A thin hard CrN coating was deposited on different substrates modified. To assess the adhesion, the scratch test and the Daimler-Benz test were used. The results are as follows:

(a) In the shot peening process, depending on the process parameters, the hardness of the material increases by about 50%. With increasing the shot peening time, the hardness also increases. Simultaneously, the surface roughness parameters R_a also significantly increases (about 20 times).

(b) The substrates subjected to a combined treatment consisting of the shot peening and nitriding show a hardness of about 50% higher than after the shot peening and about 40% higher than after the nitriding. The highest hardness was obtained for the following parameters of shot peening: pressure of 0.45 MPa and time of 360 s. Elongation in shot peening time resulted in a hardness decrease of about 50 HV0.5.

(c) Surface modification processes used increase adhesion of the coating, which is due to the increase of its hardness.

Conflicts of Interest

The authors declare that there are no conflicts of interest.

References

[1] H. Ichimura and A. Rodrigo, "The correlation of scratch adhesion with composite hardness for TiN coatings," *Surface and Coatings Technology*, vol. 126, no. 2-3, pp. 152–158, 2000.

[2] S. J. Bull, D. G Bhat, and M. H. Staia, "Properties and performance of commercial TiCN coatings. Part 2: tribological performance," *Surface and Coatings Technology*, vol. 163-164, pp. 507–514, 2003.

[3] A. Nakonieczny, "Forecasting and testing of mechanical properties of surface layers of metals (in Polish)," *Surface Engineering*, vol. 6, no. 1, pp. 3–16, 2001.

[4] T. Burakowski and T. Wierzchoń, *Surface Engineering of Metals*, CRC Press, Boca Raton, FL, USA, 1999.

[5] A. Nakonieczny, *The Performance Features of Metal Products Heat-Treated (in Polish)*, Institute of Precision Mechanics, Warsaw, Poland, 1999.

[6] L. Chekour, C. Nouveau, A. Chala, and M. A. Djouadi, "Duplex treatment of 32CrMoV13 steel by ionic nitriding and triode sputtering: application to wood machining," *Wear*, vol. 255, no. 7–12, pp. 1438–1443, 2003.

[7] R. Hoy, J.-D. Kamminga, and G. C. A. M. Janssen, "Scratch resistance of CrN coatings on nitrided steel," *Surface and Coatings Technology*, vol. 200, no. 12-13, pp. 3856–3860, 2006.

[8] H. Paschke, M. Weber, P. Kaestner, and G. Braeuer, "Influence of different plasma nitriding treatments on the wear and crack behavior of forging tools evaluated by Rockwell indentation and scratch test," *Surface and Coatings Technology*, vol. 205, no. 5, pp. 1465–1469, 2010.

[9] Ö. Alpaslan, E. Atar, and H. Çimenoğlu, "Tribological behaviour of duplex treated AISI 4140 steel," *JESTECH*, vol. 15, no. 1, pp. 39–43, 2012.

[10] D. Liu, X. Zhu, B. Tang, and J. He, "Fretting fatigue improvement of Ti6Al14V by coating and shot peening," *Journal of Materials Science & Technology*, vol. 21, pp. 246–250, 2005.

[11] Y. Fu, N. L. Loh, A. W. Batchelor et al., "Improvement in fretting wear and fatigue resistance of Ti-6Al-4V by application of several surface treatments and coatings," *Surface and Coatings Technology*, vol. 106, no. 2-3, pp. 193–197, 1998.

[12] J. Walkowicz, J. Smolik, and K. Miernik, "Investigation of the influence of ion etching parameters on the structure of nitrided case in hot working steel," *Surface and Coatings Technology*, vol. 116-119, pp. 361–366, 1999.

[13] S. M. Aouadi, D. M. Schultze, S. L. Rohde, K. C. Wong, and K. A. R. Mitchell, "Growth and characterization of Cr2N/CrN multilayer coatings," *Surface and Coatings Technology*, vol. 140, no. 3, pp. 269–277, 2001.

[14] S. C. Lee, W. Y. Ho, and F. D. Lai, "Effect of substrate surface roughness on the characteristics of CrN hard films," *Materials Chemistry and Physics*, vol. 43, no. 3, pp. 266–273, 1996.

[15] B. D. Cullity, *Elements of X-ray Diffraction*, Addison-Wesley Publishing Company Inc., London, UK, 2nd edition, 1978.

[16] J. Ratajski, "Relation between phase composition of compound zone and growth kinetics of diffusion zone during nitriding of steel," *Surface & Coatings Technology*, vol. 203, no. 16, pp. 2300–2306, 2009.

[17] G. G. Stoney, "The tension of metallic films deposited by electrolysis," *Proceedings of the Royal Society of London. Series A, Containing Papers of a Mathematical and Physical Character*, vol. 82, no. 507, pp. 172–175, 1909.

[18] W. Przybylski, *The Processing Technology of Burnishing (in Polish)*, Publisher of Scientific and Technical, Warsaw, Poland, 1987.

[19] C. Nouveau, M. A. Djouadi, and C. Decès-Petit, "The influence of deposition parameters on the wear resistance of Cr_xN_y magnetron sputtering coatings in routing of oriented strand board," *Surface and Coatings Technology*, vol. 174-175, pp. 455–460, 2003.

[20] B. Warcholinski and A. Gilewicz, "Effect of substrate bias voltage on the properties of CrCN and CrN coatings deposited by cathodic arc evaporation," *Vacuum*, vol. 90, pp. 145–150, 2013.

[21] M. Odén, C. Ericsson, G. Hakansson, and H. Ljungcrantz, "Microstructure and mechanical behavior of arc-evaporated Cr–N coatings," *Surface and Coatings Technology*, vol. 114, no. 1, pp. 39–51, 1999.

[22] X. S. Wan, S. S. Zhao, Y. Yang, J. Gong, and C. Sun, "Effects of nitrogen pressure and pulse bias voltage on the properties of Cr-N coatings deposited by arc ion plating," *Surface and Coatings Technology*, vol. 204, no. 11, pp. 1800–1810, 2010.

[23] A. Gilewicz, B. Warcholinski, P. Myslinski, and W. Szymański, "Anti-wear multilayer coatings based on chromium nitride for wood machining tools," *Wear*, vol. 270, no. 1-2, pp. 32–38, 2010.

[24] M. Odén, J. Almer, and G. Håkansson, "The effects of bias voltage and annealing on the microstructure and residual stress of arc-evaporated Cr–N coatings," *Surface and Coatings Technology*, vol. 120–121, pp. 272–276, 1999.

[25] V. D. Ovcharenko, A. S. Kuprin, G. N. Tolmachova et al., "Deposition of chromium nitride coatings using vacuum arc plasma in increased negative substrate bias voltage," *Vacuum*, vol. 117, pp. 27–34, 2015.

[26] V. D. Ovcharenko, A. S. Kuprin, G. N. Tolmachova et al., "Deposition of chromium nitride coatings from vacuum arc plasma in increased nitrogen pressure," *Problems of Atomic Science and Technology*, vol. 6, no. 20, pp. 204–207, 2014.

[27] J. W. Seok, N. M. Jadeed, and R. Y. Lin, "Sputter deposited nanocrystalline Cr and CrN coatings on steels," *Surface and Coatings Technology*, vol. 138, no. 1, pp. 14–22, 2001.

[28] J. Kusinski, M. Rozmus, and J. Bujak, "Investigation of the life-time of drills covered with the anti-wear Cr(C,N) complex coatings, deposited by means of Arc-PVD technique," *Journal of Achievements in Materials and Manufacturing Engineering*, vol. 33, no. 1, pp. 86–93, 2009.

[29] J. Esteve, J. Romero, M. Gómez, and A. Lousa, "Cathode chromium carbide coatings for molding die applications," *Surface and Coatings Technology*, vol. 188-189, pp. 506–510, 2004.

[30] M. Čekada, P. Panjan, M. Maček, and P. Šmid, "Comparison of structural and chemical properties of Cr-based hard coatings," *Surface and Coatings Technology*, vol. 151-152, pp. 31–35, 2002.

[31] J. Romero, E. Martinez, J. Esteve, and A. Lousa, "Nanometric chromium nitride/chromium carbide multilayers by r.f. magnetron sputtering," *Surface and Coatings Technology*, vol. 180-181, pp. 335–340, 2004.

[32] C. S. Barret and T. B. Massalski, *Structure of Metals*, Pergamon Press, Oxford, UK, 1980.

[33] S. J. Bull, "Failure mode maps in the thin film scratch adhesion test," *Tribology International*, vol. 30, no. 7, pp. 491–498, 1997.

[34] Y. Y. Chang, D. Y. Wang, and C. Y. Hung, "Structural and mechanical properties of nanolayered TiAlN/CrN coatings synthesized by a cathodic arc deposition process," *Surface and Coatings Technology*, vol. 200, no. 5-6, pp. 1702–1708, 2005.

15

Pull-Out Resistance Capacity of a New Perfobond Shear Connector for Steel Pile Cap Strengthening

Young-Ho Kim,[1] Jae-Yoon Kang,[2] Hyun-Bon Koo,[2] and Dae-Jin Kim[3]

[1] *Institute of R&D, Ji Seung Consultant Co., Ltd., Seoul 06128, Republic of Korea*
[2] *Korea Institute of Civil Engineering and Building Technology, Goyang 10223, Republic of Korea*
[3] *Department of Architectural Engineering, Kyung Hee University, Yongin 17104, Republic of Korea*

Correspondence should be addressed to Dae-Jin Kim; djkim@khu.ac.kr

Academic Editor: Akihiko Kimura

This study proposes a new type of the perfobond shear connector, which can be used to strengthen the steel pile cap embedded into the structure foundation, and evaluates its pull-out resistance capacity by performing a test on ten specimens. Test parameters include the embedment length of the shear connector, existence of transverse rebars passing through holes in the shear connector, and their shape, size, and number. The pull-out load versus slip curve is plotted for all specimens, and their failure modes are identified. The effects of the test parameters on the peak pull-out load are examined in this work. The test results show that the perfobond shear connector proposed in this study can retain the peak pull-out load up to 6 times higher than the one without any holes. This indicates that the existence of holes in the shear connector enables the dowel action of concrete inside the hole, resulting in the improvement of the shear resistance capacity of the connector.

1. Introduction

At the foundation to support an upper structure, there exists a connection between piles and concrete footing as illustrated in Figure 1. The integration of these two structural components is very important to guarantee the safe load transfer from the upper structure to the supporting layer of rocks. In general, this can be achieved by the use of steel reinforcing bars embedded into the concrete footing. In this case, the bond between the steel rebars and concrete is activated by the mechanical bearing of bar ribs, and it is essential to retain a sufficient level of their bond strength. However, since their rib size is limited and they have relatively small cross-sectional area, their bond strength is also limited to a small value as indicated in various concrete design code provisions such as ACI building code [1] and Eurocode 2 [2].

In order to handle this issue, we propose a new perfobond shear connector illustrated in Figure 2, which can guarantee an excellent composite behavior among the structural components at the steel pile cap. Originally, this type of perfobond shear connectors was first developed by Zellner as a remedy to the fatigue problem of steel studs in the steel composite bridge [3]. This device is basically a steel plate with holes and embedded into concrete to ensure the perfect integration of steel members and the surrounding concrete. As external loads are applied, the dowel action of concrete inside the hole is activated and enables the load transfer between the structural components made of different materials. This perfobond shear connector has been known to retain higher shear strength, better fatigue performance, and more enhanced constructability than existing shear studs [4].

Several publications have investigated the performance of the perfobond shear connector. Veldanda and Hosain [5] compared the shear strengths of the perfobond shear connector and shear studs by performing a pull-out test on specimens with these types of connections. Oguejiofor and Hosain [6–8] performed a bending test on steel-concrete composite beams with perfobond shear connectors and proposed its shear strength equation. Sara and Bahram [4] modified the shear strength equation proposed by Oguejiofor and Hosain with the consideration of the bond strength at the steel-concrete interface. More recently, Machacek and Studnicka

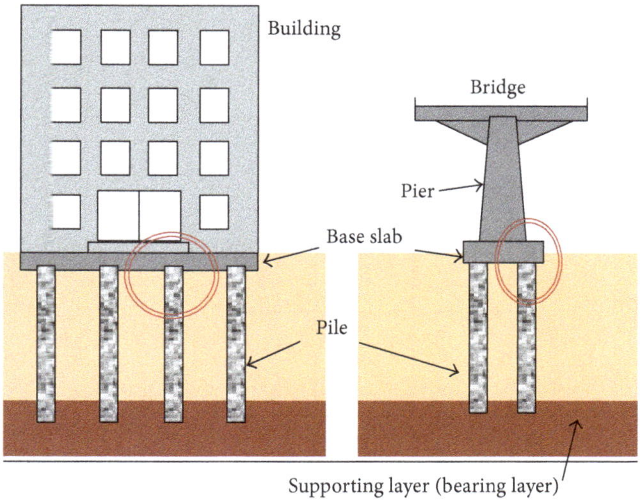

FIGURE 1: Connection between structural foundation and steel pile caps.

[9] and Valente and Cruz [10] examined the structural performance of the perfobond shear connector embedded into lightweight concrete, and still various kinds of research are performed on the development of its new details and application to different types of structural members.

This paper experimentally evaluates the shear strength of the proposed perfobond connector by performing a pull-out test on ten specimens by considering the embedment length of the shear connector, existence of transverse rebars passing through holes in the shear connector, and their shape, size, and number as test parameters. The failure characteristics of the test specimens are examined, and their load-displacement curves are analyzed. We also investigate the effects of the test variables on the pull-out strengths of the test specimens in order to demonstrate the effectiveness of the proposed perfobond shear connector.

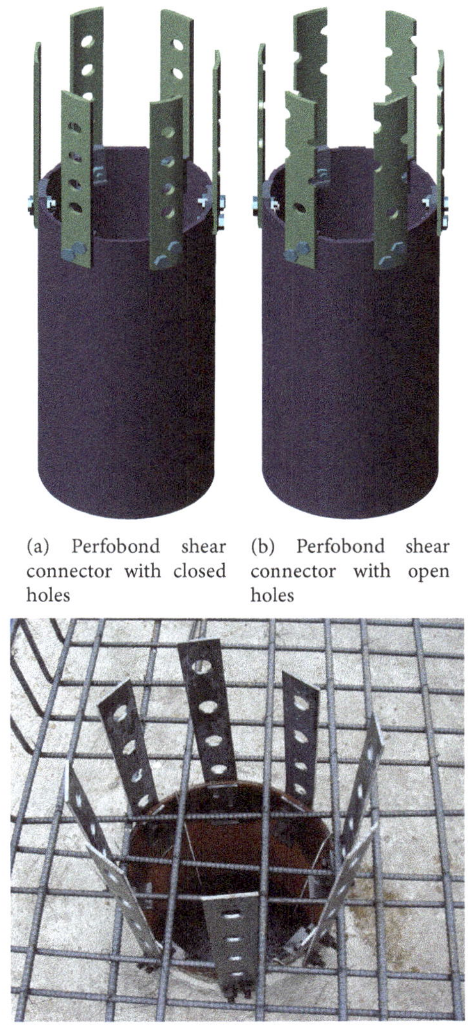

(a) Perfobond shear connector with closed holes

(b) Perfobond shear connector with open holes

(c) Construction example of the proposed perfobond shear connector for steel pile cap strengthening

FIGURE 2: Strengthening of steel pile caps using the perfobond shear connectors.

2. Experimental Program

2.1. Test Specimens. For this study, ten specimens totally were manufactured and tested to investigate the effectiveness of the newly proposed perfobond rib shear connector. In all specimens, a perforated steel flat-plate with a sectional dimension of 100 mm by 8 mm was embedded in the concrete block. In six of the test specimens, the embedment length of the perforated flat-plate was 150 mm, while it was 270 mm in other four flat-plates. The center of the perforated flat-plate coincides with that of the concrete block in all specimens. Figure 3 illustrates the geometrical configurations of the specimens described above.

Test parameters include the type (no hole, closed hole, and open hole), size (40 mm, 50 mm, and 60 mm), number (0, 1, and 2) of holes, embedment length of the shear connector (150 mm and 270 mm), and existence of transverse rebars passing through holes. These parameters were determined by considering that, in general, the most effective hole size and distance between holes are approximately 0.45 and two times the width of the perfobond connector, respectively [11]. The shapes of the perfobond connectors used for the test specimens are provided in Figure 4. The details of the ten test specimens are summarized in Table 1. For convenience, the test specimens in the table are grouped into the four categories such as Series I, II, III, and IV. From the results of Series I and II, the effects of hole size and transverse rebars passing through holes on the shear resistance of the perfobond connector can be discussed, respectively. Similarly, the effects of the embedment length of the shear connector and types of holes can be analyzed from those of Series III and IV.

The compressive strength of concrete was measured in accordance with the standards of ASTM C39 9 [12] and the measured average strength was found to be 27.30 (MPa). The yield strength, ultimate strength, and maximum strain at fracture of rebars are 490.1 MPa, 584.9 MPa, and 18.1%,

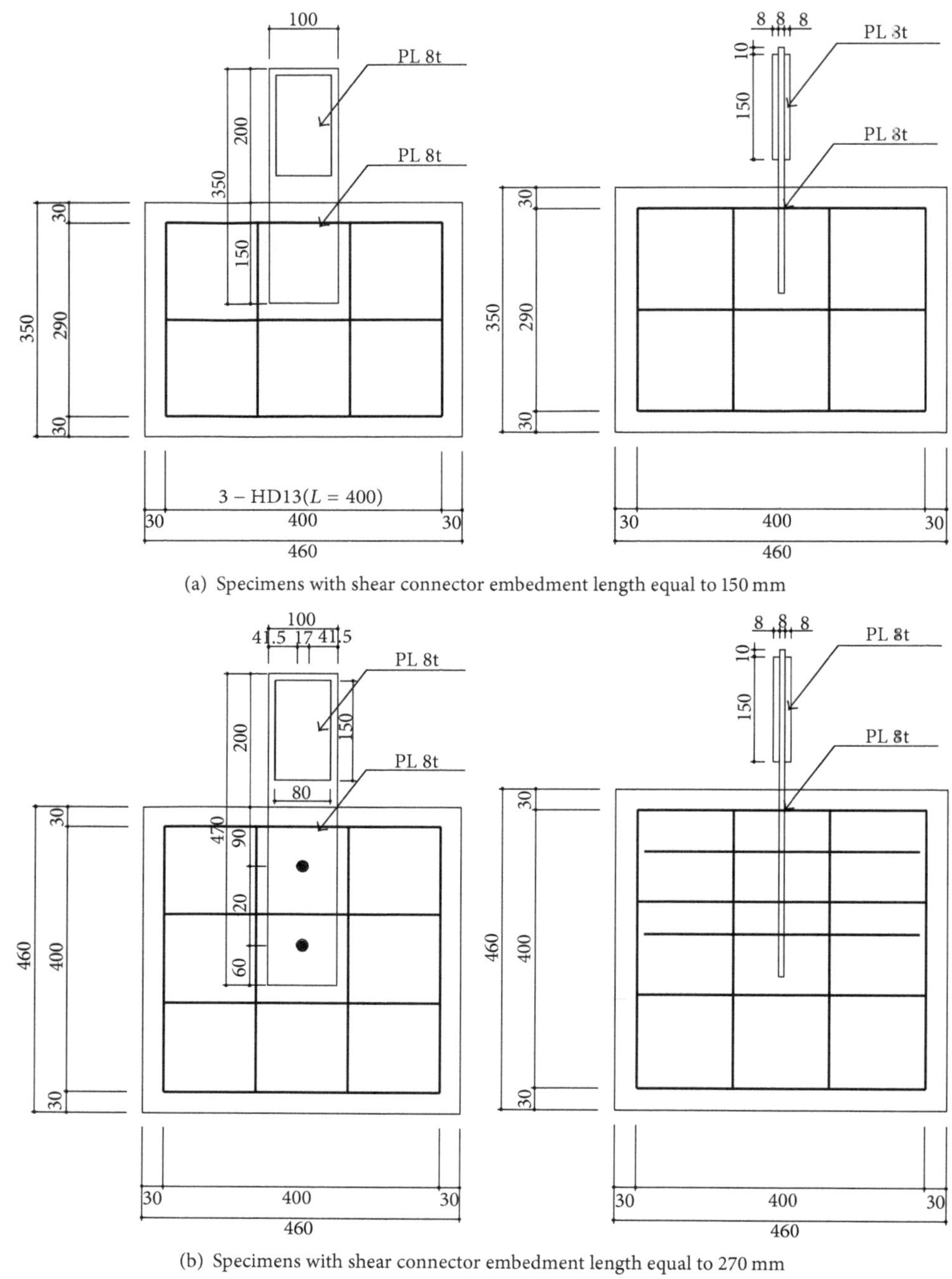

Figure 3: Size and rebar details of test specimens (unit: mm).

respectively. The corresponding values of the steel plates are 321.4 MPa, 441.5 MPa, and 33.7%, respectively. These properties are summarized in Table 2.

2.2. Testing Equipment and Procedure. Pull-out tests were conducted on the ten specimens to estimate the strength of the new perfobond shear connectors and to identify their failure modes. The test setup is shown in Figure 5. The test was conducted on a universal testing machine (UTM), which allows displacement-based load control. The maximum capacity of the UTM is 250 kN and the load was applied to the specimen at a rate of 0.5 mm/min. The magnitude of loading was measured by the load cell attached at the bottom of the actuator. The vertical displacement was monitored by two linear variable differential transducers (LVDTs) installed near the specimen, and the concrete cube of the test specimen was fixed onto the UTM base by putting additional steel frames on the top of the concrete cube and connecting it to the UTM base using long bolts, as shown in Figure 5. The load-versus-displacement data were recorded

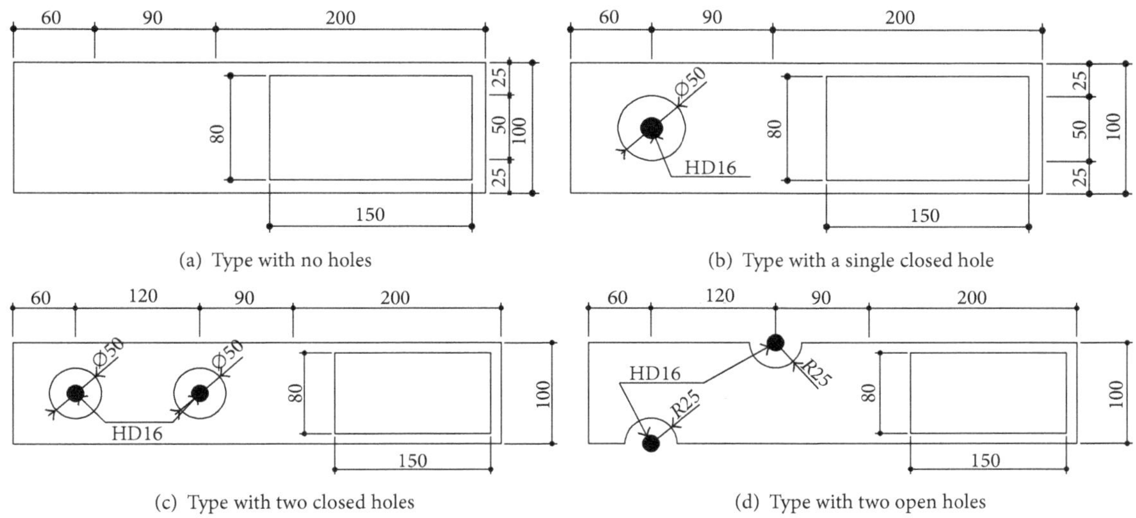

FIGURE 4: Types of perfobond shear connectors (unit: mm).

TABLE 1: Details of test specimens.

	Specimen	Type of holes	Diameter of holes (unit: mm)	Number of holes	Embedment length of shear connector (unit: mm)	Total number of transverse rebars passing through holes
Series I	PN-0015-N	No hole	N/A	0	150	0
	PO-5015-N	Closed	50	1		
	PO-6015-N		60			
Series II	PO-4015-R	Closed	40	1	150	1
	PO-5015-R		50			
	PO-6015-R		60			
Series III	PO-5027-N	Closed	50	2	270	0
	PO-5027-R					2
Series IV	PU-5027-N	Open	50	2	270	0
	PU-5027-R					2

throughout the entire loading history using a computer-aided data acquisition system.

3. Pull-Out Resistance of the Perfobond Shear Connector

3.1. General Behavior and Failure Modes. Table 3 summarizes the test results such as peak loads, displacements at peak loads, and loads corresponding to several crack propagation phenomena occurring in the concrete block. Figure 6 plots the load-displacement curves of the four series of the test specimens listed in Table 1. Crack propagation patterns were basically similar in all of the test specimens, but the amount and range of cracking were different in each specimen. The failure characteristics of the test specimens include the yielding and fracture of the perforated flat-plate, its pull-out failure, and concrete delamination and are summarized in Table 4. Depending on the three characteristics, the ultimate failure modes of the test specimens can be categorized into five groups (Types I to V).

Figure 7 shows the cracking pattern of specimen PN-0015-N on its top and side faces. It indicates that the specimen failed by bond-slip failure between the flat-plate steel shear connector and the surrounding concrete without any delamination or cracking of concrete. It seems to happen due to insufficient bond strength capacity of the steel shear connector without any holes, and this type of failure mode is categorized as Type I.

The cracking pattern of specimen PO-6015-N is given in Figure 8. In this specimen, the pull-out force applied was initially resisted by the dowel action of the concrete inside the hole of the perfobond shear connector. However, as the load increases, the local bearing pressure on the concrete inside hole also increases, and this finally resulted in its crushing failure. Similar to the case of specimen PN-0015-N, the fracture of the steel perfobond shear connector did not occur, and no significant cracking of concrete was observed outside

TABLE 2: Material properties of test specimens.

Concrete			Rebars (H16)			Steel plate (SS400)		
Compressive strength (MPa)								
Number 1	Number 2	Number 3	Yield strength (MPa)	Ultimate strength (MPa)	Maximum strain at fracture (%)	Yield strength (MPa)	Ultimate strength (MPa)	Maximum strain at fracture (%)
27.77	27.96	26.16	490.1	584.9	18.1	312.4	441.5	33.7

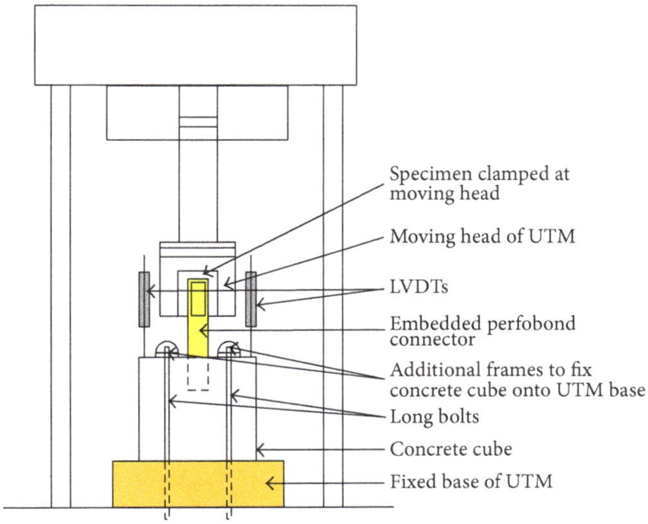

(a)

(b)

FIGURE 5: Test setup.

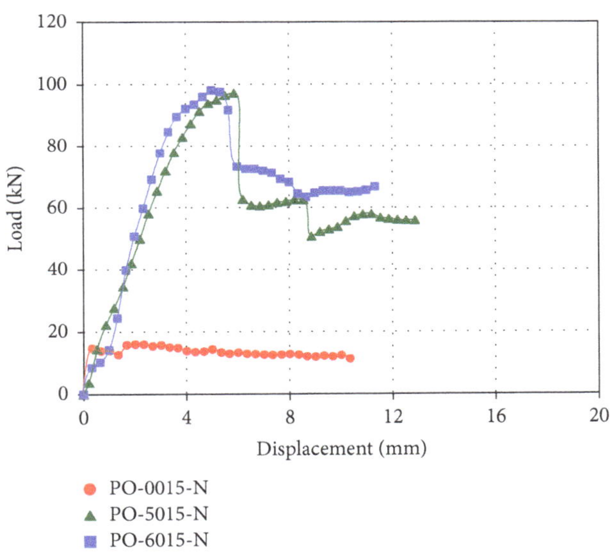

(a) Series I

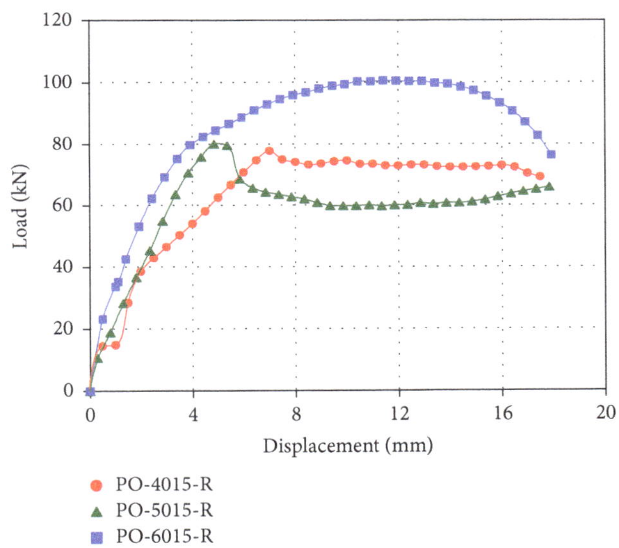

(b) Series II

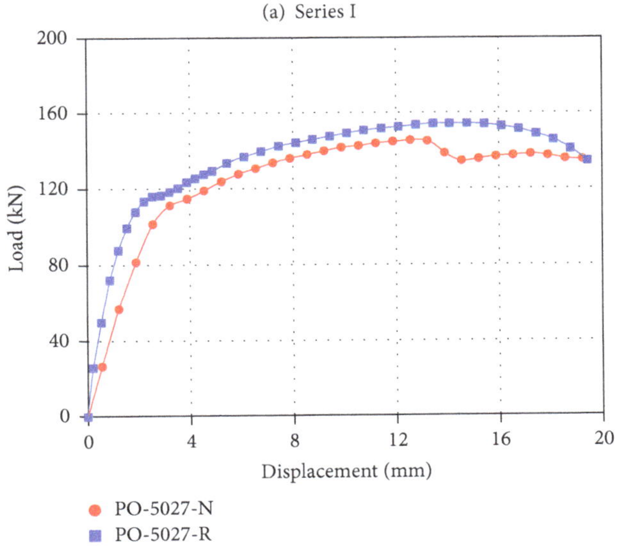

(c) Series III

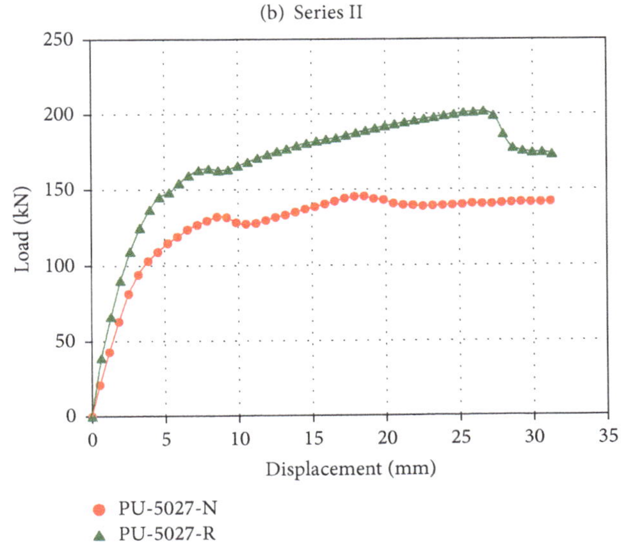

(d) Series IV

FIGURE 6: Load-displacement curves of test specimens.

TABLE 3: Results of the pull-out test.

Specimen	Peak load (kN)	Displacement at peak load (mm)	Cracking load (kN)		
			Cracking on top of the concrete block	Vertical cracking on the side of the concrete block	Horizontal cracking on the side of the concrete block
PN-0015-N	16.6	1.95	15	N/A	N/A
PO-5015-N	97.4	5.94	48	97	97
PO-6015-N	99.2	5.20	76	88	94
PO-4015-R	78.3	7.13	64	78	N/A
PO-5015-R	81.3	5.21	33	69	69
PO-6015-R	100.8	12.02	36	N/A	N/A
PO-5027-N	145.6	12.73	85	141	N/A
PO-5027-R	154.7	14.40	130	135	N/A
PU-5027-N	146.1	37.48	118	136	N/A
PU-5027-R	202.2	26.76	148	153	153

TABLE 4: Failure modes of test specimens.

Specimen	Failure characteristics			Ultimate failure mode
	Concrete delamination and cracking	Yielding and fracture of the steel flat-plate shear connector	Pull-out of the perforated flat-plate	
PN-0015-N	X	X	O	Type I
PO-5015-N	X	X	O	Type II
PO-6015-N	X	X	O	
PO-4015-R	O	X	O	Type III
PO-5015-R	O	X	O	
PO-6015-R	X	O	X	Type IV
PO-5027-N	X	O	X	
PO-5027-R	X	O	X	
PU-5027-N	X	O	X	
PU-5027-R	O	O	O	Type V

the specimen. This type of failure occurred in specimens PO-5015-N and PO-6015-N, where no transverse rebars into the hole of the perfobond shear connector exist, and it is categorized as Type II.

The failure mode of specimen PO-5015-R is shown in Figure 9. In contrast to the case of Type II, the pull-out load is sustained by the dowel action of both the transverse rebar and the concrete inside the hole in this case. As a result, instead of exhibiting local crushing failure in the concrete inside the hole, a significant amount of cracking was observed outside the specimen at its ultimate stage. No yielding and fracture of the steel perfobond shear connector happened. This type of failure was observed in specimens PO-4015-R and PO-5015-R and is categorized as Type III.

Figure 10 shows the cracking pattern and failure mode of specimen PO-6015-R. In this specimen, the perfect bond between the shear connector and surrounding concrete was achieved mainly due to the shear resistance by the shear connector with sufficient embedment length and dowel action by both the transverse rebars and the concrete inside the holes of the shear connector. As a result, its failure mode at ultimate stage was characterized by the yielding and fracture of the steel perfobond shear connector, and no significant delamination or cracking of concrete was observed during the entire stages of the test. In addition, as discussed in more details later, a ductile behavior was exhibited in the load-displacement curve even after reaching the peak load. This type of failure mode is denoted as Type IV and was observed in specimens PO-6015-R, PO-5027-N, PO-5027-R, and PU-5027-N.

Lastly, the cracking pattern of specimen PU-5027-R is exhibited in Figure 11. In contrast to Types III and IV, the ultimate failure mode of this specimen is not localized failures such as local crushing of concrete near the holes of the shear connector and yielding/fracture of the steel shear connector. Instead, the significant amounts of delamination and

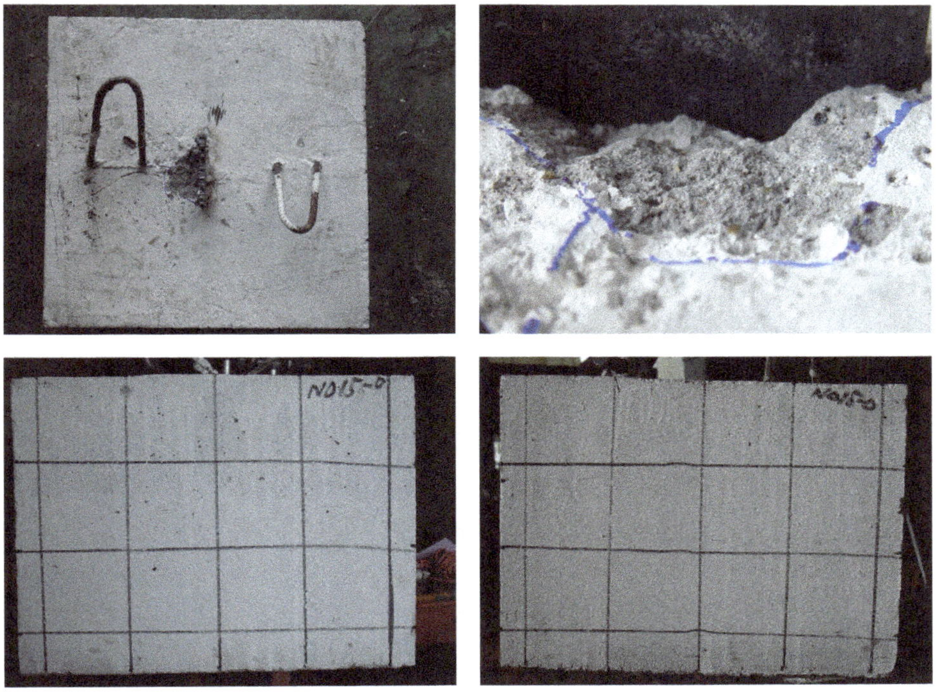

FIGURE 7: Failure mode of specimen PN-0015-N (Type I).

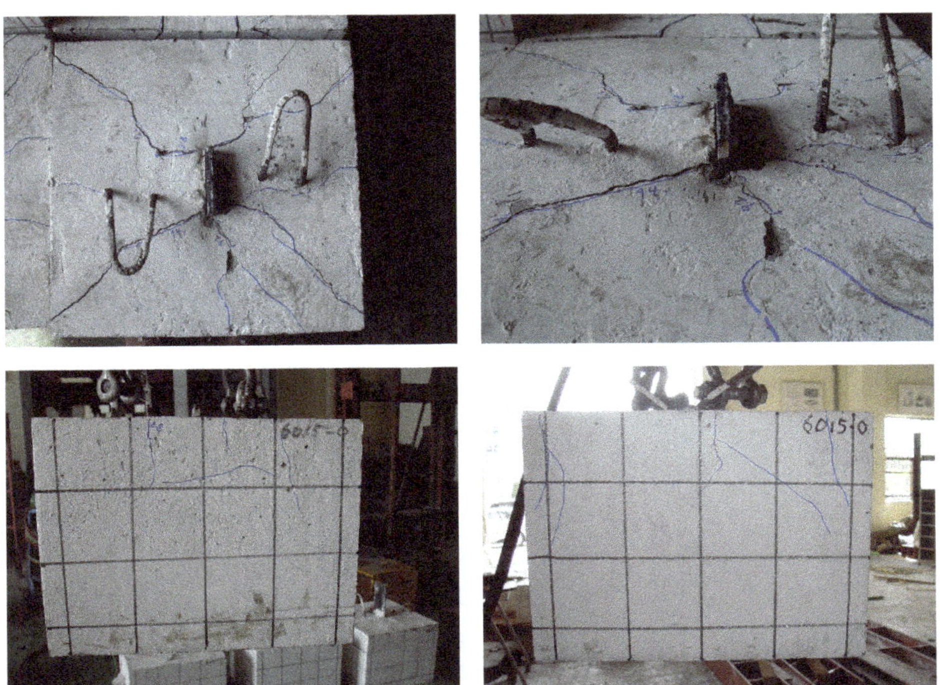

FIGURE 8: Failure mode of specimen PO-6015-N (Type II).

cracking of concrete were found in the entire specimen at ultimate stage as shown in the figure. This type of failure mode is categorized as Type V and was observed in specimen PU-5027-R.

3.2. Effects of Test Parameters on the Shear Strength of Perfobond Shear Connectors. This section discusses the effects of test parameters such as the size of connector holes, existence of transverse rebars into the holes, embedment length of the

FIGURE 9: Failure mode of specimen PO-5015-R (Type III).

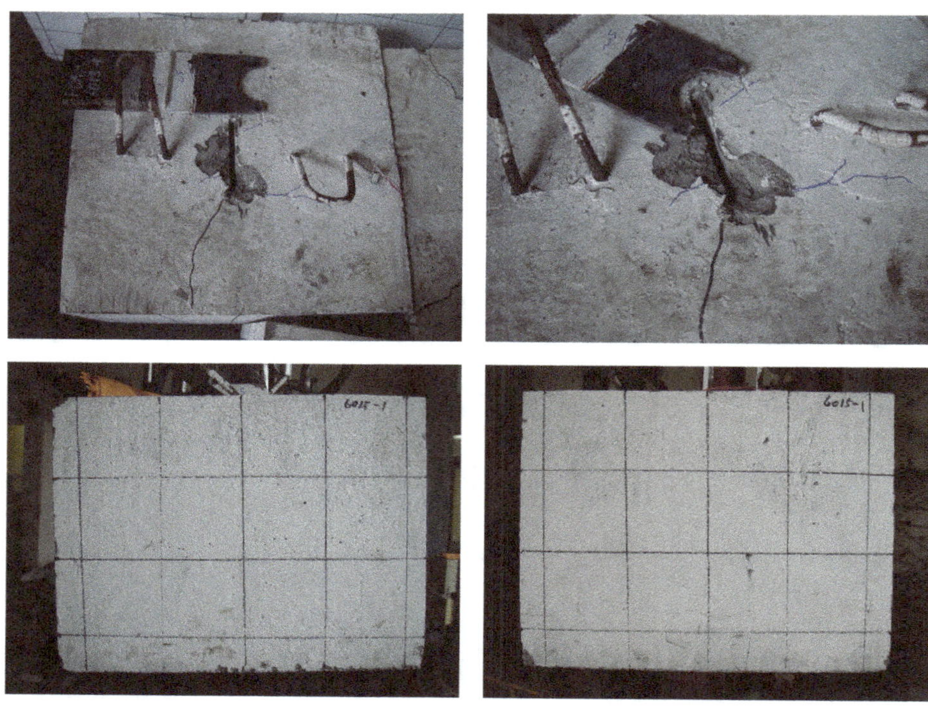

FIGURE 10: Failure mode of specimen PO-6015-R (Type IV).

shear connector, and shape of the connector hole on the shear strength of the perfobond connector.

Figure 12(a) shows the peak pull-out loads of the two sets of test specimens with different hole diameters, which are Series I and II listed in Table 3. The three test specimens PN-0015-N, PO-5015-N, and PO-6015-N in Series I do not have any transverse rebars passing through the connector holes. It can be noticed from the results of these specimens that two test specimens with 50 and 60 mm hole diameters (PO-5015-N and PO-6015-N, resp.) retain the shear strength

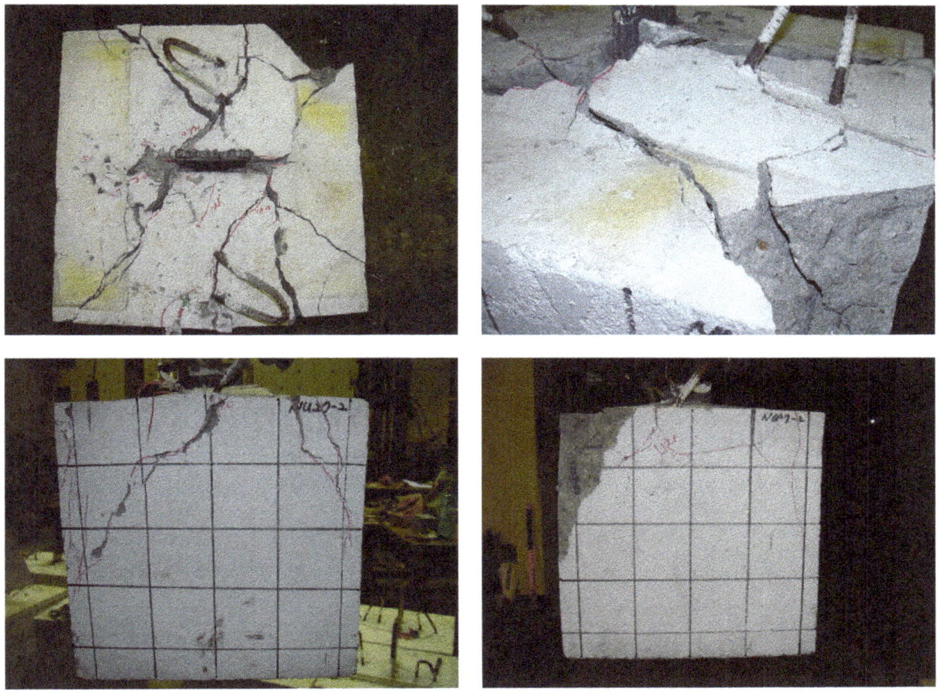

FIGURE 11: Failure mode of specimen PU-5027-R (Type V).

approximately 6 times higher than the one without any holes (PN-0015-N). This indicates that the existence of holes in the shear connector enables the dowel action of concrete inside the hole, resulting in the improvement of the shear resistance capacity of the connector. However, the peak pull-out loads of specimens PO-5015-N and PO-6015-N are almost equal to each other; thus, there seems to be a limit in the shear resistance capacity of the perfobond connector that can be improved by increasing the hole size.

The three test specimens in Series 2, which are PO-4015-R, PO-5015-R, and PO-6015-R, have a transverse rebar passing through the connector hole. Their results in the figure show that the peak pull-out loads of the two test specimens with 40 and 50 mm hole diameter (PO-4015-R and PO-5015-R, resp.) are almost equal and the specimen with 60 mm hole diameter (PO-6015-R) retains shear resistance approximately 25% higher than those of the first two specimens. This may be attributed to the fact that a relatively large hole size of the specimen PO-6015-R helps the transverse rebar to be effectively integrated into the surrounding concrete, which eventually leads to the enhancement of the shear resistance capacity. In addition, it can be observed from Figure 6(b) that the initial stiffness of Series II specimens is increased with increasing the hole diameter.

The effect of the transverse rebar on the shear strength of the perfobond connector can be identified from the results of Figure 12(b). It plots the peak pull-out loads of the three sets of specimens reinforced with transverse rebars, which have different hole diameters (d_{hole}) and embedment lengths (l_e) such as PO-5015-N and PO-5015-R, PO-6015-N and PO-6015-R, and PO-5027-N and PO-5027-R. The results of the figure indicate that the existence of the transverse rebars enhances the peak pull-out load except the set of PO-5015-N and PO-5015-R. In addition, as can be seen from the comparison between Figures 6(a) and 6(b), it is helpful for retaining a significant level of ductility after reaching the peak load. The decrease of the peak pull-out load shown in the set of PO-5015-N and PO-5015-R seems to happen since specimen does not have sufficient space around the transverse rebar and the integration between the rebar and concrete is not fully achieved.

Figure 12(c) shows the effect of embedment length on the peak pull-out strength for the two sets of test specimens with and without transverse rebars, which are PO-5015-N and PO-5027-N and PO-5015-R and PO-5027-R. The results of the figure indicate that, in both cases of the specimen sets, the peak pull-out load is increased at least by 50% by increasing the embedment length from 150 mm to 270 mm. This confirms that the increase of embedment length is greatly helpful to improve the shear resistance capacity of the perfobond connector. In addition, the test results show that the peak pull-out loads of the specimens with transverse rebars are not always greater than those of the ones without transverse rebars. This may be attributed to the fact that only a single specimen was tested for each test parameter; thus, more tests may be required for the same test parameter to clarify this issue.

The effect of the shape of perfobond holes on the peak pull-out load can be evaluated from Figure 12(d). It plots the peak pull-out loads of the two specimen sets, which are PO-5027-N and PO-5027-R and PU-5027-N and PU-5027-R. In the case of the second set of which specimens have open holes in the perfobond connector the peak pull-out load of

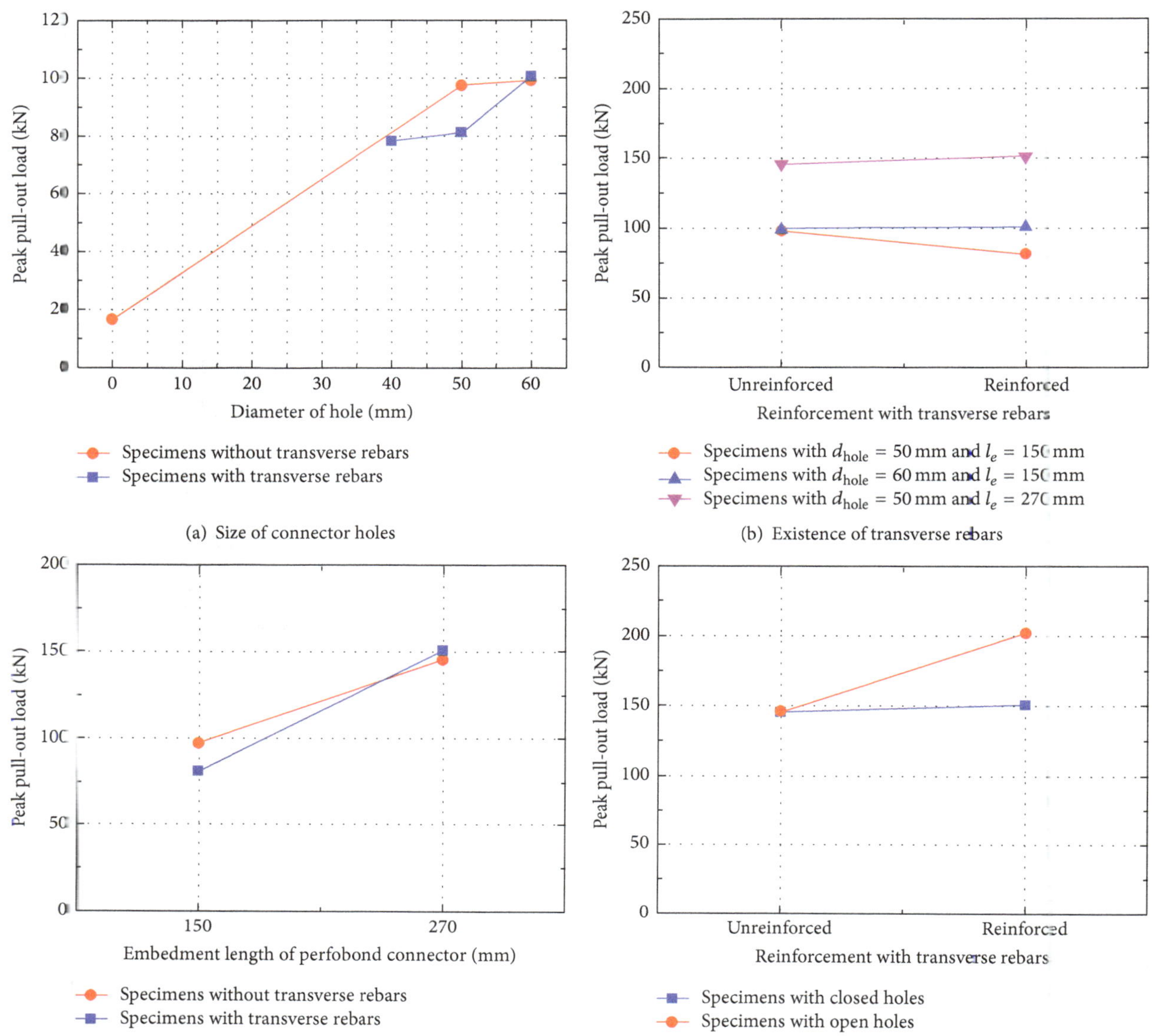

(a) Size of connector holes

(b) Existence of transverse rebars

(c) Embedment length of perfobond connector

(d) Shape of connector hole

FIGURE 12: Effects of test parameters on the shear strength of perfobond shear connectors.

specimen PU-5027-R is approximately 30% greater than that of specimen PU-5027-R, while this kind of a phenomenon is not observed in the case of the first set. This seems to happen because the test specimen with open holes has a larger cross-sectional area as shown in Figure 4; thus, it can retain a higher shear resistance capacity than the one with closed holes if its perfobond connector is fully anchored into the concrete cube with the help of transverse rebars.

4. Conclusions

In this study, we proposed a new type of the perfobond shear connector, which can be used to strengthen the steel pile cap embedded into the structure foundation, and evaluated its pull-out resistance capacity by performing a test on ten specimens. The main conclusions of this paper are as follows.

(1) The perfobond shear connector proposed in this study can retain the peak pull-out load up to 6 times higher than the one without any holes. This indicates that the existence of holes in the shear connector enables the dowel action of concrete inside the hole, resulting in the improvement of the shear resistance capacity of the connector.

(2) The size of the holes in the perfobond shear connector should be sufficiently large enough to guarantee that transverse rebars are integrated into the surrounding

concrete, leading to the enhancement of the shear resistance capacity.

(3) In general, the existence of the transverse rebars enhances the peak pull-out load and is helpful for retaining a significant level of ductility after reaching the peak load.

(4) The increase of embedment length is greatly helpful to improve the shear resistance capacity of the perfobond connector.

(5) The perfobond shear connector with open holes may retain higher shear strength capacity than the one with closed holes if it is fully anchored into the concrete cube with the help of transverse rebars.

Conflict of Interests

The authors declare that there is no conflict of interests regarding the publication of this paper.

Acknowledgment

The corresponding author of this paper acknowledges the support from the Basic Science Research Program through the National Research Foundation of Korea (NRF) funded by the Ministry of Science, ICT & Future Planning (Grant no. 2013R1A1A1076011).

References

[1] ACI Committee 318, *Building Code Requirements for Structural Concrete (ACI 318-11) and Commentary*, American Concrete Institute, Farmington Hills, Mich, USA, 2011.

[2] *Eurocode 2: Design of Concrete Structures—Part 1-1: General Rules and Rules for Buildings*, BS EN 1992-1-1, British Standard Institution, London, UK, 2004.

[3] W. Zellner, "Recent designs of composite bridges and a new type of shear connectors," in *Proceedings of the ASCE/IABSE Engineering Foundation Conference on Composite Construction*, pp. 240–252, Henniker, NH, USA, 1987.

[4] B. M. Sara and M. S. Bahram, "Perfobond shear connector for composite construction," *Engineering Journal*, vol. 39, no. 1, pp. 2–12, 2002.

[5] M. R. Veldanda and M. U. Hosain, "Behaviour of perfobond rib shear connectors: push-out tests," *Canadian Journal of Civil Engineering*, vol. 19, no. 1, pp. 1–10, 1992.

[6] E. C. Oguejiofor and M. U. Hosain, "Behaviour of perfobond rib shear connectors in composite beams: full-size tests," *Canadian Journal of Civil Engineering*, vol. 19, no. 2, pp. 224–235, 1992.

[7] E. C. Oguejiofor and M. U. Hosain, "A paramatic study of perfobond rib shear connectors," *Canadian Journal of Civil Engineering*, vol. 21, pp. 614–625, 1994.

[8] E. C. Oguejiofor and M. U. Hosain, "Numerical analysis of push-out specimens with perfobond rib connectors," *Computers & Structures*, vol. 62, no. 4, pp. 617–624, 1997.

[9] J. Machacek and J. Studnicka, "Perforated shear connectors," *Steel and Composite Structures*, vol. 2, no. 1, pp. 51–66, 2002.

[10] I. Valente and P. J. S. Cruz, "Experimental analysis of Perfobond shear connection between steel and lightweight concrete," *Journal of Constructional Steel Research*, vol. 60, no. 3–5, pp. 465–479, 2004.

[11] H.-Y. Kim and H.-B. Koo, "Composite behavior of perfobond rib shear connector for steel-concrete decks," *Journal of the Korean Society of Civil Engineers*, vol. 26, no. 1, pp. 91–97, 2006 (Korean).

[12] ASTM International, "Standard test method for compressive strength of cylindrical concrete specimens," ASTM Standard C39/C39M-11, ASTM International, 2011.

16

Stress-Strain Relation of Steel-Polypropylene-Blended Fiber-Reinforced Concrete under Uniaxial Cyclic Compression

Lihua Xu, Biao Li, Yin Chi, Changning Li, Biao Huang, and Yuchuan Shi

School of Civil Engineering, Wuhan University, Wuhan, China

Correspondence should be addressed to Yin Chi; yin.chi@whu.edu.cn

Academic Editor: Antonio Caggiano

This paper investigates the cyclic stress-strain behavior of steel-polypropylene-blended fiber-reinforced concrete (BFRC) under uniaxial cyclic compression. A total of 48 prism specimens were tested for different fiber volume fractions and aspect ratios. The results show that the introduction of blended fibers has synergetic effects on improving the cyclic behavior of concrete in terms of peak strength, postpeak ductility, hysteretic energy dissipation, and stiffness degradation. Moreover, the increase in the volume fractions of both steel and polypropylene fibers can lead to a remarkable decrease in plastic strain accumulation. Furthermore, the stiffness degradation ratio as well as the stress deterioration ratio of BFRC can be significantly alleviated in comparison with those of plain concrete, notwithstanding that the degradation amount is insensitive to the variations of fiber parameters. Subsequently, based on the test results, a constitutive model is developed to generalize the cyclic stress-strain responses of BFRC, with the contributions of blended fibers taken into account. The developed model is then verified by independent experimental results and other test data reported in the literature. It is observed that the prediction yields a close estimation of the cyclic compressive behavior of BFRC with varying fiber parameters.

1. Introduction

Concrete has been well acknowledged as one of the most widely used construction and building materials in civil engineering field. It is witnessed nowadays that a number of practical applications for large-scale concrete structures have emerged, such as high-rise buildings, large-span bridges, and nuclear power plants. However, plain concrete demonstrates explicitly its disadvantages that hinder its further application, e.g. susceptibility to cracking, low deformation threshold, and limited ductility, which at times have been the key determinants resulting in the selection of other alternative materials [1]. In an attempt to tackle the limitations, while reducing the dependency on steel reinforcement, extensive research studies have been undertaken, and to date, the majority of innovations published have succeeded in utilizing fiber-reinforced concrete (FRC) to effectively enhance the concrete properties [2–5]. In recent years, the concrete technology has resulted in a combination of different kinds of fibers, i.e., blended fiber-reinforced concrete (BFRC) materials [6–10], in order to achieve a synergetic mechanical behavior.

It has to be mentioned that the cyclic stress-strain relation of BFRC plays an important role in the elastoplastic analysis of concrete members and directly determines the ductile and hysteretic behaviors of concrete structures subjected to seismic, explosive, and fatigue loads [11, 12]. Considerable efforts have been made to investigate the mechanical behaviors of plain and confined concrete under uniaxial cyclic loading [13–25]. The corresponding empirical constitutive models and formulae for generalizing the cyclic stress-strain responses of plain and confined concrete have been well established [26–28]. Moreover, some others have succeeded in the incorporation of fracture energy and damage-plastic concepts for cyclic constitutive modeling [29–32]. Those research outcomes have led to a great success in the prediction of concrete cyclic loading responses and serve as a base for the nonlinear analysis of concrete structural performance. In comparison with plain concrete, research studies on the cyclic mechanical behavior of FRC are rarely reported in the literature, with BFRC in particular [33–38]. Of the limited tests for FRC under cyclic loading, the study conducted by Otter and Naaman is worth quoting. They claimed that steel fiber-reinforced concrete (SFRC) behaves similarly as plain concrete except that the envelope varies with addition of fibers. The conclusion has been verified by other researchers. Thereafter, based on the test results, mathematical expressions analogous to plain concrete for predicting the cyclic stress-strain curve of SFRC under uniaxial compression were developed. With respect to BFRC, experimental investigations are found to be primarily concentrated on the performance of BFRC structural members [39, 40]. Despite that, some constitutive formulations for modeling the stress-strain behaviors of BFRC under uniaxial and true triaxial loads have been established by a small amount of experimental investigations; however, the proposed models for BFRC are noted to be only suitable for monotonic loading cases [7, 8, 10]. The stress-strain relationship of BFRC under cyclic compressive load has not been investigated yet according to the author's knowledge. The effects of blended fiber and the multiple fiber-reinforcing mechanism as well as the underlying blended effects on the concrete performance degradation and energy dissipation are rarely reported. The mechanical behavior and cyclic constitutive model of BFRC, which are of critical importance for civil engineering design and BFRC material applications, remain to be further elucidated.

The objective of this paper is to study the stress-strain relation of steel-polypropylene-blended fiber-reinforced concrete (BFRC) subjected to uniaxial cyclic compression. The influences of volume fraction and aspect ratio of blended fiber on the stress-strain responses and parameters of cyclic mechanical behavior were studied. Based on the experimental results, an empirical constitutive model was developed to generalize the cyclic compressive stress-strain responses of BFRC, with the effects of blended fiber taken into consideration.

2. Experimental Program

2.1. Materials and Specimens Preparation. Prismatic specimens with a dimension of 150 mm × 150 mm × 300 mm were employed in this study. The mix proportions of plain concrete designed according to the code JGJ 55-2011 [41] are given in Table 1, which is the same in [37]. Ordinary Portland cement type P.O. 42.5 was used as the binder. Crushed granitic rocks of sizes between 5 mm and 20 mm and normal river sands with a fineness modulus of 2.7 were used as the coarse aggregates and fine aggregates, respectively. The aggregate gradations of them are presented in Figure 1. A water reducer with a reducing rate of about 20% was adopted.

Two different types of fibers were considered in this study, namely, steel fiber (SF) and polypropylene fiber (PF), as shown in Figure 2. The corresponding properties are given in Table 2. According to the previous studies [7, 37], the volume fraction of SF ranging from 0.5% to 2.0% and the aspect ratio between 30 and 80 are recommended. In this work, the volume fractions of SF were selected as 1.0%, 1.5%, and 2.0%, and the aspect ratios of 30, 60, and 80 were adopted. In addition, monofilament PFs with low volume fractions of 0.1%, 0.15%, and 0.2% were, respectively, used in order to avoid the balling effect due to excessive dosage and ensure the even distribution of fibers. The lengths of PF having a fixed diameter of 48 μm were selected to be 8 mm,

TABLE 1: Designed concrete mix proportions (kg/m^3) [37].

Cement	Sand	Gravel	Water	Superplasticizer	Water-cement ratio
417	724	1086	175	2.1	0.42

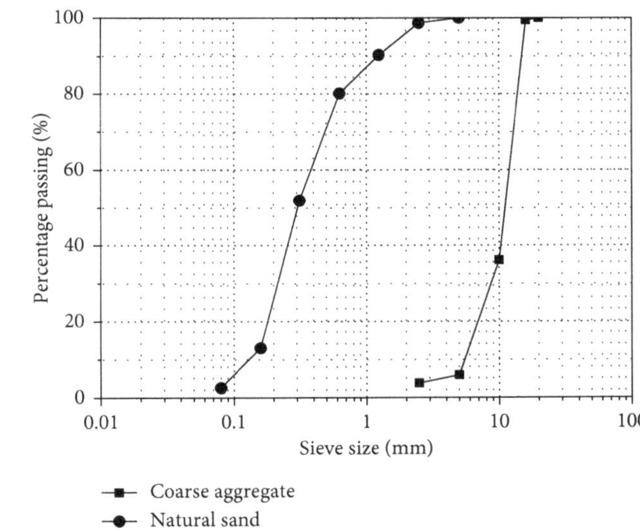

FIGURE 1: Grading curves for fine and coarse aggregates.

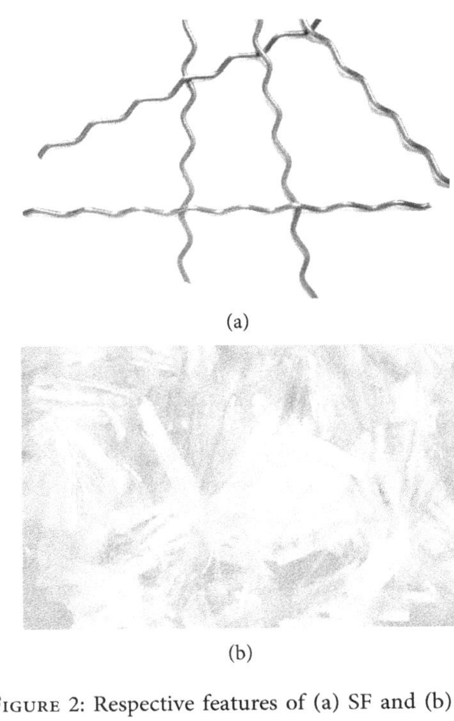

(a)

(b)

FIGURE 2: Respective features of (a) SF and (b) PF.

Table 2: Major properties of fibers.

Fiber species	No.	Aspect ratio	Equivalent diameter (mm)	Fiber type	Density (g/cm^3)	Tensile strength (MPa)
Steel fiber	SA	30	0.55	Corrugated type	7.8	≥600
	SB	60				
	SC	80				
Polypropylene fiber	PA	167	0.048	Monofilament type	0.91	≥400
	PB	280				
	PC	396				

13.6 mm, and 19 mm, corresponding to the aspect ratios of 167, 280, and 396, respectively.

All the specimens were fabricated following the Chinese Standard CSCE 38:2004 [42]. After a uniform mix, the concrete was cast into plastic forms. After 24 hours, the specimens were demoulded carefully and stored in a standard curing room with a constant temperature of 20°Cand humidity of 95% until the 28-day strength was achieved. Then the specimens were removed from the curing room, and the surfaces were cleaned with water. In addition, for each mixture, six cubes of 150 mm side length were prepared for compressive strength (f_{cu}) and splitting tensile strength (f_{st}) tests. The test results from averaging a group of three are listed in Table 3.

2.2. Test Setup. The cyclic tests were performed on a universal electrohydraulic servo-rock testing machine-INSTRON-1346 with a 2000 kN load capacity, as shown in Figure 3. One linear variable displacement transducer (1#LVDT) with a maximal range of 5 mm was used to measure the vertical displacement, and the lateral displacements were monitored by 2#, 3#LVDT during the loading process. The axial loads and displacements were automatically recorded to the data acquisition system with a synchronized frequency of 50 Hz. The test setup is schematically shown in Figure 3.

2.3. Loading Scheme. A preload of approximately 10% of the ultimate strength was applied at the beginning of each test. A hierarchical loading method with a controlled displacement load was used, as shown in Figure 4. In the prepeak region (stageI), the displacement increment in each step was 0.15 mm with a loading speed of 0.01 mm/s. In the postpeak region (stage II), the displacement increment was then set to 0.3 mm. The load-controlled method with a speed of 10 kN/s was used in the unloading process.

3. Experimental Results

3.1. Failure Pattern. Figure 5 shows the typical failure modes of concrete specimens. It is found that there are distinct differences in the failure modes between plain concrete and FRC. The failure of FRC is dominated by shear cracks, whilst that of plain concrete is determined by tensile cracks. For plain concrete (S000P000), the concrete specimen collapses into fine columns sharing the axial load at the end of loading process. Nearly vertical macrocracks that coalesced from microcracks and initial defects can be observed. In comparison with plain concrete, diagonal cracks with an angle ranging from 45° to 60° can be seen obviously in FRC specimens. For polypropylene fiber-reinforced concrete (PFRC), the specimens demonstrate a brittle failure pattern, indicating that PF induces sudden energy dissipation due to fiber-fracturing events. For SFRC, the cracks are finer and smaller than those of plain concrete, which is mainly owing to the fiber crack-arresting effect. Moreover, it is evidenced that BFRC specimen behaves more ductile, with better specimen integrity.

Table 3: Details of specimens.

No.	Specimens	SF		PF		f_{cu} (MPa)	f_{st} (MPa)
		V_{sf}	l_{sf}/d_{sf}	V_{pf}	l_{pf}/d_{pf}		
1	S000P000	—	—	—	—	47.53	2.75
2	SB10P000	1.00%	60	—	—	51.75	3.96
3	SB15P000	1.50%	60	—	—	52.91	4.47
4	S000PA15	—	—	0.15%	167	48.80	3.47
5	S000PC15	—	—	0.15%	396	50.65	3.58
6	SB10PA10	1.00%	60	0.10%	167	53.71	3.99
7	SB10PA15	1.00%	60	0.15%	167	55.2	4.13
8	SB15PA10	1.50%	60	0.10%	167	58.05	4.61
9	SA15PB15	1.50%	30	0.15%	280	54.65	4.19
10	SB15PA15	1.50%	60	0.15%	167	53.35	4.25
11	SB15PB15	1.50%	60	0.15%	280	52.35	4.36
12	SB15PC15	1.50%	60	0.15%	396	51.15	4.50
13	SC15PA15	1.50%	80	0.15%	167	53.70	4.76
14	SB15PA20	1.50%	60	0.20%	167	50.91	3.76
15	SB20PA15	2.00%	60	0.15%	167	56.22	5.36
16	SB20PA20	2.00%	60	0.20%	167	54.65	5.07

3.2. Cyclic Stress-Strain Response. The typical axial stress-strain relations of BFRC for various fiber volume fractions and aspect ratios under cyclic compression are illustrated in Figure 6. The cyclic stress-strain responses can be summarized as follows:

(i) *Distinct Unloading and Reloading Paths.* A full cycle of the stress-strain response is composed of two distinct and well-defined paths: unloading branch and reloading branch. The unloading branch typically exhibits a curvilinear feature that is characterized by a progressive slope reduction. The path eventually intersects the horizontal strain axis with an unrecoverable strain that is referred to as plastic strain. Immediately after, the reloading path is approximately linear until the intersection with unloading path, which is followed by a significant nonlinearity once the unloading strain, is exceeded.

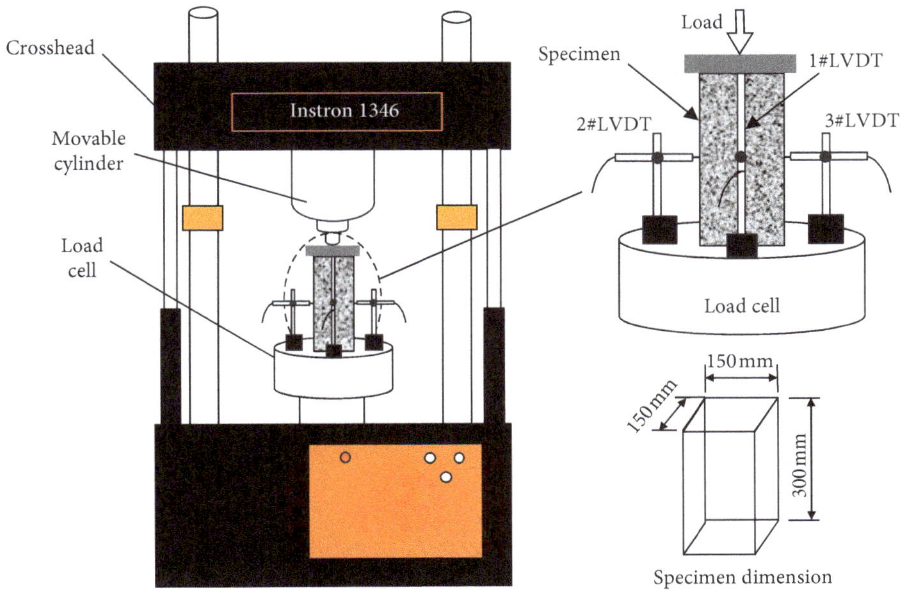

FIGURE 3: Schematic diagram of experimental setups.

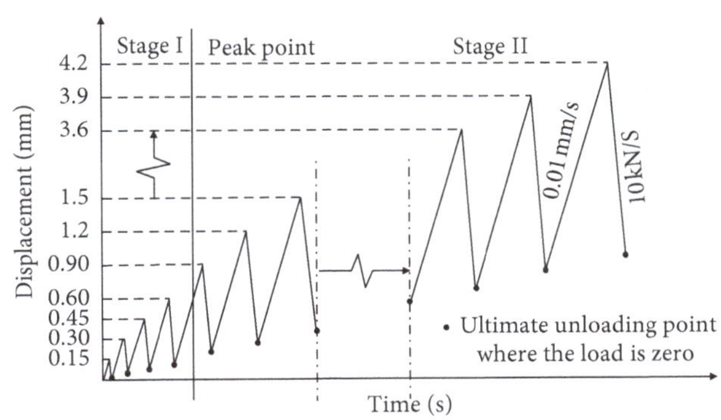

FIGURE 4: Schematic diagram of cyclic loading procedure [37].

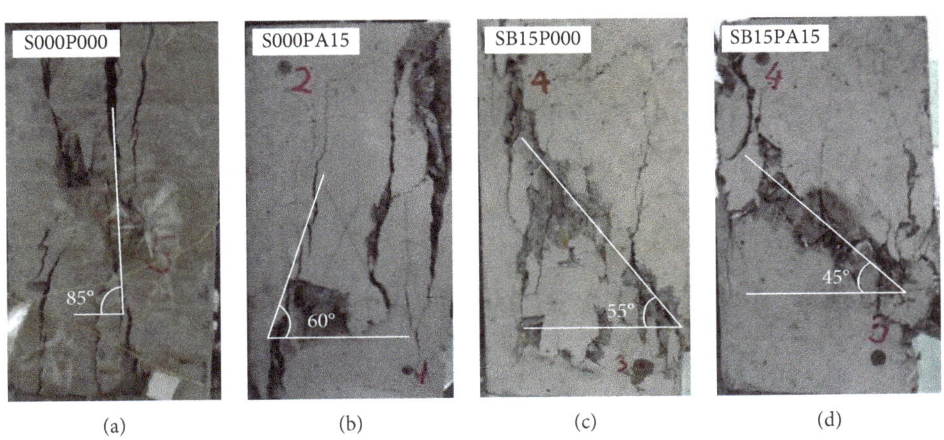

FIGURE 5: Typical failure modes for concrete specimens.

(ii) *Remarkable Performance Degradation*. Remarkable degradation in elastic stiffness and stress can be observed for all the stress-strain curves, which differ widely for different fiber parameters. The initial segment for each unloading path is almost vertical with an abrupt stress drop for a small strain change;

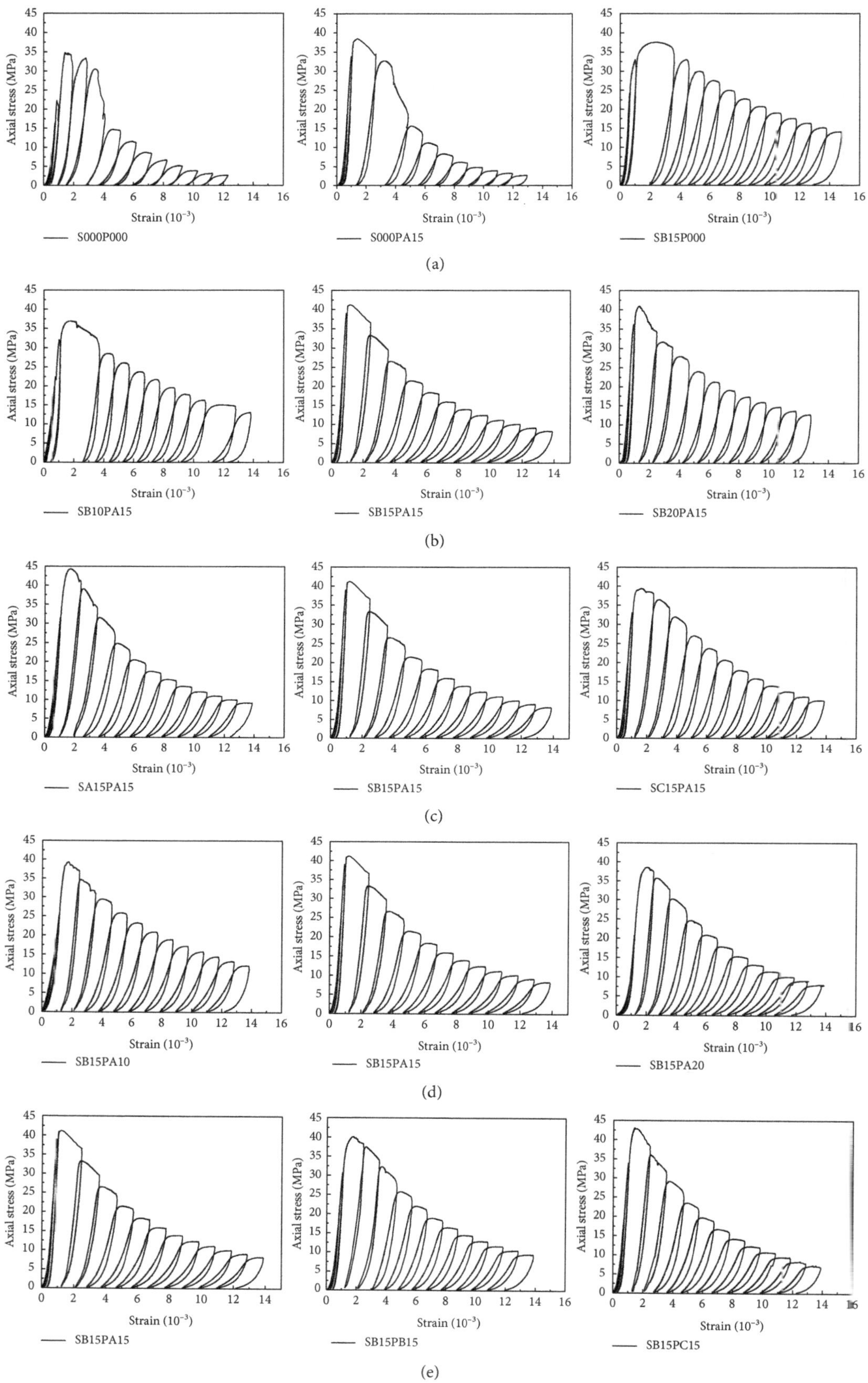

FIGURE 6: Cyclic stress-strain relations of BFRC specimen for various fiber volume fractions and aspect ratios. (a) Effect of fiber types. (b) Effect of volume fraction of SF. (c) Effect of aspect ratio of SF. (d) Effect of volume fraction of PF. (e) Effect of aspect ratio of PF.

however, when approaching the strain axis, the curvature of the unloading path turns larger. It is noted that in comparison with the envelope stress at the same unloading strain, significant stress deterioration can be observed after unloading and reloading, which is caused by the crack propagation and damage accumulation.

(iii) *Hysteretic Energy Dissipation Capacity.* Prior to the peak strength, the unloading path is almost overlapping with the loading path because the concrete specimen almost remains elastic, where the hysteretic energy dissipation is not evident. However, in the postpeak region, the hysteretic loop becomes more obvious with increasing loading cycles. In addition, the capacity of FRC in hysteretic energy dissipation is stronger than that of plain concrete, and for plain concrete, a small hysteretic loop after the peak stress is observed, and it fades away with increasing loading cycles. The phenomenon is attributed to the debonding between aggregate and cement matrix in plain concrete, and the interfacial cracks propagate along with the principle macrocracks. With respect to SFRC, the energy dissipation remains active. At the failure stage, fiber sliding and pull-out mechanism are the main contributors to energy dissipation.

(iv) *Envelope Curve.* In general, the stress-strain envelope curves of FRC are smoother than those of plain concrete. For plain concrete, a sharp load drop after the peak point due to damage accumulation can be observed. For FRC, the load drop is slighter than that of plain concrete owing to fiber crack-bridging effect, especially for SFRC and BFRC. Furthermore, a slower rate of elastic stiffness degradation is observed for SFRC specimens, which is consistent with the conclusions in the literature [34, 37].

Figure 7 depicts a schematic diagram of a cyclic stress-strain curve extracted from the current results, which is believed to be well suited to reproduce the actual cyclic compressive response. The corresponding morphologies of typical specimen surfaces for each loading stage are also shown in Figure 7. It can be observed from the test results that the cyclic compressive process can be divided into six stages:

(i) Elastic stage (Path OA). The initial microcracks are restrained by the randomly distributed fibers. The contributions of blended fiber on the mechanical behavior of BFRC at this stage are negligible.

(ii) Initial crack propagating stage (Path AB). Visible fine cracks can be observed on the specimen at this stage. The internal crack distributions and the degree of crack propagation are determined by the volume fraction of PF [10].

(iii) Stable crack propagating stage (Path BC). Macrocracks can be seen when the load reaches the ultimate strength, and at the peak stress, a sudden increase in the width of cracks can be observed, which is induced by the events of steel fiber pull-out, matrix cracking, and debonding between the aggregate and matrix.

(iv) Fracture-occurring stage (Path CD). The cracks in specimens propagate and coalesce into a main crack. The deformation of PF increases continuously or PF fractures. At this stage, steel fiber debonding occurs continuously with increasing loads and loading cycles.

(v) Continuous failure stage (Path DE). The width of the main cracks becomes larger, and the partial SFs are completely pulled out, with small pieces of concrete peeled off.

(vi) Convergence stage (Path EF). The tensile stress between the two sides of a crack is undertaken by SFs, and when the increasing amount of fibers is pulled out, the concrete specimen collapses into failure.

In addition, Path GH is the unloading stage, where the stress is diminishing and the elastic deformation is recovered, with partial cracks closure. Path HI is the first reloading stage. The axial stress has not reached the last unloading stress, and therefore, the width of cracks almost remains as a constant. Path IJ is the second reloading stage. At this stage, the cracks propagate continually until the next unloading strain.

4. Analysis and Discussion

4.1. Envelope Curve. In order to analyze the effects of blended fiber on the mechanical behavior of concrete, the envelope curves referenced as the upper boundary of a cyclic response are plotted for each specimen, as shown in Figure 8. The results indicate that the fibers mainly affect the postpeak response of concrete; the envelope curves in the prepeak region for different fiber parameters are almost identical with the lone exception of peak strength. Figures 8(a) and 8(b) show that when adding SF into concrete matrix, an enhancement in peak stress is observed, and the degradations in concrete mechanical properties in terms of strength and initial elastic stiffness are slighter than those of PFRC specimens (S000PA15). The peak strength of concrete increases with an increase in the fiber volume fraction and decreases with increasing fiber aspect ratio. In addition, the effect of SF on the ductile behavior has a certain fluctuation under cyclic loads. With respect to PF, it can be observed from Figures 8(c) and 8(d) that the addition of PF has a slight improvement on the peak strength of BFRC specimens; however, the degradation in strength is much more significant than that of SFRC specimens (SB15P000).

Based on the above descriptions, it can be concluded that the synergistic effect of blended fiber is positive on the concrete strength, despite that the effect of PF has a small fluctuation. However, adding PF into concrete matrix has insignificant effects on the ductility of concrete specimens

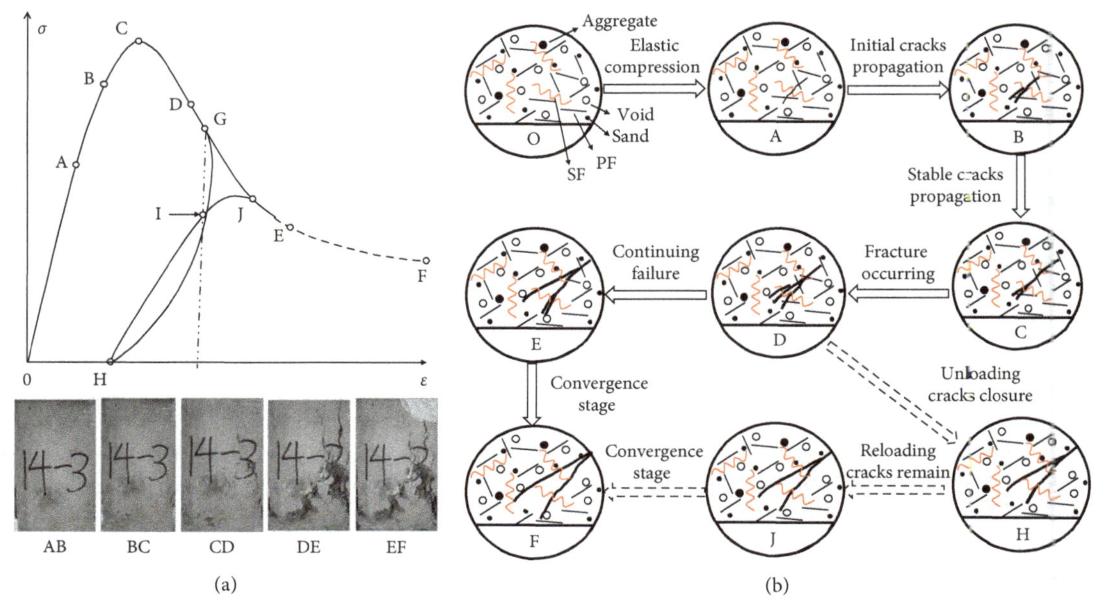

FIGURE 7: Ideal schematic of cyclic compressive process of BFRC specimen. (a) Schematic of stress-strain curve and the corresponding morphologies of typical specimen surface for each loading stage. (b) The schematic drawing showing propagation of a crack.

when subjected to cyclic loading in this study, which can be interpreted that initial damages such as microcracks and weak interfaces increase with increasing volume fraction of PF.

4.2. Plastic Strain. The plastic strain is the accumulated residual axial strain of concrete when the load is unloaded to zero [37]. At present, various formulae for the relationship between plastic strain and envelope unloading strain in the literature have been proposed [18–20]. In this work, from a regression analysis, the power function ($y = px^q$) proposed by Bahn and Hsu [18] is used, where the parameter p is a coefficient of plastic strain and the parameter q is an optimum order which controls the curvature of the relationship curve. Figure 9 shows the relations of plastic strain (ε_{pl}) and envelope unloading strain (ε_{unl}) of BFRC specimens for different fiber volume fractions and aspect ratios subjected to uniaxial cyclic compression. The best fitting parameters are listed in Table 4. The comparisons of parameter p between the fitting and predicted results are shown in Figure 9(f).

It is seen from Figure 9(a) that the plastic strain of FRC specimen is smaller than that of plain concrete at the same envelope unloading strain. It can be interpreted that fibers can bridge microcracks that induce permanent strain due to crack propagation and damage accumulation. Moreover, the crack-arresting capacity of PF is smaller than that of SF. In addition, in most cases, as the fiber volume fraction (for both PF and SF) increases (see in Figures 9(b) and 9(d)), the plastic strain of BFRC specimen decreases in comparison with SFRC and PFRC. Despite that the effect of fiber aspect ratio is fluctuating, all the plastic strains of BFRC are smaller than SFRC and PFRC (see in Table 4), resulted in a notable synergistic effect.

The above analysis could be explained by the effective fiber bridging effect of fiber reinforcing index, which is proportional to the fiber volume fraction and aspect ratio [39]. The format of fiber characteristic parameter λ_{sf} can be given as follows:

$$\lambda_{\text{f}} = \frac{V_{\text{f}} l_{\text{f}}}{d_{\text{f}}}. \tag{1}$$

In this work, two reinforcing indexes η_1 and η_2 are introduced, which are functions of fiber characteristic parameters. However, due to the lack of research on the synergetic effect of blended fibers on the concrete performance, the positive synergetic effect is neglected in this study. Conservatively, the contributions of blended fiber are only numerically superimposed, i.e., the factor η_i is defined as follows:

$$\eta_i = 1 + \alpha_i \lambda_{\text{sf}} + \beta_i \lambda_{\text{pf}}. \tag{2}$$

Thus, the fitting parameters p, q can be expressed as $p = a_0 \eta_1, q = b_0 \eta_2$, where p_0 and q_0 are the fitting parameters for plain concrete. Based on the least-squares analysis, rounded coefficients $\alpha_1 = -0.111$, $\beta_1 = -0.281$ and $\alpha_2 = -0.07$, $\beta_2 = 0.08$ are suggested to evaluate the contributions of SF and PF for the reinforcing indexes η_1 and η_2, with the goodness of fit being 0.978 and 0.998, respectively. However, because the effects of fibers on the shapes of the curve are insignificant, the influences can be ignored [33–37]. Herein, $q = 1.276$ is suggested in this paper. The analytical formula of p is shown as follows:

$$p = 0.589\left(1 - 0.111\lambda_{\text{sf}} - 0.281\lambda_{\text{pf}}\right) \tag{3}$$

Figure 9(f) illustrates the comparisons of the predicted and experimental fitting results for the coefficient a. It is observed that a reasonable agreement is obtained. Therefore,

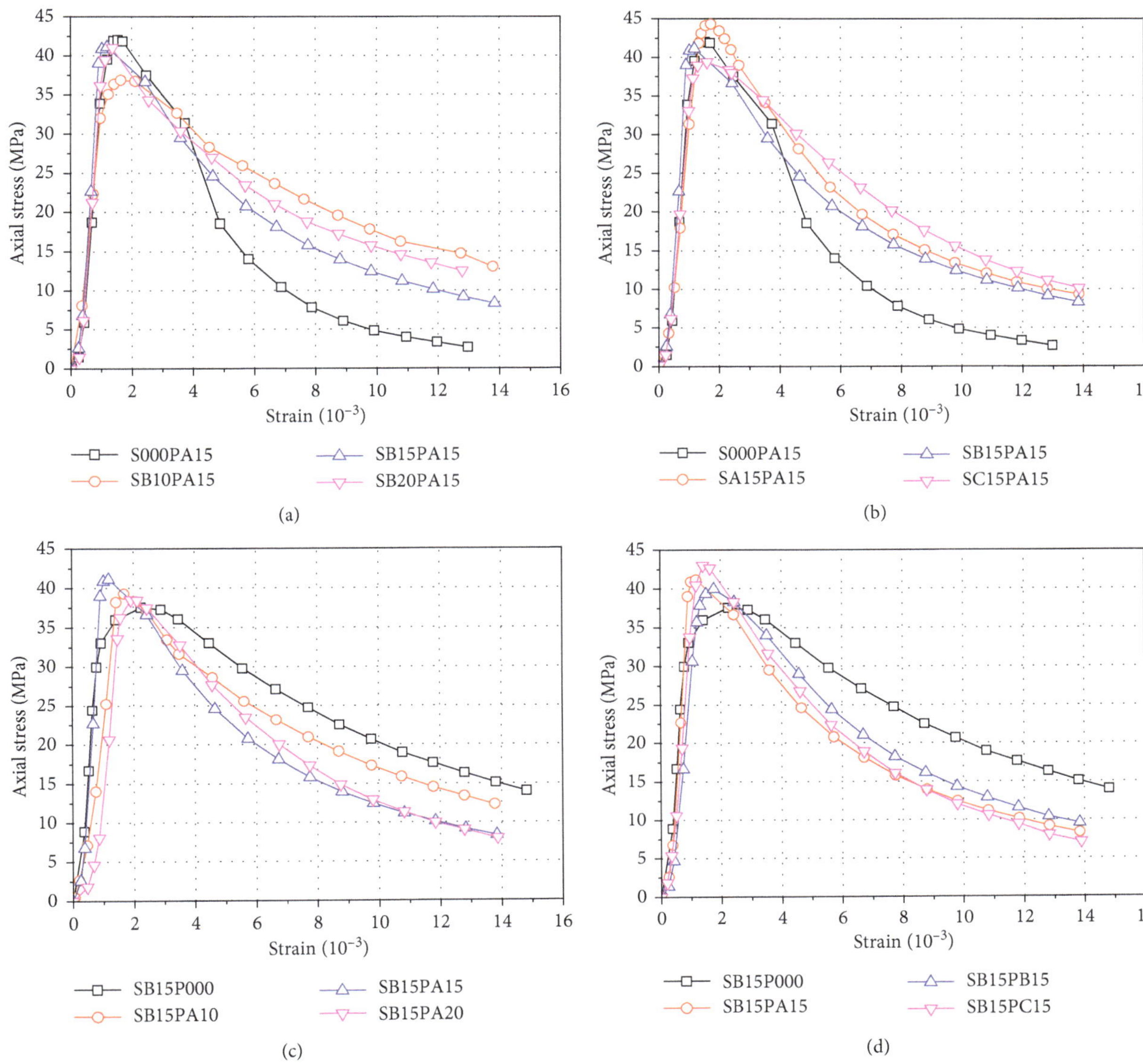

FIGURE 8: Envelope curves of BFRC specimens for different fiber volume fractions and aspect ratios. (a) Effect of volume fraction of SF. (b) Effect of aspect ratio of SF. (c) Effect of volume fraction of PF. (d) Effect of aspect ratio of PF.

the relation between the plastic strain ratio $\varepsilon_{pl}/\varepsilon_{cu}$ and envelope unloading strain ratio $\varepsilon_{unl}/\varepsilon_{cu}$ is expressed as follows:

$$\frac{\varepsilon_{pl}}{\varepsilon_{cu}} = 0.589\left(1 - 0.111\lambda_{sf} - 0.281\lambda_{pf}\right)\left(\frac{\varepsilon_{unl}}{\varepsilon_{cu}}\right)^{1.276}. \quad (4)$$

4.3. Elastic Stiffness Degradation. Elastic stiffness degradation is a key aspect to reflect the damage evolution, which can be quantified by the variations of elastic reloading modulus during cyclic loading. In the present study, as a simplified calculation rule, the slope between the unloading point and ultimate unloading point is defined as the unloading elastic modulus, as shown in Figure 10. The calculated elastic stiffness values of each cycle as well as the initial modulus of the undamaged material E_0 are listed in Table 5.

Figure 11 shows the elastic stiffness degradation process with increasing unloading strain for BFRC specimens. Distinct elastic stiffness degradation with increasing loading cycles can be observed from Figure 11(f). In general, the process can be divided into two stages: descending stage and stable stage. At the descending stage, an obvious drop in the unloading modulus can be observed. The elastic stiffness ratio $(E_{unl,i}/E_0)$ decreases with increasing unloading strain which may due to initial defect propagation as the loads and loading cycles increased. The microcracks fall into macrocracks, and the damages in specimens accumulate gradually. Thereafter, the elastic modulus of a specimen almost remains constant until the concrete specimen collapses into failure.

In comparison with plain concrete, the elastic stiffness degradation of FRC is moderate. The stiffness ratios of FRC are larger than that of plain concrete (see in Figure 11(a)).

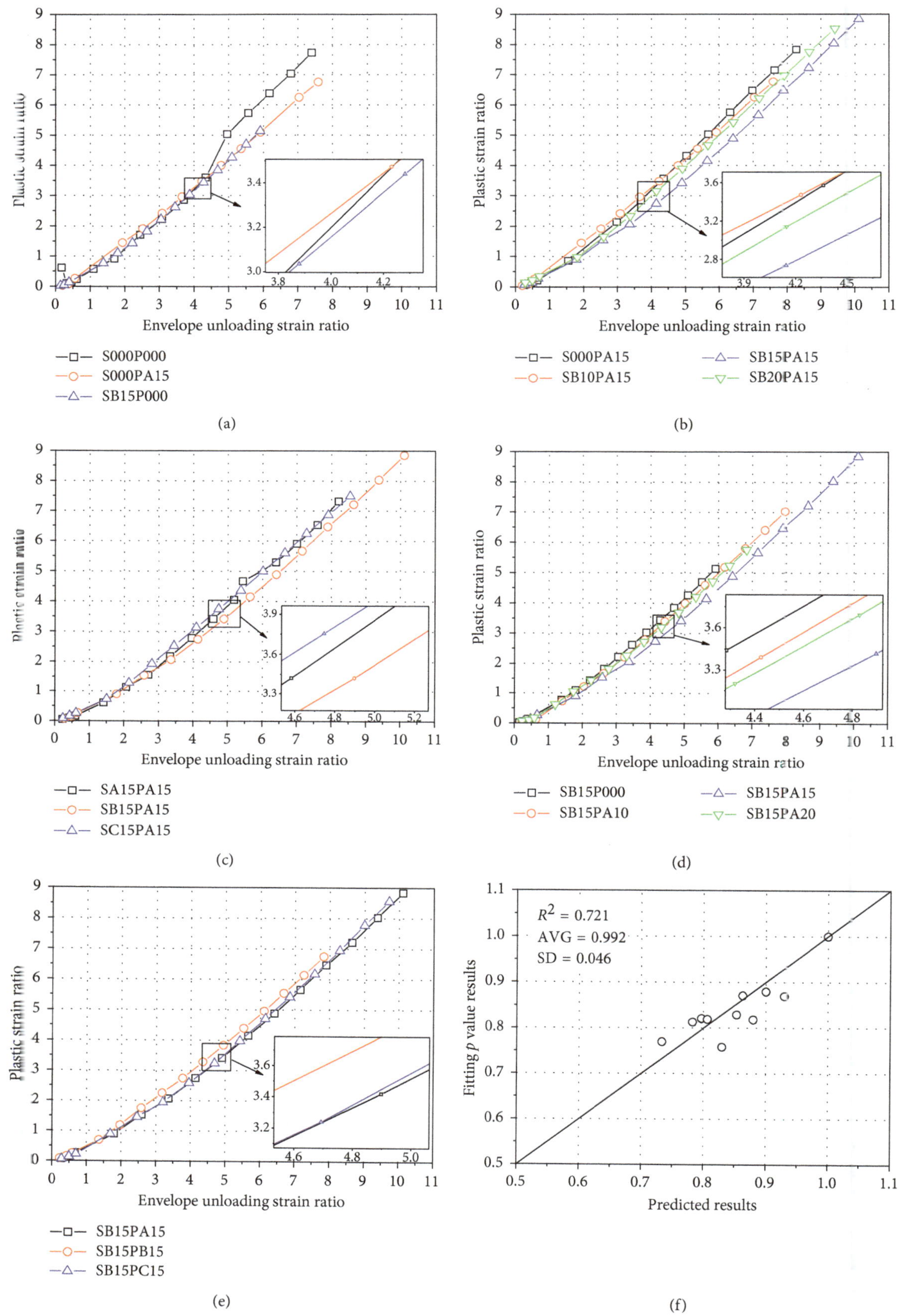

FIGURE 9: Relations between plastic strain and envelope unloading strain. (a) Effect of fiber type. (b) Effect of volume fraction of SF. (c) Effect of aspect ratio of SF. (d) Effect of volume fraction of PF. (e) Effect of aspect ratio of PF. (f) Comparisons of plastic strain between fitting and predicted results.

Table 4: Fitting parameters for the relation between plastic strain and envelope unloading strain.

Specimens	λ_{sf}	λ_{pf}	a	R_t^a
S000P000	0.00	0.000	0.589	1.000
S000PA15	0.00	0.251	0.527	0.870
SB15P000	0.90	0.000	0.533	0.880
SB10PA15	0.60	0.251	0.528	0.871
SB15PA15	0.90	0.251	0.459	0.758
SB20PA15	1.20	0.251	0.497	0.820
SA15PA15	0.45	0.251	0.496	0.818
SC15PA15	1.20	0.251	0.497	0.821
SB15PA10	0.90	0.167	0.502	0.829
SB15PA20	0.90	0.334	0.496	0.819
SB15PB15	0.90	0.420	0.493	0.813
SB15PC15	0.90	0.594	0.467	0.770

Note: a is the basic characteristic parameters of fitting results; R_t^a represents the ratio of plastic strain fitting parameters related to that of the control concrete specimen.

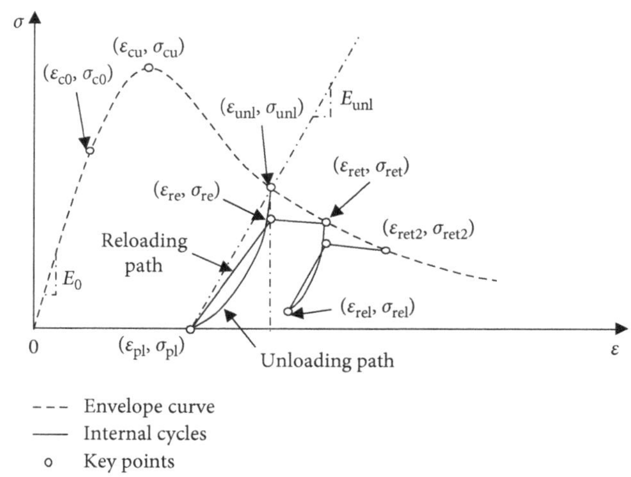

Figure 10: Key parameters' definition of cyclic stress-strain of BFRC.

A sharp drop in elastic stiffness of plain concrete specimen is observed when the load reaches the peak strength, while that of FRC specimen is smoother [37]. The results indicate that fibers can effectively alleviate the speed of elastic stiffness degradation, especially for SF. Moreover, for plain concrete, the specimens are sliced into small columns sharing the axial load at the end of loading process, with the stiffness ratio below 0.03. While for FRC, the stiffness ratio is higher than 0.06 accounted for that the loads are carried by both fibers and concrete. With respect to the effects of fiber volume fraction (see in Figures 11(b) and 11(d)), an increase in the volume fraction of SF induces an increase in the elastic stiffness ratio, while an opposite trend can be observed as the volume fraction of PF increased. In terms of fiber aspect ratio, the influences on elastic stiffness degradation are insignificant (see in Figures 11(c) and 11(e)).

4.4. Stress Deterioration. Research results on the stress deterioration of uniaxial cyclic compression have shown that the stress of a reloading branch is less than the envelope unloading stress at the same unloading strain [14–20]. In this work, a parameter φ is introduced to quantify the stress deterioration ratio [43] shown in Figure 10 and defined as the following:

$$\varphi = \frac{\sigma_{re}}{\sigma_{unl}}, \tag{5}$$

where σ_{re} is the stress on the reloading path at a strain level equal to the envelope unloading strain (ε_{unl}).

The relations between the stress deterioration ratio and nondimensionalized envelope unloading strain ($\varepsilon_c/\varepsilon_{cu}$) are shown in Figure 12. The results indicate that the stress deterioration ratio accesses to 1.0 when the matrix is still elastic. A sharp drop in the stress deterioration ratio is followed when the load reaches the ultimate strength. Thereafter, the stress deterioration ratio remains stable in the subsequent loading cycles.

With respect to the fiber effects, the results indicate that the influences are insignificant, especially when the nondimensionalized unloading strain exceeds 3.0. However, a fluctuant fiber effect can still be seen when the nondimensionalized unloading strain is less than 3.0, which can be ignored according to the conclusions in the literature [25–28]. In this study, based on the analysis of test results, a critical point where $\varepsilon_{unl}/\varepsilon_{cu} = 3$ is postulated, and the constant coefficient 0.85 is obtained for the stress deterioration ratio at the stable stage. For simplicity, the following model for stress deterioration process is proposed, as shown in Figure 13 (black straight line) and expressed as the following:

$$\begin{aligned} &\varphi = 1 - 0.05\frac{\varepsilon_{unl}}{\varepsilon_{cu}} \quad 0 < \frac{\varepsilon_{unl}}{\varepsilon_{un}} < 3, \\ &\varphi = 0.85 \quad \frac{\varepsilon_{unl}}{\varepsilon_{un}} > 3. \end{aligned} \tag{6}$$

5. Cyclic Constitutive Model

This section presents a cyclic constitutive model for BFRC based on the experimental results in the current study. The proposed model consists of three components: (1) a monotonic stress-strain model to describe the envelope curve, (2) a power expression for the unloading path, and (3) a broken line for the reloading path, as shown in Figure 10. The key parameters of the cyclic constitutive model are defined in Figure 10.

5.1. Cyclic Envelope Curve. Previous experimental studies have shown that the monotonic stress-strain curve can be generally assumed as the envelope curve of concrete material under cyclic compression [13–19, 21–27]. The model, initially proposed by Guo and Zhang [44], is further developed and amended that accounts for the presence of blended fiber. It involves a parabola consisting of a curvilinear ascending portion followed by a descending branch, as shown in the following equation.

TABLE 5: Elastic stiffness versus unloading strain for concrete specimens.

ε_{unl}	E_{unl}	ε_{unl}	E_{unl}	ε_{unl}	E_{unl}	ε_{unl}	E_{unl}	ε_{unl}	E_{unl}	ε_{unl}	E_{unl}
S000P000		*S000PA15*		*SB15P000*		*SB10PA15*		*SB15PA15*		*SB20PA15*	
0.00	53.86	0.00	71.35	0.00	67.62	0.00	75.34	0.00	75.06	0.00	84.00
1.02	33.28	0.99	50.20	1.01	50.13	1.03	57.82	0.92	70.85	0.96	74.21
1.82	22.70	2.43	34.50	3.46	20.60	3.51	36.46	2.46	29.99	2.48	29.92
2.82	15.92	4.68	15.30	4.50	19.16	4.58	25.53	3.54	20.57	3.54	23.46
4.05	8.51	5.79	10.49	5.54	15.57	5.61	21.26	4.62	13.71	4.60	19.02
5.10	6.34	6.84	8.35	6.58	13.82	6.65	18.42	5.68	10.77	5.65	17.22
6.15	4.54	7.88	7.00	7.62	12.35	7.68	15.59	6.72	8.86	6.69	15.15
7.20	3.97	8.90	5.83	8.65	11.14	8.69	13.62	7.75	7.66	7.72	13.78
8.23	3.03	9.91	5.18	9.69	10.20	9.73	12.01	8.80	6.62	8.74	12.78
9.24	2.56	10.94	4.86	10.69	9.37	10.75	10.82	9.81	6.05	9.77	12.08
10.26	2.25	11.95	4.29	11.73	8.67	12.78	10.22	10.82	5.69	10.78	11.35
11.28	1.96	12.95	3.85	12.74	8.13	13.78	8.56	11.85	5.15	11.79	10.85
12.27	1.89	—	—	13.77	7.81	—	—	12.86	4.91	12.81	10.46
SA15PA15		*SC15PA15*		*SB15PA10*		*SB15PA20*		*SB15PB15*		*SB15PC15*	
0.00	61.29	0.00	66.04	0.00	59.00	0.00	58.07	0.00	68.49	0.00	65.02
1.02	42.40	1.00	57.39	1.10	27.48	1.19	33.48	1.04	51.33	1.00	49.74
2.38	30.29	2.44	30.67	2.44	31.26	2.44	31.98	2.42	32.29	2.42	33.70
3.49	21.85	3.48	24.72	3.54	21.77	3.52	23.46	3.48	24.52	3.51	22.21
4.56	14.59	4.55	21.30	4.56	17.08	4.58	16.44	4.56	19.84	4.59	14.79
5.64	11.68	5.60	17.48	5.62	14.26	5.64	12.06	5.63	14.71	5.67	11.26
6.70	9.87	6.64	14.79	6.65	12.70	6.70	9.50	6.69	11.40	6.71	9.06
7.75	8.67	7.69	12.53	7.68	11.59	7.73	7.79	7.71	9.51	7.75	7.79
8.76	7.85	8.73	10.65	8.71	10.62	8.77	6.52	8.74	8.15	8.79	6.72
9.20	10.21	9.76	9.38	9.74	9.70	9.80	5.58	9.77	7.10	9.80	5.89
10.82	6.45	10.79	8.04	10.74	9.07	10.82	4.91	10.81	6.29	10.83	5.27
11.84	5.92	11.80	7.31	11.79	8.44	11.84	4.40	11.82	5.73	11.85	4.78
12.84	5.52	12.82	6.47	12.79	7.92	12.85	4.07	12.84	5.21	12.88	4.55
13.86	5.45	—	—	13.82	7.46	13.87	3.64	13.85	4.90	13.89	4.21

$$y = ax + (3-2a)x^2 + (a-2)x^3 \quad 0 < x < 1,$$
$$y = \frac{x}{b(x-1)^2 + x} \quad x > 1, \tag{7}$$

where $x = \varepsilon_c/\varepsilon_{cu}$, $y = \sigma_c/\sigma_{cu}$, and σ_c and ε_c represent the axial stress and strain in which ε_{cu} and σ_{cu} are the axial strain and stress at peak point (transition point), as shown in Figure 10. The physical meaning of the parameter a, which controls the parabolic ascending portion of the stress-strain curve before the transition point reached, is the ratio of initial elastic modulus and tangential modulus at peak stress. From a regression analysis of the results in the current study and literature [7], the parameter a is determined as the following:

$$a = 1.64 + 0.4772\lambda_{sf} - 0.4917\lambda_{pf}. \tag{8}$$

The a value ranges from 1.5 to 3.0. When the calculated value is less than 1.5, it is set equal to 1.5, and as the calculated value is larger than 3.0, it is equal to 3.0. The parameter b that determines the slope of the descending branch is also related to the fiber reinforcing indexes. It enables to simulate the varying softening behaviors as the fiber parameter changes. The calculation of b value is developed using Equation (9). It is noted that the effects of concrete strength are neglected during the evaluation of a and b values. The comparisons of parameters a and b between experimental and predicted results are shown in Figure 14, with reasonable agreements observed.

$$b = 1.185 - 0.501\lambda_{sf} - 0.307\lambda_{pf}. \tag{9}$$

In addition, the ε_{cu} and σ_{cu} are also calculated as a function of fiber characteristic parameters from experimental results in the literature, as shown in Equation (10). The comparisons of peak stress and peak strain between experimental and predicted results are shown in Figure 15.

$$\sigma_{cu} = \sigma_{cu0}\left(1 + 0.179\lambda_{sf} + 0.341\lambda_{pf}\right),$$
$$\varepsilon_{cu} = \varepsilon_{cu0}\left(1 + 0.55\lambda_{sf} + 0.303\lambda_{pf}\right), \tag{10}$$

where σ_{cu0} and ε_{cu0} represent the peak stress and peak strain for plain concrete. As a simplified format, the relationship between σ_{cu0} and ε_{cu0} is developed according to the literature [44] for convenient use, given as

$$\varepsilon_{cu0} = 263.3\sqrt{\sigma_{cu0}} \cdot 10^{-6} \tag{11}$$

5.2. Unloading Path. Considerable research results have shown that the unloading path from the envelope curve to the plastic strain point is nonlinear, which becomes significant when the unloading path intersects the strain axis. In addition, the shape of the unloading curve depends on the initial unloading point and ultimate unloading point [24, 26]. In order to capture the aforementioned features in concrete cyclic response, many formulae have been developed [26–31, 43]. Among them, the power type of

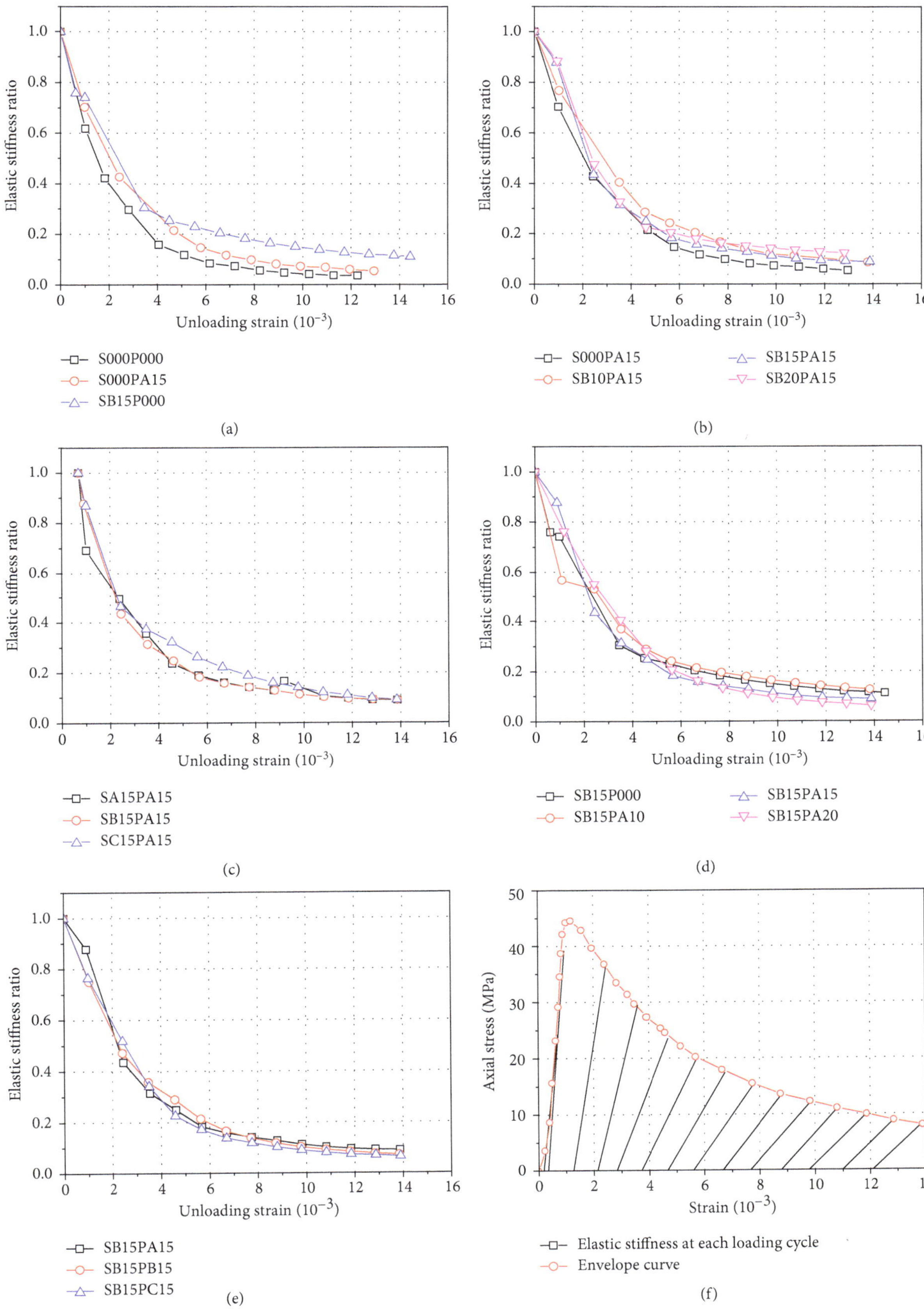

FIGURE 11: Relations between elastic stiffness degradation and envelope unloading strain. (a) Effect of fiber type; (b) effect of volume fraction of SF; (c) effect of aspect ratio of SF; (d) effect of volume fraction of PF; (e) effect of aspect ratio of PF; and (f) schematic diagram of typical elastic stiffness degradation process of BFRC.

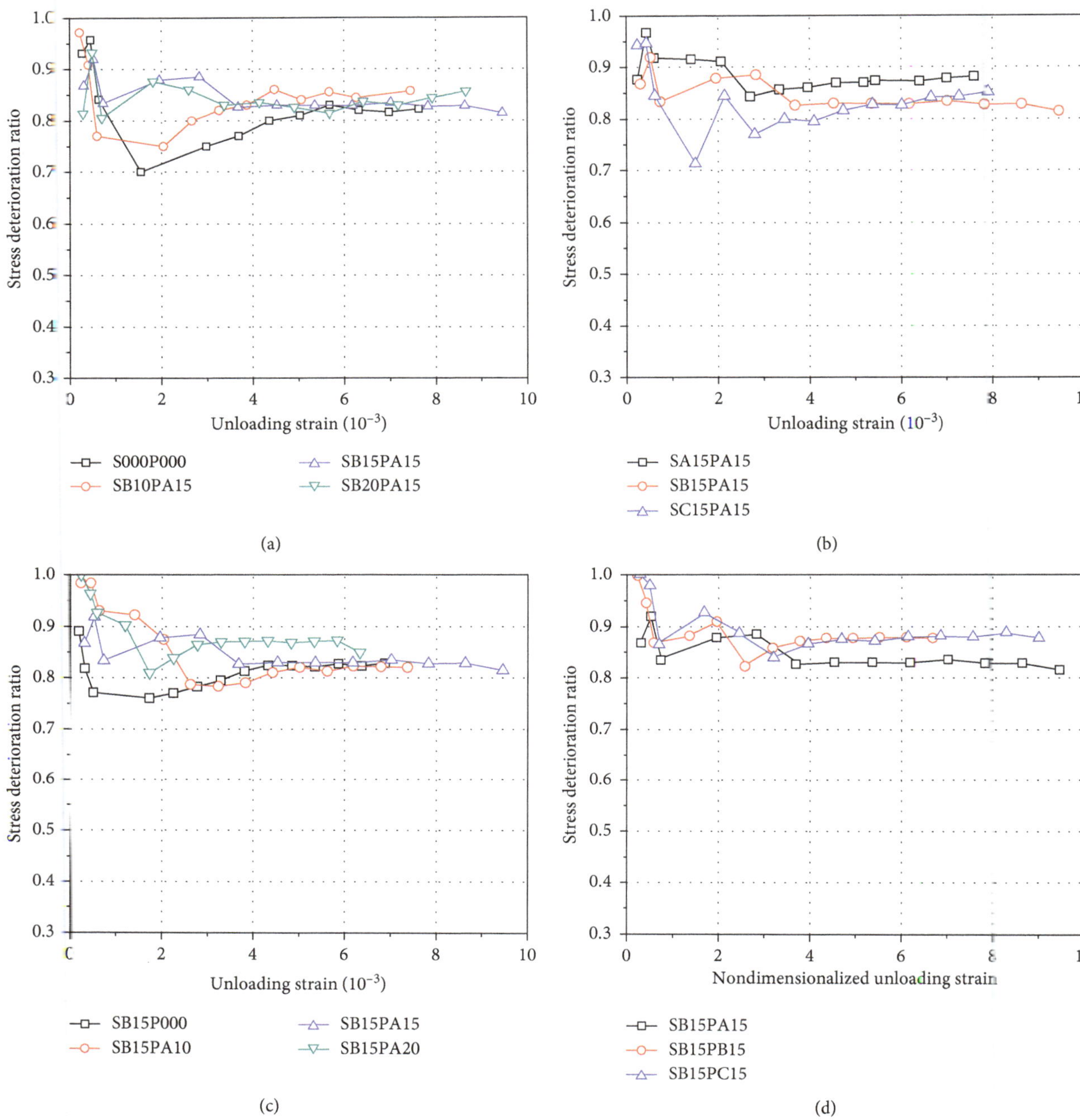

FIGURE 12: Relations between stress deterioration ratio and envelope unloading strain. (a) Effect of steel fiber volume fraction; (b) effect of steel fiber aspect ratio; (c) effect of polypropylene fiber volume fraction; and (d) effect of polypropylene fiber aspect ratio.

equation proposed by Bahn and Hsu [18] for modeling the behavior of unloading path can provide best estimation of the unloading curve. The equation reads as follows:

$$\text{for linear unloading: } \frac{\sigma_c}{\sigma_{unl}} = \frac{\varepsilon_c - \varepsilon_{pl}}{\varepsilon_{unl} - \varepsilon_{pl}}, \quad (12)$$

$$\text{for nonlinear unloading: } \frac{\sigma_c}{\sigma_{unl}} = m\left(\frac{\varepsilon_c - \varepsilon_{pl}}{\varepsilon_{unl} - \varepsilon_{pl}}\right)^n, \quad (13)$$

where m and n are the parameters of the unloading curve. m is a coefficient of unloading curve which can be obtained by fitting the experimental results, with the value ranging between 0.95 and 1.0. In general, $m = 1$ is assumed and adopted for the concrete material. The parameter n controls the curvature of unloading path that relates to the envelope unloading strain closely [24–27]. By means of a regression analysis of the current test results, the formula is written as Equation (14), with the best fitting result shown in Figure 16.

$$n = 1 + 1.132\left(\frac{\varepsilon_{unl}}{\varepsilon_{cu}}\right)^{0.436} \quad (14)$$

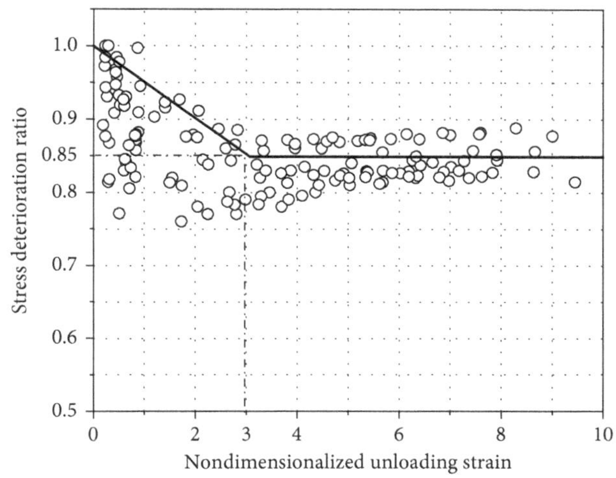

FIGURE 13: Relation between stress deterioration ratio and nondimensionalized unloading strain for BFRC.

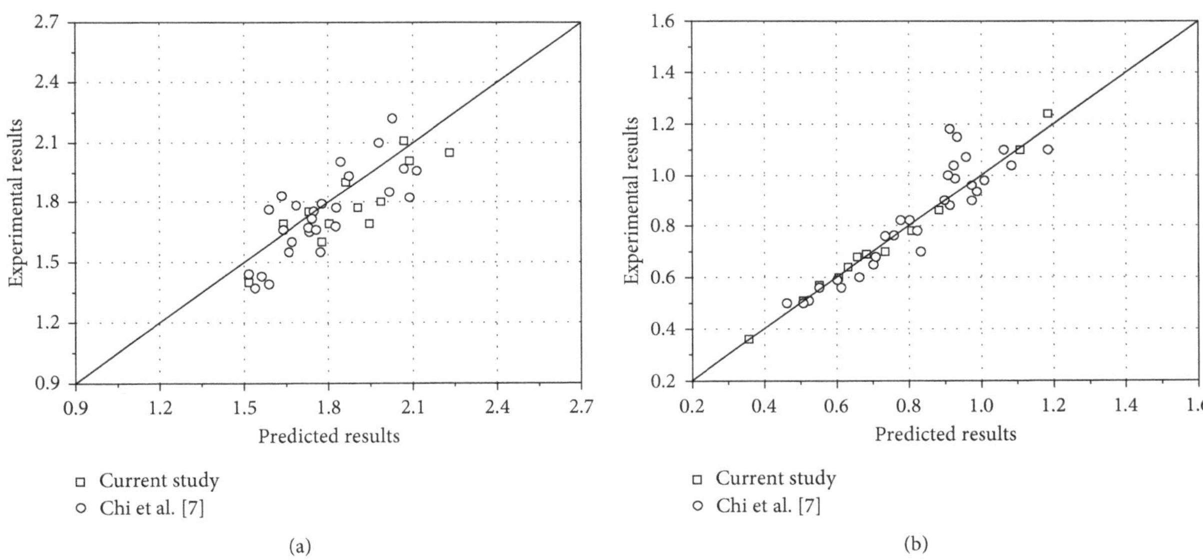

FIGURE 14: Comparisons of parameters between experimental and predicted results. (a) Parameter a; (b) parameter b.

5.3. Reloading Path. As aforementioned, the reloading path of a full cycle consists of a linear segment and a parabolic segment. At present, many equations have been proposed to simulate the reloading path, from linear function to nonlinear approximation. However, the predictions exhibit remarkable deviations in particular when the reloading path is approaching to the intersection point with the unloading curve. To tackle this problem, a sectional model [26, 43] is adopted and described by a broken line, as shown in Figure 10.

5.3.1. The First Linear Portion. The first linear portion of the reloading path is defined by the following equation:

$$\sigma_c = \frac{\sigma_{re}}{\varepsilon_{unl} - \varepsilon_{pl}} \left(\varepsilon_c - \varepsilon_{pl}\right), \quad \varepsilon_c < \varepsilon_{unl}, \tag{15}$$

where σ_{re} is the reloading transition point stress when the strain reaches the unloading strain ε_{unl}. According to Equation (5), σ_{re} can be expressed as follows:

$$\sigma_{re} = \varphi \sigma_{unl} \tag{16}$$

5.3.2. The Second Linear Portion. The second linear portion of the reloading path is expressed by the following equation:

$$\sigma_c = \frac{\sigma_{ret} - \sigma_{re}}{\varepsilon_{ret} - \varepsilon_{unl}} \left(\varepsilon_c - \varepsilon_{unl}\right) + \sigma_{re}, \quad \varepsilon_c > \varepsilon_{unl}, \tag{17}$$

where the point $(\sigma_{ret}, \varepsilon_{ret})$ is the intersection point of the reloading path and the envelope curve. The relationship between σ_{ret} and ε_{ret} can be determined by Equation (7).

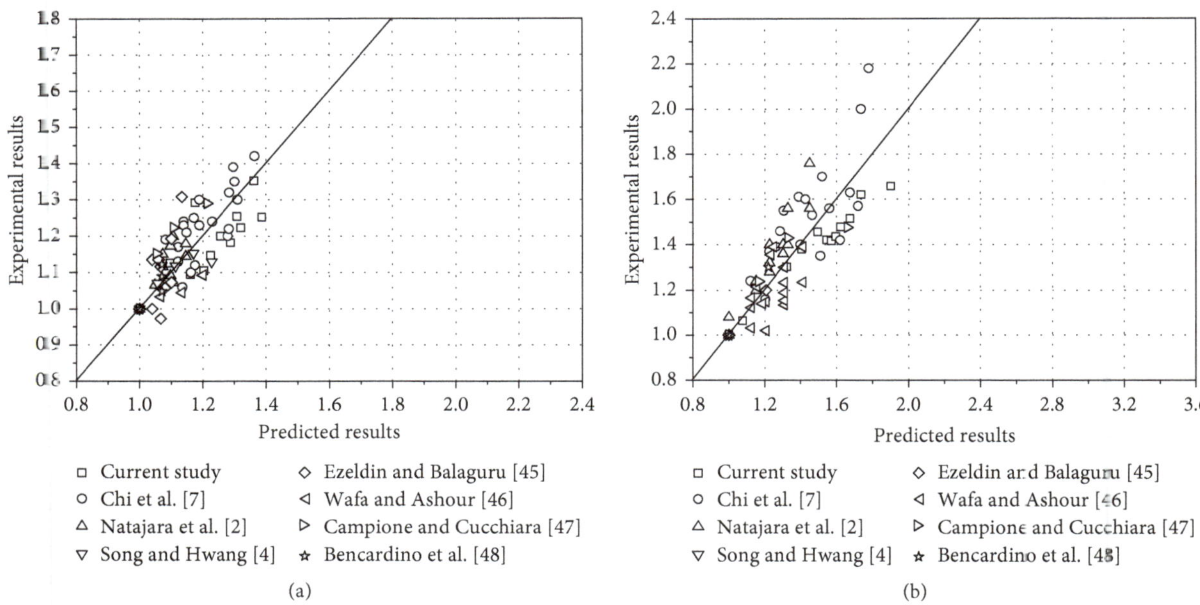

FIGURE 15: Comparisons of mechanical parameters between experimental [2, 4, 7, 45–48] and predicted results. (a) Peak stress; (b) peak strain.

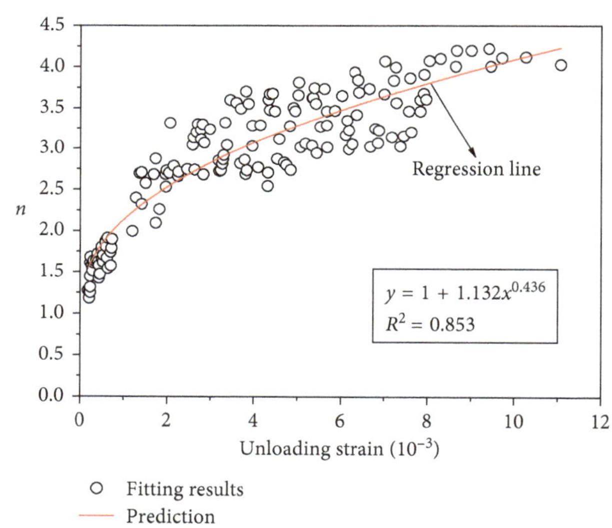

FIGURE 16: Variation of n with $\varepsilon_{unl}/\varepsilon_{co}$.

5.4. Partial Unloading and Reloading Path. For incomplete unloading, the unloading path of the stress-strain curve is interrupted at the point $(\varepsilon_{rel}, \sigma_{rel})$ (Figure 10). The partial unloading path followed by a reloading path can be assumed as a portion of the complete unloading curve. The model equation can be expressed as Equation (13). Assuming that no damage occurs during the unloading process [28–30], the reloading path can also use the broken line model as shown in Equations (15) and (17).

6. Comparison of Experimental Tests and Proposed Models

In order to facilitate better understanding and implementation, a step-by-step process for generating the stress-strain curves of BFRC under cyclic compression is summarized in Figure 17. For a given BFRC specimen, in addition to the concrete compressive strength (σ_{un}) and fiber characteristic parameters of SF (λ_{sf}) and PF (λ_{pf}), only the experimental values of the unloading strain (ε_{unl}) and the

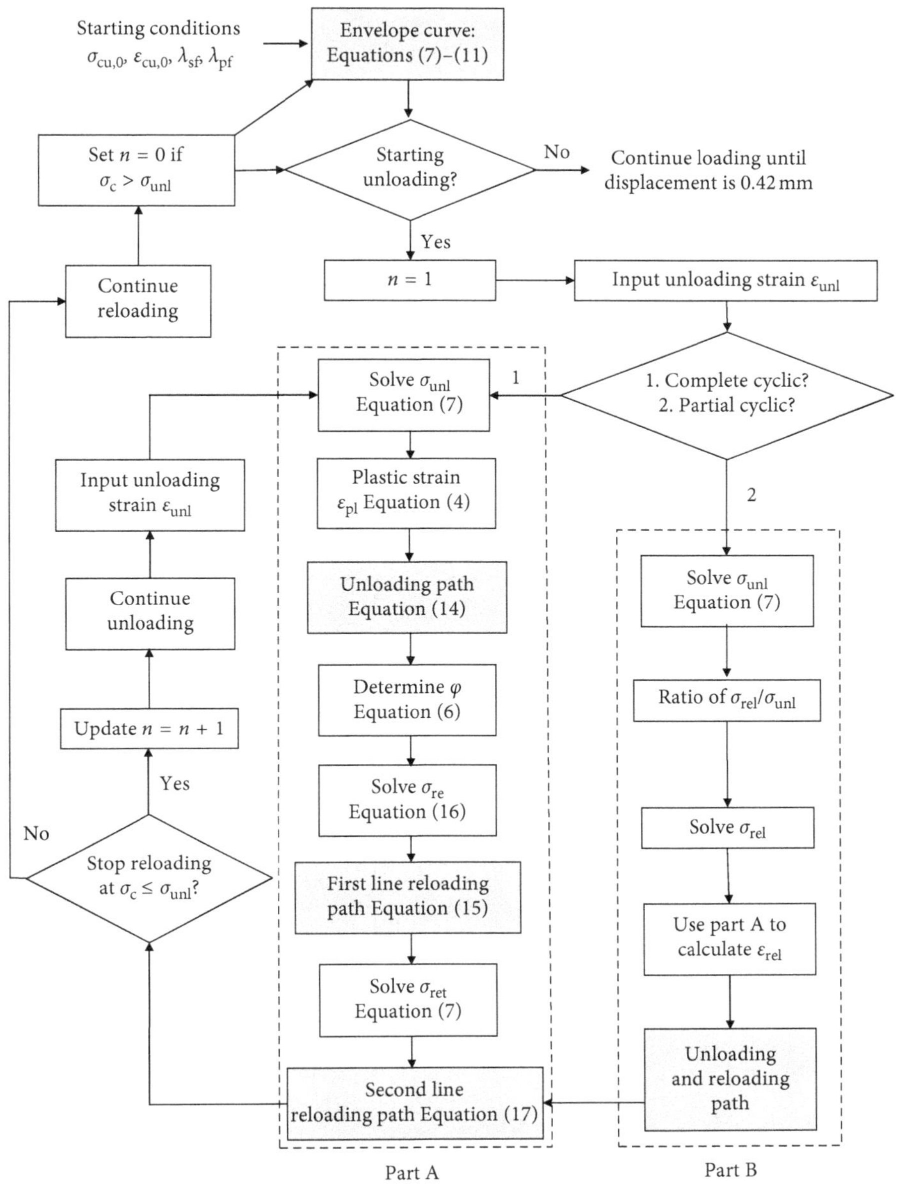

FIGURE 17: Flowchart describing the generation of stress-strain curves under cyclic compression.

partial unloading stress (σ_{rel}) are used as the input parameters. The unloading stress (σ_{unl}), the plastic strain (ε_{pl}), the stress and strain at the transition point (ε_{re}, σ_{re}), and intersection point (ε_{ret}, σ_{ret}) are all calculated from the above-proposed formulae.

Figure 18 shows the typical comparisons of the models' predictions with test results in this study and open literature for plain concrete, PFRC, SFRC, and BFRC, respectively. It is evident that despite slight discrepancies, the proposed models can provide fairly good estimation of cyclic stress-strain response of FRC.

However, it should be noted that the model for PFRC and BFRC is only validated based on the current test results due to the lack of test data in the literature.

7. Conclusions

(i) The introduction of blended fibers into concrete matrix has positive influences on improving the mechanical behavior of concrete under cyclic loading, reflected in that an improvement on peak strength and postpeak ductility can be observed, notwithstanding that PF has an insignificant effect for ductility. The failure pattern of BFRC is ductile, while that of plain concrete is brittle.

(ii) Obvious degradation on elastic stiffness and strength with increasing loading cycles is observed, especially on the postpeak region. The capacity of FRC in hysteretic energy dissipation is stronger than

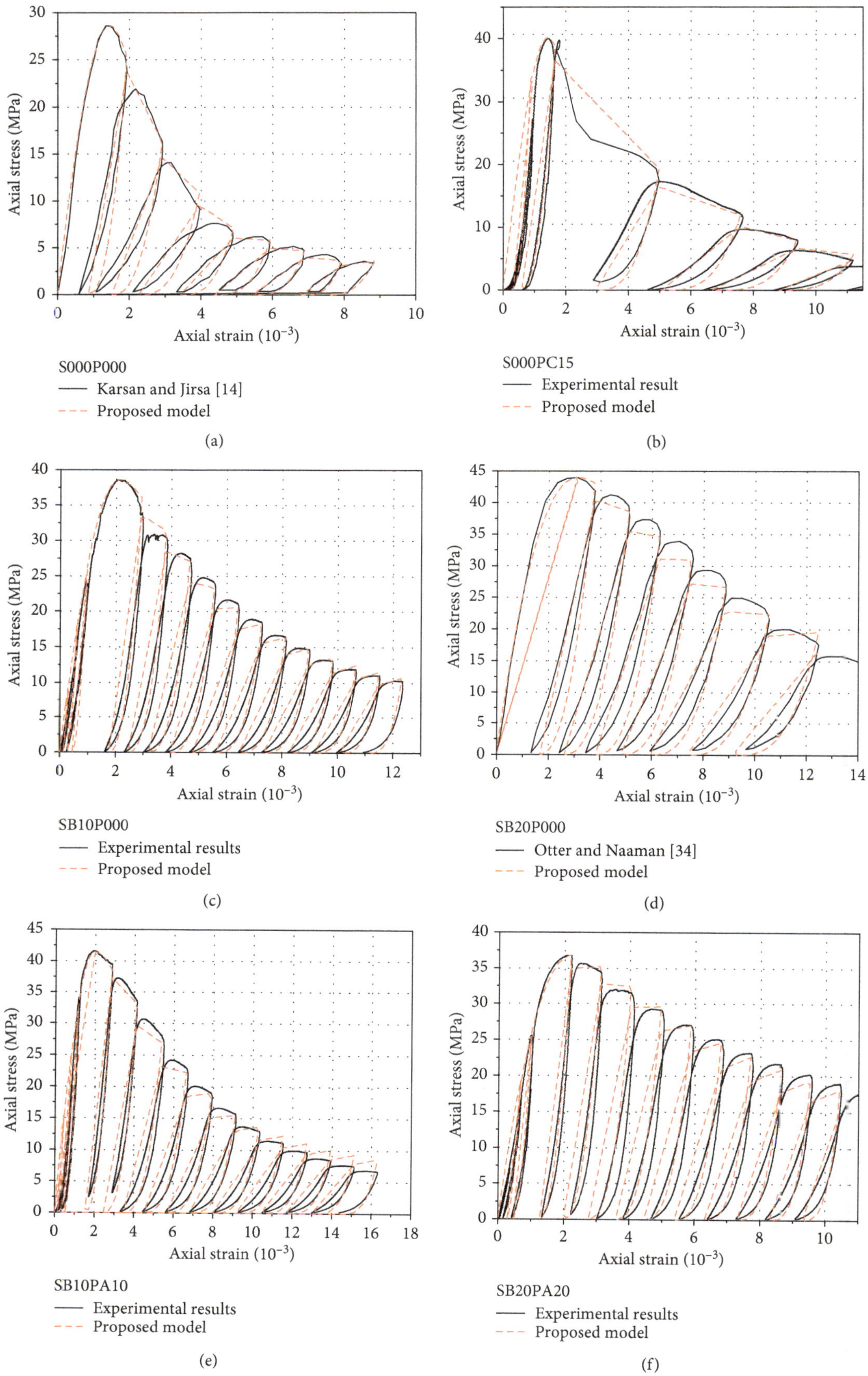

FIGURE 18: Typical comparisons of experimental results and proposed models for concrete.

that of plain concrete. With respect to SFRC and BFRC, the energy dissipation remains active owing to the steel fiber sliding and pull-out mechanisms.

(iii) In comparison with plain concrete, the increase in volume fractions of both fibers can lead to a remarkable decrease in plastic strain accumulation. The stiffness degradation ratio as well as the stress deterioration ratio of BFRC is significantly alleviated in comparison with those of plain concrete, despite the degradation degrees being insensitive to the variations of fiber parameters.

(iv) A semiempirical constitutive model for BFRC under uniaxial cyclic compression is developed to generalize the cyclic stress-strain response, with the contributions of blended fiber taken into consideration. The predictions are found in satisfactory agreement with the results in this study and other literature.

Conflicts of Interest

The authors declare that they have no conflicts of interest.

Acknowledgments

The present work was supported by the Chinese National Natural Science Foundation (Grant nos. 51608397 and 51478367). The financial support is gratefully acknowledged.

References

[1] T. H. Wee, M. S. Chin, and M. A. Mansur, "Stress-strain relationship of high-strength concrete in compression," *Journal of Materials in Civil Engineering*, vol. 8, no. 2, pp. 70–76, 1996.

[2] M. C. Nataraja, N. Dhang, and A. P. Gupta, "Stress-strain curves for steel-fiber reinforced concrete under compression," *Cement & Concrete Composites*, vol. 21, no. 5-6, pp. 383–390, 1999.

[3] M. H. Zhang, L. Li, and P Paramasivam, "Flexural toughness and impact resistance of steel-fibre-reinforced lightweight concrete," *Magazine of Concrete Research*, vol. 56, no. 5, pp. 251–262, 2004.

[4] P. S. Song and S. Hwang, "Mechanical properties of high-strength steel fiber-reinforced concrete," *Construction and Building Materials*, vol. 18, no. 9, pp. 669–673, 2004.

[5] A. M. Brandt, "Fibre reinforced cement-based (FRC) composites after over 40 years of development in building and civil engineering," *Composite Structures*, vol. 86, no. 1–3, pp. 3–9, 2008.

[6] H. D. Yun, F. Fukuyama, E. Jeon, I. S. Yang, and C. S. Choi, "Mechanical properties of high-performance hybrid-fibre-reinforced cementitious composites (HPHFRCCs)," *Magazine of Concrete Research*, vol. 59, no. 4, pp. 257–271, 2007.

[7] Y. Chi, L. H. Xu, and Y. Zhang, "Experimental study on hybrid fiber-reinforced concrete subject to uniaxial compression," *Journal of Materials in Civil Engineering*, vol. 26, no. 2, pp. 211–218, 2014.

[8] Y. Chi, L. H. Xu, and H. S. Yu, "Constitutive modeling of steel-polypropylene hybrid fiber reinforced concrete using a non-associated plasticity and its numerical implementation," *Composite Structures*, vol. 111, no. 1, pp. 497–509, 2014.

[9] S. B. Daneti and T. H. Wee, "Behaviour of hybrid fibre reinforced high-strength lightweight aggregate concrete," *Magazine of Concrete Research*, vol. 63, no. 11, pp. 785–796, 2011.

[10] L. H. Xu, L. Huang, Y. Chi, and G. D. Mei, "Tensile behavior of steel-polypropylene hybrid fiber-reinforced concrete," *ACI Materials Journal*, vol. 2, no. 113, pp. 219–229, 2016.

[11] H. Fathi and H. Dabbagh, "Damage mechanism of SCC under cyclic loading with different speed," *Construction and Building Materials*, vol. 101, no. 1, pp. 252–259, 2015.

[12] M. Nehdi, A. Said, and M. Elmorsi, "Non-linear model for reinforced concrete under cyclic loading," *Magazine of Concrete Research*, vol. 57, no. 4, pp. 211–224, 2005.

[13] B. P. Sinha, K. H. Gerstle, and L. G. Tulin, "Stress-strain relations for concrete under cyclic loading," *ACI Structural Journal*, vol. 61, no. 2, pp. 195–211, 1964.

[14] L. D. Karsan and J. O. Jirsa, "Behavior of concrete under compression loadings," *Journal of Structural Division*, vol. 95, no. 12, pp. 2543–2563, 1969.

[15] D. Z. Yankelevsky and H. W. Rainhardt, "Uniaxial behavior of concrete under cyclic loading," *Journal of Structural Engineering*, vol. 115, no. 1, pp. 166–182, 1989.

[16] Y. H. Lee and K. J. Willan, "Mechanical properties of concrete under cyclic compression," *ACI Materials Journal*, vol. 94, pp. 457–467, 1997.

[17] K. Maekawa and K. F. El-Kashif, "Cyclic cumulative damaging of reinforced concrete in post-peak regions," *Journal of Advanced Concrete Technology*, vol. 2, no. 2, pp. 257–271, 2004.

[18] B. Y. Bahn and C. T. T. Hsu, "Stress-strain behavior of concrete under cyclic loading," *ACI Materials Journal*, vol. 95, no. 2, pp. 178–193, 1988.

[19] E. Osorio, J. M. Bairan, and A. R. Mari, "Lateral behavior of concrete under uniaxial compressive cyclic loading," *Materials and Structures*, vol. 46, no. 5, pp. 709–724, 2013.

[20] S. Sinaie, A. Heidarpour, X. L. Zhao, and J. G. Sanjayan, "Effect of size on the response of cylindrical concrete samples under cyclic loading," *Construction and Building Materials*, vol. 84, no. 1, pp. 399–408, 2015.

[21] H. P. Lee, A. Z. Awang, and W. Omar, "Steel strap confined high strength concrete under uniaxial cyclic compression," *Construction and Building Materials*, vol. 72, no. 1, pp. 48–55, 2014.

[22] H. K. Cheong and S. H. Perry, "Cyclic loading of laterally confined concrete columns," *Materials and Structures*, vol. 26, no. 9, pp. 557–562, 1993.

[23] R. Abbasnia and H. Ziaadiny, "Behavior of concrete prisms confined with FRP composites under axial cyclic compression," *Engineering Structures*, vol. 32, no. 3, pp. 648–655, 2010.

[24] Z. Y. Wang, D. Y. Wang, S. T. Smith, and D. G. Lu, "CFRP-confined square RC columns. I: experimental investigation," *Journal of Composites for Construction*, vol. 16, no. 2, pp. 150–160, 2012.

[25] L. Lam, J. G. Teng, C. H. Cheung, and Y. Xiao, "FRP-confined concrete under axial cyclic compression," *Cement & Concrete Composites*, vol. 28, no. 10, pp. 949–958, 2006.

[26] L. Lam and J. G. Teng, "Stress-strain model for FRP-confined concrete under cyclic axial compression," *Engineering Structures*, vol. 31, no. 2, pp. 308–321, 2009.

[27] N. F. Hany, E. G. Hantouche, and M. H. Harajli, "Axial stress-strain model of CFRP-confined concrete under monotonic and cyclic loading," *Journal of Composites for Construction*, vol. 19, no. 6, pp. 1–16, 2015.

[28] Z. Y. Wang, D. Y. Wang, S. T. Smith, and D. G. Lu, "CFRP-confined square RC columns. II: cyclic axial compression stress-strain model," *Journal of Composites for Construction*, vol. 16, no. 2, pp. 161–170, 2012.

[29] J. F. Sima, P. Roca, and C. Molins, "Cyclic constitutive model for concrete," *Engineering Structures*, vol. 30, no. 3, pp. 695–706, 2008.

[30] M. Breccolotti, M. F. Bonfigli, A. D'Alessandro, and A. L. Materazzi, "Constitutive modeling of plain concrete subjected to cyclic uniaxial compressive loading," *Construction and Building Materials*, vol. 94, pp. 172–180, 2015.

[31] W. He, Y. F. Wu, and K. M. Liew, "A fracture energy based constitutive model for the analysis of reinforced concrete structures under cyclic loading," *Computer Methods in Applied Mechanics & Engineering*, vol. 197, no. 51-52, pp. 4745–4762, 2008.

[32] M. Neuenschwander, M. Knobloch, and M. Fontana, "Suitability of the damage-plasticity modeling concept for concrete at elevated temperatures: experimental validation with uniaxial cyclic compression tests," *Cement and Concrete Research*, vol. 79, no. 2, pp. 57–75, 2016.

[33] D. E. Otter and A. E. Naaman, "Model for response of concrete to random compressive loads," *Journal of Structural Engineering*, vol. 115, no. 11, pp. 2794–2809, 1989.

[34] D. E. Otter and A. E. Naaman, "Properties of steel fiber reinforced concrete under cyclic load," *ACI Materials Journal*, vol. 85, no. 4, pp. 254–261, 1988.

[35] K. Komlos and B. Babal, "Hybrid fibre-reinforced concrete under repeated loading," *Nuclear Engineering Design*, vol. 156, no. 1, pp. 195–200, 1995.

[36] G. Campione and L. L. Mendola, "Stress-strain behavior in compression of lightweight fiber reinforced concrete under monotonic and cyclic loads," *Earthquake Resitant Engineering Structures III*, vol. 57, no. 9, pp. 387–396, 2001.

[37] B. Li, L. Xu, Y. Chi, B. Huang, and C. Li, "Experimental investigation on the stress-strain behavior of steel fiber reinforced concrete subjected to uniaxial cyclic compression," *Construction and Building Materials*, vol. 140, no. 1, pp. 109–118, 2017.

[38] Y. Tanigawa and S. Hatanaka, "Stress-strain relations of steel fiber reinforced concrete under repeated compressive load," *Cement and Concrete Research*, vol. 13, no. 6, pp. 801–808, 1983.

[39] L. Huang, L. H. Xu, Y. Chi, and H. R. Xu, "Experimental investigation on the seismic performance of steel-polypropylene hybrid fiber reinforced concrete columns," *Construction and Building Materials*, vol. 87, no. 1-2, pp. 16–27, 2015.

[40] N. Ganesan, P. V. Indira, and M. V. Sabeena, "Behavior of hybrid fibre reinforced concrete beam-column joints under reverse cyclic loads," *Materials & Design*, vol. 54, no. 2, pp. 685–693, 2014.

[41] JGJ 55-2011, *Specification for Mix Proportion Design of Ordinary Concrete*, Ministry of House and Urban-Rural Development of People's Republic of China, Beijing, China, 2011.

[42] CSCE 38, *Technical Specification for Fiber Reinforced Concrete Structures*, China Association for Engineering Construction Standardization, China, Beijing, China, 2004.

[43] J. Saka and K. Kawashima, "Unloading and reloading stress-strain model for confined concrete," *Journal of Structural Engineering*, vol. 132, no. 1, pp. 112–22, 2006.

[44] Z. H. Guo and X. Q. Zhang, "Experimental investigation of the complete stress-strain curve of concrete," *Journal of Building Structures*, vol. 3, no. 1, pp. 1–12, 1982, in Chinese.

[45] A. S. Ezeldin and P. N. Balaguru, "Normal- and high-strength fiber reinforced concrete under compression," *Journal of Materials in Civil Engineering*, vol. 4, no. 4, pp. 415–429, 1992.

[46] F. F. Wafa and S. A. Ashour, "Mechanical properties of high-strength fiber reinforced concrete," *ACI Materials Journal*, vol. 88, no. 6, pp. 449–455, 1992.

[47] G. Campione and C. Cucchiara, "Steel-concrete bond in lightweight fiber reinforced concrete under monotonic and cyclic actions," *Engineering Structures*, vol. 27, no. 6, pp. 881–890, 2005.

[48] F. Bencardino, L. Rizzuti, and G. Spadea, "Experimental tests v/s theoretical modeling for FRC in compression," in *Proceedings of 6th International Conference on Fracture Mechanics of Concrete and Concrete Structures-FraMCoS-6*, vol. 3, pp. 1473–1480, Catania, Italty, June 2007.

17

An Investigation of the Effect of Parameters and Chip Slenderness Ratio on Drilling Process Quality of AISI 1050 Steel

Zülküf Demir and Rifat Yakut

Mechanical and Manufacturing Engineering of Technology, Faculty of Batman University, 72100 Batman, Turkey

Correspondence should be addressed to Rifat Yakut; rifat45@gmail.com

Academic Editor: Wei Zhou

The chip slenderness ratio is a vital parameter in theoretical and applicable machining operations. In predrilled drilling operations of AISI 1050 steel alloy, HSS drills were employed, and the effect of the selected parameters on the chip slenderness ratio and also the effect of the chip slenderness ratio on the thrust force, surface roughness, drilled hole delamination, tool wear, and chip morphology were investigated. The major parameters, influential on the chip slenderness ratio, were feed rate and point angle, while spindle speed was too small to be negligible. With increasing the chip slenderness ratio, the thrust force and the tool wear decreased, which resulted in appropriate chip morphology, but there were increases in surface roughness. However, the chip slenderness ratio had no effect on the drilled hole delamination.

1. Introduction

Drilling process is a metal cutting method, which processes metal pieces in required geometry and dimension, in manufacturing area [1]. This manufacturing method has an important place as it is used about 33% among traditional manufacturing methods [2]. In the drilling process, researchers conducted many works after having completed complicated operations and manufacturing processes in order to obtain the required highest sensibility and best quality [3]. A lot of parameters such as drill point angle, material used for drilling, and properties of drilling equipment affect the drilling process. The tool equipment affects the demented of the drilled hole, the tool wear, the surface quality, and the chip length and thus the chip morphology [4]. As the feed rate increases, the surface quality is influenced inversely, and in other words, the surface roughness increases. Because the thrust force and the tool wear decrease at high spindle speeds and drill point angles, the surface roughness values decrease; thus, the surface quality improves. However, the process temperature and drilled hole delamination increase at high spindle speeds and drill point angles [5]. The delamination size of the drilled hole depends on the thrust force and feed rate in drilling processes [6, 7]. Besides, thrust force increases the drilled hole delamination and the tool wear [8]. While the numbers of the drilled hole increase, the tool wear ratio increases as well [9]. As the chip height and thrust force increase due to increase in the feed rate, the chip height and the thrust force decrease with increase in spindle speed [10, 11].

Furthermore, the feed rate is a significant parameter, affecting the process quality and the surface roughness, more than the spindle speed in the drilling process [12–20]. With sufficient increase in the thrust force in the drilling process, the web thickness of the drill and the relief angle increase; otherwise, the drill point angle and the helical angle decrease [21, 22]. However, with increase in the feed rate and point angle, the thrust force and surface roughness increase [23]. In the drilling process, vibration, tool wear, and dry process conditions cause changes in the diameter of the drilled hole [24]. At low cutting speeds and feed rates, due to the requested chip morphology availability, a better performance is obtainable in drilling processes. In addition, the tool point angle does not significantly affect the process temperature [25]. During the process, discontinuous and short shaped chips are obtained, due to the intermittent plastic deformation at low spindle speeds. With increasing the drill point angle and the spindle speed, the thrust force, resulting from the applied pressure on workpiece material by the drill point, decreases. Chip thickness varies considerably on all along the cutting edge during a cycle of the drill point

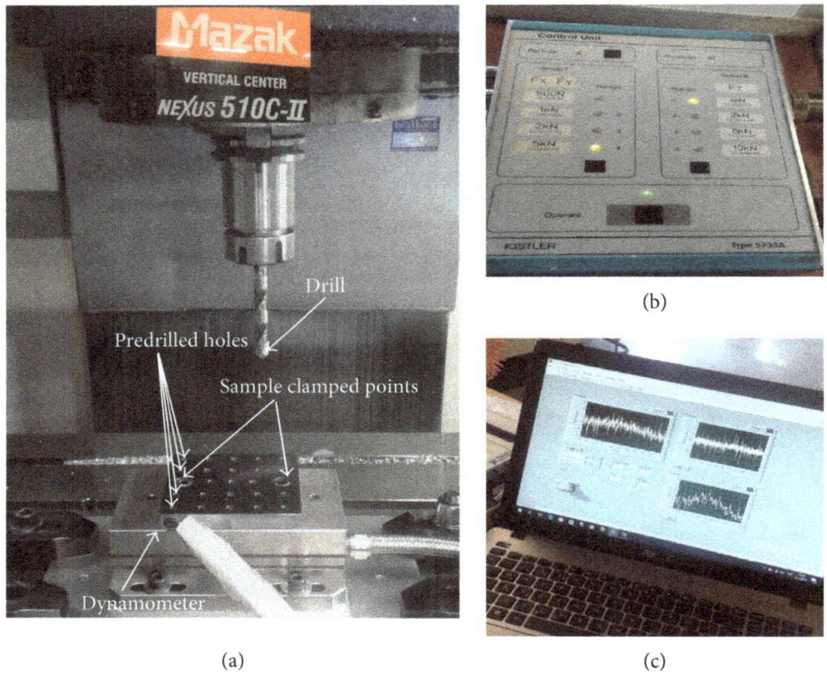

FIGURE 1: Experimental setup: (a) CNC milling machine, (b) 5233A-type dynamometer, and (c) computer.

[26]. With increasing the cutting conditions, the morphology of chip structure changes into a brittle shape from ductile. With enhancing the feed rate, the chip height increases, and also the thrust force, the drilled hole delamination, the surface roughness, and the tool wear increase as well [11]. Conical cutting edge, starting from the drill point, causes changes in cutting force all along the cutting edge of the drill [27]. Because the conical geometry is formed by the point angle of drill, the main feature of the drilling process shows differences from all along the cutting edge. While the feed rate gets the maximum ratio at the outmost diameter of the drill, it approaches to zero value toward the point axis of the drill where it constitutes core of the drill because of decrease in the diameter [28].

The chip slenderness ratio is the proportion of cutting depth to feed rate, which is the vital parameter to affect the outcomes, in machining. Therefore, the investigation of the chip slenderness ratio is the main phenomenon. Furthermore, the goal of this study is to investigate the effect of the chip slenderness ratio on the predrilled drilling process. In the literature, it is identified as a ratio of the depth of cutting to the feed rate, constituting an important parameter in machining operations. In drilling processes, due to the fact that the depth of the chip depends on the diameter of the drilled hole and the drill point, the chip slenderness ratio was calculated by the ratio of the square of the feed rate to the tool chip contact area. Considering the geometry of the drill point, the chip area was calculated mathematically depending on the drill point angle. The chip areas were calculated by crossing these two measured geometric dimensions. The effects of the drill point angle, spindle speed, and feed rate on the chip slenderness ratio were investigated during the study. In addition, with the help of the calculated chip area from this study, theoretical chip slenderness ratios were calculated.

TABLE 1: Mechanical characteristics and chemical components of AISI 1050 alloy steel.

Mechanical characteristics	
Intensity	7.85 g/cm^3
Stress strength	690 MPa
Tensile strength	580 MPa
Slip modulus	80 GPa
Elastic modulus	190–210 GPa
Poisson's ratio	0.27–0.30
Brinell hardness value	197
Heat conductivity	49.8 W/mK
Chemical components	
Name of the element in composition	Percentage (%) ratio of composition
C	0.47–0.55
Mn	0.6–0.90
P	≤0.040
S	≤0.050
Fe	98.46–98.92

2. Experimental Procedure

The experiments were conducted on Mazak Vertical Center CNC Milling Machine, which is the NEXUS 510 C-II model as shown in Figure 1. AISI 1050 alloy steel samples were prepared as $100 \times 100 \times 15\,mm^3$ dimensions. The samples were predrilled 5 mm in diameter before the experiment. The distance was arranged as 20 mm between the centers of

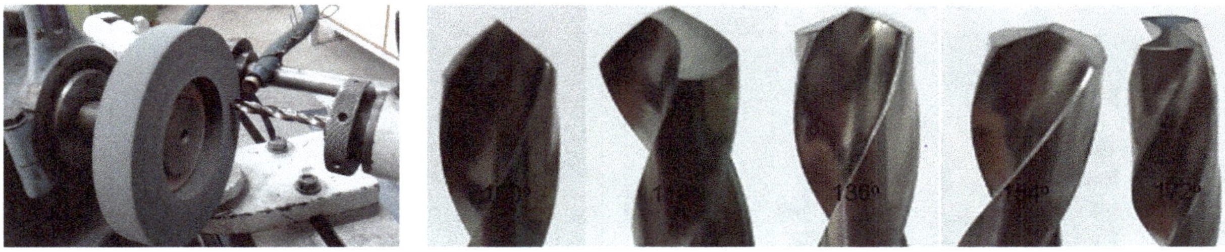

(a) (b)

FIGURE 2: (a) Tool grinding machine and (b) drills with different point angles.

TABLE 2: Characteristics of HSS drill points.

Drill diameter (mm)	10 ± 0.01
Type of drill	2-Helical channel, HSS
Drill point angle (°)	100, 118, 136, 154
Helical angle (°)	30
Handle length (mm)	46
Type of handle	Cylindrical
Total length (mm)	133

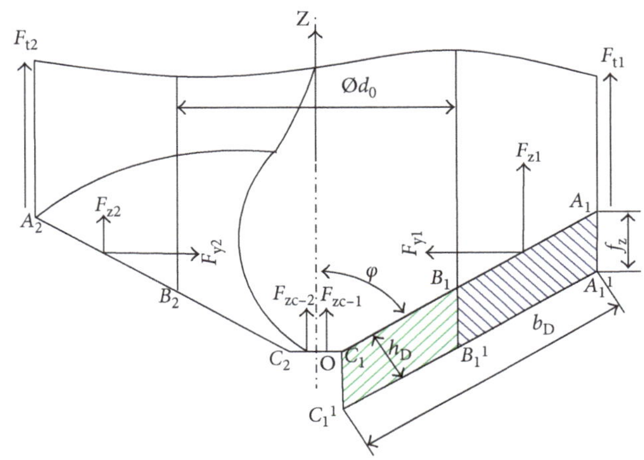

FIGURE 3: Geometry of the drill point [29].

the drilled holes on the samples. The chemical components and mechanical characteristics of the AISI 1050 alloy steel are shown in Table 1. During the drilling process, the thrust force was measured with Kistler 5233A Dynamometer which is in Newton unit.

With the help of the depth micrometer, the drilled hole delamination was measured as μm unit at 4 different points, which is 5 mm in depth, and the arithmetic average of these values was taken into account. Wear areas of the drill points were monitored with Stereo LEICA Z16 APO microscope, and images were recorded. Surface roughness values were measured as 0.8 mm by using TR 200 TIME SIRF, a surface roughness value device.

During the experimental study, 4 different drill point angle values were selected: 100°, 118°, 136°, and 154°. The drill points were grinded at selected point angles by using the tool grinding machine, as shown in Figure 2(a). Pictures of the grinded drill points can also be seen in Figure 2(b). The geometrical properties of HSS drill points are shown in Table 2.

3. Theory of Cutting Edge Geometry

3.1. Geometry of Chip Analyses. The length of the cutting edge is formed depending on the angle of the drill point during the drilling process. Therefore, the geometry of drill is vital for drilling processes because of changing the cutting speed and showing differences throughout the cutting edge. Cutting speed gets the maximum ratio at the outmost diameter of the drill, and it decreases and approaches to zero toward the axis of the drill where the drill constitutes core [28]. The angle of the drill point has an important influence on the length of the cutting edge in the drilling process.

The geometry of the drill point is shown in Figure 3. In this figure, $Ød_o$ is the diameter of the predrilled hole of the workpiece sample before experimental processes. Effective forces on 1st and 2nd cutting edges of the drill were shown, respectively, as F_{t1} and F_{t2} tangential forces, F_{z1} and F_{z2} thrust forces, and F_{y1} and F_{y2} radial forces in the revolving direction. $F_{zc\text{-}1}$ and $F_{zc\text{-}2}$ are the effective thrust forces on the core of the drill point. $\Phi/2 = \varphi$ is half of the value of the drill point angle. Besides, the values of the point angles used in this study are Φ_1, Φ_2, Φ_3, and Φ_4, respectively, and they are equal to 100°, 118°, 136°, and 154°, respectively.

In this experimental study, in order to eliminate the effects of the drill core, holes were predrilled 5 mm in diameter. The dimensions of the cutting edges of the drills, used in experimental, drilling processes, in which the surplus were not predrilled, are calculable as seen in (1), depending on the geometry of the tip of the drill, as shown in Figure 4 according to predrilling diameters and drill point angle:

$$l = \left|C_1 A_1\right| = b_D = \frac{d_h - d_o}{2 \cdot \sin \varphi} \text{ mm.} \tag{1}$$

The uncut chip height (h_D) can be calculated based on the feed rate (f_z) and point angle (φ) by using (2), in accordance with the geometry of the drill point, as shown in Figure 4. Since the height of the chip does not change depending on the predrilling diameter, the following equation is valid for both predrilled and without predrilled drilling processes:

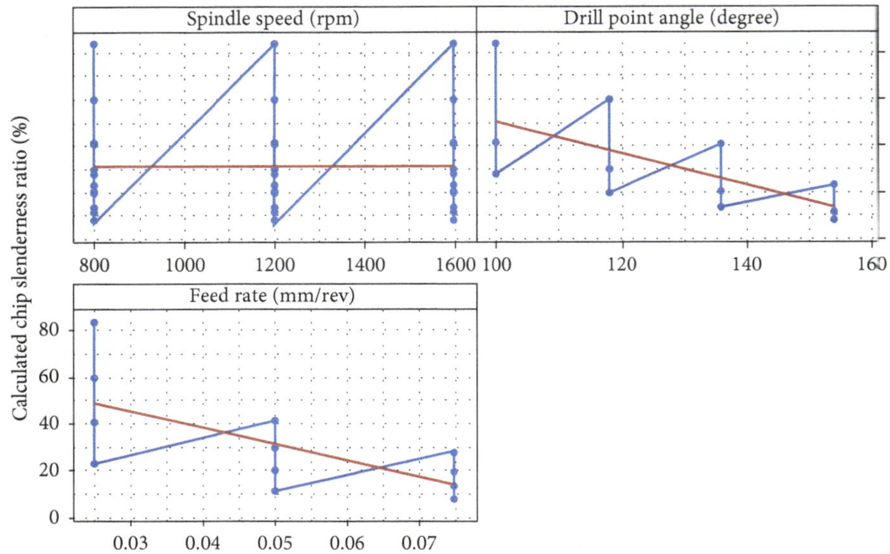

FIGURE 4: Effects of the selected parameters on the calculated chip slenderness ratios

$$h_D = h_{De} = f_z \cdot \cos\varphi \text{ mm}. \tag{2}$$

In this experimental study, as surplus were predrilled 5 mm in diameter, the influential chip width and height values were demonstrated with b_{De} and h_{De}, respectively.

$$l_e = |C_1A_1| = b_{De} = \frac{d_h - d_o}{2 \cdot \sin\varphi} \text{ mm}. \tag{3}$$

In these circumstances, the effective chip area (A_e) and chip width (b_{De}), depending on the chip height (h_{De}), can be calculated in accordance with the following equation:

$$A_{ce} = A_{ce}|B_1B_1^1 - A_1A_1^1| = h_{De} \cdot b_{De} \text{ mm}^2. \tag{4}$$

In this experimental study, 4 different drill point angles were selected. These were represented with Φ_1, Φ_2, Φ_3, and Φ_4 symbols, and they were selected as 100°, 118°, 136°, and 154°, respectively.

According to the geometry of the drill point, the length of the cutting edge without predrilled hole (l), and the length of cutting edge with predrilled hole (l_e), in other words, the uncut chip width values (b_D) and (b_{De}), showed differences depending on the change in the drill point angle.

In calculating the chip slenderness ratio, the chip width (b_{De}), the chip height (h_{De}), and the chip area (A_{ce}) values were taken into consideration in predrilled drilling operations, depending on the drill point angles by using (1)–(4).

3.2. Analysis of the Chip Slenderness Ratio. The chip slenderness ratio is an important parameter when the feed rate of cutting depth in expanded cutting speed law is considered. This ratio is named as "chip slenderness factor" or "chip morphology." For the same chip contact area, the chip slenderness ratio is high at higher selected chip cutting depths, whereas it is low at smaller selected chip cutting depths. Manufacturers investigate the effect of the cutting depth and feed rate on the chip slenderness ratio simultaneously in real applicable machining operations. The optimum chip slenderness ratio is 5 : 1 according to the results in the literature.

However, this value (5 : 1) is not valid for the turning operations because the selected feed rates are bigger than the selected cutting depths. Measuring of the chip dimensions, obtained from the experiments, such as chip thickness and chip width, is too difficult. Therefore, the calculated chip slenderness ratio derived from (5)–(7) was taken into consideration. The chip contact area can be written as demonstrated in the following equations in machining operations:

$$A = a \cdot f. \tag{5}$$

As seen in the following equation, the chip slenderness ratio (G) can be identified as the rate of the depth of cutting (a) to the feed rate (f):

$$G = \frac{a}{f}. \tag{6}$$

When the common solution of (5) and (6) is considered, the chip slenderness ratio (G) can be calculated by the ratio of the chip contact area (A) to the square of the feed rate (f^2):

$$G = \frac{A}{f^2}. \tag{7}$$

4. Results and Discussion

4.1. Effects of the Selected Parameters on the Chip Slenderness Ratio. The chip slenderness ratio is an operational parameter accepted by researchers, working in machining manufacturing area. This parameter is demonstrated as the ratio of selected chip depth (a) to the feed rate (f) [30]. Feed rate is an important result in machining manufacturing operation, and it is the most important parameter determining the quality of the operation. Due to the fact that the chip slenderness ratio is expressed as the ratio of the chip depth to the feed rate (6), the effect of this ratio for operation is important. As the chip depth can be selected by the operator in milling and turning operations, this ratio can be directly considered as the ratio of the chip depth to the feed rate. However, in drilling processes, the chip depth is based directly on the diameter of hole to be drilled; in short, it is based on the diameter of the selected drill. So, this ratio should be considered as the ratio of the chip area to the feed rate square (7), and it should also be considered for conducting investigations and analysis in this study.

In this study, chip areas were investigated as calculated (theoretical) chip areas. The calculated (theoretical) chip areas were calculated using (2)–(4) and depending on the geometry of the drill point as shown in Figure 3.

It was investigated that the effects of parameters should be calculated by chip slenderness ratios in Figure 4. The calculated chip slenderness ratio was between 7.72 and 84.00. Differences between theoretical chip slenderness ratios were numerically high, but it was determined that chip slenderness ratios are directly proportional to the change in accordance with the selected parameters. The most effective parameters on the chip slenderness ratio were feed rate and angle of the drill point. When both feed rate and point angle increase, the chip slenderness ratio decreases linearly as shown in Figure 4. However, the effect of spindle speed on the chip slenderness ratio is too little.

Effects of the spindle speed on the theoretical chip slenderness ratio should be zero according to (6) and (7).

In the circumstance, it is concluded that geometrical and mathematical equations providing the trust results, theoretically, depend on the selected parameters. The highest value of the theoretical chip slenderness ratio was 84 and the lowest 7.72.

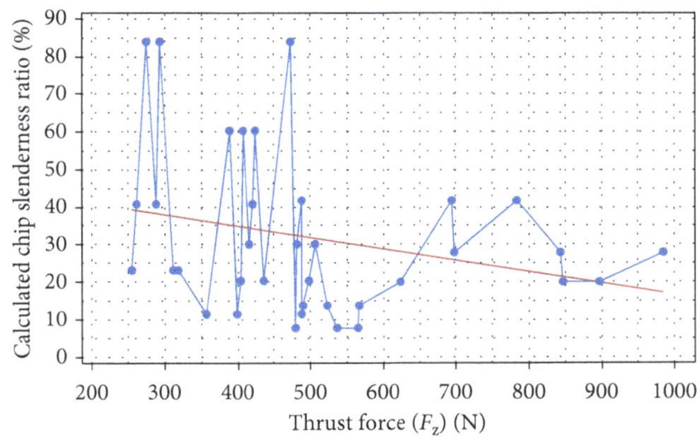

FIGURE 5: Effects of the experimental and theoretical chip slenderness ratios on the thrust force.

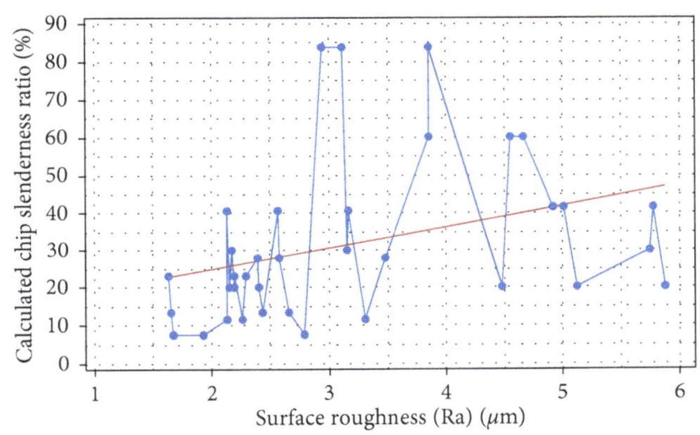

FIGURE 6: The effect of the calculated chip slenderness ratios on the surface roughness.

4.2. Effects of the Chip Slenderness Ratio on Thrust Force. During drilling operations, thrust force is one of the important factors determining the quality of process and supplying information about the results. If the thrust force is low during the operation, this means that operation quality is high. In Figure 5, it is seen that the thrust force decreased when the chip slenderness ratio increased. On the other hand, it was also seen that the chip slenderness ratio decreased collaterally with the thrust force when the drill point angle increased for the theoretical chip slenderness ratio.

However, with increase in the theoretical chip slenderness ratio, decrease in thrust force was observed.

As demonstrated in Figure 5, the lowest thrust force values were obtained as 274.70 and 292.09 N at the theoretical chip slenderness ratio of 84. However, the spindle speed had no effects on the theoretical chip slenderness ratio depending on the geometry of the drill point. On the contrary, the change resulted in only depending on the feed rate and the drill point angle.

4.3. Effects of the Chip Slenderness Ratio on Surface Roughness. Because of the chip slenderness ratio directly depends on the feed rate, the most effective parameter on surface roughness, thus on the chip slenderness ratio is the most influential parameter on surface roughness. According to the graph in Figure 6, when the chip slenderness ratio increased, the values of surface roughness also increased, but surface quality decreased.

At the highest (60–80) and the lowest (7–20) chip slenderness ratio values, the maximum and the minimum surface roughness values were 4.50 μm and 1.6 μm, respectively, as shown in Figure 6.

4.4. Effect of the Chip Slenderness Ratio on Drilled Hole Delamination. Due to the fact that the drilled hole delamination in drilling operations mostly depends on the motions of the tool and the interactions between the tool and the workpiece, it is not much affected by the chip slenderness ratio. The effect of the theoretical chip slenderness ratio on the delamination of the drilled hole is demonstrated in Figure 7. According to the results, the effect of the chip slenderness ratio on the delamination was unclear, and change in delamination values was determined depending on the changes in the chip slenderness ratio. Hole delamination cannot be investigated depending on the chip slenderness factor.

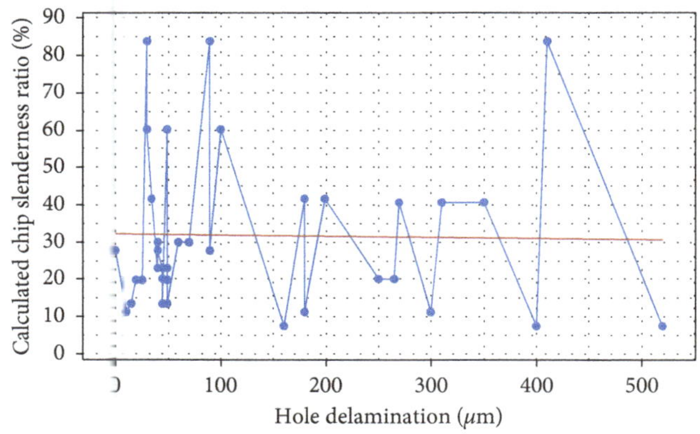

FIGURE 7: Effect of the calculated chip slenderness ratios on the drilled hole delamination.

4.5. The Effect of the Calculated Chip Slenderness Ratio on Chip Morphology. In drilling operations, morphology is the most important marker, affecting the operation quality. In this study, the effect of the chip slenderness ratio on the chip morphology was demonstrated, as shown in Figures 8(a)–8(c), for selecting 0.025, 0.050, and 0.075 mm/rev feed rates, respectively.

As shown in Figure 8(a), at 0.025 mm/rev feed rate, with increasing the drill point angle, the chip morphology changed from brittle and short forms to continuous. Furthermore, at lower point angles, especially at 100° and 118°, the serrating chip form resulted in the obtained chips, while the serrating chip formation decreased at higher point angles, as 136° and 154°. The serrated chip formation demonstrates the quantity of the vibrations, undesired phenomenon during the machining operations because of causing the severe wear and shorter tool life. However, at 0.05 mm/rev feed rate, as shown in Figure 8(b), at higher point angles, especially at 136° and 154°, and at collateral spindle speed, 1600 rpm spindle speed, the serrated chip formation decreased. At 0.075 mm/rev feed rate, as shown in Figure 8 (c), at lower point angles, 100° and 118° and at lower spindle speeds, 800 and 1200 rpm, the serrated chip formation increased linearly. However, at 136° and 154° point angles, the serrated formation in the removed chip decreased as shown in Figure 8(c), definitely.

While the requested ductile form was obtained, long and continuous chip morphology with increasing point angle, at 0.025 mm/rev feed rate conditions in machining operation also the tool wear decreased. In addition, with increasing the drill point angle, the chip slenderness ratio decreased.

Continuous, long, and ductile chip morphology with decreasing chip slenderness ratios was obtained with the selected drill point angles 136° and 154°. As shown in Figure 8(b), decrease in chip slenderness ratio, at 0.050 mm/rev feed rate, was only obtained at the drill point angle 154°. But the continuous, long, and ductile chip morphology was also obtained with other drill point angles. In Figure 8(c), while chip slenderness ratio increased under 0.075 mm/rev feed rate and 800 rpm spindle speed conditions, the value of this ratio decreased because of increasing the spindle speed.

Here, positive effects were observed when the feed rate and the spindle speed collaterally changed with each other. Also at 0.075 mm/rev feed rate, the lowest chip slenderness ratios were obtained with selective drill point angles 136° and 154°. Moreover, chips were obtained in the requested morphology, and tool wear decreased at these angles. Burning marks on the cutting edge of the tool and chips were seen because of high process temperature during the drilling operations.

4.6. The Effect of the Calculated Chip Slenderness Ratio on Tool Wear. Tool wear is an important identifier for the tool life in machining operations. Less tool wear provides longer tool life. The effect of the drill point angle, spindle speed, and feed rate on the tool wear is shown in Figure 9 in predrilled drilling operations.

At 0.025 mm/rev feed rate, with increasing drill point angle, calculated chip slenderness ratio decreased linearly. Moreover, the tool wear quantity decreased as shown in Figure 9(a).

At 0.050 mm/rev feed rate, the tool wear increased at higher spindle speeds, as 1200 and 1600 rpm, but at lower drill point angles, it decreased as 100° and 118°. However, at 500 rpm spindle speed and 100° drill point angle, the tool wear decreased. At 136° and 154° point angles, the tool wear decreased for all selected spindle speeds. However, at higher spindle speeds, due to the higher process temperature, there were burning marks on the drill cutting edges as shown in Figure 9(c).

With increasing the feed rate to 0.075 mm/rev, the tool wear increased only at lower spindle speed and 100° point angle; with increasing the point angle, the wear quantity of the tool cutting edge increased at 1600 rpm spindle speed.

5. Conclusions

In predrilled drilling operations of AISI 1050 steel alloy, HSS drills were employed, and the effect of the selected parameters on the chip slenderness ratio and also the effect of the chip slenderness ratio on the thrust force, the surface roughness, the drilled hole delamination, the tool wear, and the chip morphology were investigated. The derived results in this study were listed below.

The major parameters influential on the chip slenderness ratio were the feed rate and the drill point angle, but the effect of the spindle speed was too small to be negligible; with decreasing the chip slenderness ratio, the thrust force and tool wear decreased, and also the removed chip morphology changed from shorter, brittle, and serrated to continuous, ductile, and free of serrated form. Besides, with increasing the calculated chip slenderness ratio, the surface roughness increased collaterally. However, the chip slenderness ratio did not have any important effect to be taken into consideration on the hole delamination.

Furthermore, with changing point angles from 100° to 154°, the chip slenderness ratio decreased from 84 to 7, the

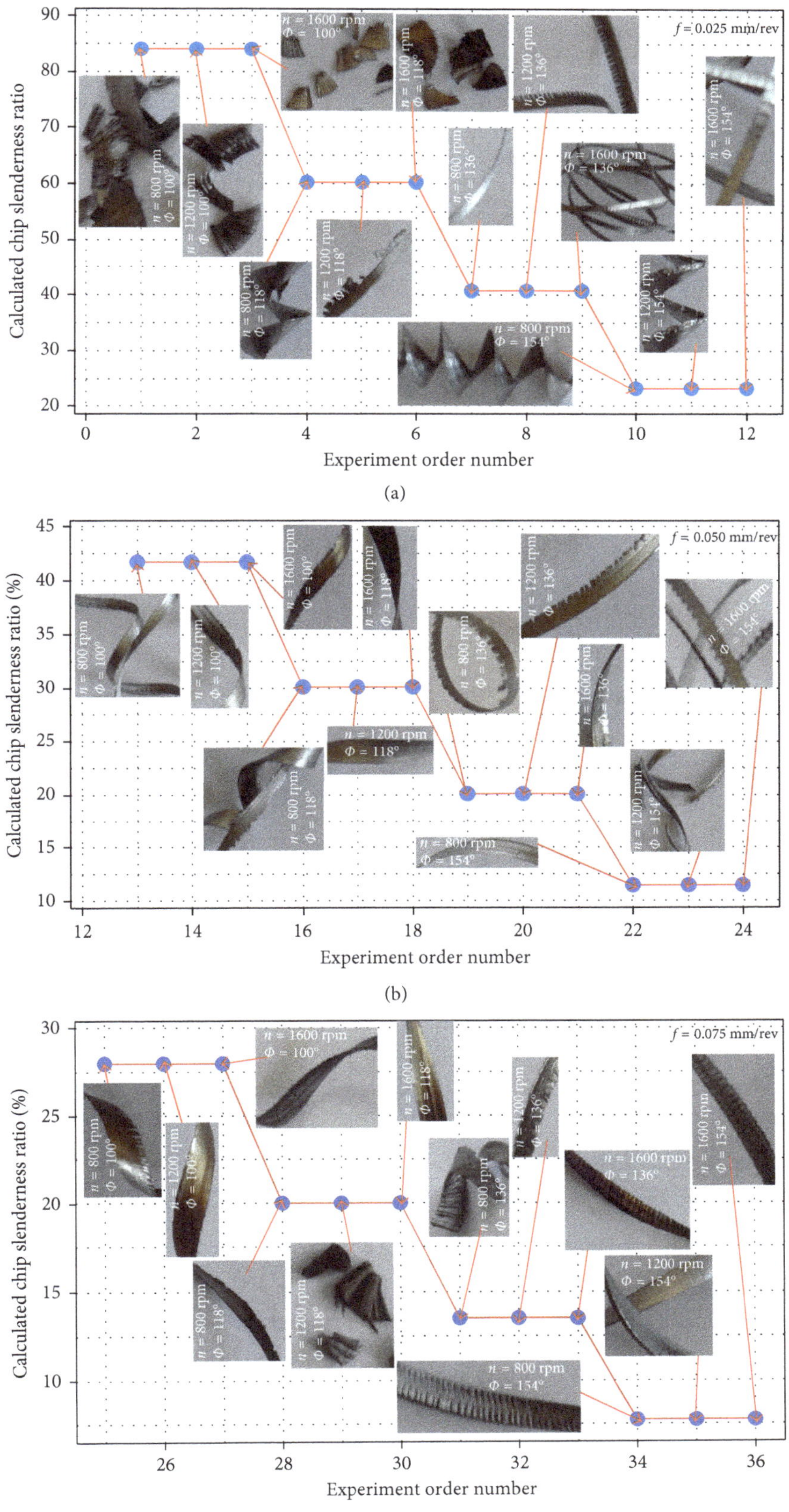

FIGURE 8: Effects of experimental and theoretical chip slenderness ratios on the chip morphology. (a) 0.025 mm/rev, (b) 0.050 mm/rev, and (c) 0.075 mm/rev feed rates.

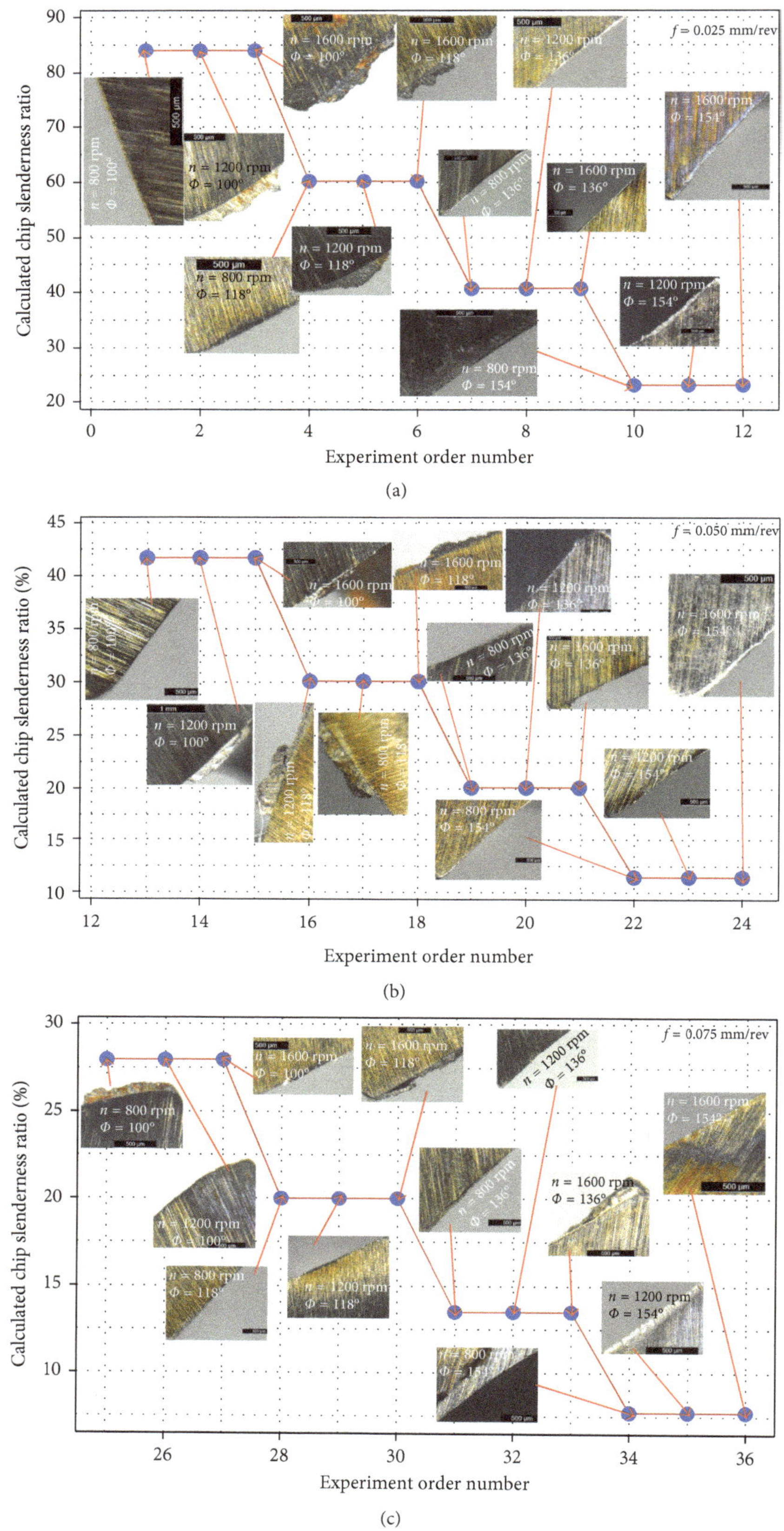

FIGURE 9: Effects of experimental and theoretical chip slenderness ratios on the tool wear. (a) 0.025 mm/rev, (b) 0.050 mm/rev, and (c) 0.075 mm/rev feed rates.

quantity of the tool wear decreased, and the removed chip morphology was changed from brittle, shorter, and serrated forms, undesired chip forms, to continuous, longer ductile, and free of serrated form, the most desired chip form in machining operations.

Conflicts of Interest

The authors declare that they have no conflicts of interest.

References

[1] A. A. Thakre and S. Soni, "Modeling of burr size in drilling of aluminum silicon carbide composites using response surface methodology," *Engineering Science and Technology an International Journal*, vol. 19, no. 3, pp. 1199–1205, 2016.

[2] W. C. Chen and C. C. Tsao, "Cutting performance of different coated twist drills," *Journal of Materials Processing Technology*, vol. 88, no. 1–3, pp. 203–207, 1999.

[3] M. B. Lazar and P. Xiraouchhakis, "Mechanical load distribution along the main cutting edges in drilling," *Journal of Materials Processing Technology*, vol. 213, no. 2, pp. 245–260, 2013.

[4] S. Kalidas, R. E. Devor, and S. G. Kapoor, "Experimental investigation of the effect of drilling coatings on hole quality under dry and wet drilling conditions," *Surface and Coatings Technology*, vol. 148, no. 2-3, pp. 117–128, 2001.

[5] G. S. Samy and K. S. Thirumalai, "Measurement and analysis of temperature, thrust force and surface roughness in drilling of AA (6351)-B_4C composite," *Measurement*, vol. 103, pp. 1–9, 2017.

[6] N. Z. Karimi, H. Heidary, and G. Minak, "Critical thrust and feed prediction models in drilling of composite laminates," *Composite Structures*, vol. 148, pp. 19–26, 2016.

[7] Z. Jia, R. Fu, B. Niu, B. Qian, Y. Bai, and F. Wang, "Novel drill structure for damage reduction in drilling CFRP composites," *International Journal of Machine Tools and Manufacture*, vol. 110, pp. 55–65, 2016.

[8] K. Kalaiselvan, N. Murugan, and S. Parameswaran, "Production and characterization of AA6061–B_4C stir cast composite," *Materials and Design*, vol. 32, no. 7, pp. 4004–4009, 2011.

[9] U. Çaydaş, A. Hasçalık, Ö. Buytoz, and A. Meyveci, "Performance evaluation of different twist drills in dry drilling of AISI 304 Austenitic stainless steel," *Materials and Manufacturing Processes*, vol. 26, no. 8, pp. 951–960, 2011.

[10] K. Palanikumar and A. Muniaraj, "Experimental investigation and analysis of thrust force in drilling cast hybrid metal matrix (Al–15%SiC–4%graphite) composites," *Measurement*, vol. 53, pp. 240–250, 2014.

[11] Y. H. Çelik, H. Yildiz, and C. Özek, "Effect of cutting parameters on workpiece and tool properties during drilling of Ti-6Al-4V," *Materials Testing*, vol. 58, no. 6, pp. 519–525, 2016.

[12] B. Ramesh, A. Elayaperumal, and S. Satishkumar, "Effect of the standard and special geometry design of a drill body on quality characteristics and multiple performance optimization in drilling of thick laminated composites," *Procedia Manufacturing*, vol. 97, pp. 390–401, 2014.

[13] P. Feng, J. Wang, J. Zhang, and J. Zheng, "Drilling induced tearing defects in rotary ultrasonic machining of C/SiC composites," *Ceramics International*, vol. 43, no. 1, pp. 791–799, 2017.

[14] J. Wang, P. Feng, J. Zheng, and J. Zhang, "Improving hole exit quality in rotary ultrasonic machining of ceramic matrix composites using a compound step-taper drill," *Ceramics International*, vol. 42, no. 12, pp. 13387–13394, 2016.

[15] M. Ramulu, P. N. Rao, and H. Kao, "Drilling of $(Al_2O_3)p$/6061 metal matrix composites," *Journal of Materials Processing Technology*, vol. 124, no. 1-2, pp. 244–254, 2002.

[16] J. P. Davim, "Study of drilling metal-matrix composites based on the Taguchi techniques," *Journal of Materials Processing Technology*, vol. 132, no. 1–3, pp. 250–254, 2003.

[17] S. Basavarajappa, G. Chandramohan, J. P. Davim et al., "Drilling of hybrid aluminium matrix composites," *International Journal of Advanced Manufacturing Technology*, vol. 35, no. 11-12, pp. 1244–1250, 2008.

[18] J. P. Davim and A. M. Baptista, "Cutting force, tool wear and surface finish in drilling metal matrix composites," *Proceedings of the Institution Mechanical Engineers, Part: E Journal of Process Mechanical Engineering*, vol. 215, no. 2, pp. 177–183, 2001.

[19] Y. Altunpak, M. Ay, and S. Aslan, "Drilling of a hybrid Al/SiC/Gr metal matrix composites," *International Journal of Advanced Manufacturing Technology*, vol. 60, no. 5–8, pp. 513–517, 2012.

[20] A. Taşkesen and K. Kütükde, "Experimental investigation and multi-objective analysis on drilling of boron carbide reinforced metal matrix composites using grey relational analysis," *Measurement*, vol. 47, pp. 321–330, 2013.

[21] S. Ema, "Effect of twist drill point geometry on torque and thrust," *Science Report of the Faculty of Education, Gifu University*, vol. 36, pp. 165–174, 2012.

[22] R. Ghosh, R. Sarkar, S. Paul, and S. K. Pal, "Biocompatibility and drilling performance of beta tricalcium phosphate: yttrium phosphate bioceramic composite," *Ceramics International*, vol. 42, no. 7, pp. 8263–8273, 2016.

[23] U. Çaydaş and M. Çelik, "AA 7075-T6 investigation effects of cutting parameters on surface roughness, tool temperature and feed force," *Polytechnic Magazine*, vol. 20, no. 2, pp. 419–425, 2017.

[24] Y. H. Çelik, "Investigating the effects of cutting parameters on the hole quality in drilling the Ti-6Al-4V alloy," *Materials and Technology*, vol. 48, no. 5, pp. 653–659, 2014.

[25] A. Z. Sultan, S. Sharif, and D. Kurniawan, "Chip formation when drilling AISI 316L stainless steel using carbide twist drill," *Procedia Manufacturing*, vol. 2, pp. 224–229, 2015.

[26] P. A. Rey, J. Ledref, J. Senatore, and Y. Landon, "Modelling of cutting forces in orbital drilling of titanium alloy Ti-6Al-4V," *International Journal of Machine Tools & Manufacture*, vol. 106, pp. 75–88, 2016.

[27] R. S. Anand and K. Patra, "Mechanistic cutting force modelling for micro-drilling of CFRP composite laminates," *CIRP Journal of Manufacturing Science and Technology*, vol. 16, pp. 55–63, 2017.

[28] E. M. Trent, *Mechanics of Drilling, Metal Cutting, Metal Cutting Operations and Terminology*, University of Birmingham, Birmingham, UK, 1984.

[29] V. P. Astakhov, *Drilling System, Drills, Science and Technology of Advanced Operations*, Taylor & Francis Group LLC, Boca Raton, FL, USA, 2014.

[30] M. Kronenberg, *The Extended Cutting Speed Law, Machining Science and Application, Theory and Practice for Operation and Development of Machining Processes*, Pergamon Press Inc., Oxford, UK, 1966.

18

Effect of Nd^{3+} Ion Concentration on the Corrosion Resistance of API X70 Steel in Chloride-Rich Environments

D. M. Martinez de la Escalera,[1] J. J. Ramos-Hernandez,[2] E. Porcayo-Palafox,[3] J. Porcayo-Calderon,[3] J. G. Gonzalez-Rodriguez,[3] and L. Martinez-Gomez[2,4]

[1]*Facultad de Ingeniería, Universidad Autónoma de Campeche, Campus V, Av. Ing. Humberto Lanz, 24085 Campeche, CAM, Mexico*
[2]*Instituto de Ciencias Físicas, Universidad Nacional Autónoma de México, Avenida Universidad s/n, 62210 Cuernavaca, MOR, Mexico*
[3]*CIICAp, Universidad Autónoma del Estado de Morelos, Avenida Universidad 1001, 62209 Cuernavaca, MOR, Mexico*
[4]*Corrosion y Protección (CyP), Buffon 46, 11590 México City, Mexico*

Correspondence should be addressed to J. Porcayo-Calderon; jporcayoc@gmail.com

Academic Editor: Yee-wen Yen

In this study, the effect of the addition of Nd^{3+} ions as a corrosion inhibitor of the API X70 steel in a medium rich in chlorides was evaluated. The performance of the Nd^{3+} ions was evaluated by means of electrochemical techniques such as potentiodynamic polarization curves, open circuit potential measurements, linear polarization resistance, and electrochemical impedance spectroscopy, as well as by means of scanning electron microscopy and EDS measurements. The results showed that Nd^{3+} ions reduce the corrosion rate of steel at concentrations as low as 0.001 M Nd^{3+}. At higher concentrations, the inhibition efficiency was only slightly affected although the concentration of chloride ions was increased by the addition of the inhibitor. The adsorption of the Nd^{3+} ions promotes the formation of a protective layer of oxides/hydroxides on the metal surface, thereby reducing the exchange rate of electrons. Nd^{3+} ions act as a mixed inhibitor with a strong predominant cathodic effect.

1. Introduction

In the last decades, there has been a great interest in the use of rare earth metals for the synthesis of compounds with inhibitory capacity of the corrosion of metals and alloys (Al, Fe, Mg, and Zn mainly) in aggressive environments. These compounds have been used as soluble elements in the electrolyte or as an additive of organic coatings. It has been recognized that rare earth salts are low-toxicity compounds whose average lethal dose (LD50) is similar to that of sodium chloride (NaCl) [1–3]; therefore, the interest in the use of rare earths has emerged as an alternative to replace those toxic and highly carcinogenic corrosion inhibitors that affect both people and the environment, and many of these dangerous inhibitors are based on chromates ($CrO^{2-}{}_4$), mercurate, nitrite, and alkynol [4]. Most of the studies have been reported based on the use of both Ce and La salts as corrosion inhibitors, in the form of either chlorides, nitrates, or other salts [2, 4–9]. Many of these studies have been carried out in chloride-rich solutions, and the excellent corrosion inhibiting capacity of rare earth salts has been attributed to the high reactivity of rare earths.

It is known that, in an electrochemical corrosion process, the cathodic reaction activates the O_2 reduction reaction causing an increase of the pH in the cathodic sites of the metallic surface [4]. It has been observed that this local increase in pH enhances the precipitation of rare earth ions, and this promotes the formation of a protective film of complex hydrated oxides on the cathode sites of the metal surface. The formed protective film reduces the rate of oxygen reduction and increases the cathodic polarization rate, thus causing a reduction in the corrosion potential as well as in the corrosion current density [2–4, 6, 7, 10].

In the literature, several studies can be found which indicate that various lanthanide base compounds possess excellent properties as corrosion inhibitors. In particular, the best anticorrosive properties have been detected mainly on light alloys, such as alloys based on aluminum, magnesium, and zinc [7–17]. However, some studies on iron-base alloys have also shown that rare earths and their compounds are capable of inhibiting the iron corrosion [4, 18–20]. Since iron is the main element of many structural alloys with applications in aqueous media rich in chlorides, then the constant search for alternatives to counteract the adverse effects of the corrosive environment is of interest.

Although the main technological applications of rare earths have been focused in the field of high technology electronics, from the point of view of their abundance, La, Ce, Pr, and Nd emerge as the main candidates to be used as corrosion inhibitors. In particular, the availability of Nd for the synthesis of corrosion inhibitors can be attractive if the new technologies for the recycling of rare earths are considered [21–24]. Therefore, in this study, the corrosion inhibitory capacity of neodymium chloride on API X70 steel in a 3.5% NaCl solution at room temperature is evaluated. However, because the neodymium chloride molecule is formed by one Nd atom, three Cl atoms, and six water molecules, in this study the inhibitory capacity is evaluated as a function of the amount of Nd^{3+} ions added instead of considering the entire molecule. The performance of the Nd^{3+} ions as a corrosion inhibitor is evaluated by means of electrochemical techniques such as potentiodynamic polarization curves, open circuit potential measurements, linear polarization resistance, and electrochemical impedance spectroscopy, and corroded specimens were analyzed by scanning electron microscopy and EDS measurements.

2. Experimental Procedure

2.1. Corrosion Inhibitor. The source of Nd^{3+} ions used in this study was neodymium chloride. This salt was obtained by dissolving neodymium oxide in a dilute solution of HCl until a clear solution was obtained. Subsequently, the solution was subjected to dryness at 60°C, and the resulting salt used in that condition was used for the corrosion tests.

$$Nd_2O_3 + 6HCl \longrightarrow 2NdCl_3 + 3H_2O \xrightarrow{\Delta} 2NdCl_3 \cdot 6H_2O \quad (1)$$

Because the neodymium chloride molecule is formed by approximately 40% of the Nd^{3+} cation, 30% of the Cl^- anion, and 30% of water (at.%), in this study and in order to avoid ambiguities, it was decided to evaluate the effect of the concentration of the rare earth cation as a corrosion inhibitor rather than the salt as a whole.

2.2. Electrochemical Evaluation. As a study material, an API X70 steel was used, from which specimens with dimensions 10*10*5 mm were obtained. A copper wire was welded by the spot-welding technique on one of the faces of the specimens, and in this condition, they were encapsulated in epoxy resin. The encapsulated specimens were roughened with abrasive paper from grade 120 to grade 1200 and subsequently polished with 0.5-micron diamond paste using ethanol as a lubricant. Subsequently, they were washed with distilled water and ethanol, dried, and used immediately in the corrosion tests.

As a corrosive medium, distilled water with 3.5% by weight of NaCl was used at room temperature under nonaerated conditions. The performance as a corrosion inhibitor of the neodymium chloride was determined, considering the concentration of Nd^{3+} ions added to the corrosive medium. In this sense, the concentrations of Nd^{3+} ions evaluated were 0.0001 M, 0.0005 M, 0.001 M, 0.005 M, and 0.01 M.

The electrochemical evaluation was performed in a three-electrode electrochemical cell using the encapsulated specimens as working electrodes, a saturated calomel electrode as the reference electrode, and a graphite bar with an area greater than that of the working electrode as a counter electrode. The electrochemical techniques used were potentiodynamic polarization curves (anodic and cathodic), open circuit potential measurements, linear polarization resistance measurements, and electrochemical impedance spectroscopy.

The anodic and cathodic branches of the polarization curves were obtained independently using a new working electrode in each case. The cathodic polarization curve was obtained by scanning cathodically the working electrode from +10 mV to −300 mV, and the anodic curves were obtained in a similar way but from −10 mV to 500 mV, in both cases with respect to its open circuit potential. In both cases, before starting the test, the electrochemical cell was allowed to stabilize for 24 hours before starting the scan at a speed of $0.1667\ mV{\cdot}s^{-1}$. The open circuit potential (OCP) measurements were carried out by recording the potential of the working electrode with respect to the reference electrode at intervals of one hour. For linear polarization resistance (RPL) measurements, the working electrode was polarized at ±10 mV with respect to its open circuit potential, the value of the polarization resistance was determined from the slope of the obtained E-I ratio, and measurements were made at one-hour intervals. The electrochemical impedance spectra (EIS) were obtained by applying an amplitude perturbation of 10 mV with respect to the open circuit potential of the working electrode in a frequency range from 100 kHz to 0.01 Hz. OCP, LPR, and EIS measurements were performed for 100 hours.

3. Results and Discussion

3.1. Potentiodynamic Polarization Curves. Figures 1 and 2 show the anodic and cathodic branches, respectively, of API X70 steel in 3.5% NaCl solution with the addition of Nd^{3+} ions at different concentrations. From both figures can be observed a cathodic direction shift of the corrosion potential, with respect to the blank, at concentrations greater than 0.0001 M Nd^{3+}, and this behavior is indicative of the existence of a cathodic inhibition process. In a similar way, a displacement is observed, with respect to the blank, of both anodic and cathodic branches towards lower current

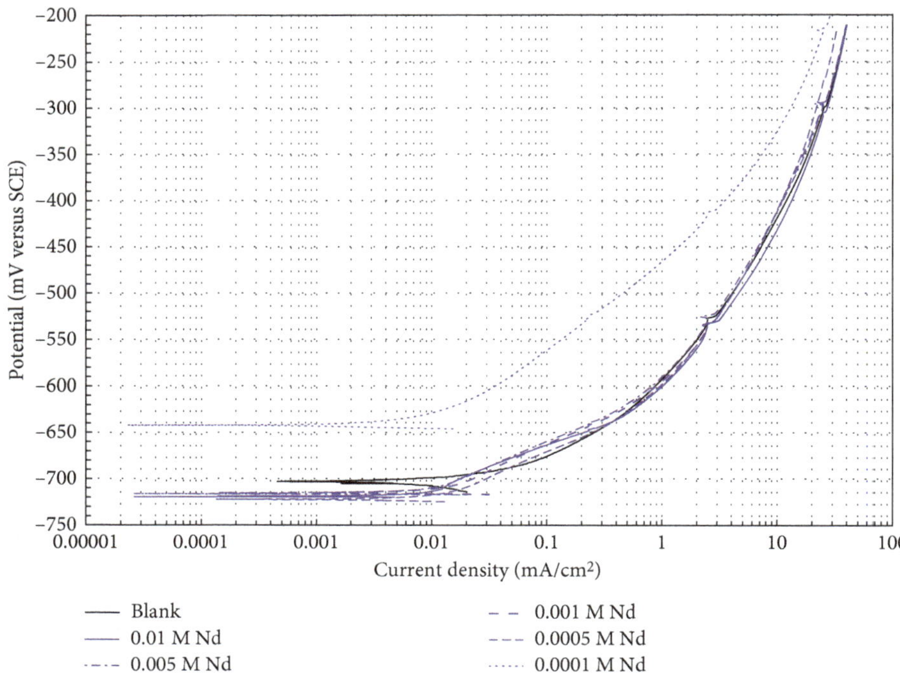

FIGURE 1: Anodic polarization curves for API X70 steel in 3.5% NaCl solution at different concentrations of Nd^{3+} ions ($0.1667\,mV{\cdot}s^{-1}$).

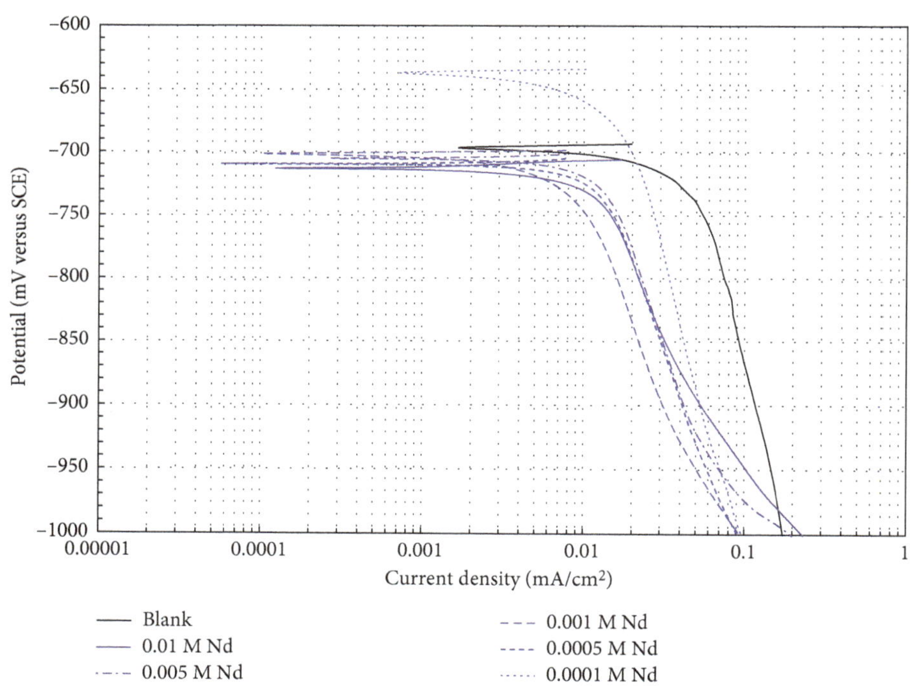

FIGURE 2: Cathodic polarization curves for API X70 steel in 3.5% NaCl solution at different concentrations of Nd^{3+} ions ($0.1667\,mV{\cdot}s^{-1}$).

densities. From the anodic branches, it is possible to observe a change in its slope to potentials slightly above the corrosion potential; this suggests that the presence of Nd^{3+} ions also affects the anodic reaction of the electrochemical corrosion process of the API X70 steel.

From the cathodic branch, it is possible to observe that the addition of Nd^{3+} ions caused a decrease in the cathodic current regardless of the concentration of added inhibitor. The greatest reduction in the cathodic current is observed at the concentration of 0.001 M Nd^{3+}. It has been suggested that the reduction observed in the cathodic current is due to the high concentration of OH^- ions in the cathode sites and to the high reactivity of the lanthanide ions, which by reacting form an insoluble oxide/hydroxide layer on the cathodic

TABLE 1: Electrochemical parameters of the polarization curves in the presence of Nd^{3+} ions.

(Nd^{3+}) M	Cathodic E_{corr} (mV)	Anodic E_{corr} (mV)	Ba (mV/Dec)	Bc (mV/Dec)	Anodic I_{corr} (mA/cm^2)	Cathodic I_{corr} (mA/cm^2)
0	−696	−703	72	517	0.0466173	0.047242
0.01	−713	−720	53	308	0.0107901	0.011432
0.005	−704	−716	56	380	0.0095416	0.0129336
0.001	−707	−714	57	346	0.0124705	0.0083153
0.0005	−713	−720	56	371	0.0120315	0.0126311
0.0001	−638	−642	92	645	0.0056876	0.0251889

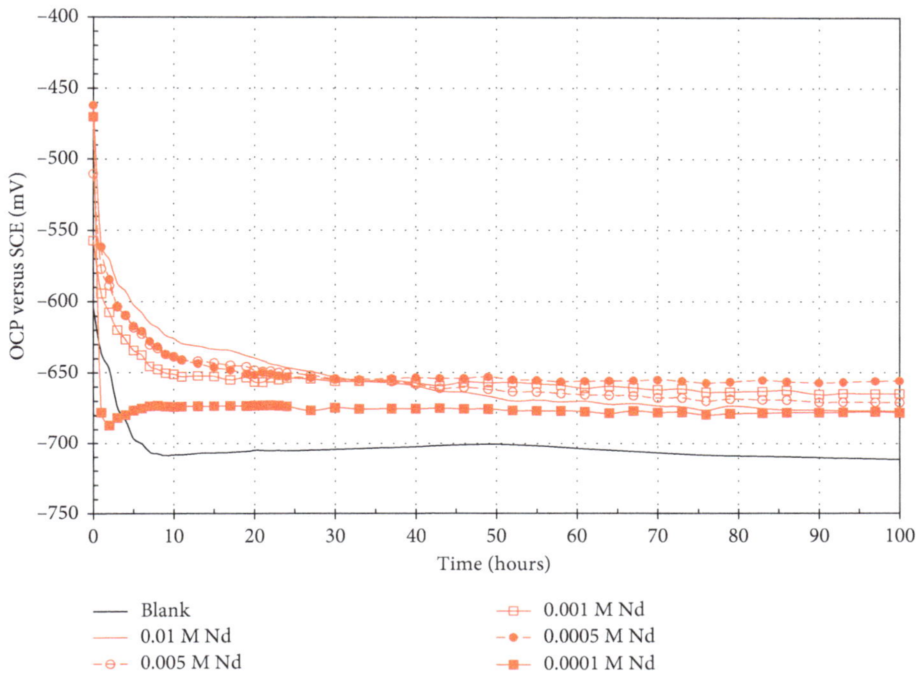

FIGURE 3: Variation in OCP values for API X70 steel in 3.5% NaCl solution at different concentrations of Nd^{3+} ions.

sites, reducing the cathodic reaction rate and the localized corrosion processes [2–4, 6, 7, 10].

Table 1 shows the electrochemical parameters extracted from the anodic and cathodic polarization curves. In general, it can be seen that there is a reduction in both anodic and cathodic slopes. This is due to the displacement towards lower current densities of both branches ($\Delta E/\Delta I$). The greatest reduction is observed in the cathodic slopes, which suggests a modification in the oxygen reduction reaction [7]. Only at the lowest concentration of Nd^{3+}, an increase in the cathodic slope was observed. On the other hand, although there was an increase in the Cl^- ion concentration due to the addition of the inhibitor, the anodic current density showed a decrease close to an order of magnitude. However, in the case of cathodic current density, the greatest decrease occurred at concentrations up to 0.001 M Nd^{3+}. Notwithstanding the beneficial effect of the rare earth chlorides on the corrosion inhibition process, their addition in high concentrations will cause a decrease in their inhibitory capacity due to the inherent increase of the chloride ions in the electrolyte [6].

It is known that the addition of an inhibitor can affect the overall corrosion process by affecting the anodic dissolution, the oxygen reduction, or both. In all three cases, a decrease in the corrosion current density is observed; however, a displacement of the corrosion potential towards more active potentials characterizes a cathodic inhibitor and in the opposite case to an anodic type. When an inhibitor affects both processes (corrosion current density and corrosion potential), it is considered as a mixed inhibitor, and its net effect in the polarization curves will be determined by the inhibition magnitude of the anodic and cathodic reactions [6]. The previous analysis shows that the addition of Nd^{3+} ions changed the corrosion process of the API X70 steel, affecting both the anodic and cathodic branches (different current densities); this suggests that both the local anodic dissolution and the cathodic reduction of oxygen are affected simultaneously [2, 4]. However, based on the negative displacement of the corrosion potential and the decrease of the cathodic current, it can be said that the Nd^{3+} ions act as mixed inhibitors with a predominant cathodic effect.

3.2. Open Circuit Potential Measurements. Figure 3 shows the variation in OCP values of API X70 steel in 3.5% NaCl solution as a function of time at different concentrations of

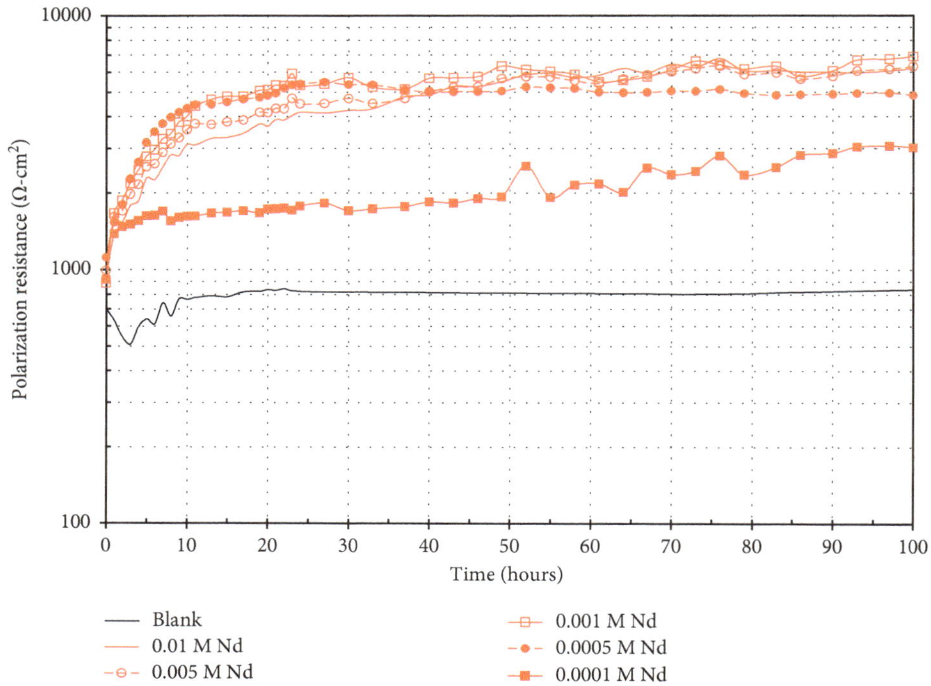

FIGURE 4: Variation in polarization resistance values for API X70 steel in 3.5% NaCl solution at different concentrations of Nd^{3+} ions.

Nd^{3+} ions. It is observed that, in the absence of Nd^{3+} ions, the steel showed an abrupt drop in its OCP values from approximately −600 mV to around −705 mV in the first 6 hours of immersion; afterwards, it showed a slight increase until 50 hours of immersion followed by a slight decrease in the rest of the test. It has been suggested that abrupt variations in OCP values at the beginning of the test may be a consequence of a reorganization of the surface layers due to the acceleration of the anodic process or a decrease in the activity of the cathodic process [4]. In the absence of Nd^{3+} ions, the most active OCP values were obtained. On the other hand, in the presence of Nd^{3+} ions, it is also possible to observe an abrupt drop in the OCP values at the beginning of the test, and subsequently after 50 hours of immersion, it is possible to observe a quasistationary state. It is known that the displacement of the OCP values to more active values besides being associated with an increase in the anodic reaction rate (metallic solution) can also correspond to the presence of an inhibition process of the cathodic reaction [2]. In the case of rare earth-based corrosion inhibitors in chloride-rich solutions, this displacement has been associated with the formation of a layer of rare earth products onto cathodic sites of the metal surface, blocking the reduction of dissolved oxygen [2, 23]. It is possible to observe that, regardless of the concentration of Nd^{3+} ions, the OCP values are within a band of 25 mV.

It has been suggested that a displacement in the OCP values greater than ±85 mV, with respect to the blank, is typical of anodic or cathodic inhibitors depending on the direction of displacement; otherwise, they are mixed inhibitors [24–26]. However, the results of this study show a variation in the OCP values lower than 85 mV; this reaffirms that the Nd^{3+} ions act as a mixed inhibitor with a strong predominant cathodic effect as previously established.

3.3. Linear Polarization Resistance Measurements. Figure 4 shows the variation in the polarization resistance values of API X70 steel as a function of time at different concentrations of Nd^{3+} ions. In the absence of inhibitor, oscillations in the polarization resistance values are observed in the first 10 hours of immersion, and a quasistationary state is subsequently observed. These variations are consistent with the observations established in the OCP measurements; that is, the initial behavior corresponds to a reorganization of the surface layer due to the acceleration of the anodic dissolution process [2, 4]. In the presence of Nd^{3+} ions in all cases, a significant increase in the polarization resistance values was observed in the first 10 hours of immersion and, subsequently, a less pronounced ascendant trend was observed with slight oscillations. These trends demonstrate the corrosion inhibitory capacity of Nd^{3+} ion. According to the graph at concentrations higher than 0.0005 M Nd^{3+}, similar polarization resistance values are observed.

Figure 5 shows the inhibition efficiency as a function of the concentration of Nd^{3+} ions. The calculation of the inhibition efficiency was according to the following expression:

$$E(\%) = \left(\frac{Rp_i - Rp_b}{Rp_i}\right) * 100, \quad (2)$$

where Rp_i corresponds to the polarization resistance in the presence of inhibitor and Rp_b corresponds to the polarization resistance in the absence of inhibitor. In each case, the

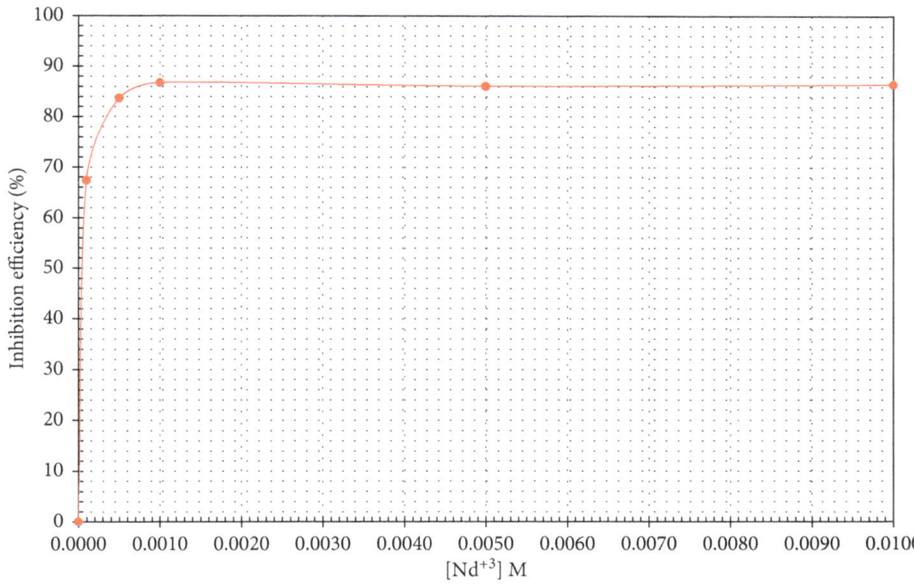

FIGURE 5: Effect of the concentration of Nd^{3+} ions on the corrosion inhibition efficiency of API X70 steel in 3.5% NaCl solution.

polarization resistance values used correspond to the average of the last 50 hours of the values reported in the previous figure. According to the graph, it is observed that the inhibition efficiency is close to 90% at concentrations higher than 0.0005 M Nd^{3+}. The highest inhibition efficiency was obtained with 0.001 M Nd^{3+}; at higher concentrations, the inhibition efficiency tends to decrease slightly, possibly due to the increase in the concentration of chloride ions.

3.4. Electrochemical Impedance Spectroscopy Measurements. Figure 6 shows the Nyquist and Bode diagrams for API X70 steel in 3.5% NaCl solution at different concentrations of Nd^{3+} ions after 100 hours of immersion. According to the Nyquist diagram, in the absence of inhibitor, the steel shows the presence of a depressed capacitive semicircle with a diameter less than 1000 Ω-cm^2. This is confirmed from the Bode diagram which shows the presence of a single time constant at frequencies around 10 Hz with a maximum phase angle close to 54°. From the evolution of the impedance module, $|Z|$, it is possible to define the high-frequency plateau (≥1000 Hz) which indicates the absence of some layer of corrosion products on its surface with viscous characteristics that affect the diffusion of species; however, the low-frequency plateau is not fully defined (≤1 Hz). On the other hand, in the presence of inhibitor, regardless of its concentration from the Nyquist diagram, the apparent presence of a single capacitive semicircle is also visible; however, the analysis of the Bode diagrams reveals the existence of three time constants. The fact that these are not visible in the Nyquist diagram is due to that the first two time constants emerge from the high-frequency region to the intermediate-frequency region, and the experimental points mostly visible correspond to the low-frequency region. According to the Bode diagram in its impedance module format, the first time constant is located in the high-frequency region around 10,000 Hz, the second time constant in the intermediate-frequency region from 10 to 200 Hz, and the third time constant at a frequency less than 1 Hz. Correspondingly, the three time constants are associated with a change in slope of the relation $\log f$ versus $\log |Z|$ at the indicated frequencies. The maximum values of impedance modulus are in agreement with the results of polarization resistance and inhibition efficiency that indicate that, with the addition of 0.001 M Nd^{3+}, the greater inhibition of the corrosion process is obtained.

Figure 7 shows the evolution of the impedance spectra as a function of time for API X70 steel with the addition of 0.001 M Nd^{3+}. Similar characteristics were observed at other concentrations. In general, from the Nyquist diagrams, the apparent presence of a single capacitive semicircle whose diameter increases as a function of time is observed. However, again from the Bode diagram in its phase angle format, the presence of three time constants is observed. It has been suggested that this is because in passive systems, the corrosion process is controlled by the reactions occurring at the interface with the electrolyte, and these are detected in the low-frequency region; therefore, the predominant signal in the Nyquist diagrams corresponds to response of these processes [4].

In the high-frequency region, the evolution (around 10,000 Hz) of the first time constant with a maximum phase angle of 28° is observed. In the intermediate-frequency region, the presence of the second time constant is observed around 5 Hz with a maximum phase angle of 60°, and in the low-frequency region (1 ≤ Hz), the third time constant is located with a maximum phase angle around 50°. In the impedance module format, $|Z|$, a slope change of the relation $\log f$ versus $\log |Z|$ associated with the presence of the three time constants is observed. It is also evident that the absence of the plateaus of both high- and low-frequency regions is due to the presence of a time constant in those regions. It is

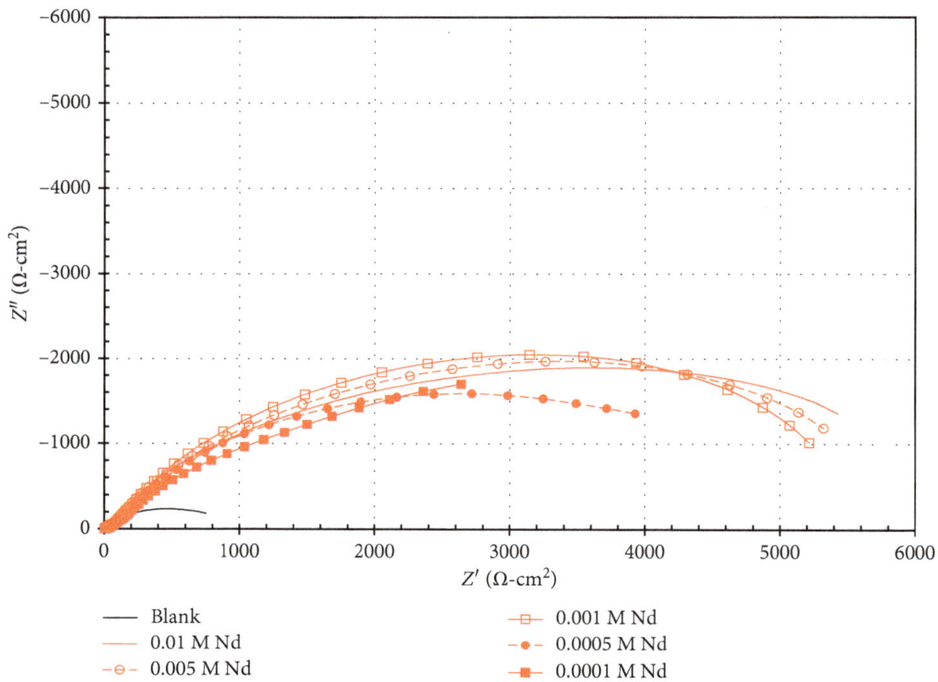

(a)

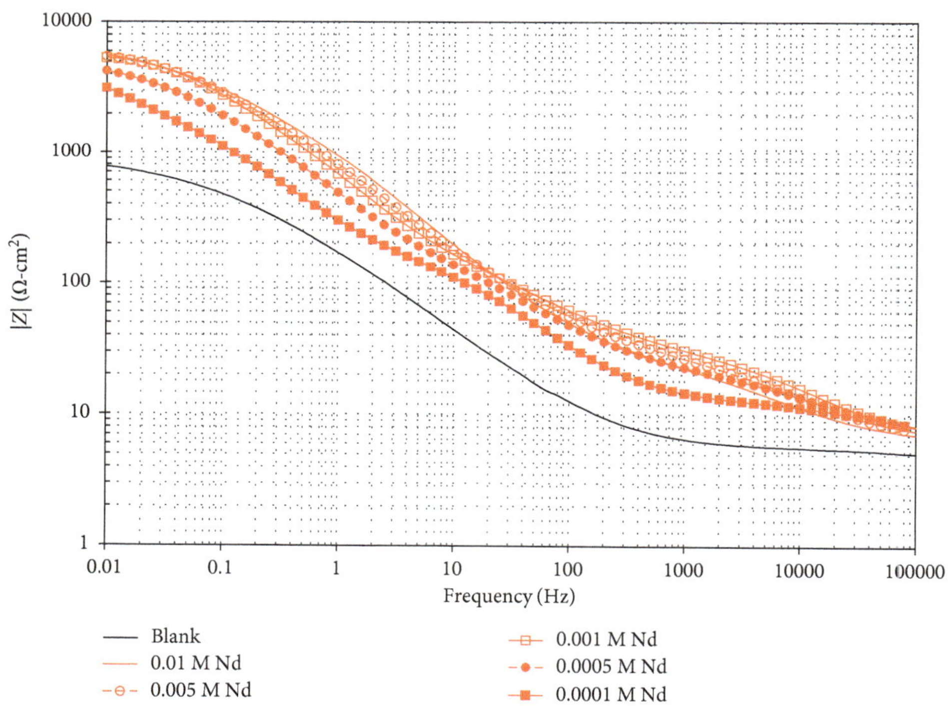

(b)

FIGURE 6: Continued.

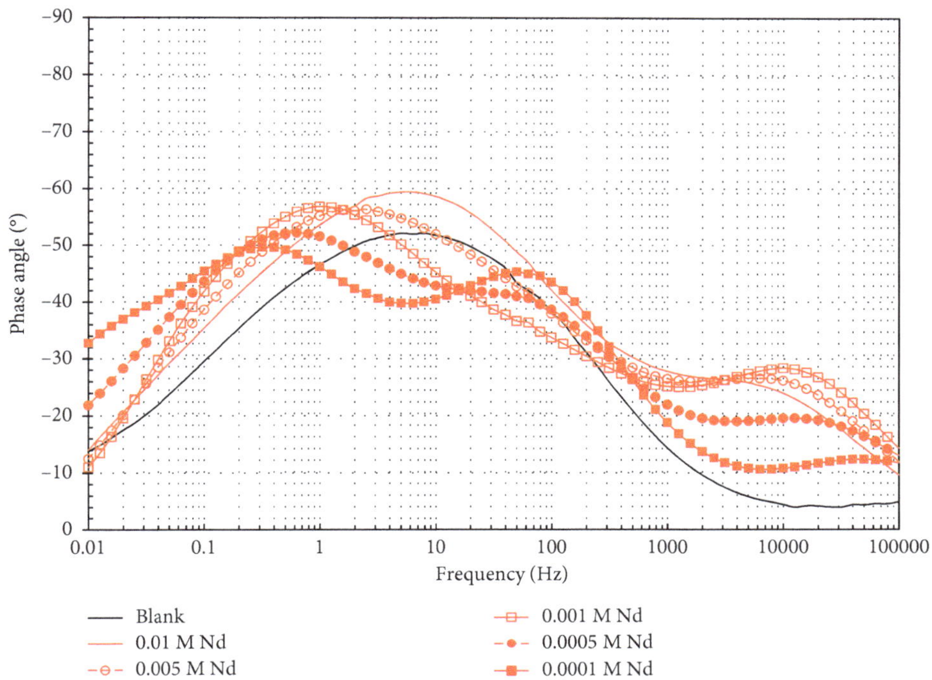

(c)

FIGURE 6: Nyquist and Bode diagrams for API X70 steel with and without the addition of Nd^{3+} ions after 100 hours of immersion in 3.5% NaCl solution.

possible that the absence of the high-frequency plateau is due to the presence of a surface layer on the working electrode, and the absence of the low-frequency plateau indicates that the maximum impedance module is greater than the last registered value, as well as the existence of a process of diffusion-adsorption of species on the surface of the working electrode.

It is evident that, in the presence of Nd^{3+} ions, there was a change in the mechanism of the corrosion process, and this can be attributed to the formation or deposition of an Nd oxide film on the metal surface. In general, the precipitation of the oxide layer is due to hydrolysis reactions of the metal cations with the OH^- ions produced at the cathodic sites. The increase in pH of the cathodic sites causes the precipitation of compounds of the type $Me(OH)_3$, which subsequently are transformed to more stable metal oxides (Me_2O_3). The mechanism of precipitation of both rare earth oxides and hydroxides has been cited in the literature previously [7, 27–30]. In general, the hydrolysis reaction of lanthanide cations (Ln) can occur according to the following reactions:

$$\begin{aligned} O_2 + 2H_2O + 4e^- &\longrightarrow 4OH^- \\ Ln^{3+} + OH^- &\longrightarrow Ln(OH)^{2+} \\ Ln^{3+} + 2OH^- &\longrightarrow Ln(OH)^{+} \\ Ln^{3+} + 3OH^- &\longrightarrow Ln(OH)_3\downarrow \end{aligned} \quad (3)$$

The hydrolysis reaction of the rare earth cations can be associated with the presence and evolution of the time constant observed in the high-frequency region. Because the rare earth hydroxides are only stable in alkaline environments, their stability will depend on the continuous production of OH^- ions, and this will allow their lateral growth by forming a three-dimensional film that inhibits the corrosion process. Because a decrease in pH will cause its dissolution, then its transformation to oxide can occur almost simultaneously with the formation of complex hydroxides.

$$2Ln(OH)_3 \longrightarrow Ln_2O_3 + 3H_2O \quad (4)$$

This reaction can be associated with the presence and evolution of the second time constant due to the formation of a protective film with hydrophobic characteristics. The hydrophobic nature of rare earth oxides has been associated with its unique electronic configuration characterized by the presence of nonaccessible metallic orbitals that reduce its affinity for water, and only the oxygen atoms attached directly to the rare earth are able to form hydrogen bonds with water molecules [31, 32]. Similar observations have been reported, indicating that, by increasing the atomic number of the lanthanides, their reactivity decreases due to the fact that the external electrons have more nuclear attraction [4].

The presence of the third time constant observed in the low-frequency region (the only visible capacitive semicircle in the Nyquist diagram) represents the capacitive response of the electrochemical double layer and the charge transfer resistance, and its evolution is due to the presence of the first two time constants.

Figure 8 shows the equivalent circuits used to fit the impedance spectra. To model the behavior of API X70 steel

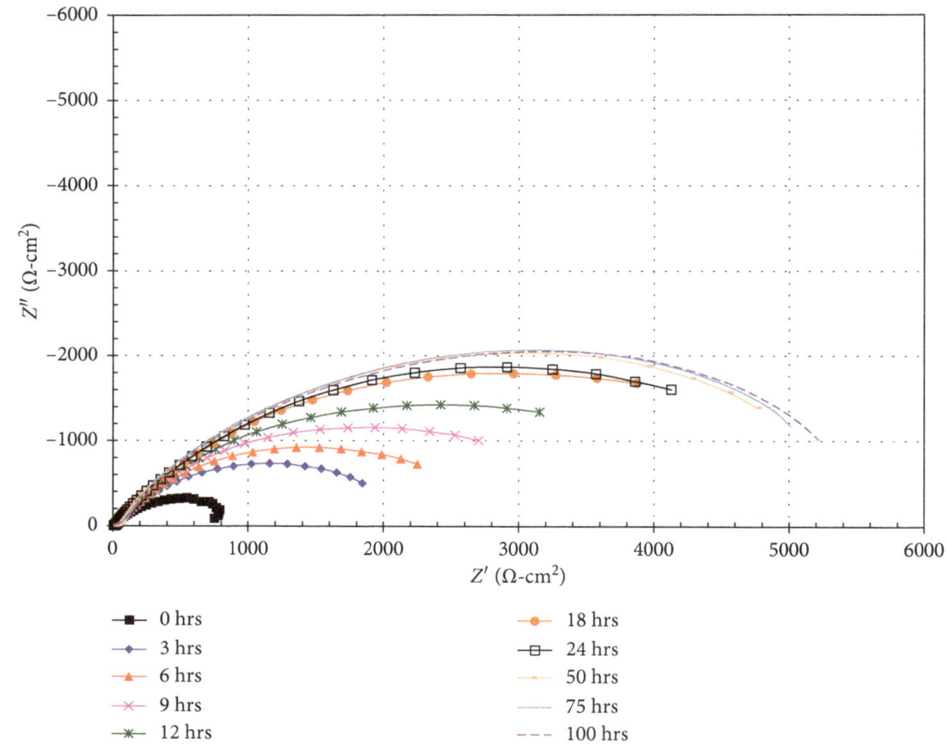

(a)

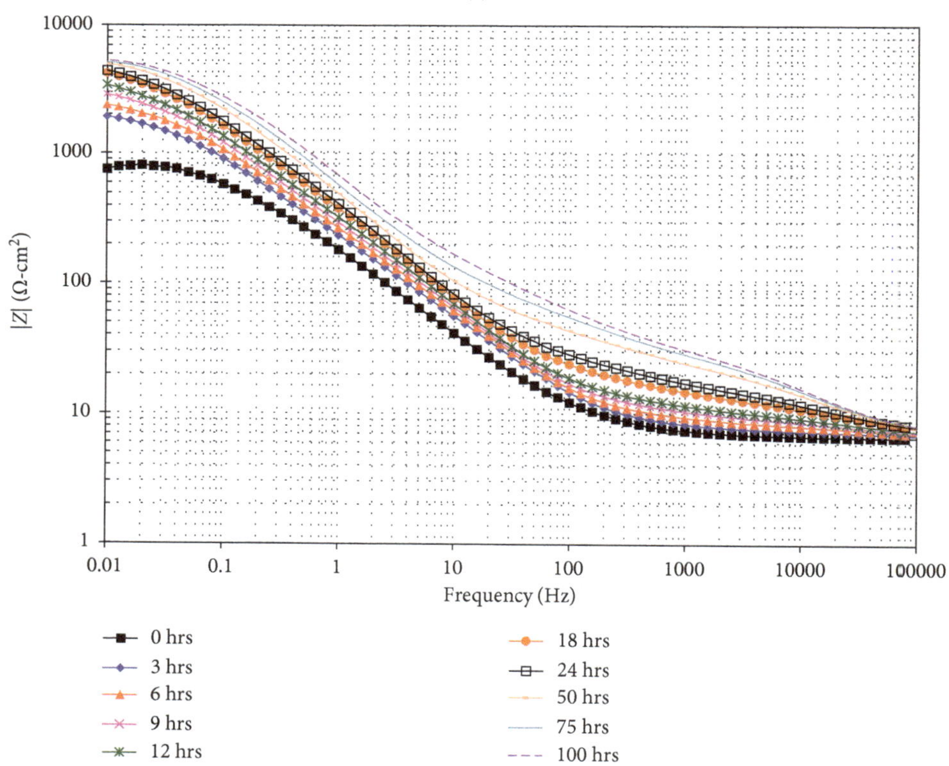

(b)

Figure 7: Continued.

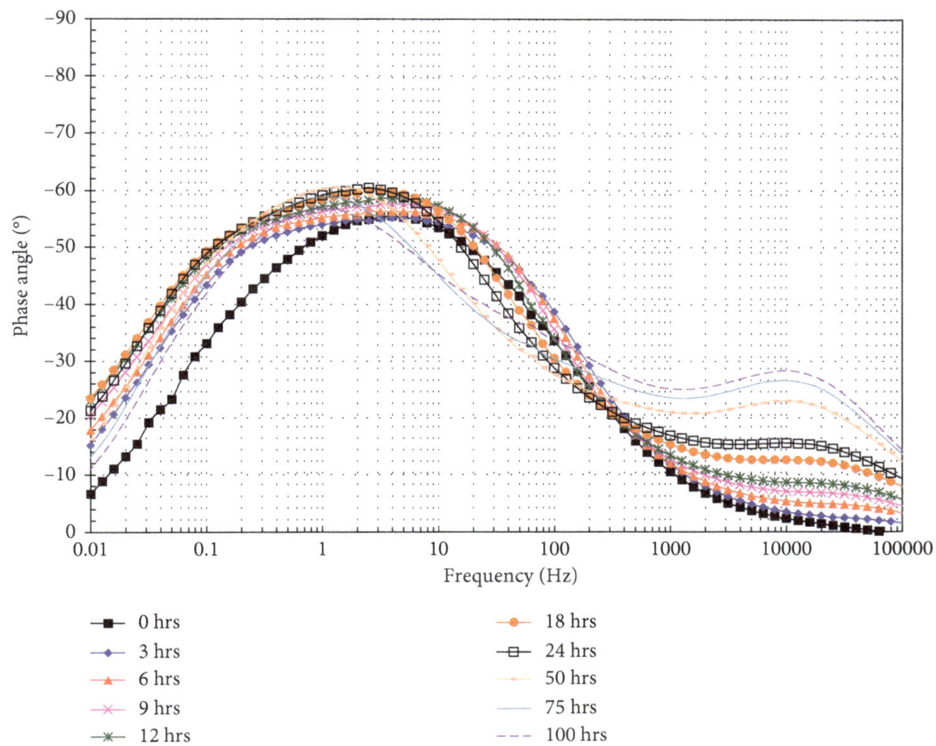

(c)

FIGURE 7: Evolution of the Nyquist and Bode diagrams for API X70 steel in 3.5% NaCl solution with the addition of 0.001 M Nd^{3+}.

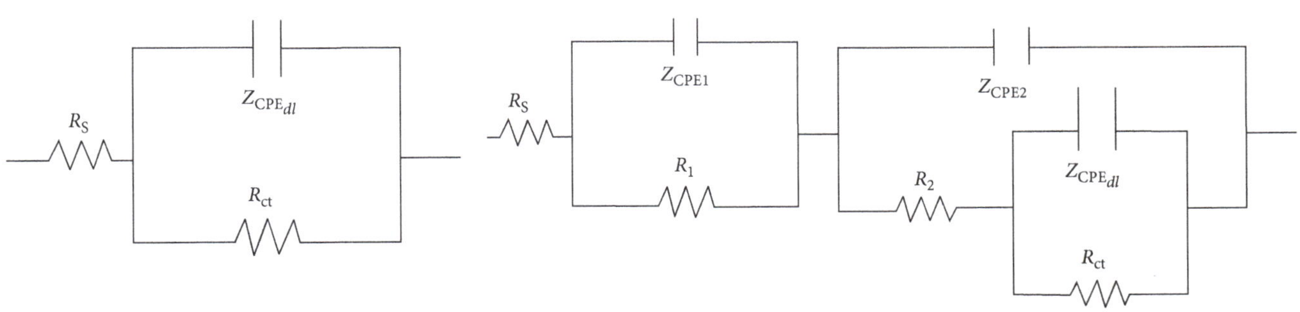

(a) (b)

FIGURE 8: Proposed equivalent circuits for modeling the impedance spectra: (a) without the addition of Nd^{3+} ions and (b) with the addition of Nd^{3+} ions.

without the addition of Nd^{3+} ions, the equivalent circuit of Figure 8(a) was used; in this case, R_s is the electrolyte resistance, R_{ct} the charge transfer resistance, and $Z_{CPE_{dl}}$ the impedance of the double-layer capacitance. The constant phase element (CPE) was used in order to compensate the inhomogeneity due to the irregularities of the surface. The impedance of the CPE is defined as

$$Z_{CPE} = \frac{1}{Y_0}(i\omega)^{-n}, \quad (5)$$

where Y_0 is a proportional factor, i is the imaginary number, ω is the angular frequency ($\omega = 2\pi f$), f is the frequency, and n has the meaning of a phase shift.

On the other hand, with the addition of Nd^{3+} ions, the impedance spectra were modeled with the equivalent circuit of Figure 8(b). In this case, R_1 represents the resistance of the $Nd(OH)_3$ film adsorbed on the metal surface, R_2 represents the resistance of the Nd_2O_3 film precipitated on the metal surface, and Z_{CPE_1} and Z_{CPE_2} are the impedances of the constant phase elements of the $Nd(OH)_3$ film and of the Nd_2O_3 film, respectively.

The evolution of the fitting parameters of the impedance spectra after 100 hours of immersion (Figure 6) according the equivalent circuits of Figure 8 is shown in Table 2. From the R_{ct} data, the same tendency obtained in the other electrochemical measurements carried out can be observed;

TABLE 2: Evolution of electrochemical parameters of the EIS spectra after 100 hours of immersion in 3.5% NaCl solution.

Nd^{3+} ion concentration	R_1 (Ω-cm^2)	Y_{0_1} (μF·cm^{-2})	n_1	R_2 (Ω-cm^2)	Y_{0_2} (μF·cm^{-2})	n_2	R_{ct} (Ω-cm^2)	$Y_{0_{dl}}$ (μF·cm^{-2})	n_{ct}
0.0 M	—	—	—	—	—	—	848	$1.63E-3$	0.65
0.0001 M	15.92	$1.09E-3$	0.30	100.8	$1.31E-4$	0.87	4677	$1.06E-3$	0.63
0.0005 M	16.66	$1.49E-3$	0.59	141.0	$1.81E-4$	0.78	4971	$4.14E-4$	0.69
0.001 M	22.85	$3.98E-5$	0.69	161.4	$1.82E-4$	0.74	5950	$2.02E-4$	0.76
0.005 M	19.87	$4.80E-5$	0.69	41.0	$5.03E-5$	0.90	5886	$2.94E-4$	0.66
0.01 M	22.24	$2.03E-3$	0.60	7.9	$1.59E-4$	0.81	5770	$1.71E-4$	0.42

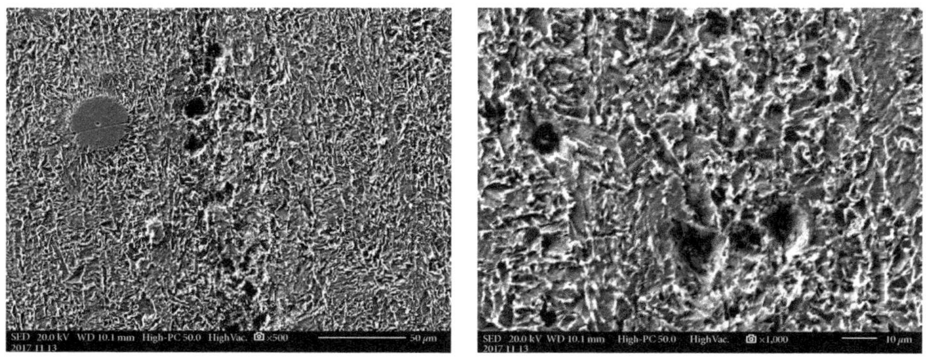

(a) (b)

FIGURE 9: Morphological aspect of API X70 steel after 100 hours of immersion in 3.5% NaCl solution.

that is, increasing the concentration of Nd^{3+} ions increases the corrosion resistance of the API X70 steel. The maximum charge transfer resistance is observed at a concentration of 0.001 M Nd^{3+}, and at higher concentrations, it decreases slightly. In addition, it can also be observed that, at this concentration, the resistance values of R_1 and R_2 are the highest, which also contribute to increase the corrosion resistance of the steel. This contribution can be observed by the values of n_{ct}, where the maximum value is also observed for the concentration of 0.001 M Nd^{3+}, which indicates a more capacitive surface.

3.5. Morphological Analysis of Corroded Surfaces. Figure 9 shows morphological aspects of the surface of the API X70 steel corroded for 100 hours in 3.5% NaCl solution in the absence of Nd^{3+} ions. The micrographs clearly show that the steel underwent a nonuniform corrosion process where pitting attack is evident. The absence of corrosion products indicates that the steel was not able to form a layer of protective corrosion products, suggesting the formation of soluble corrosion products.

The degradation of the steel occurred according to the basic reactions of iron oxidation and water reduction:

$$\begin{aligned} &Fe \longleftrightarrow Fe^{2+} + 2e^- \\ &O_2 + 2H_2O + 4e^- \longleftrightarrow 4OH^- \end{aligned} \tag{6}$$

In addition to the possible formation of nonprotective iron compounds such as hydroxides or oxyhydroxides, in addition to iron oxides,

$$\begin{aligned} Fe^{2+} + 2OH^- &\longleftrightarrow Fe(OH)_2 + H_2O \\ 4Fe(OH)_2 + O_2 + 2H_2O &\longleftrightarrow 4Fe(OH)_3 \\ 4Fe(OH)_2 + O_2 &\longleftrightarrow 2Fe_2O_3 + 4H_2O \\ Fe(OH)_3 + O_2 &\longleftrightarrow FeOOH + H_2O \\ 4Fe(OH)_2 + O_2 &\longleftrightarrow 4FeOOH + 2H_2O \\ 2FeOOH + H_2O &\longleftrightarrow Fe_2O_3 + H_2O \end{aligned} \tag{7}$$

Additionally, the presence of chloride ions promotes the formation of highly soluble iron chlorides and oxychlorides:

$$\begin{aligned} Fe^{2+} + 2Cl^- &\longleftrightarrow FeCl_2 \\ Fe^{2+} + Cl^- + H_2O &\longleftrightarrow FeOCl + H_2\uparrow \end{aligned} \tag{8}$$

The attack experienced by the API X70 steel is correlated with the microstructure of the steel. It is known that depending on the manufacturing conditions and thermal treatments to which an X70 steel is subjected, its microstructure can be formed by the presence of different phases, the main being the polygonal ferrite, quasipolygonal ferrite, Widmanstätten ferrite, acicular ferrite, bainitic granular ferrite, bainite ferrite, martensite in addition to the presence of precipitates (Nb(CN), Fe_3C, MnS, and SiO_2) [33–35]. This combination of phases can favor a galvanic effect between them, causing the corrosion of the electrochemically less stable phase, generally the ferrite [36], in addition to a nonuniform distribution of the local anodic/cathodic ratio that favors localized corrosion [37]. The microstructure of the API X70 steel (Figure 10) was revealed by attacking the

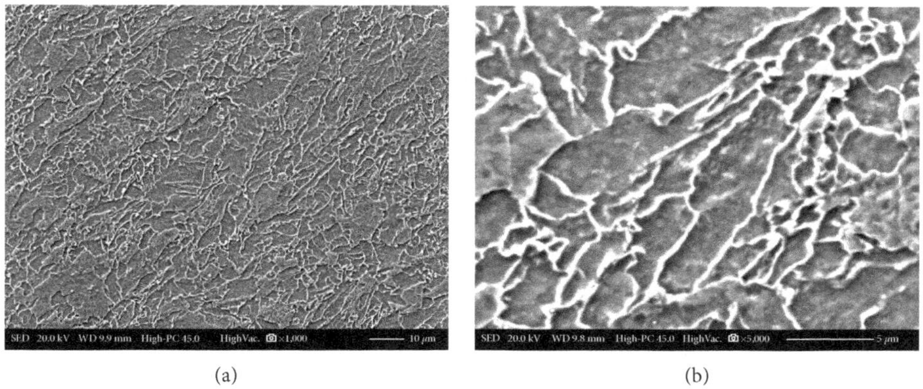

FIGURE 10: Microstructure of the API X70 steel as received at different magnifications.

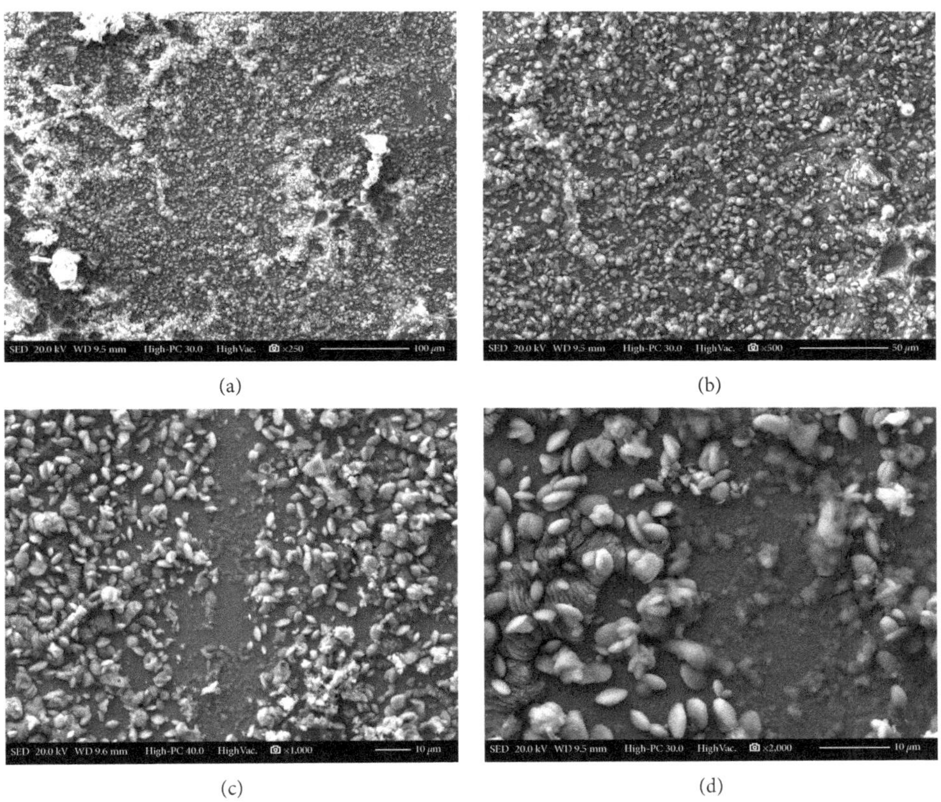

FIGURE 11: Surface appearance of the API X70 steel after 100 hours of immersion in 3.5% NaCl solution with the addition of 0.001 M Nd^{3+}.

steel with a 2% Nital solution (2 ml HNO_3 and 98 ml ethanol). According to the micrographs, it can be observed that the microstructure is composed of quasipolygonal ferrite grains, acicular ferrite, granular bainite, and perlite. Comparing the microstructure of the steel with the morphology of its surface after the corrosion process, it is possible to observe a great similarity between them. This similarity is in agreement with that mentioned above; the presence of multiple phases favors the appearance of galvanic pairs that reduce the resistance to corrosion of steel.

Figure 11 shows the morphological aspects of the surface of the API X70 steel corroded for 100 hours in 3.5% NaCl solution in the presence of 0.001 M Nd^{3+}. It is clearly observed that the surface appearance is completely different from that observed in the absence of inhibitor (Figure 9). The surface appearance shows the presence of a surface layer with protuberances, both morphologies possibly composed of Nd compounds.

A view at higher magnification (Figure 12) allows to define a thin layer adhered to the metal surface as well as

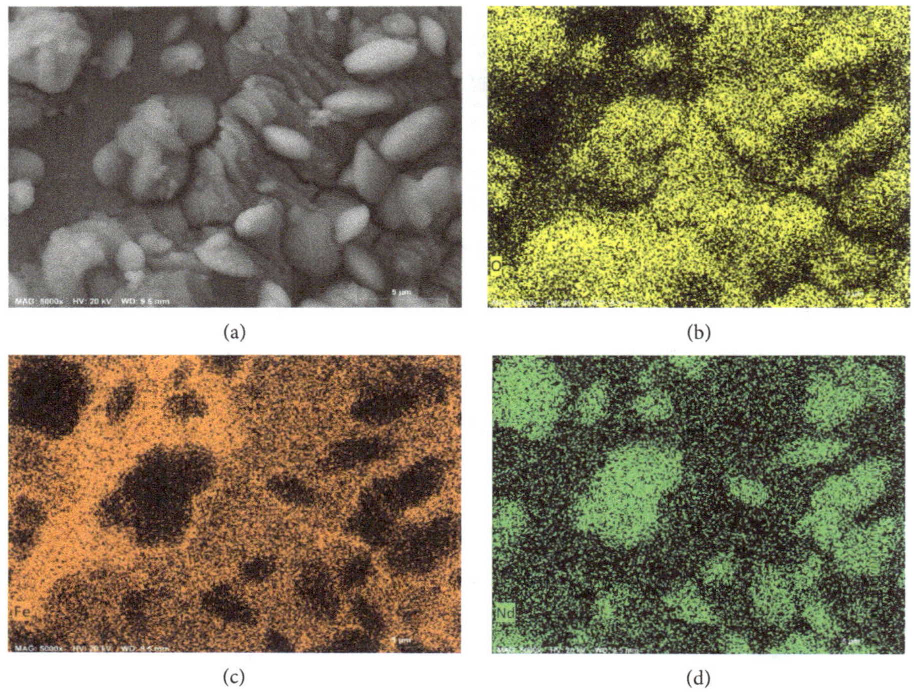

FIGURE 12: View to higher magnification to the corrosion product layer developed on the steel surface and mapping of elements.

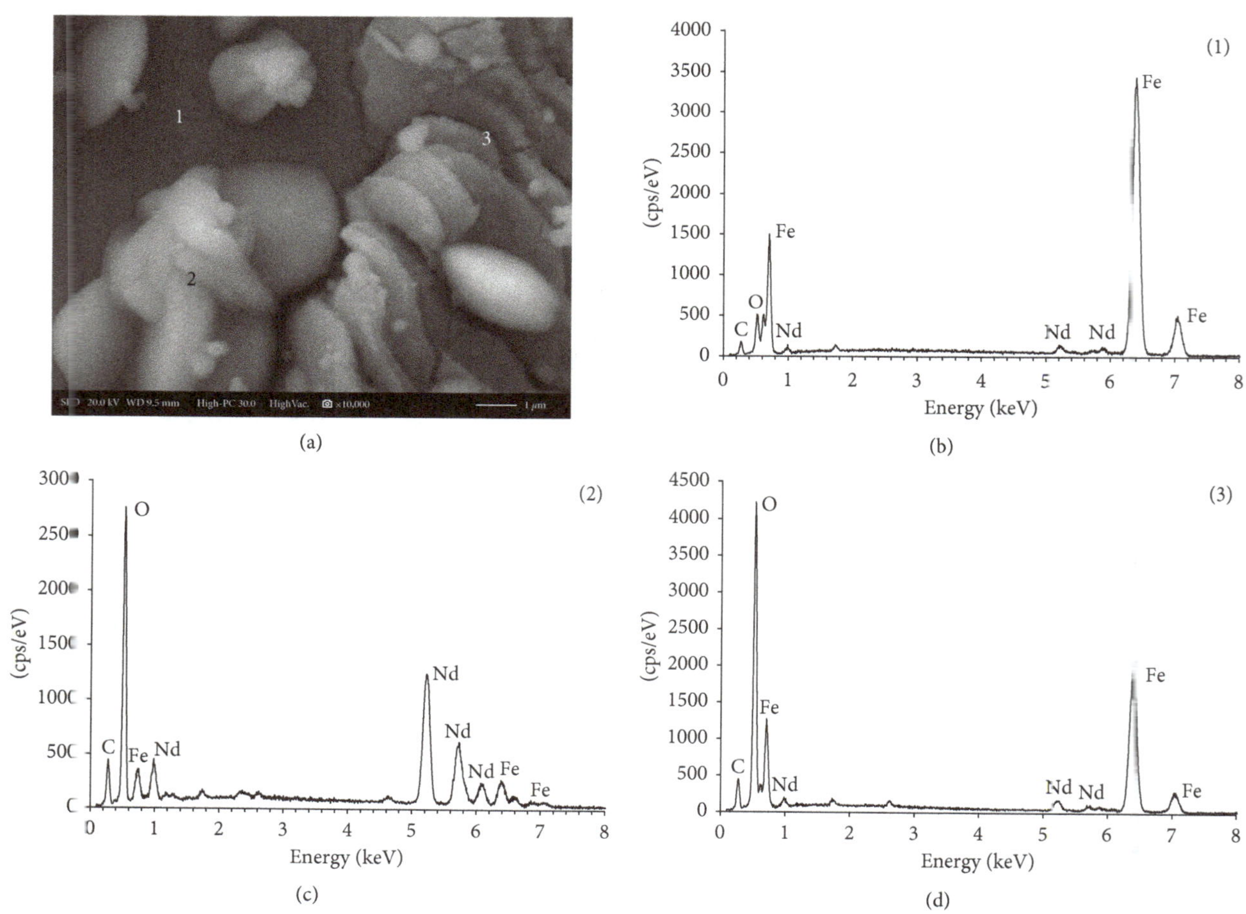

FIGURE 13: Elemental EDS analysis to the different morphologies observed on the API X70 steel surface.

the presence of multiple crystals in apparent cathodic sites on the surface of the steel. The mapping of elements shows the presence of Nd over the all surface, and its greater presence is detected in the crystals that protrude from the surface (Figure 12). A point analysis of the main morphologies identified (Figure 13) showed that the layer adhered to the surface (point 1) had an elemental composition of Fe = 92%, Nd = 3%, and O = 5% (% by weight), the crystals that protrude from the surface (point 2) had an elemental composition of Fe = 7%, Nd = 60%, and O = 33%, and the morphologies with the appearance of stacked layers (point 3) have an elemental composition of Fe = 54%, Nd = 7%, and O = 39%. According to the morphology of point 3, it is possible to deduce that the layer adhered to the metal surface had a thickness between approximately 200 and 400 nm.

It has been indicated that although the formation of a protective layer limits the exchange of electrons with the oxidizing species, the anodic and cathodic reactions are carried out through microdefects in the protective film [7]. Based on the observed evidence, it is possible to suggest that the Nd^{3+} ions protect the steel due to its adsorption on the metal surface forming a thin layer of oxyhydroxides. However, possibly the presence of imperfections causes the cracking and detachment of the protective layer forming the morphologies of stacked layers observed. In the case of small imperfections such as pores (cathode sites), the Nd^{3+} ions are able to precipitate and seal them, causing the formation of the small crystals identified as protuberances.

4. Conclusions

Measurements by means of different electrochemical techniques showed that Nd^{3+} ions are able to reduce the corrosion rate of API X70 steel in chloride-rich environments. Maximum protection is achieved in concentrations as low as 0.001 M Nd^{3+}. Regardless of the fact that, at higher concentrations, the concentration of Cl^- ions is increased and the protection efficiency is only slightly affected. Polarization curves and OCP measurements suggest that Nd^{3+} ions act as a mixed inhibitor with a strong predominant cathodic effect. According to the EIS measurements and morphological analysis, the presence of a protective layer composed of Nd oxides/hydroxides was observed. Although the thickness of the protective layer was 200–400 nm, its presence was able to reduce the exchange of electrons with the oxidizing species. However, the presence of microdefects on its surface caused the formation of morphologies such as stacked layers and protuberances due to the formation of active sites.

Conflicts of Interest

The authors declare that there are no conflicts of interest regarding the publication of this paper.

Acknowledgments

Financial support from the Consejo Nacional de Ciencia y Tecnología (CONACYT, México) (Project 232611 "Laboratorio Nacional de Materias Primas, Metalurgia y Aleaciones Estratégicas Basadas en Tierras Raras Orientadas a Fortalecer la Sustentabilidad de los Sectores Energía, Transporte y Comunicaciones") is gratefully acknowledged. The authors are grateful for the comments and support provided by Dra M. Casales-Diaz.

References

[1] M. Bethencourt, F. J. Botana, J. J. Calvino, M. Marcos, and M. A. Rodriguez-Chacon, "Lanthanide compounds as environmentally-friendly corrosion inhibitors of aluminium alloys: a review," *Corrosion Science*, vol. 40, no. 11, pp. 1803–1819, 1998.

[2] D. Mohammedi, F. Ismail, R. Rehamnia, R. Bensalem, and O. Savadogo, "Corrosion behaviour of steel in the presence of rare earth salts: synergistic effect, corrosion engineering," *Science and Technology*, vol. 50, no. 8, pp. 633–638, 2015.

[3] M. A. Arenas, A. Conde, and J. J. de Damborenea, "Cerium: a suitable green corrosion inhibitor for Tinplate," *Corrosion Science*, vol. 44, no. 3, pp. 511–520, 2002.

[4] Z. Yanhua, Z. Jia, Y. Yongsheng, and Z. Xianguang, "Research on anti-corrosion property of rare earth inhibitor for X70 steel," *Journal of Rare Earths*, vol. 31, no. 7, pp. 734–740, 2013.

[5] H. Allachi, F. Chaouket, and K. Draoui, "Corrosion inhibition of AA6060 aluminium alloy by lanthanide salts in chloride solution," *Journal of Alloys and Compounds*, vol. 475, no. 1-2, pp. 300–303, 2009.

[6] S. Bernal, F. J. Botana, J. J. Calvino, M. Marcos, J. A. Perez-Omil, and H. Vidal, "Lanthanide salts as alternative corrosion inhibitors," *Journal of Alloys and Compounds*, vol. 225, no. 1-2, pp. 638–641, 1995.

[7] B. Davó and J. J. De Damborenea, "Use of rare earth salts as electrochemical corrosion inhibitors for an Al–Li–Cu (8090) alloy in 3.56% NaCl," *Electrochimica Acta*, vol. 49, no. 27, pp. 4957–4965, 2004.

[8] D. Ho, N. Brack, J. Scully, T. Markley, M. Forsyth, and B. Hinton, "Cerium dibutylphosphate as a corrosion inhibitor for AA2024-T3 aluminum alloys," *Journal of The Electrochemical Society*, vol. 153, no. 9, pp. B392–B401, 2006.

[9] H. Allachi, F. Chaouket, and K. Draoui, "Protection against corrosion in marine environments of AA6060 aluminium alloy by cerium chlorides," *Journal of Alloys and Compounds*, vol. 491, no. 1-2, pp. 223–229, 2010.

[10] X.-B. Chen, T. Cain, J. R. Scully, and N. Birbilis, "Technical note: experimental survey of corrosion potentials for rare earth metals Ce, Er, Gd, La, and Nd as a function of pH and chloride concentration," *Corrosion*, vol. 70, no. 4, pp. 323–328, 2013.

[11] D. R. Arnott, B. R. W. Hinton, and N. E. Ryan, "Cationic-film-forming inhibitors for the protection of the AA 7075 aluminum alloy against corrosion in aqueous chloride solution," *Corrosion*, vol. 45, no. 1, pp. 12–18, 1989.

[12] Y. Xingwen, C. Chunan, Y. Zhiming, Z. Derui, and Y. Zhongda, "Study of double layer rare earth metal conversion coating on aluminum alloy LY12," *Corrosion Science*, vol. 43, no. 7, pp. 1283–1294, 2001.

[13] M. Forsyth, T. Markley, D. Ho et al., "Inhibition of corrosion on AA2024-T3 by new environmentally friendly rare earth

organophosphate compounds," *Corrosion*, vol. 64, no. 3, pp. 191–197, 2008.

[14] D. Zhao, J. Sun, L. Zhang, Y. Tan, and J. Li, "Corrosion behavior of rare earth cerium based conversion coating on aluminum alloy," *Journal of Rare Earths*, vol. 28, pp. 371–374, 2010.

[15] Y. Shengxue, L. Qiaoyan, H. Jing, Z. Zhanwei, and Z. Qianyun, "Preparation and performance of rare earths chemical conversion film on magnesium alloy," *Journal of Rare Earths*, vol. 24, no. 1, pp. 397–400, 2006.

[16] L. Li, J. Lei, S. Yu, Y. Tian, Q. Jiang, and F. Pan, "Formation and characterization of cerium conversion coatings on magnesium alloy," *Journal of Rare Earths*, vol. 26, no. 3, pp. 383–387, 2008.

[17] S.-X. Yu, J.-Y. Cao, L. Chen, J. Han, and R.-J. Zhang, "Corrosion resistance, composition and structure of RE chemical conversion coating on magnesium alloy," *Transactions of Nonferrous Metals Society of China*, vol. 18, pp. s349–s353, 2008.

[18] F. Blin, S. G. Leary, G. B. Deacon, P. C. Junk, and M. Forsyth, "The nature of the surface film on steel treated with cerium and lanthanum cinnamate based corrosion inhibitors," *Corrosion Science*, vol. 48, no. 2, pp. 404–419, 2006.

[19] X. Li, S. Deng, H. Fu, G. Mu, and N. Zhao, "Synergism between rare earth cerium (IV) ion and vanillin on the corrosion of steel in H_2SO_4 solution: weight loss, electrochemical, UV–vis, FTIR, XPS, and AFM approaches," *Applied Surface Science*, vol. 254, no. 17, pp. 5574–5586, 2008.

[20] X. Cui, Y. Yang, E. Liu, G. Jin, J. Zhong, and Q. Li, "Corrosion behaviors in physiological solution of cerium conversion coatings on AZ31 magnesium alloy," *Applied Surface Science*, vol. 257, no. 23, pp. 9703–9709, 2011.

[21] G. G. Zaimes, B. J. Hubler, S. Wang, and V. Khanna, "Environmental life cycle perspective on rare earth oxide production," *ACS Sustainable Chemistry & Engineering*, vol. 3, no. 2, pp. 237–244, 2015.

[22] Z. Sun, Y. Xiao, H. Agterhuis, J. Sietsma, and Y. Yang, "Recycling of metals from urban mines–a strategic evaluation," *Journal of Cleaner Production*, vol. 112, pp. 2977–2987, 2016.

[23] M. A. Arenas and J. J. De Damborenea, "Growth mechanisms of cerium layers on galvanised steel," *Electrochimica Acta*, vol. 48, no. 24, pp. 3693–3698, 2003.

[24] O. L. Riggs Jr., Corrosion Inhibition, C. C. Nathan, Ed., National Association of Corrosion Engineers, Houston, TX, USA, 2nd edition, 1973.

[25] M. A. Amin, M. A. Ahmed, H. A. Arida et al., "Monitoring corrosion and corrosion control of iron in HCl by non-ionic surfactants of the TRITON-X series–Part III. Immersion time effects and theoretical studies," *Corrosion Science*, vol. 53, no. 5, pp. 1895–1909, 2011.

[26] M. S. Al-Otaibi, A. M. Al-Mayouf, M. Khan, A. A. Mousa, S. A. Al-Mazroa, and H. Z. Alkhathlan, "Corrosion inhibitory action of some plant extracts on the corrosion of mild steel in acidic media," *Arabian Journal of Chemistry*, vol. 7, no. 3, pp. 340–346, 2014.

[27] K. A. Yasakau, M. L. Zheludkevich, S. V. Lamaka, and M. G. S. Ferreira, "Mechanism of corrosion inhibition of AA2024 by rare-earth compounds," *Journal of Physical Chemistry B*, vol. 110, no. 11, pp. 5515–5528, 2006.

[28] E. A. Matter, S. Kozhukharov, M. Machkova, and V. Kozhukharov, "Comparison between the inhibition efficiencies of Ce(III) and Ce(IV) ammonium nitrates against corrosion of AA2024 aluminum alloy in solutions of low chloride concentration," *Corrosion Science*, vol. 62, pp. 22–33, 2012.

[29] T. Hu, H. Shi, T. Wei, F. Liu, and S. Fan, "En-Hou Han, cerium tartrate as a corrosion inhibitor for AA 2024-T3," *Corrosion Science*, vol. 95, pp. 152–161, 2015.

[30] L. J. Zhang, Z. Zhang, J. Q. Zhang, and C. N. Cao, "The study of the $La(NO_3)_3$ inhibition on X70 pipeline steel in 3.0(wt.%) NaCl solution," *Materials and Corrosion*, vol. 56, no. 9, pp. 630–635, 2005.

[31] G. Carchini, M. García-Melchor, Z. Łodziana, and N. Lopez, "Understanding and tuning the intrinsic hydrophobicity of rare-earth oxides: A DFT+U study," *ACS Applied Materials & Interfaces*, vol. 8, no. 1, pp. 152–160, 2016.

[32] R. Lundy, C. Byrne, J. Bogan et al., "Exploring the role of adsorption and surface state on the hydrophobicity of rare earth oxides," *ACS Applied Materials & Interfaces*, vol. 9, no. 15, pp. 13751–13760, 2017.

[33] B. Hwang, Y. M. Kim, S. Lee, N. J. Kim, and S. S. Ahn, "Correlation of microstructure and fracture properties of API X70 pipeline steels," *Metallurgical and Materials Transactions A*, vol. 36, no. 3, pp. 725–739, 2005.

[34] Y.-B. Guo, C. Li, Y.-C. Liu et al., "Effect of microstructure variation on the corrosion behavior of high-strength low-alloy steel in 3.5wt% NaCl solution," *International Journal of Minerals, Metallurgy, and Materials*, vol. 22, no. 6, pp. 604–612, 2015.

[35] T. V. Shibaeva, V. K. Laurinavichyute, G. A. Tsirlina, A. M. Arsenkin, and K. V. Grigorovich, "The effect of microstructure and non-metallic inclusions on corrosion behavior of low carbon steel in chloride containing solutions,," *Corrosion Science*, vol. 80, pp. 299–308, 2014.

[36] C. W. Du, X. G. Li, P. Liang, Z. Y. Liu, G. F. Jia, and Y. F. Cheng, "Effects of microstructure on corrosion of X70 pipe steel in an alkaline soil," *Journal of Materials Engineering and Performance*, vol. 18, no. 2, pp. 216–220, 2009.

[37] M. Alizadeh and S. Bordbar, "The influence of microstructure on the protective properties of the corrosion product layer generated on the welded API X70 steel in chloride solution,," *Corrosion Science*, vol. 70, pp. 170–179, 2013.

19

Development of an Efficient Steel Beam Section for Modular Construction Based on Six-Sigma

Tae-Hyu Ha,[1] Bong-Ho Cho,[2] Hongjin Kim,[3] and Dae-Jin Kim[4]

[1] *POSCO, Seoul, Republic of Korea*
[2] *Department of Architecture, Ajou University, Suwon, Republic of Korea*
[3] *Department of Architectural Engineering, Kyungpook National University, Daegu, Republic of Korea*
[4] *Department of Architectural Engineering, Kyung Hee University, Yongin, Republic of Korea*

Correspondence should be addressed to Bong-Ho Cho; bhcho@ajou.ac.kr

Academic Editor: Jae-Han Lim

This study presents a systematic approach for the development of an efficient steel beam section for modular construction based on Six-Sigma. Although the Six-Sigma is frequently implemented in manufacturing and other service industries, it is a relatively new concept in the area of building design and construction. As a first step in this approach, market studies and surveys are conducted to obtain the opinions of potential customers. Then the opinions of customers are converted into quality characteristics for the steel beam using the quality function deployment methodology. A steel hollow flanged channel is chosen as the main modular beam shape, and the design concept is derived and developed by applying the Pugh matrix methodology. A pilot test was performed to validate the effectiveness of the developed beam section. The results indicated that the developed channel beam section showed excellent performance and retained high accuracy in fabrication, thus resulting in a significant reduction of steel consumption.

1. Introduction

Modular construction is an industrialized construction system, with a process that mainly consists of the factory production of preengineered building units, the delivery of the fabricated units to the construction site, and their fast assembly [1]. The structural system of steel frame modular buildings can be largely classified into two types. The first is the beam-column frame system and is suitable for use in school buildings or military barracks, which require relatively large open space [2]. The second is the load-bearing wall system, and dormitories or bachelor officer quarters are good applications of this type as they can be built with a series of transportable unit room modules [3, 4].

In general, the modular unit of the beam-column frame system is constructed by connecting structural members such as corner columns, internal columns, floor beams, and ceiling beams by bolting or welding. Figure 1 illustrates an example of standard modules for elementary schools in Korea. The dimensions are 9.8 m (length), 7.5 m (column to column span), 3 m (width), and 3.2 m (height). Table 1 provides the composition ratio of the main structural members used in this standard module. In the beam-column frame system, the upper beam of the lower level unit is connected with the lower beam of the upper level unit. This connection scheme is called the double beam system. As a result, beam members make up a larger portion of the structural components of the standard module, which is 81% as shown in Table 1. The beam members play an important role as they not only contribute greatly to the overall construction cost of modular buildings, but also may affect various factors such as story height and floor vibration. Therefore, the selection of adequate beam members is essential to improving the cost efficiency and performance of modular buildings.

Six-Sigma was first developed by Motorola in the 1980s [5] and became well known in the 1990s when Jack Welch made it central to his business strategy at General Electric. Since the late 1990s, most manufacturing companies in Korea

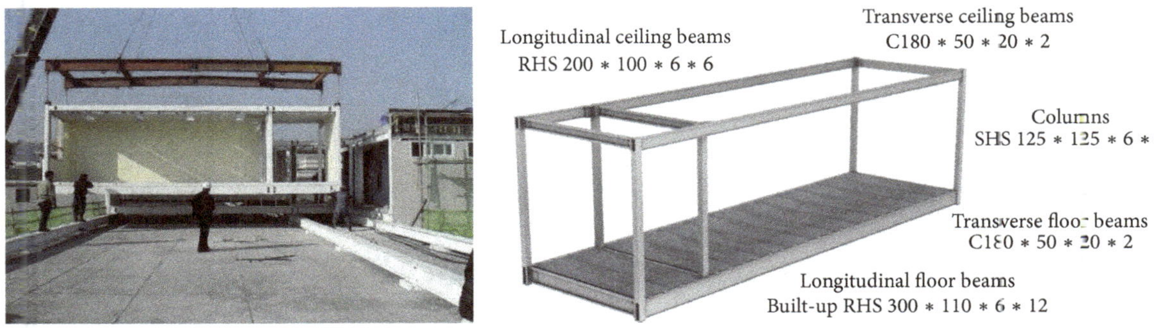

(a) Framed modular unit (b) Structural members of the modular unit

FIGURE 1: Standard modular unit and structural members.

TABLE 1: Composition ratio of main structural members used in the standard module.

Component	Shape	Weight ratio (%)		
Longitudinal floor beams	Built-up rectangular hollow section	41	73	81
Longitudinal ceiling beams	Rectangular hollow section	32		
Transverse floor and ceiling beams	Roll-formed C shape		7	
Columns	Square hollow section		19	

have applied the Six-Sigma for productivity enhancement, process improvement, and new product development. While the Six-Sigma has been applied to construction management and housing design on several occasions such as [5, 6], this is one of very few studies where the Six-Sigma concept is employed to develop new steel structural members for modular construction.

In general, Six-Sigma projects follow either the DMAIC or DFSS methodology. The DMAIC focuses on improving an existing business process, while the DFSS aims at creating new products or process designs. Based on this discussion, this study proposes a systematic approach for the development of a steel modular beam section that can be used in the beam-column frame system and validates its effectiveness through a pilot test.

2. QFD Methodology and Derivation of CTQ

2.1. QFD Methodology. The philosophy of the Six-Sigma approach emphasizes the reflection of customers' needs in the development of production processes. In order to achieve this, the identification and categorization of potential customers are required, and the importance of each customer group needs to be evaluated. In the Six-Sigma methodology, these customer requirements are referred to as the voice of customers (VOCs), which are obtained through surveys and interviews and utilized as basic data for product development. According to the results of the interviews, module manufacturers require a decrease in manufacturing cost as well as a shortening of order-delivery period, while designers expect various construction-related requirements to be satisfied. The VOC of the building owner is mainly focused on the excellent performance of the manufactured modules.

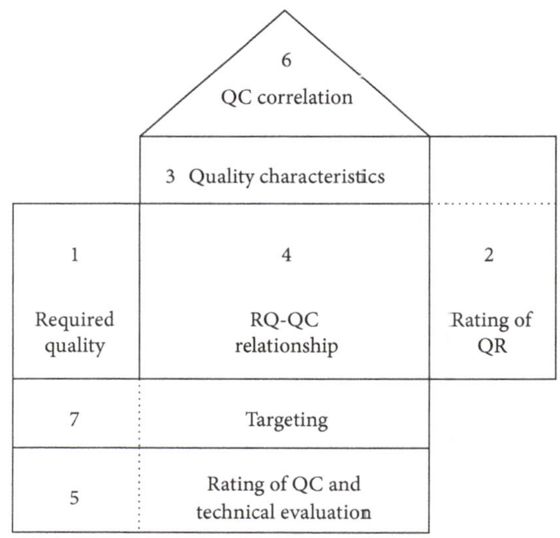

FIGURE 2: HOQ required to conduct QFD.

This study adopts the quality function deployment (QFD) methodology to convert customers' needs into quality characteristics for the products to be developed. The QFD methodology, first proposed by Akao in the 1960s [7], is a tool that can provide a quantitative evaluation on the relation between customer requirements and product characteristics using a correlation matrix [8, 9]. Figure 2 illustrates the house of quality (HOQ), which is required to conduct QFD.

In general, the QFD methodology follows a number of procedures: (1) determination of required quality; (2) importance rating for customers' needs; (3) identification of quality characteristics; (4) correlation analysis between the required quality and quality characteristics; (5) technical evaluation; (6) interrelation analysis of the quality characteristics if required; (7) target specification setup for the quality characteristics. In this study, the required quality is

TABLE 2: Priority rating for the customer requirements.

Customers' required quality	(1) Importance rating	(2) Satisfaction rating	(3) Possible quality levels	(4) Level-up ratio	(5) Sales Point	(6) Absolute priority	Priority Ratios	Ratings
Small steel quantity	4.6	2.8	4	1.4	○	7.7	10%	2
Low manufacturing cost	4.6	2.6	5	1.9	◎	13.1	17%	1
Easy structural design	3.6	3.4	4	1.2		4.3	5%	7
Easy architectural design	3.6	3.4	4	1.2		4.3	5%	7
Easy connection	3.6	3.4	4	1.2		4.3	5%	7
Transportation with low-bed trailer	3.6	3.2	4	1.3		4.7	6%	5
Transportation with cargo truck	3	3	3	1		3	4%	15
Easy installation of plumbing equipment	3.4	3.2	4	1.3		4.4	6%	6
A short time for stocking after order	3.6	2.8	4	1.4	○	6	8%	4
The stock is secured in advance	3	2.6	3	1.2	○	4.3	5%	7
Small manufacturing deformation	3.8	3	4	1.3	◎	7.4	9%	3
Less surface defects	3.6	2.8	3	1.1		4	5%	12
Easy application of high-strength steel	3	2.8	3	1.1		3.3	4%	14
Structural safety	3.6	3.6	4	1.1		4	5%	12
Less floor vibration	4.2	4	4	1		4.2	5%	11

(1) Importance rating: values obtained from customer survey (1~5).
(2) Satisfaction rating: values obtained from customer survey (1~5).
(3) Quality levels that may be obtained from the viewpoint of developers (1~5).
(4) Level up ratio: (3)/(2).
(5) Sales point from the viewpoint of developers (◎: 1.5, ○: 1.2).
(6) Absolute priority: (1)∗(4)∗(5).

determined from the VOCs obtained from the first round of interviews, and then the quality characteristics are prioritized based on the evaluations of customers and experts. Finally, the target values and importance ratings for the quality characteristics are assigned.

In this study, QFD methodology is used for the development of a new steel modular beam section. The results of priority ratings for the VOCs, where the requirements of both customers and product developers are considered, are provided in Table 2. The item labeled "customers' required quality" in Table 2 is a reorganization of the VOCs of each customer group for QDF analysis, and its priority ratings are determined based on the results of the customer surveys. From the results of the table, items with high priority are found to have "low manufacturing cost," "small steel quantity," "small manufacturing deformation," and "short time for stocking after order." In the determination of the priority rating in Table 2, the importance of customers' requirements and satisfaction with existing products are determined from the surveys. The "planning quality" and "sales point" in Table 2 are the indicators for the attainable quality and sales quantity from the perspective of developers, respectively.

2.2. Derivation of CTQ. From a correlation analysis between the required qualities and quality characteristics, parameters that are critical to quality (CTQ) from the customers' perspective can be derived. The quality characteristics are given different weight values (◎: 5 points, ○: 3 points, and △: 1 point) depending on the relationship with the corresponding customer requirements. The overall importance can be calculated by summing up the multiplications of the weighted value and absolute importance in Table 3. The quality characteristics with high importance score can be selected as the potential CTQs.

This process of deriving CTQs is important in the Six-Sigma methodology, which reflects the customers' needs as much as possible. The VOC does not contain any direct statements on the quality characteristic, but it can be converted into designable quality characteristics through QFD analysis. Table 3 lists the importance ratings derived from the correlation analysis on the required quality and quality characteristics.

From the QFD analysis, the quality characteristics with high importance are found to be "weld length per unit length," "moment of inertia per sectional area of beam," and "weight of floor beams per unit area of plan." These items

TABLE 3: Correlation analysis between the required qualities and quality characteristics.

Customers' required quality	Quality characteristic															Absolute priority	Priority ratios	Ratings
	Price of beams			Contribution to competitiveness of modules							Structural performance of bema			Quality of manufacturing				
	Surface area per unit length	Welding length per unit length	Span length	Thickness of webs	Diameter of web opening	Number of webs	Thickness of flanges	Number of openings	Height of ceiling and floor beams	Weight of floor beams per unit area of plan	Moment of inertia per sectional area of beam	Section modulus per sectional area of beam	Lever arm of bolts at the end plate	Number of axes of symmetry	Radius of curved section			
Low cost of beams																		
Small steel quantity				○		○	△			◎	◎	○				7.7	10%	2
Low manufacturing cost	○	◎			○			○								13.1	17%	1
Easy design of modular building																		
Easy structural design			◎								◎	○	◎			4.3	5%	7
Easy architectural design	◎		◎		○			○	○	◎				○	○	4.3	5%	7
Easy fabrication and transportation																		
Easy connection				◎			◎						◎		○	4.3	5%	7
Transportation with low-bed trailer									◎				○			4.7	6%	5
Transportation with cargo truck									◎				○			3	4%	15
Easy installation of plumbing equipment					◎			◎								4.4	6%	6
Short time for stocking																		
A short time for stocking after order	○	◎						○								6	8%	4
The stock is secured in advance			◎													4.3	5%	7
High quality of manufacturing																		
Small manufacturing deformation		○				◎								◎		7.4	9%	3
Less surface defects															○	4	5%	12
Easy application of high-strength steel	◎									◎					○	3.3	4%	14
Structural performance																		
Structural safety											◎	◎				4	5%	12
Small floor vibration									○		◎					4.2	5%	11
Absolute priority	126	194	116	62	92	90	46	110	95	138	182	72	101	80	48		100%	
Priority ratios	8.1%	12.5%	7.5%	4.0%	5.9%	5.8%	3.0%	7.1%	6.1%	8.9%	11.7%	4.6%	6.5%	5.1%	3.1%			
Ratings	4	1	5	13	9	10	15	6	8	3	2	12	7	11	14			

TABLE 4: CTQ definitions and target levels.

CTQ	Definition	Measuring method	Current level Built-up RHS	Current level Channel	Target level
Welding length per unit length	Welding length per 1 m beam length	Measuring welding length	4	0	Less than 2
Moment of inertia per sectional area of beam	Moment of inertia divided by sectional area of beam	Calculation of moment of inertia and sectional area of beams	127 cm^2 (7.5 m span)	132 cm^2 (7.5 m span)	Larger than 150 cm^2 (7.5 m span)
Weight of floor beams per unit area of plan	Total weight of beams used for modules divided by plan area of module	Calculation of weight of beams after redesign of modules	0.30 kN/m^2 (7.5 m span)	0.25 kN/m^2 (7.5 m span)	Less than 0.28 kN/m^2 (7.5 m span)

TABLE 5: Classification of key functions for the steel modular beam section.

Main process	Manufacturing beams	Assembly of modules	Transportation and installing	Ensuring structural performance
Must be functions	—	Welding end-plate Connecting beam and column Connecting beam and slab Connecting beam and ceiling track Connecting beam and wall track	Transportation of modules	Flexural strength of beam Shear strength of beam Minimize deflection of beam Minimize floor vibration Earthquake and wind resistance Fire resistance
Delighter functions	—	—	Plumbing equipment penetration	—
Primary satisfier functions	Minimize a space for stock	—	—	—

are selected as CTQs in the following step and expressed quantitatively. Among the 15 quality characteristics, the top 3 items are selected as CTQs and defined as shown in Table 4, which shows their definitions, measurement methods, current, and target levels. The target levels are determined using the existing built-up RHS (rectangular hollow sections) and channels as references.

3. Product Design Using Pugh Matrix Methodology

3.1. Function Analysis and Pugh Matrix. The core functions of the steel modular beam section are determined and its optimal design is developed utilizing the CTQs derived in the previous phase. Table 5 describes the main processes required for the manufacturing of the steel modular beam section and classifies the functions required in each process into one of three functions, namely, the basic function (necessary), attractive function (delighter), and discriminative function (primary satisfier). The basic functions must be satisfied, while the attractive and differentiated functions can be optionally considered in the design process.

Key functions can be determined from a correlation analysis between the quality characteristics and associated functions by applying the same method used to create CTQs in Table 3. Table 6 shows the relationship between the quality characteristics and the key functions. From this relationship, "minimize deflection and vibration of beam," "flexural strength of beam," "earthquake and wind resistance," and "piping equipment penetration" are selected as key functions.

After derivation of the key functions, an optimal design concept satisfying the derived key functions needs to be developed. In order to achieve this, various methods can be used such as the benchmarking of advanced technologies, brainstorming by experts, TRIZ, and mind mapping. In this study, various solutions that can satisfy the key functions are created, and several possible design options are obtained by combining these solutions. Finally, the Pugh matrix is employed to select the optimal design among the designs obtained.

Table 7 lists the available solutions for each of the key functions of the steel modular beam section. The table indicates that there are two solutions for "piping equipment penetration" and "connecting beam and slab," four solutions for structural functions, specifically "flexural strength of beam," "minimize deflection and vibration of beam," and "earthquake and wind resistance," and one solution for all of the remaining functions. These solutions can be combined to produce an optimal design concept. By combining the solutions listed in Table 7, seven design concepts are derived. They satisfy various functions required for the floor and ceiling

Table 6: Correlation between the quality characteristics and key functions.

Quality characteristic	Key function										
	Assembly of modules					Ensuring structural performance				Secondary function	
	Welding end-plate	Connecting beam and column	Connecting beam and slab	Connecting beam and ceiling track	Connecting beam and wall track	Flexural strength of beam	Minimize deflection and vibration of beam	Earthquake and wind resistance	Fire resistance	Plumbing equipment penetration	Transportation of modules
Price of beam											
Surface area per unit length									◎		
Welding length per unit length						◎		○			
Contribution to competitiveness of module											
Span length							◎				
Thickness of webs			◎	◎							
Diameter of web opening						○				◎	
Number of webs	○	◎	○	○							
Thickness of flanges					◎						
Number of openings						○				◎	
Height of ceiling and floor beams							◎				◎
Weight of floor beams per unit area of plan											
Structural performance of beam											
Moment of inertia per sectional area of beam							◎				
Section modulus per sectional area of beams						◎		◎			
Lever arm of bolts at the end plate	◎	◎						◎			
Quality of manufacturing											
Number of axes of symmetry			◎	○							
Radius of curved section				○	◎						
Priority ratio	6.5%	9.5%	8.6%	6.7%	4.7%	16.7%	19.7%	11.9%	0.6%	10.1%	4.8%
Rating	8	5	6	7	10	2	1	3	11	4	9

Table 7: Available solutions for each of the key functions.

Key functions	Solutions			
Piping equipment penetration	Install reinforcing device at penetration Solution A		Penetrate without reinforcing Solution B	
Transportation of modules	Limiting module height less than 3.3 m			
Welding end-plate	Limiting flange width less than 120 mm			
Connecting beam and column	Ensuring bolting space at end-plate			
Connecting beam and slab	Integrate beam and slab support Solution a		Separate beam and slab support Solution b	
Connecting beam and ceiling track	Connect track to web of beam with screw			
Connecting beam and wall track	Ensuring flange width of 60 mm			
Flexural strength of beam Minimize deflection and vibration of beam Earthquake and wind resistance	Use hollow flange (flange thickness = web thickness) Solution I	Use hollow flange (flange thickness > web thickness) Solution II	Use open flange (flange thickness = web thickness) Solution III	Use open flange (flange thickness > web thickness) Solution IV

Table 8: Design concepts obtained by combinations of the available solutions.

Combination of solutions	Solutions A + a + I	Solutions A + b + I	Solutions A + a + II	Solutions A + b + II	Solutions B + a + III	Solutions B + b + III	Solutions B + b + IV
Design concepts	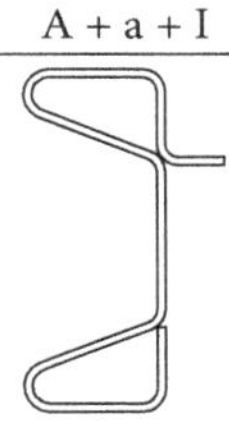	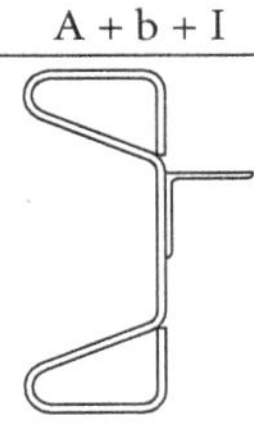	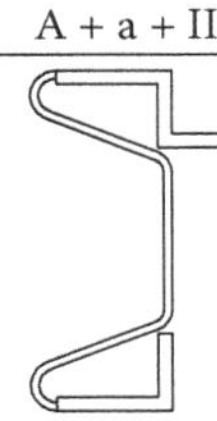	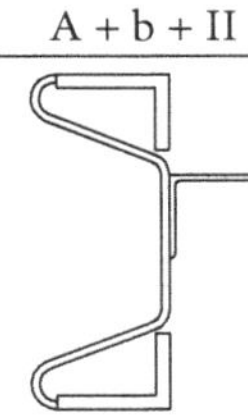		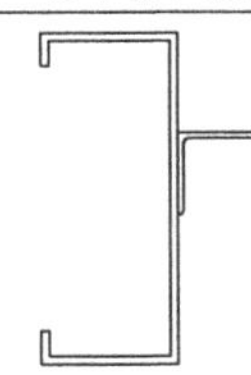	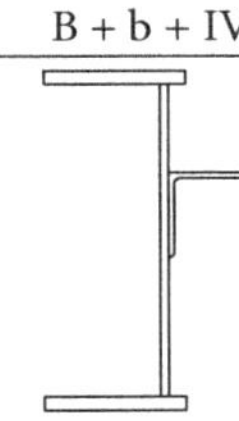
Manufacturing cost	Very high	High	Very high	Very high	Middle	Middle	Very high
Steel quantity	Very low	Low	Very low	Very low	Low	High	Low
Compatibility between floor and ceiling beam	Not compatible	Compatible	Not compatible	Compatible	Not compatible	Compatible	Compatible

beams of the modular beam-column frame system and show different characteristics with regard to manufacturing cost, steel consumption, and availability for both of ceiling and floor beams. Their characteristics are summarized in Table 8.

In order to select the optimal design among the candidates listed in Table 8, a Pugh matrix is created in Table 9. In this methodology, a relative evaluation is performed on each candidate by comparing it with the reference design for each of the key criteria. The built-up RHS is chosen as the reference design for this comparison. Thus, if a candidate shows better performance than the reference design, it is denoted as +. If the candidate is worse than or is equal to the reference design, it is denoted as − or S, respectively. The candidate with the greatest difference between the sum of pluses and that of minuses is chosen as the optimal design. The results in Table 9 indicate that design concept 2 is the optimal design. Thus, it is selected as the prototype for the steel modular beam section.

The prototype has hollow flanges in both the upper and lower parts of the section as shown in Table 9. It can be manufactured by a roll forming method, in which thin steel sheets are passed through several rolls and are consecutively bent at room temperature. The cross-sectional shape is illustrated in Figure 3. The steel beam member with the cross-sectional shape shown in Figure 3 is hereafter referred to as the modular construction optimized (MCO) beam for convenience.

3.2. Detailed Design of MCO Beams. In this section, a detailed design of the MCO beam is discussed considering the manufacturing process. With the detailed design components, the cross-section has a variable portion as well as an invariable portion as shown in Figure 4. Although MCO beams manufactured through the roll forming process require smaller manufacturing costs than general built-up beams, a variety in depth is also necessary for their applications to modular

TABLE 9: Selection of the optimal design using the Pugh matrix methodology.

Better + Same S Worse −		Reference	Concept 1	Concept 2	Concept 3	Concept 4	Concept 5	Concept 6	Concept 7
Key criteria	Weight	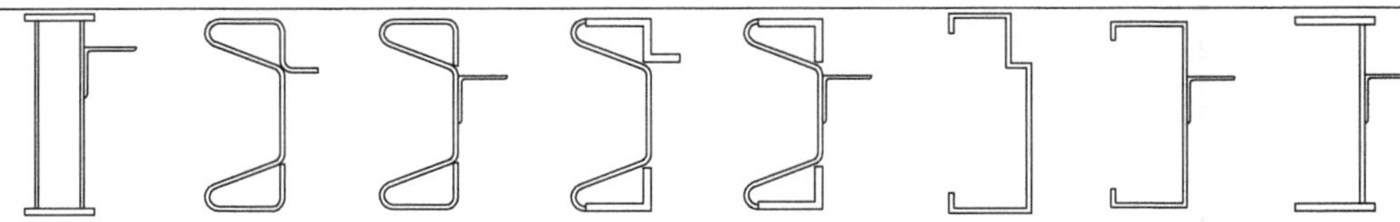							
Manufacturing cost	5	S	+	+	S	S	+	+	−
Fire proofing cost	4	S	−	−	−	−	−	−	−
Steel quantity	4	S	+	+	+	+	−	−	+
Easy fabrication	3	S	+	+	+	+	+	+	+
New investment	5	S	−	−	−	−	+	+	−
Compatibility	4	S	−	+	−	+	−	+	−
Creativity	3	S	+	+	+	+	S	S	S
Sum of +		0	4	5	3	4	3	4	2
Sum of −		0	3	2	3	2	3	2	4
Sum of S		7	0	0	1	1	1	1	1
Weighted sum of +		0	15	19	10	14	13	17	7
Weighted sum of −		0	13	9	13	9	12	8	18

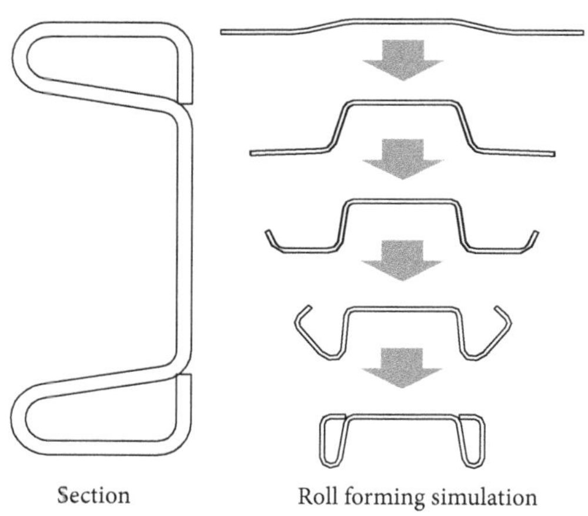

FIGURE 3: Cross-sectional shape of the MCO beam and its roll forming simulation.

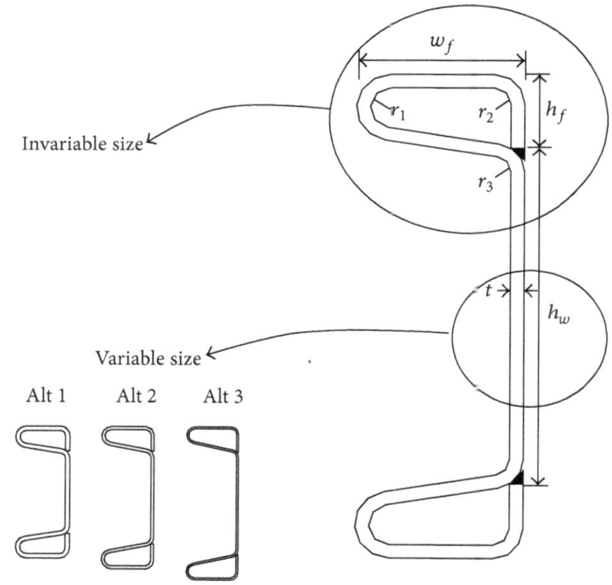

FIGURE 4: Cross-sectional design of the MCO beam.

buildings with diverse spans. Therefore, the upper and lower hollow flange parts are designed as the invariable portion and produced using a single roll. In contrast, the thickness and width of the web section are variable. The design requirements for each design factor of the MCO beam are listed in Table 10. SPA-H steel is used for manufacturing the MCO beams since it has high durability as well as a high tensile strength of 490 MPa.

The design procedures for the MCO beam with invariable cross-section (hollow flange) can be summarized as follows.

Step 1 (setting up the range of the flange width as shown in Figure 5). (i) If the minimum width of the column is 125 mm, the flange width is in the range of 100~120 mm.

(ii) The minimum width of the horizontal flat flange portion must be at least 50 mm.

Step 2 (determination of r_1 and r_2 shown in Figure 6). These parameters are constant regardless of the steel plate thickness if the same roll forming process is used.

(i) Bending capacity of SPA-H steel: if the bending angle is 180 degrees, the inner radius must be at least 1.5 times the flange thickness.

TABLE 10: Design requirements for each design factor of the MCO beam

Design factor	Design requirement
Invariable section	
Flange width	(i) Limit the flange width to less than the column width (column width: 125 mm or 150 mm) (ii) Require the minimum flat flange width (50 mm) to be connected to the wall stud
Radius of curved section	(i) Minimum radius of curved section less likely to cause a crack
Flange height	(i) Flange to thickness ratio is larger than 61 ($h_f > (h - 61t)/2$)
Welding length	(i) Prevent local buckling
Variable section	
Thickness of steel plate	(i) Variable sections for variable floor loads and spans
Web height	(i) Variable sections for variable floor loads and spans (ii) Able to penetrate pipes

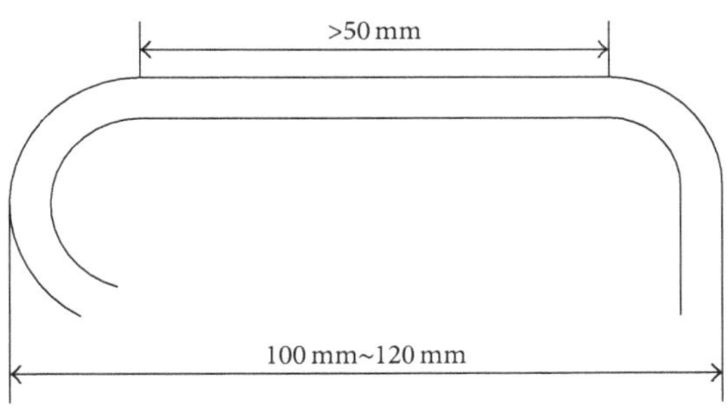

FIGURE 5: Range of the flange width.

(ii) The maximum steel plate thickness is limited to 10 mm.

(iii) If the bending angle of r_1 is in the range of 90 to 180 degrees, $r_1 = 15$ mm.

(iv) If the bending angle of r_2 is 90 degrees, $r_2 = 12$ mm.

Step 3 (determination of the flange width). (i) If the steel plate thickness is 10 mm and r_1 and r_2 are 15 and 12 mm, respectively, the flat flange width is in the range of 53 to 73 mm, thus satisfying the condition in Step 1.

(ii) The flange width is determined as 120 mm, which is the maximum value allowed by the condition in Step 1, considering its easy attachment onto ceilings and floor tracks.

Step 4 (determination of r_3). This value is constant if the same roll forming process is used regardless of the flange thickness.

(i) If bending angle of r_3 is less than 90 degrees, r_3 = 12 mm, which is the same as r_2.

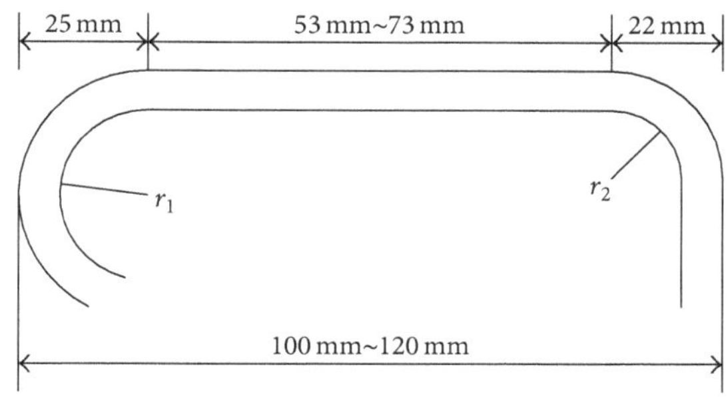

FIGURE 6: Determination of r_1 and r_2.

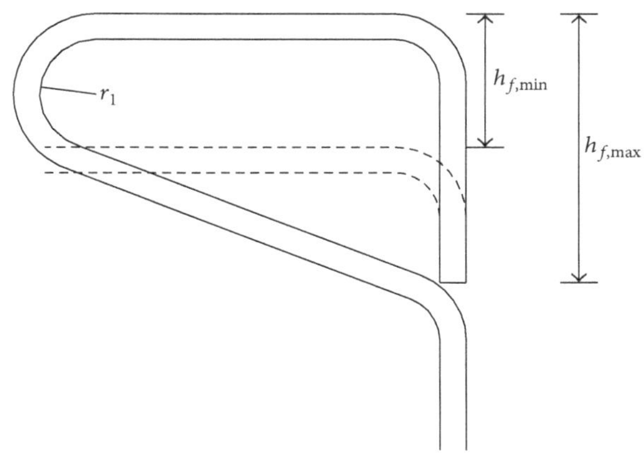

FIGURE 7: Determination of the flange height.

Step 5 (setting up the range of the flange height shown in Figure 7). (i) If it is assumed that the upper and lower flanges are parallel to each other and the flange thickness is 10 mm, $h_{f,\min1} = 10 + 15 + 15 + 10 = 50$ mm.

(ii) The flange height can be calculated by considering the maximum beam height of 400 mm and minimum steel plate thickness of 4.5 mm and applying the maximum width-thickness ratio of 61 to prevent elastic local buckling as $h_{f,\min2} = (400 - 61 \times 4.5)/2 = 63$ mm.

Step 6 (final selection of invariable dimensions of the upper hollow flange). (i) Front welding is performed on the steel plate connection as shown in Figure 8.

The following design conditions for the variable portion of the MCS beam cross-section are assumed for three module spans of 6, 7.5, and 8.1 m: floor beam load in the range of 2~5 kN/m^2, floor beam live load in the range of 2~5 kN/m^2, ceiling beam load in the range of 0.3~0.5 kN/m^2, and ceiling beam live load in the range of 0.5~1 kN/m^2. Various cross-sectional shapes for the MCO beam are designed by changing the steel plate thickness and web height as design variables, while satisfying the requirements on the dimensions of the cross-section discussed above. As shown in Table 11, 6 steel plate thickness values and 7 web heights are used as design values. The effective moment of inertia for the cross-sections

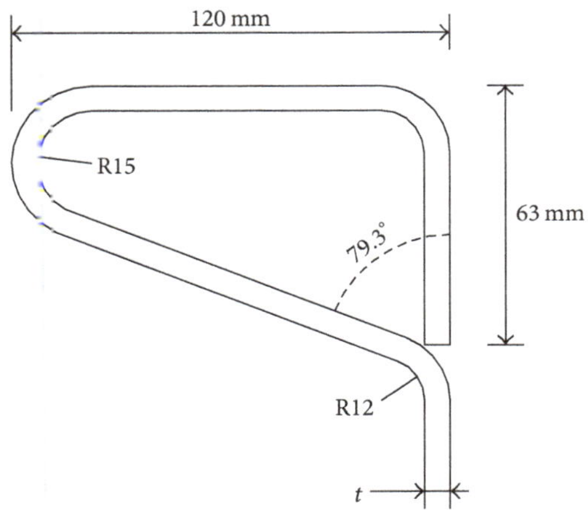

Figure 8: Invariable dimensions of the upper hollow flange of the MCO beam.

Table 11: Design requirements and values on the variable sectional portion of the MCO beam.

Design factors	Design requirements	Design values
Thickness of steel plate	Use standard steel products	(i) 4.5 mm, 6 mm, 7 mm, 8 mm, 9 mm, 10 mm
Depth of beam	Apply modular coordinate	(i) Floor beam: 270 mm, 300 mm, 330 mm, 360 mm, 390 mm (ii) Ceiling beam: 200 mm, 240 mm

in the table ranges from 1,530 to 18,218 cm^4. A total of 42 of the designed MCO beam cross-sectional shapes are given in Figure 9.

4. Design Verification and Pilot Test

As the final stage of the DFSS methodology, MCO beam rolls are produced using the roll forming facilities of a local company in Korea, and their formability and welding workability are assessed. Figure 10 shows the procedure for a pilot production of the MCO beam sections developed. A long strip in coiled steel is passed through sets of rolls until a shape with upper and lower hollow flanges is obtained. Then the production of the MCO beam section is completed by performing welding on the roll formed shape. Figure 11 shows an example of the assembled modular unit with the produced MCO beams.

The results of CTQ and productivity assessments on the final products are summarized in Table 12. The deformation level of the MCO beam that occurs during its production is less than 5 mm per 10 m. While some deformation may take place during secondary machining such as welding of brackets and end plates, additional correction can be performed in the process of module assembly, resulting in an error less than 3 mm per unit module. To confirm that the produced beams have a sufficient moment of inertia per unit cross-sectional area, a total of 150 cases are tested for various floor spans and loading conditions. The results of the test show that the average floor span is 183.38 cm^2 and the standard deviation is 16 cm^2. Thus, the average is well above the design target, which is 160 cm^2. For three spans of 6, 7.5, and 8.1 m, floor beams with optimal cross-sections are selected, and then the weight of the floor beams per unit area is evaluated for each case. It can be seen from the results of the table that target values are met for all of the three cases.

5. Concluding Remarks

In this study, a systematic approach was presented for the development of an efficient steel beam section for modular construction based on Six-Sigma. As the first step of this approach, market studies and surveys are conducted to obtain the opinions of potential customers. Then the opinions of the customers were converted into the quality characteristics for the channel-shaped steel beam using the quality function deployment methodology. A steel hollow flanged channel was chosen as the main modular beam shape, and the design concept was derived and developed by applying the Pugh matrix methodology. A pilot test was performed to validate the effectiveness of the developed beam section. The main conclusions of this study are as follows:

(1) To reflect customers' needs in the development of production processes, the QFD methodology was applied, and "welding length per unit length," "moment of inertia per sectional area of beam," and "weight of floor beams per unit area of plan" were selected as CTQs. Quantitative target values for these items were also set up.

(2) Seven design concepts were developed from an analysis on the key functions of the steel modular beam to satisfy the derived CTQs. The Pugh matrix methodology was applied to derive the optimal design of the steel modular beam, which has hollow flanges at the top and bottom.

(3) In the final product design, the cross-section of the MCO beam has the upper and lower hollow flange parts as the invariable portion and the web section as the variable portion. The results of the pilot production of the MCO beam indicate that the weight of the floor beams per unit area can be reduced to an average of 0.25 kN/m^2 for the standard 7.5 m span, which is 18% less than the corresponding value for the built-up RHS beam. All the modules assembled with the MCO beam products show an error less than 3 mm per unit module, thus satisfying the target value.

(4) Based on the DFSS methodology of Six-Sigma, this study proposed a methodology which can effectively reflect customers' needs, such as the reduction of manufacturing cost, less material usage, and reduced structural deformation of modular units, in the product design. It is expected that the proposed method

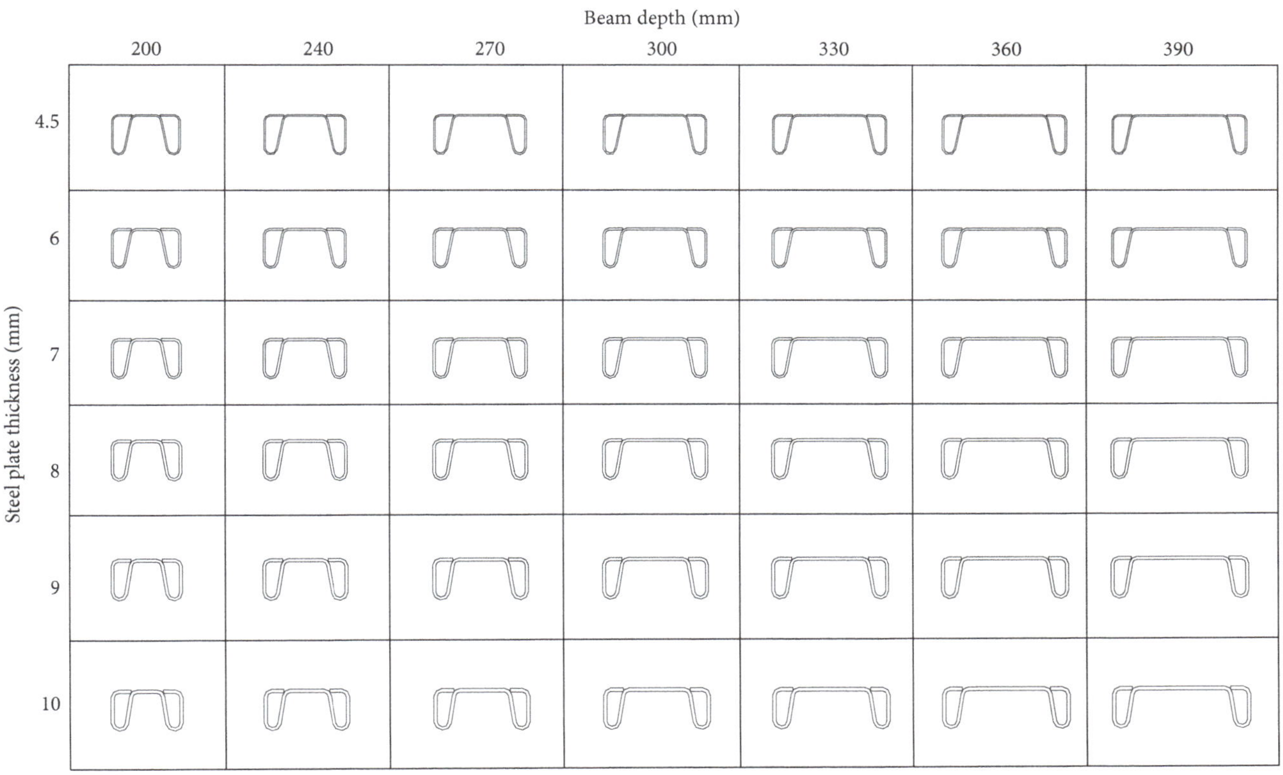

FIGURE 9: Developed cross-sectional shapes for the MCO beam.

FIGURE 10: Pilot production process for the developed MCO beam.

FIGURE 11: Assembled modular unit with the produced MCO beams.

TABLE 12: Design targets and test results for the produced MCO beams.

Design factors	Design targets	Test results
Verification of CTQ		
Welding length per unit length (m/m)	Less than 2	2
Moment of inertia per unit sectional area of beam (cm^2)	Larger than 150	Average 183.86
Weight of floor beams per unit area (kN/m^2)		
6 m span	Less than 0.25	0.18
7.5 m span	Less than 0.28	0.25
8.1 m span	Less than 0.35	0.32
Pilot test		
Pilot production of beam	Flexural deformation is less than 0.05%	Less than 0.05%
Pilot assembly of module	Tolerance of 3 mm	Less than 3 mm

can be utilized to develop other types of structural modular beams and can satisfy the diverse requirements of different customer groups.

Competing Interests

The authors declare that there is no conflict of interests regarding the publication of this article.

Acknowledgments

This research was supported by a grant from the National Research Foundation of Korea (Grant no. NRF-2016 R1D1A1B01010615).

References

[1] R. M. Lawson, R. G. Ogden, and R. Bergin, "Application of modular construction in high-rise buildings," *Journal of Architectural Engineering*, vol. 18, no. 2, pp. 148–154, 2012.

[2] S.-G. Hong, B.-H. Cho, K.-S. Chung, and J.-H. Moon, "Behavior of framed modular building system with double skin steel panels," *Journal of Constructional Steel Research*, vol. 67, no. 6, pp. 936–946, 2011.

[3] H.-J. Kim, J.-S. Lee, H.-Y. Kim, B.-H. Cho, Y. Xi, and K.-H. Kwon, "An experimental study on Fire resistance of medical modular block," *Steel and Composite Structures*, vol. 15, no. 1, pp. 103–130, 2013.

[4] T.-S. Eom, T.-H. Ha, B.-H. Cho, and T.-H. Kim, "Cyclic loading tests on framed stud walls with strap braces and steel sheathing,"

Journal of Structural Engineering, vol. 141, no. 7, Article ID 04014173, 2015.

[5] L. S. Pheng and M. S. Hui, "Implementing and applying six sigma in construction," *Journal of Construction Engineering and Management*, vol. 130, no. 4, pp. 482–489, 2004.

[6] J. A. Johnson, H. Gitlow, S. Widener, and E. Popovich, "Designing new housing at the University of Miami: a 'six sigma'© DMADV/DFSS case study," *Quality Engineering*, vol. 18, no. 3, pp. 299–323, 2006.

[7] Y. Akao, "Development history of quality function deployment," *The Customer Driven Approach to Quality Planning and Deployment*, 1994.

[8] A. Hassan, A. Siadat, J.-Y. Dantan, and P. Martin, "Conceptual process planning an improvement approach using QFD, FMEA, and ABC methods," *Robotics and Computer-Integrated Manufacturing*, vol. 26, no. 4, pp. 392–401, 2010.

[9] J. Chun and J. Cho, "QFD model based on a suitability assessment for the reduction of design changes in unsatisfactory quality," *Journal of Asian Architecture and Building Engineering*, vol. 14, no. 1, pp. 113–120, 2015.

Experimental Study on Reinforced Concrete Column Incased in Prefabricated Permanent Thin-Walled Steel Form

Jae Yuel Oh,[1] Deuck Hang Lee,[2] Jungmin Lee,[1] Kang Su Kim,[1] and Sung-Bae Kim[3]

[1] *Department of Architectural Engineering, University of Seoul, 163 Seoulsiripdae-ro, Dongdaemun-gu, Seoul 02504, Republic of Korea*
[2] *Department of Civil and Environmental Engineering, University of Illinois at Urbana-Champaign, 205 N. Mathews Ave., Urbana, IL 61801, USA*
[3] *Department of Structural Design, The Naeun Structural Engineering, 22 Yangpyeong-ro, Yeongdeungpo-gu, Seoul 07205, Republic of Korea*

Correspondence should be addressed to Kang Su Kim; kangkim@uos.ac.kr

Academic Editor: Ana S. Guimarães

Conventional construction methods of reinforced concrete (RC) structures generally require a long construction period and high costs due to many on-site temporary form works. In this study, a prefabricated permanent thin-walled steel form integrated with reinforcement cage (PPSFRC) was developed, and it makes for a fast-built construction by reducing the temporary form works. Axial compression tests were conducted on a total of 9 test specimens to investigate the structural performances of the newly developed columns. The proposed column construction method utilized relatively thinner steel plates compared to conventional concrete-filled tube (CFT) columns, but it was designed to have sufficient resistance performances against the lateral pressure of fresh concrete and to prevent the buckling of the thin plates by utilizing the steel angles and channel stiffeners prefabricated in the permanent thin-walled steel form. The experimental results showed that the column specimens fabricated by the PPSFRC method had better local buckling resistance and behaved in a more ductile manner compared to the conventional CFT columns. In addition, the axial strengths of the test specimens were compared with those estimated by design provisions, and the flexural moments induced by initial imperfection or accidental eccentricity of axial loads were also discussed in detail.

1. Introduction

Reinforced concrete (RC) structures have been widely used in constructions of buildings and infrastructures based on technology and know-how accumulated over the last several decades as well as excellent cost competitiveness of concrete material. However, recently with a sharp increase in labor costs, the on-site temporary works including concrete formworks require high costs in construction markets. Especially for the cases of bridge and building structures with high story heights, the construction time and costs for the temporary works have significantly increased, and thus the competitiveness of the RC construction method has been greatly fallen off [1–3]. In typical RC frame constructions, the construction process of columns is one of the critical paths, and it is, therefore, absolutely advantageous to minimize the on-site temporary works for reduction of the construction period. In this study, a prefabricated permanent steel form integrated with reinforcing cage (PPSFRC) was developed to overcome such problems inherent in the typical RC construction method, and the structural performance of the columns produced by utilizing the new fast-built construction method was also investigated.

Figure 1 shows the prefabricated permanent steel form integrated with reinforcing cage (PPSFRC) developed in this study according to the fabrication process. The thin-walled steel plate tube was reinforced by the multiple steel stiffener angles and channels in the longitudinal and transverse directions, and longitudinal reinforcements and shear reinforcements were also placed inside the permanent steel form. After the fabrication of a half part of steel cage, the longitudinal and transverse reinforcement were placed in the cage, and the multiple steel stiffener angles were then welded to the longitudinal channels in the longitudinal and

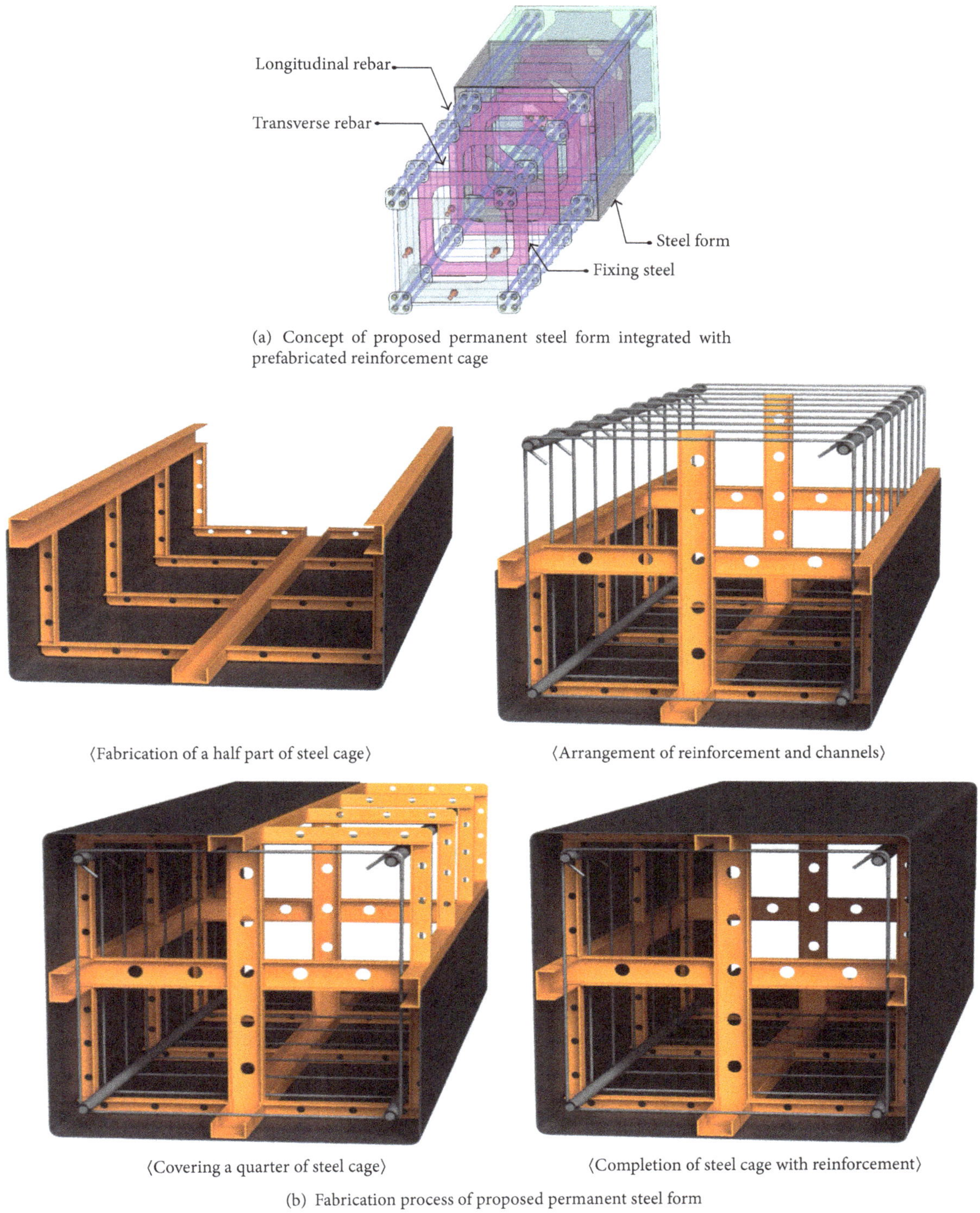

Figure 1: Permanent thin-walled steel form integrated with prefabricated reinforcement cage.

transverse directions. Then, after a quarter of the steel cage was covered by welding it to the longitudinal channels, the last quarter of the steel cage was covered by welding it to the longitudinal channels as well. The thin-walled steel form was designed to prevent excessive deformations that can be possibly induced during the transportation and temporary works and by the lateral pressure of fresh concrete. Since the reinforcement cages are prefabricated inside the permanent steel form, the concrete can be cast right after installations of the steel form, and thus construction time can be significantly reduced. The permanent steel form has quite a thin thickness, and thus the strength contribution of the steel form on

Table 1: Summary of test specimens.

Number	Specimen	Compressive strength of concrete (MPa)	Steel form	Yield strength of reinforcing bars (MPa)	Gap
1	R21-S600	21	X	600	X
2	P21-S600	21	O	600	X
3	P21-S400	21	O	400	X
4	P21-S600G	21	O	600	O
5	R30-S600	30	X	600	X
6	P30-S600	30	O	600	X
7	P30-S400	30	O	400	X
8	P30-S600G	30	O	600	O
9	CFT	21	X	—	X

the column is neglected, and only the concrete and rebar placed inside the steel form are considered in the strength calculation just like a typical RC column. If the column with PPSFRC is considered as a composite member, the minimum steel area ratio ($A_{s,\min}$) and the width-to-thickness ratio ($(B$ or $D)/t_{\max}$) should satisfy the design criteria required for steel-concrete composite columns. In the design of the columns with PPSFRC, however, the strength contribution of the steel form on the column strength is neglected, which means that the requirements of the minimum steel area ratio and width-to-thickness ratio can be neglected in their design. For this reason, an efficient design can be possible by employing very thin steel plates. The PPSFRC columns can also secure good fire resistance that is equal to or higher than that of conventional RC members, and thus no additional fire protection is required for their applications [4, 5].

2. Experimental Program

2.1. Test Specimens. In order to investigate the structural performances of the PPSFRC columns under axial loads, a total of nine specimens was fabricated, among which two specimens were conventional RC columns (R-series specimens), six specimens were PRSFRC columns (P-series specimens), and one specimen was concrete-filled tube column (CFT column). As summarized in Table 1, the types of column, the compressive strengths of concrete, and the strength grades of reinforcing bar were considered as the key test parameters, and, especially in the P21-S600G and P30-S400G specimens, 10 mm gaps were intentionally introduced between the steel form and the load-transfer plate on the top to prevent the local buckling (see Figure 2(b)) [6–9]. As shown in Figure 2, all the test specimens were 1480 mm in height and 350 mm in width (B). The top and bottom of the test specimens were reinforced with the load-transfer steel plates with a thickness of 40 mm to avoid stress concentrations at loading points and to introduce uniformly distributed compressive stresses over the cross section of the test specimens. In addition, two stiffeners with 12 mm and 75 mm in thickness and length, respectively, were provided on each side of the top and bottom of the specimens to prevent the local buckling (see Figure 2(b)). For better workability of fresh concrete, a hole with a diameter of 150 mm was placed at the center of the top load-transfer plate, and four air-vent holes with a diameter of 30 mm were placed on each corner of the column specimen. The P-series specimens and the CFT specimen had a hole with a 20 mm in diameter at 300 mm away from the bottom face of the test specimens so that the concrete filling can be visually checked during the concrete casting. In the R-series and P-series specimens, four screw-ribbed reinforcing bars with a diameter of 25.4 mm were used in the longitudinal direction, and hoops with a diameter of 9.5 mm were provided at 150 mm spacing. The reason for the use of the screw-ribbed reinforcing bar with the large diameter is to facilitate simple mechanical splice connections of the longitudinal reinforcements [10]. In the steel plates of the P-series specimens, the steel tube was welded with groove in the longitudinal direction. C-40 × 25 × 4 steel channels were attached on four inner sides of the steel form in the longitudinal direction by fillet welding with 3 mm wide and 100 mm long at every 300 mm to prevent excessive deformations induced by the lateral pressure of fresh concrete, and L-25 × 25 × 3.2 steel angles were also placed at 600 mm spacing in the transverse direction and welded to the longitudinal channels using fillet welding with a width of 5 mm.

Table 2 shows a summary of the current design code provisions related to CFT columns, specified in the American Institute of Steel Construction (AISC), Eurocode 4, and Korean Building Code (KBC) [11–13]. The revised KBC and AISC alleviated the limits of the width-to-thickness ratio (B/t_w) and the minimum area ratio of a steel material ($A_{s,\min}$) and expanded the application ranges of the material strengths of concrete and steel plates (f_c' and $F_{y,\max}$). For the P-series specimens and the CFT specimen tested in this study, 3.2 mm and 7.0 mm steel plates were used, respectively, and corresponding steel ratios (A_s/A_g) were 3.6% and 7.0%, respectively, which satisfy the minimum steel ratio ($A_{s,\min}$) specified in the design provisions shown in Table 2. The width-to-thickness ratio (B/t_w) of the CFT specimen was 50.0, which was designed to meet the provisions specified in KBC, AISC, and Eurocode 4, whereas that of the P-series specimens was 109.0, which was intentionally designed not to satisfy the current design provisions. Since the main feature of the proposed method is to use thin-walled steel plates as the permanent steel form and not to account for its strength contribution in the structural design, the width-to-thickness

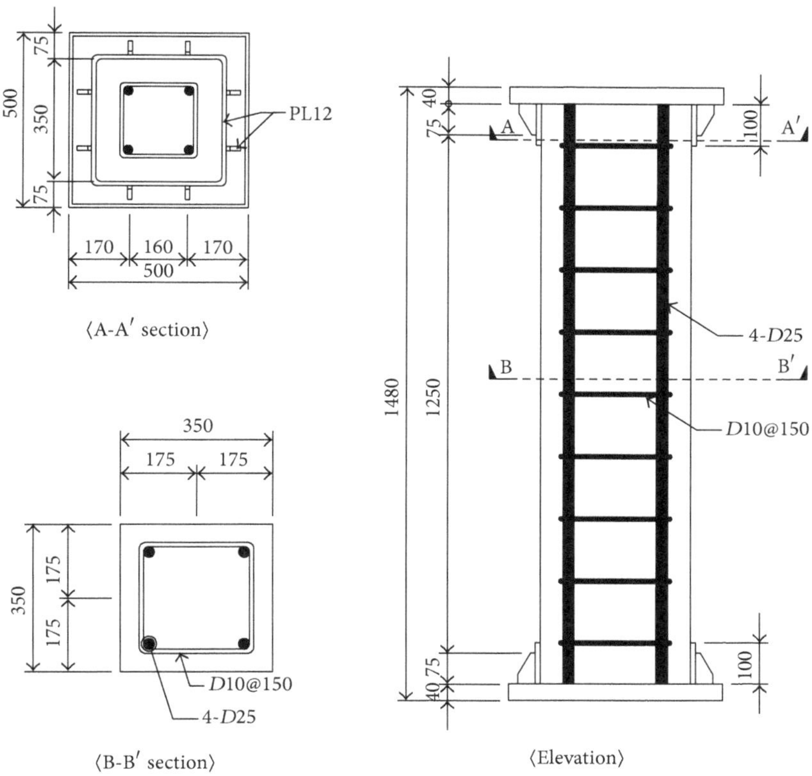

(a) R series specimens

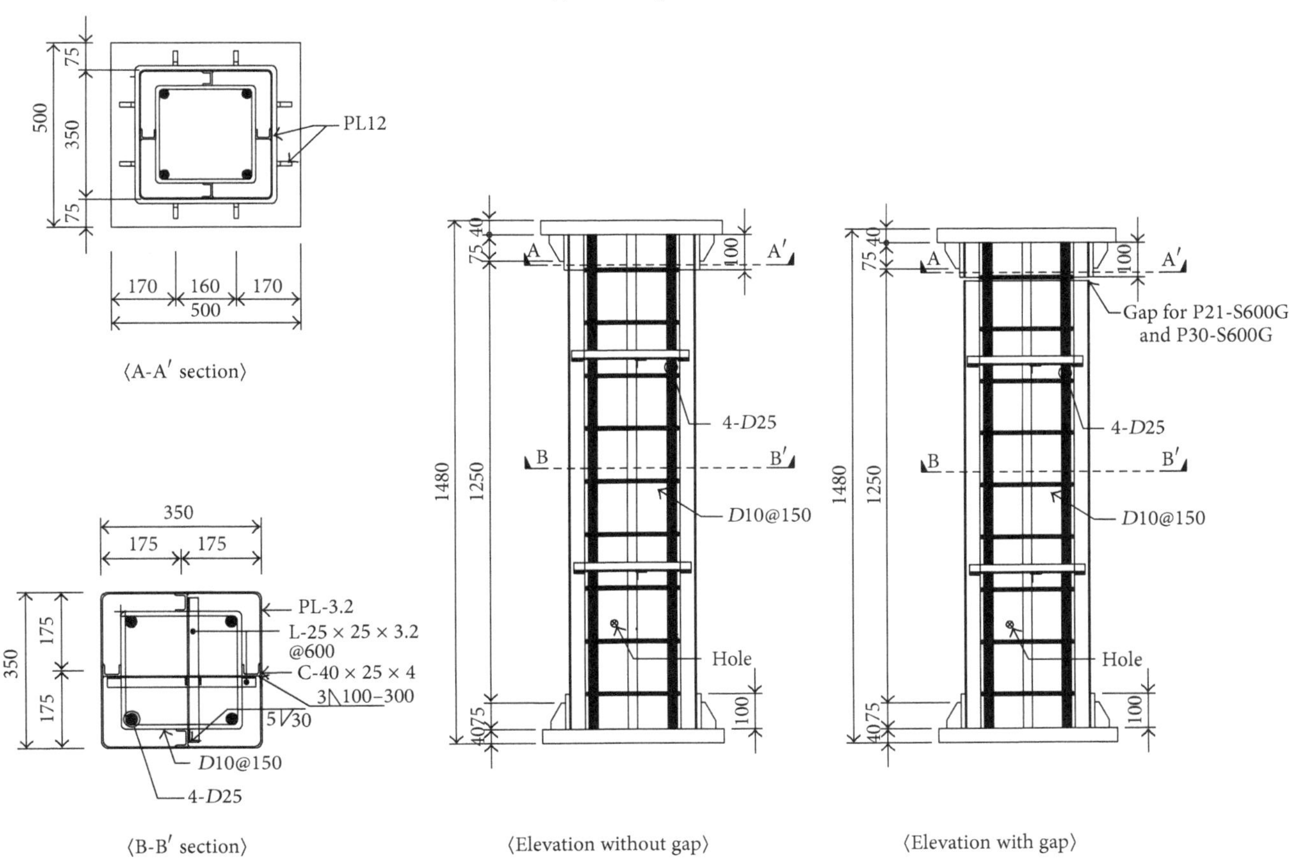

(b) P series specimens

Figure 2: Continued.

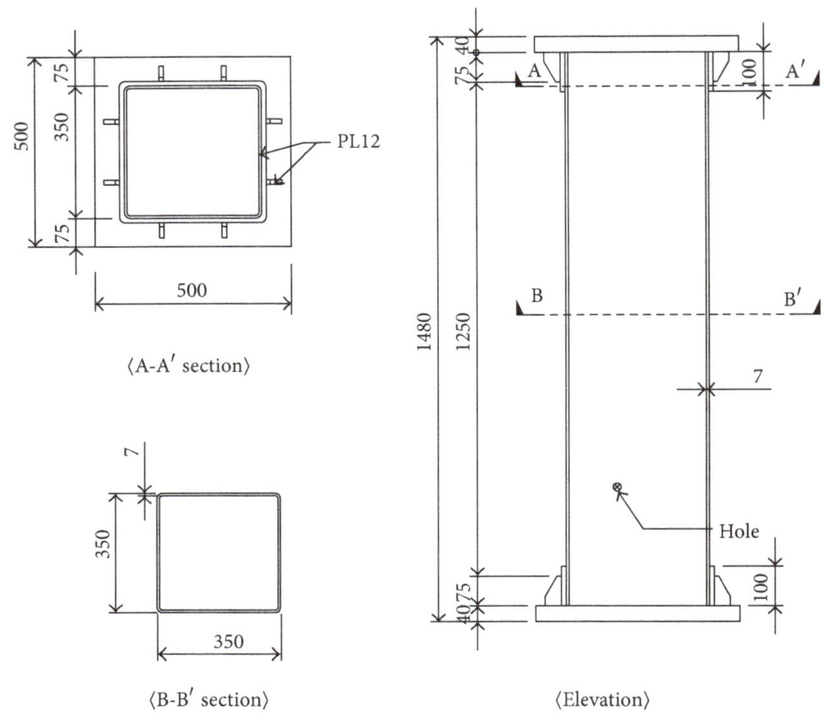

(c) CFT specimen

FIGURE 2: Dimensional details of test specimens.

TABLE 2: Specifications regarding CFT in national design codes.

Code	KBC 2005		KBC 2009		AISC 1999		AISC 2005		Eurocode 4	
A_s/A_g	3% or above		1% or above		4% or above		1% or above		2%–9%	
f_c'	20.5 MPa or greater		21–70 MPa		21–55 MPa		21–70 MPa		20–50 MPa	
$F_{y,max}$	415 MPa		440 MPa		415 MPa		440 MPa		450 MPa	
Shape	□	○	□	○	□	○	□	○	□	○
$(B \text{ or } D)/t_w$	$\sqrt{\frac{3E}{F_y}}$	$\sqrt{\frac{8E}{F_y}}$	$\sqrt{\frac{5.1E}{F_y}}$	$\frac{0.15E}{F_y}$	$\sqrt{\frac{3E}{F_y}}$	$\sqrt{\frac{8E}{F_y}}$	$\sqrt{\frac{5.1E}{F_y}}$	$\frac{0.15E}{F_y}$	$52\sqrt{\frac{235}{F_y}}$	$90\frac{235}{F_y}$

ratio (B/t_w) was, therefore, intentionally increased for the P-series specimens.

2.2. Material Properties. The concrete mix proportions used in the test specimens are summarized in Table 3. The design compressive strengths were 21.0 MPa and 30.0 MPa, respectively, and those obtained from material tests were 26.3 MPa and 44.2 MPa, respectively. The yield strengths of SD400 and SD600 threaded bars with a diameter of 25.0 mm were measured as 460 MPa and 690 MPa, respectively. For the P-series specimens and the CFT specimen, grade SS400 steel plates with 3.2 mm and 7.0 mm in thickness were used, respectively, and their yield strengths obtained from material tests were 350 MPa and 388 MPa, respectively, as summarized in Table 4. As shown in Figure 3, the universal testing

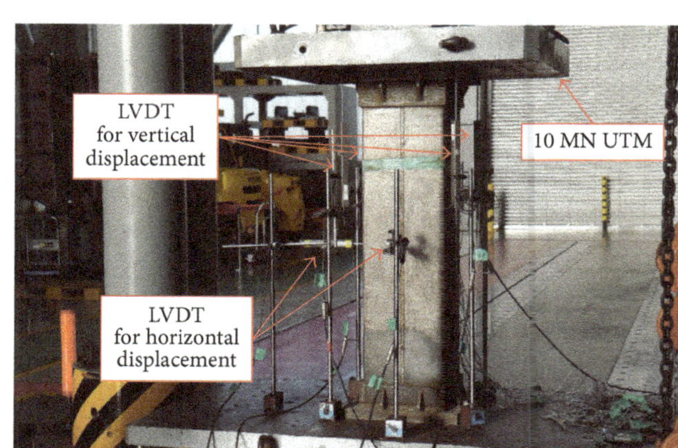

FIGURE 3: Test setup and measurement devices.

TABLE 3: Concrete mix proportion.

Design strength (MPa)	G_{max} (mm)	W/C (%)	S/A (%)	Unit volume weight (kg/m^3)				
				W	C	S	G	AD_2
21	25	56.1	48.8	179	319	856	915	1.60
30	25	42.1	46.9	120	284	844	974	3.25

TABLE 4: Material properties of concrete, rebars, and steel plates.

Type	Concrete		Rebar				Steel plate			
	21 MPa	30 MPa	400 MPa		600 MPa		3.2 mm		7 mm	
			Yield	Ultimate	Yield	Ultimate	Yield	Ultimate	Yield	Ultimate
Strength (MPa)	26.3	44.2	463	596.6	691.8	847.0	344.1	423.9	367.5	411.1

machine of 10.0 MN capacity was used to apply axial loads on the column specimens with a constant loading rate of 0.02 mm/sec. In order to measure the vertical displacements of the specimens during testing, linear variable differential transformers (LVDTs) were installed on four corners of each specimen, and two LVDTs were additionally installed on two sides to measure the lateral displacements.

2.3. Test Results. Figures 4(a) and 4(b) show the axial load-deformation responses of the test specimens with the design compressive strengths of 21.0 MPa and 30 MPa, respectively. The average values of the axial displacements measured from the four LVDTs were used for the axial deformation, and the loading was terminated when reduced to 80.0% of the maximum load. The R21-S600 and R30-S600 specimens were conventional RC column specimens without the external steel form. Their maximum loads were 3,042 kN and 3,782 kN, respectively, and the maximum displacements were 7.9 mm and 6.8 mm, respectively.

As shown in Figure 5(a), the stiffness of the P21-S600 specimen slightly reduced at a load of about 3,000 kN due to the initiation of local buckling, which was observed at the section by $B/2$ (175 mm) away from the top of the column specimen, and then the local buckling was gradually propagated downwards. As shown in Figure 5(b), at the maximum load of about 5,396 kN, the local buckling occurred at a spacing of about 300 mm, which was almost equal to the spacing of the transverse steel stiffener channels. After the maximum load, a heavy local buckling occurred in the upper part of the column as shown in Figure 5(c), and the displacement reached 29.4 mm at final stage as shown in Figure 5(d). Other P-series specimens showed similar behavioral characteristics to the P21-S600 specimen. The P21-S600G and P30-S600G specimens with 10 mm gap between the steel form and the load-transfer steel plate on the top of the column showed the same axial stiffness with the RC specimens (R21-S600 and R30-S600 specimens) as shown in Figures 4(a) and 4(b), respectively, which is because the axial load was transferred only through the concrete due to the gap of the steel form during the initial loading stage. As the applied load increased, the gap was gradually reduced and eventually closed, and then their behavior appeared to be very similar to those of the specimens P21-S600 and P30-S600. According to Johansson and Gylltoft [6], the concrete-filled composite columns with gaps between load-transfer plate and circular steel tube showed more ductile behavior compared to the one without gap. This is because when the steel plate without the gap was used, local tensile deformation occurred due to the confinement effect of the circular steel tube and led to the yielding or local buckling of the steel plates near the connection between the load-transfer plate and the steel form that occurred earlier than those without the gap. In this study, however, it was not observed because the confinement effect was relatively small due to the use of the thin steel plates with rectangular shaped sections.

In the case of the CFT specimen, the axial loads increased with a constant initial stiffness, and the stiffness decreased gradually just before the local buckling, which was observed at the section by $B/2$ away from the top end of the column at a load of about 4,500 kN. At the ultimate load of 4,700 kN with the maximum displacement of 12.0 mm, a compressive failure occurred with heavy local buckling. Compared to the P-series specimens, the CFT specimen showed a very low deformation capability as shown in Figure 4(a), which can be inferred from the failure patterns shown in Figure 6. As shown in Figure 6(b), in the case of the P-series specimens except the P21-S600G and P30-S600G specimens, relatively small local buckling was distributed rather than being concentrated at a section due to the steel stiffener angles provided inside the column specimens.

As shown in Figure 6(c), in the case of the P21-S600G specimen, which had a 10 mm gap on the top of the steel tube, the local buckling was delayed by the gap, and the specimen had a convex radial deformation over a wide region of the steel tube before closing of the gap because of the Poisson strains of the concrete. Then, after closing of the gap, small local buckling concentrated at the top and middle of the column length occurred. The P30-S600G specimen, which also had a 10 mm gap on the top of the steel tube, had a very similar buckling behavior to the P21-S600G specimen. On the other hand, in the CFT specimen, as shown in Figure 6(d), the local buckling was concentrated at the section by $B/2$ from the top of the column, and thus the deformation capability of the CFT column was relatively low compared to the P-series specimens.

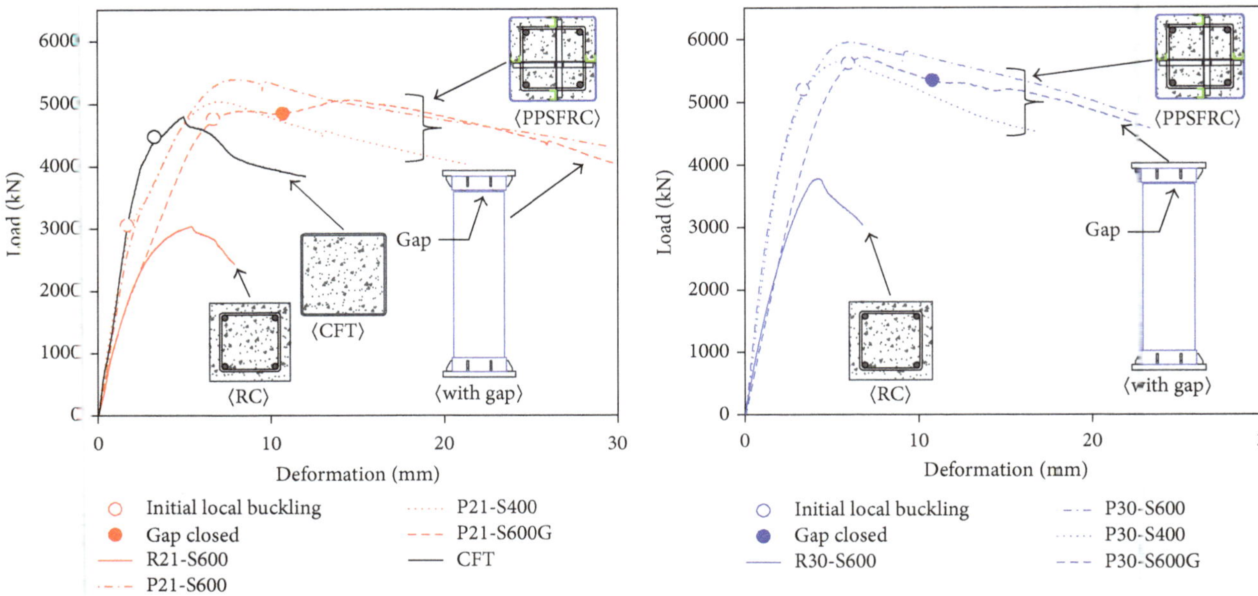

(a) R21 and P21 series specimens

(b) R30 and P30 series specimens

FIGURE 4: Axial load-deformation responses of test specimens.

TABLE 5: Comparison of test results and calculated strengths.

Number	Specimen	P_u (kN)	P_0 (kN)	P_n (kN)	e_m (mm)	e_i (mm)	e_{tot} (mm)	P_e (kN)	P_u/P_0	P_u/P_n	P_u/P_e
1	R21-S600	3,042	4,011	3,209	13	7	20	3,354	0.76	0.95	0.91
2	P21-S600	5,396	5,748	3,109	7	7	14	4,895	0.94	1.74	1.10
3	P21-S400	5,041	5,282	2,736	7	7	14	4,895	0.95	1.84	1.03
4	P21-S600G	5,062	5,748	3,109	20	7	27	4,507	0.88	1.63	1.12
5	R30-S600	3,782	5,789	4,631	16	7	23	5,313	0.65	0.82	0.71
6	P30-S600	5,964	7,571	4,567	20	7	27	6,185	0.79	1.31	0.96
7	P30-S400	5,660	7,104	4,194	18	7	25	6,138	0.80	1.35	0.92
8	P30-S600G	5,721	7,571	4,567	26	7	33	5,864	0.76	1.25	0.98
9	CFT	4,797	6,250	5,000	27	7	34	4,933	0.77	0.96	0.97
				Average					0.81	1.32	0.97

3. Analytical Study

The ACI318-14 presents the axial capacity estimation equation of RC columns based on 567 column test results conducted at the University of Illinois and Leigh University from 1927 to 1933 [14], as follows:

$$P_0 = f_y A_{st} + (0.85 f_c')(A_g - A_s), \quad (1)$$

where f_y and A_{st} are the yield strength and the sectional area of reinforcements, respectively, and f_c' and A_g are the compressive strength and the gross sectional area of concrete, respectively. Figure 7 and Table 5 show the comparison between the axial strengths calculated by (1) (P_o) and those measured through the experiments (P_u). The ratios of the axial strengths of the test specimens to those estimated by (1) (P_u/P_o) were distributed between 0.65 and 0.95, which indicates that the axial capacities estimated by (1) overestimate the test results. This would be most likely because the initial imperfection (e_i) inevitably occurring during the production process and the minimum accidental eccentricity (e_m) caused by the test setup of the specimen were not taken into consideration. For this reason, ACI318-14 [14] adopted a factor of 0.8 for RC columns with rectangular cross sections. In this study, the eccentricity induced by the initial imperfection (e_i) was assumed to be $L/200$ as presented in Eurocode 4 [12]. Since the length of the column specimens tested in this study was 1400 mm, the initial imperfection (e_i) of the specimens was estimated to be 7.0 mm. In addition, the minimum accidental eccentricity (e_m) that could possibly occur when axial load was applied to the column specimens needs to be considered to analyze the test results. As shown in Figure 8, the eccentricity of the loading causes bending moment, and the rotation angle (θ) caused by the bending moment can be calculated, as follows:

$$\tan\theta \approx \theta = \frac{L}{D}, \quad (2)$$

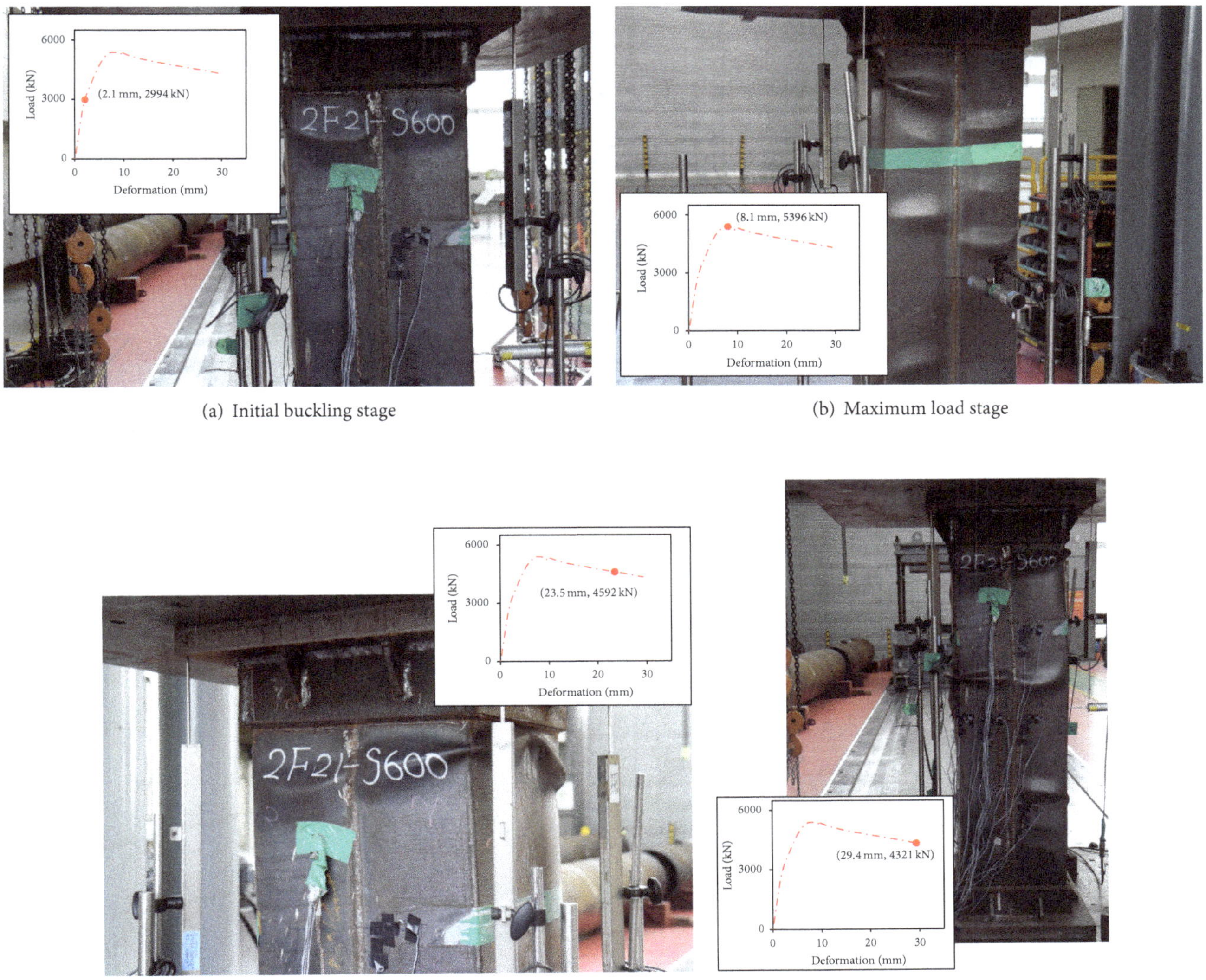

(a) Initial buckling stage

(b) Maximum load stage

(c) Postpeak stage

(d) Final stage

FIGURE 5: Detailed behavioral characteristics of P21-S600 specimen.

where L is taken to be $B/2$ and D is the axial deformation at the extreme fiber of the cross section, which was measured from the LVDTs installed during the all loading stages. If the bending moment was constant along the member axis, the relationship between the rotation angle of the member (θ) and the bending moment (M) can be expressed, as follows:

$$M = \theta \frac{EI}{l}, \tag{3}$$

where E and I are the modulus of elasticity and the moment of inertia. As aforementioned, although the columns with PPSFRC are not considered as composite members for the design, the reinforcing bars and steel plates need to be considered for the calculation of E and I in (3) that are needed to estimate the minimum accidental eccentricity (e_m), which was done by the transformed section method [15]. The bending moment (M) can be estimated by substituting the measured rotation angle (θ) into (3), and the magnitude of eccentricity (e_m) can be calculated, as follows:

$$e_m = \frac{M}{P}. \tag{4}$$

As summarized in Table 5, the minimum accidental eccentricities (e_m) of the specimens estimated from (4) were distributed between 7.0 mm and 27.0 mm. Considering the eccentricity (e_m) of axial loads and the initial imperfection (e_i), the estimated total eccentricities (e_{tot}) of the test specimens can be estimated, which were ranged from 14.0 mm to 34.0 mm. Accordingly, as shown in Figure 9(a), the axial strength (P_o) was reduced to P_e by considering the effect of the eccentricity (e_{tot}), and Figure 9(b) shows the analysis results of the P21-S600 specimen considering the influence of the eccentricity (e_{tot}). The effective axial strengths of the column specimens (P_e) were calculated by considering the effects of

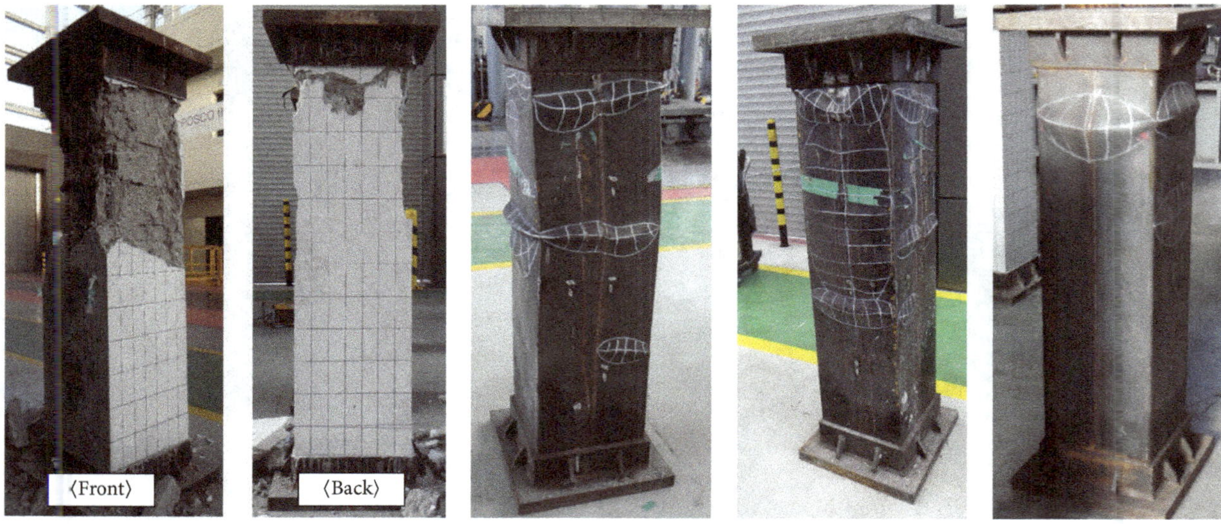

(a) R21-S600 (b) P21-S600 (c) P21-S600G (d) CFT

FIGURE 6: Failure patterns of test specimens.

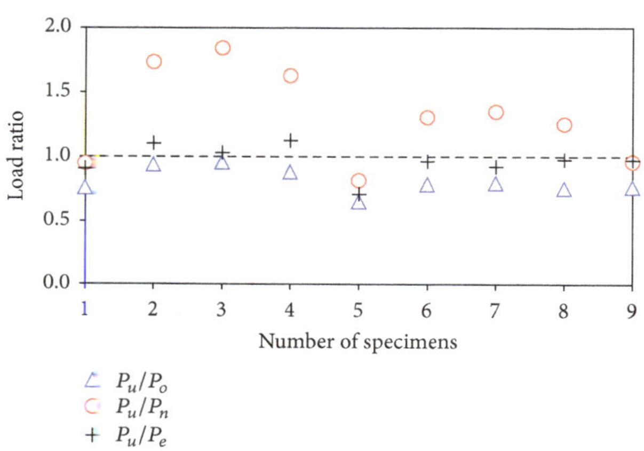

FIGURE 7: Comparisons of observed and calculated axial strengths of test specimens.

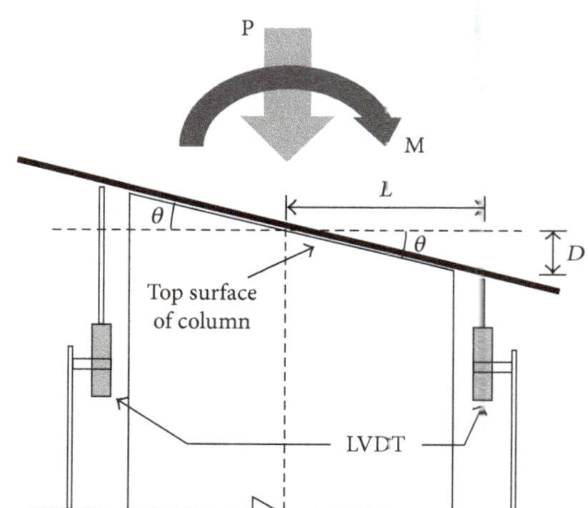

FIGURE 8: Axial force and flexural moment induced by minimum eccentricity.

the flexural moment caused by the total eccentricity (e_{tot}), and the estimated results are shown in Table 5 and Figure 7. The average of the ratios between the test results and the effective axial strengths (P_u/P_e) for all the specimens was 0.97, and the effective axial strength calculated by considering the additional eccentricity (P_e) agreed closely with the experimental results. As mentioned above, in the ACI318-14, the nominal axial strength of RC column (P_n) is limited to $0.8P_o$, and this corresponds to the total eccentricity (e_{tot}) of 35.0 mm according the estimation method shown in Figure 9(a), which is also equivalent to about 10.0% of the cross-sectional width (B).

As shown in Table 5 and Figure 7, the average of ratio between the nominal strengths and ultimate strength of the test specimens (P_u/P_n) was 1.32, where it is noted that the contribution of the permanent steel form was not considered in calculating the axial strengths (P_n), as mentioned earlier. It is, therefore, expected that a sufficient margin of structural safety can be secured when the existing RC design equation is applied to the column members fabricated by the PPSFRC method proposed in this study.

4. Conclusions

In this study, the prefabricated permanent steel form integrated with reinforcing cage (PPSFRC) was developed, and the structural performance of the reinforced concrete column specimens fabricated by the PPSFRC method was investigated through the axial compression testing on a total of the nine specimens. In addition, the applicability of the current axial strength estimation models specified in national design

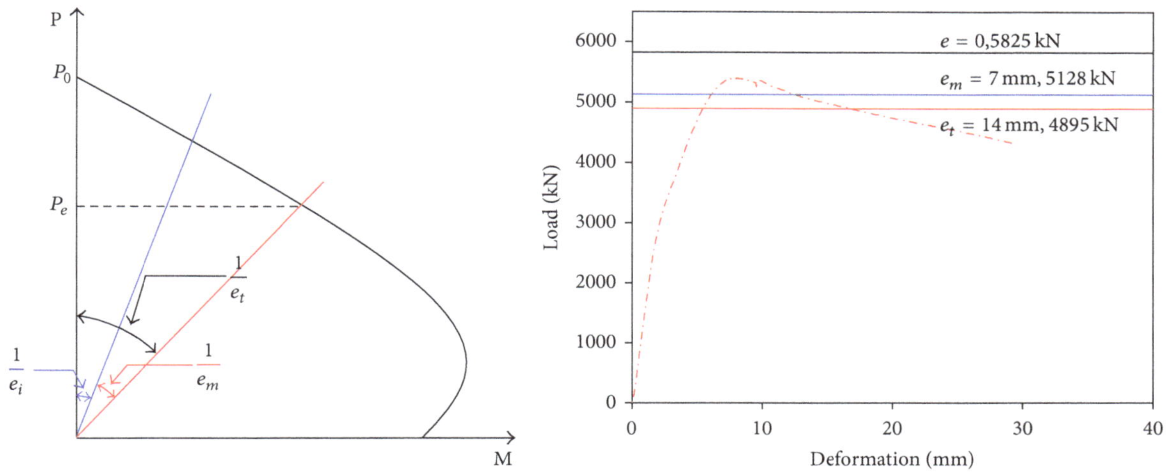

(a) Eccentricity effect in P-M relation curve

(b) Strength of P21-S600 considering eccentricity

FIGURE 9: P-M interaction model considering the effect of eccentricity.

codes on the PPSFRC columns was also evaluated. From this study, the following conclusions can be obtained:

(1) It was observed that local buckling was well controlled, not concentrated at a section, by the steel channels and angles provided in the longitudinal and transverse direction of the PPSFRC columns, and the PPSFRC columns showed more ductile behavioral characteristics compared to the conventional CFT column.

(2) The axial strength of the cross section without a consideration of the initial imperfection and the minimum eccentricity of applied loads overestimated the measured axial strengths of the specimens (P_u) by up to 25%, and the effective axial strength of the column (P_e) considering the influence of the minimum eccentricity and the initial imperfection provided a good analysis accuracy.

(3) It turned out that $B/10$ considered as the minimum eccentricity of the column in ACI318-14 was very close to the total eccentricity (e_{tot}), which is a sum of the loading eccentricity measured in this study (e_m) and the eccentricity induced by the initial imperfection (e_i).

(4) The PPSFRC column specimens (P_u) showed enhanced axial strengths (P_n) with sufficient reserved strengths by more than 30%, compared to the design strength of the conventional RC column. Therefore, it is expected that the PPSFRC columns can be designed with a sufficient margin of safety by the conventional RC design method.

Competing Interests

The authors declare that there is no conflict of interests regarding the publication of this paper.

Acknowledgments

This research was supported by the Research Grant from University of Seoul through the Korea Agency for Infrastructure Technology Advancement funded by the Ministry of Land, Infrastructure and Transport of the Korean government (Project no. 15TBIP-C092932-02).

References

[1] A. Mwafy, N. Hussain, and K. El-Sawy, "Seismic performance and cost-effectiveness of high-rise buildings with increasing concrete strength," *The Structural Design of Tall and Special Buildings*, vol. 24, no. 4, pp. 257–279, 2015.

[2] H. Sezen and M. Shamsai, "High-strength concrete columns reinforced with prefabricated cage system," *Journal of Structural Engineering*, vol. 134, no. 5, pp. 750–757, 2008.

[3] M. Shamsai, E. Whitlatch, and H. Sezen, "Economic evaluation of reinforced concrete structures with columns reinforced with prefabricated cage system," *Journal of Construction Engineering and Management*, vol. 133, no. 11, pp. 864–870, 2007.

[4] T. T. Lie and T. D. Lin, "Fire performance of steel reinforced concrete columns," *Journal of Structural Engineering*, vol. 141, no. 4, 2014.

[5] T. T. Lie and T. D. Lin, "Influence of restraint on fire performance of reinforced concrete columns," *Fire Safety Science*, vol. 1, pp. 291–300, 1986.

[6] M. Johansson and K. Gylltoft, "Mechanical behavior of circular steel-concrete composite stub columns," *Journal of Structural Engineering*, vol. 128, no. 8, pp. 1073–1081, 2002.

[7] J. Liu, X. Wang, and S. Zhang, "Behavior of square tubed reinforced-concrete short columns subjected to eccentric compression," *Thin-Walled Structures*, vol. 91, pp. 108–115, 2015.

[8] X. Zhou and J. Liu, "Seismic behavior and strength of tubed steel reinforced concrete (SRC) short columns," *Journal of Constructional Steel Research*, vol. 66, no. 7, pp. 885–896, 2010.

[9] X. Zhou, B. Yan, and J. Liu, "Behavior of square tubed steel reinforced-concrete (SRC) columns under eccentric compression," *Thin-Walled Structures*, vol. 91, pp. 129–138, 2015.

[10] S.-J. Lee, D. H. Lee, K. S. Kim, J.-Y. Oh, M.-K. Park, and I.-S. Yang, "Seismic performances of RC columns reinforced with screw ribbed reinforcements connected by mechanical splice," *Computers and Concrete*, vol. 12, no. 2, pp. 131–149, 2013.

[11] AISC-LRFD, *Load and Resistance Factor Design Specification for Structural Steel Buildings*, AISC, Chicago, Ill, USA, 1999.

[12] Eurocode4, Design of steel and composite structures, Part 1-1 General rules and rules for buildings, 1994.

[13] KBC, *Korean Building Code*, Architectural Institute of Korea, Seoul, Republic of Korea, 2009.

[14] ACI-318, *Building Code Requirements for Reinforced Concrete*, ACI, Detroit, Mich, USA, 2014.

[15] J. G. MacGregor, J. K. Wight, S. Teng, and P. Irawan, *Reinforced Concrete: Mechanics and Design*, Prentice Hall, Upper Saddle River, NJ, USA, 1997.

21

Study on Seismic Behavior of Steel Frame-Steel Shear Wall with Assembled Two-Side Connections

Yuben Zhang[1] **and Xun Zhan**[2]

[1]*Zhejiang Industry Polytechnic College, Shaoxing 312000, China*
[2]*Tongji University, Shanghai 200092, China*

Correspondence should be addressed to Yuben Zhang; zhangyuben@zjipc.com

Academic Editor: Miguel Castro

Low-cycle reciprocating loading tests were carried out on a steel frame with prefabricated beam-only connected steel plate shear wall (specimen A) and a steel frame with welded beam-only connected steel plate shear wall (specimen B). The seismic performance of the two different types of steel frame-steel plate shear wall specimens was studied and the failure modes, hysteresis curve, skeleton curve, and seismic performance index, etc of two groups of specimens were obtained, together with the studies of failure characteristics, ductility, energy dissipation, and stiffness degradation of the two specimens. The results showed that the assembled steel frame-steel plate shear wall with connection joints of steel shear wall with discontinuous cover plate connected on both sides (DCPC) have good seismic performance. On the basis of no loss of seismic performance, DCPC joints can provide better energy dissipation capacity than traditional welded steel plate shear wall structures and ensure the good postearthquake repair function.

1. Introduction

Assembled steel structure is made of factory-processed modules spliced by field bolts, and it has the advantages of fast construction speed, good seismic performance, and easy quality assurance [1]. The development of assembled steel structure buildings can effectively resolve the excess capacity of iron and steel and make the construction industry return to industrialization [2]. As a new type of lateral resistance structure system, steel plate shear wall can bear most of the lateral loads in the structure, improve the initial lateral stiffness and horizontal bearing capacity of the structure, and act as the first line of defence before the main frame is destroyed so as to reduce the damage degree of the main frame and improve the collapse resistance of the structure [3, 4]. It consists of embedded steel plate and edge member, in which the embedded steel plate can be connected with the edge member by bolting or welding. In order to make the steel plate shear wall system more suitable for the fabricated high-rise steel structure system, it is very important to study the influence of the fabricated connection form of the steel plate shear wall-embedded wall panel and the edge member on the lateral resistance of the structure.

In the existing engineering, the connection mode of quadrilateral welding is widely used [5], which leads to problems such as excessive welding quantity, difficult to guarantee the quality of welding seam, high cost of the structure, and long time limit. The bolted steel plate shear wall is mainly used in the structural test [6–10], and its purpose is to facilitate the connection of the test device and the repeated use of some components, but in the finite element numerical analysis, welding is still used to simulate the connection between the wallboard and the frame. The existing bolted joints are densely arranged and require high-processing accuracy, which not only makes the practical difficulty of the bolted joints increase greatly, but also does not meet the energy-saving and high-efficiency characteristics of the assembly structure. Therefore, while guaranteeing the transmission performance, it is necessary to effectively reduce the number of bolts and avoid and prevent the impact of processing errors on the structure to realize the assembly of the steel plate shear wall.

Rong and Rusli [11, 12] analyzed and studied the influence of the connection mode between the embedded wallboard and the surrounding frame on the mechanical properties of the steel plate shear wall system and proposed that when the embedded wallboard is connected with the surrounding frame, the action of the steel plate on the frame column may lead to premature failure of the frame column; the steel plate shear wall connected only with the frame beam can avoid premature failure of the column. The use of steel plate shear walls connected on both sides can effectively reduce the number of connecting bolts, facilitate structural opening, and stiffness adjustment and avoid the adverse effects of buckling of steel plate shear walls on columns. Considering the characteristics of the assembled steel structure and steel plate shear wall system, a new type of discontinuous cover-plate shear wall joint (DCPC) is proposed, which is suitable for assembling high-rise steel structure Low cyclic loading tests are carried out on one specimen of the welded steel frame-steel plate shear wall on both sides and one specimen of the steel frame-steel plate shear wall with DCPC on both sides.

2. Design Principle, Construction, and Characteristics of DCPC Node

The purpose of this joint is to effectively connect the embedded wallboard with the frame beam by using high-strength bolts. The design principle of joints is to avoid sliding between the embedded wallboard and the connecting plate. Therefore, the maximum shear force that a single high-strength bolt can bear should be guaranteed not less than the ultimate tensile strength of the tension band it shares. The advantage of this joint is that it can not only ensure the load transfer performance of the joint and give full play to the good lateral resistance of the steel plate shear wall, but also effectively reduce the accumulated error between the holes and is easy to assemble. DCPC can not only increase the number of effective friction surfaces and critical friction, but also avoid the phenomenon of the "unbutton."

As shown in Figure 1, the joint structure is characterized by connecting thin steel plate shear walls at both ends to facilitate structural opening and stiffness adjustment, avoiding adverse effects of buckling of steel plate shear walls on columns and opening large bolt holes in embedded steel plates and flanges of beams to increase the allowable value of processing cumulative errors and facilitate assembly, reduce the size of a single connector, reduce the cumulative error value that may be caused, and improve the allowable value of machining accuracy of components.

The arrangement of the embedded wall panel clamped by the bottom plate and the cover plate not only increases the friction surface between the wall panel and the connecting piece, but also effectively increases the critical friction and controls the bolt slip, and the out-of-plane warping force caused by buckling of the steel plate shear wall in the joint area is well avoided; even after the bolt hole is deformed, the connecting bolt will not collapse, so as to prevent the bolt from being released one by one from the outside to the inside. The connecting base plate and frame beam are welded and stiffeners are set to ensure the strength of the joint area, so that the connecting base plate and cover plate are free of residual deformation after the structure earthquake, which is convenient for the repair and replacement of the embedded steel plate after the earthquake.

3. Testing Situation

3.1. Specimen Design. Two single span two-story specimens (1/3 scale) are designed to simulate the bottom two stories of the actual steel frame-steel plate shear wall structure. The two specimens are welded steel frame-steel plate shear wall specimens on both sides (B) and connected steel frame-steel plate shear wall specimens with DCPC on both sides (A). The specimens consist of three parts: hollow seamless square steel tube column, H-shaped steel beam, and steel plate shear wall. The dimensions of the square steel tube flange column and H-shaped steel beam of the two groups of specimens are the same, and the thickness of the inner wall plate of the steel plate shear wall is 4 mm. Because the steel plate shear walls are arranged on both sides of the middle beam, the middle beam is always alternately deformed in the equilibrium position due to the alternating influence of the tension bands on both sides during the reciprocating loading process. However, the roof beam is only affected by one side of the steel plate shear wall and the residual deformation cannot be restored. In order to avoid the failure of the test due to the premature failure of the roof beam, the section of the roof beam is larger [13]. The cross-sectional dimensions of the frame and inner steel plate are shown in Table 1, steel is Q235B, and the test results of mechanical properties are shown in Table 2.

The processing diagram of specimen A is shown in Figure 2(a), the outer frame is machined in the factory, in which the flange of the frame beam is opened with large holes and welded with the connecting bottom plate, and the beam-column joints are welded. According to Figure 3(f), the connecting cover plate is machined and large bolt holes are opened on the embedded steel plate, and then all the components are shipped to the laboratory for assembly, as shown in Figure 2(b). The wall plate is connected by 10.9 grade M24 friction-type high-strength bolts, while the cover plate is connected with the beam by 10.9 grade M20 friction-type high-strength bolts.

In specimen B, the steel plate shear wall is connected with the frame beam by traditional fishtail plate welding and only the bottom of the specimen is connected with the ground beam by DCPC. By comparing the test results, the influence of the assembled joints and the traditional welded joints on the hysteretic behavior of the steel frame-steel plate shear wall is studied. Detailed processing drawings of components and beams are shown in Figure 3.

3.2. Test Device

3.2.1. Layout of Measuring Points. Two tension line displacement meters and three displacement percentile meters

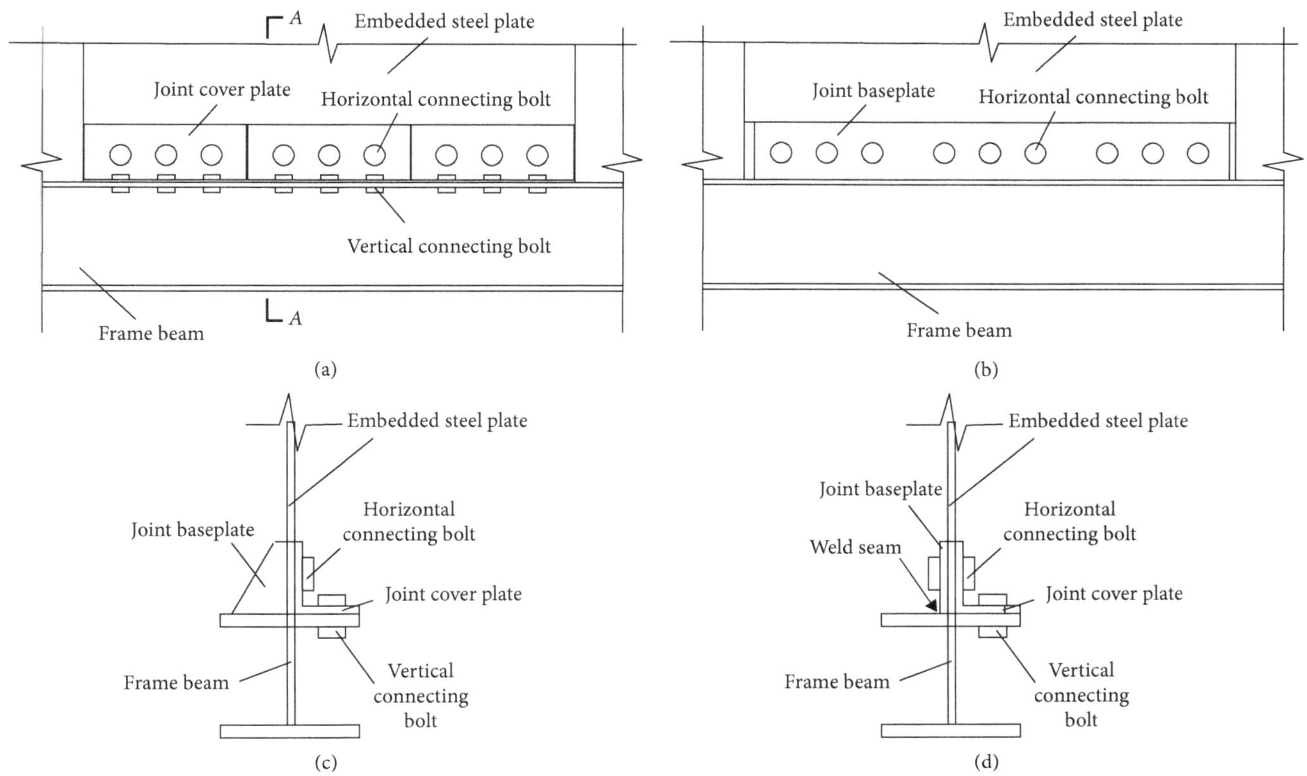

FIGURE 1: Layout of DCPC. (a) Structural map of discontinuous cover plate. (b) Structural map of baseplate. (c) Structural map of the side (d) A-A profile.

TABLE 1: Specific dimensions of specimens (mm).

Specimen number	Square steel string $D \times t_c$	First floor of steel girder $h_b \times b_f \times t_w \times t_f$	Second floor of steel girder $h_b \times b_f \times t_w \times t_f$	Embedded wall panels $t \times W \times H$
A	200 × 12	200 × 200 × 8 × 12	300 × 200 × 8 × 12	4 × 900 × 1190
B	200 × 12	200 × 200 × 8 × 12	300 × 200 × 8× 12	4 × 900 × 1190

TABLE 2: Mechanical properties of steel.

Thickness/mm	Yield strength (N/mm^2)	Tensile strength (N/mm^2)	Elongation (%)	Elastic modulus (×10^4 N/mm^2)
4	286	440	25.9	20
8	276	428	25.2	20.5
10	285	425	31.9	19.6
12	267	429	29.8	20

were set up in the experiment. Three displacement percentile meters are mainly used to measure the relative displacement between the bottom of the column and the ground beam and the absolute displacement of the ground beam. One stay wire displacement meter is set at the beam end of each floor to measure the displacement of the beam end. The displacement reading at the top beam is taken as the control index of the specimen loading. A displacement dial indicator is set at the back side of the specimen to measure the out-of-plane deformation of the joint area cover plate. In order to facilitate comparative analysis, the two components adopt basically the same strain gauge layout scheme. The only difference is that the specimen A also arranges strain gauges on the connecting cover of the wall-slab joint area, while the specimen B does not arrange strain gauges in the joint area. The layout of displacement gauge and strain gauge points is shown in Figure 4. The axial force on the top of the column and the reciprocating load on the end of the beam are output by the pressure sensor installed on the jack.

3.2.2. Loading Device. In order to ensure that the two floors of the specimens are at the same shear level, a 200 t MTS electrohydraulic servo programmable actuator is used to exert low-cycle reciprocating load at one end of the top beam of the frame. Two compression beams are used to suppress

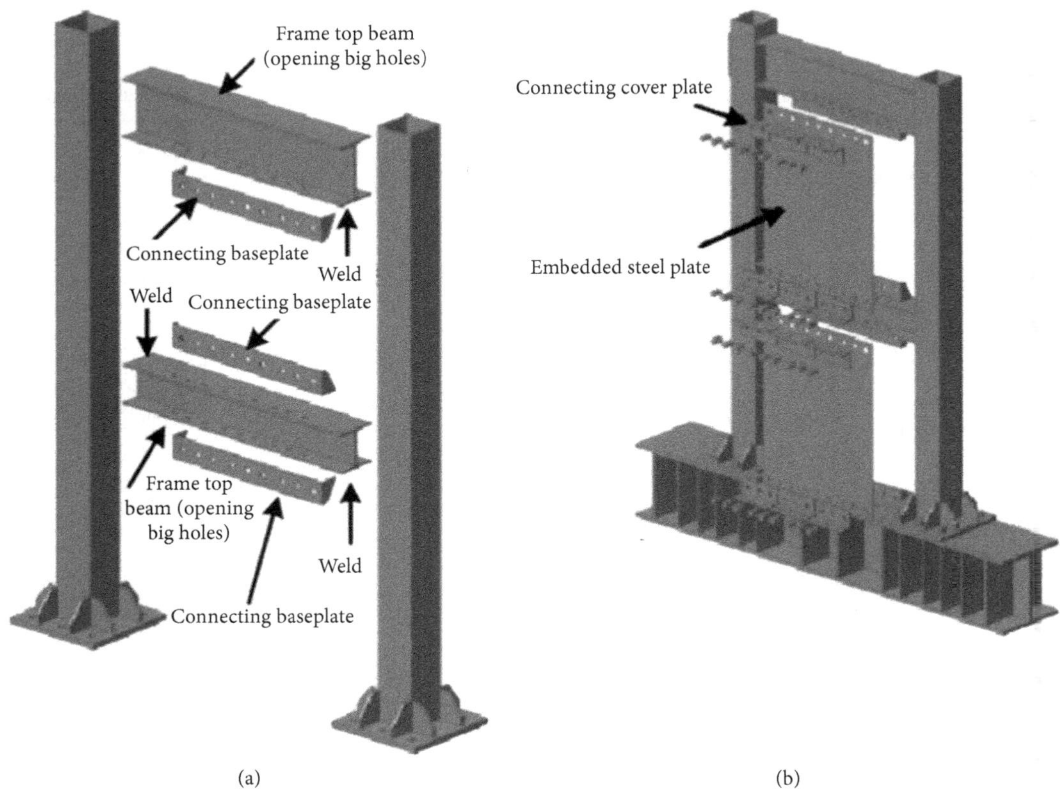

FIGURE 2: Schematic diagram of manufacture and installation for specimen A. (a) Outer frame drawing. (b) On-site assembly drawings.

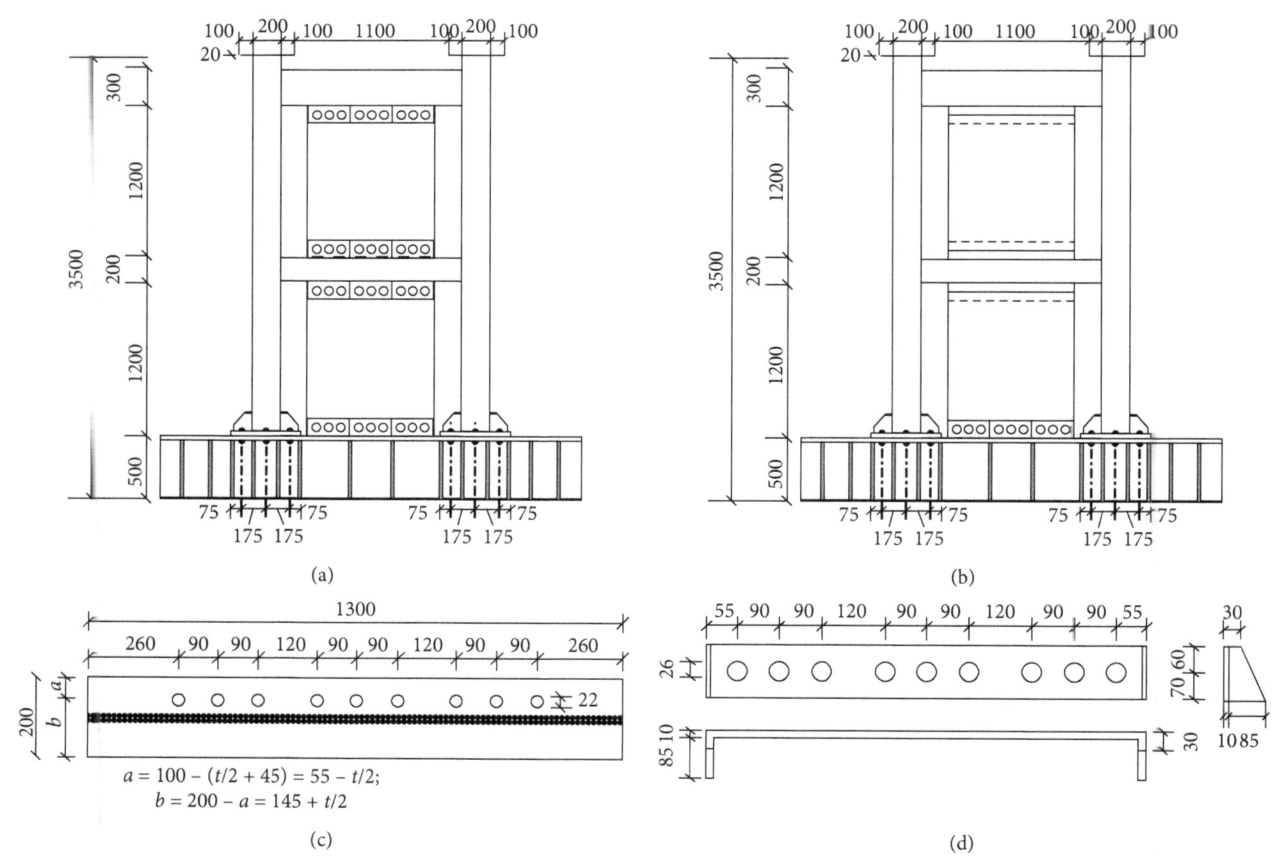

FIGURE 3: Continued.

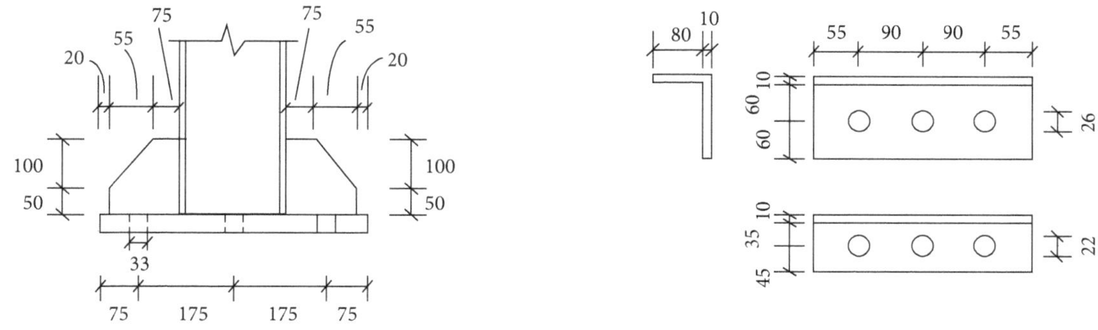

(e) (f)

FIGURE 3: Detailed drawing of specimen processing. (a) Specimen A. (b) Specimen B. (c) Beam flange opening. (d) Connecting floor. (e) Column base. (f) Connecting cover plate.

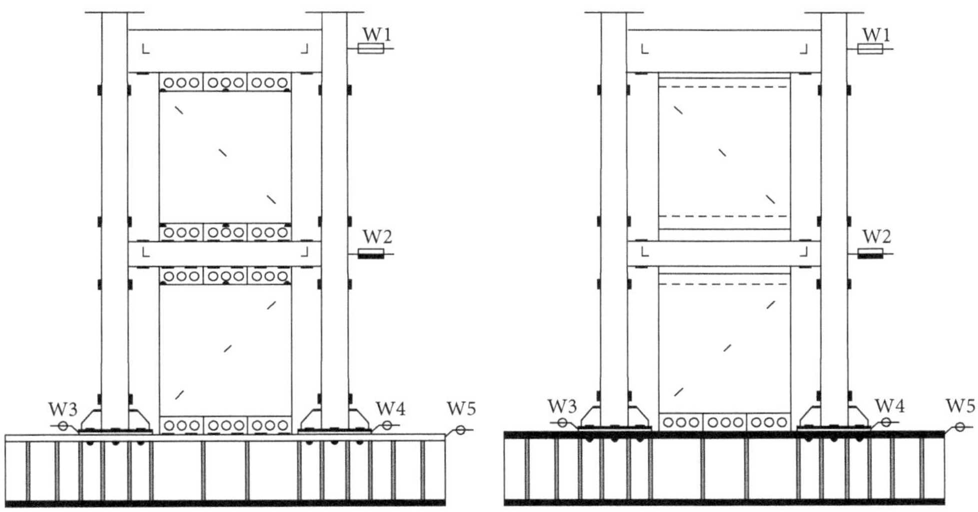

(a) (b)

FIGURE 4: Layout of measuring points for displacement and strain of specimens. (a) Specimen A. (b) Specimen B.

both ends of the ground beam to resist the overall overturning moment and to achieve the fixed end restraint at the bottom of the plane frame. The horizontal top beam is used at both ends of the ground beam to prevent the rigid body displacement of the specimen in the plane. In order to prevent the ground beam from sliding during the loading process, the end restraint shall be checked and the bolts shall be tightened from time to time during the test loading process. Vertical load is provided by two vertical loading systems, which are composed of hydraulic jack, reaction frame, and rolling guide. A spherical hinge is set between the column jack and the jack to realize the free rotation of the column jack. A sliding roller trough is set between the column jack and the steel beam of the reaction frame to realize the sliding of the jack together with the column jack and ensure that the column jack always provides vertical load. In order to avoid out-of-plane instability of the test specimen during loading, a lateral support system is set up, which consists of a portal frame and a beam supported on the portal frame. Each floor is provided with a lateral brace, and the lower floor is arranged in the frame beam position. However, in order to avoid the horizontal jack and ensure that the horizontal jack has enough loading space, the lateral support of the upper floor is arranged below the top beam.

3.2.3. Loading System. According to the American AISC seismic code [14], this test adopts a displacement-controlled loading mode and takes the interstory displacement angle as the controlling factor, as shown in Figure 5.

Firstly, the axial pressure is applied on the top of the column and the axial pressure ratio of the column is 0.3. After successful loading under axial compression, the anchor bolts and frame column foot bolts are tightened again to eliminate the influence of the anchor bolts, the ground flatness, and the bottom flatness of the specimens on the test and to check whether the test device and the measuring equipment are working properly. Unload the jack on the top of the column, rebalance the acquisition system, and prepare for official loading. A predetermined axial load is applied to the top of the column, and the axial load is adjusted during the whole test process to ensure that the axial load is basically

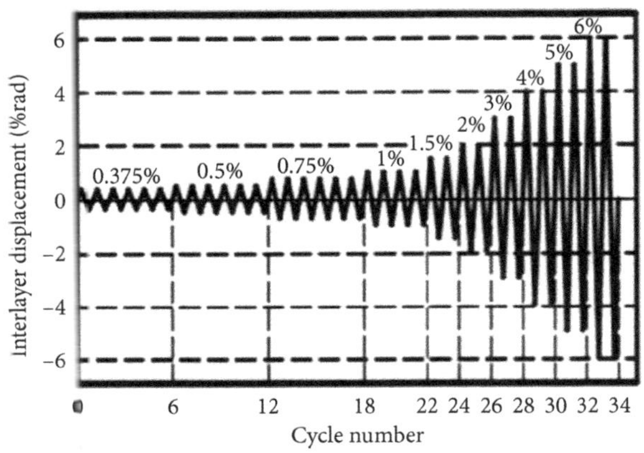

Figure 5: Loading protocol.

constant during the whole test process. Lower cyclic loads are applied at the end of the top beam, and the number of cycles and the corresponding interstory displacement angle of each stage load are shown in Figure 5.

3.3. *Test Phenomena*

3.3.1. Specimen B. According to the test scheme of specimen B, 640 kN vertical loads were applied on the top of the two columns, respectively. After checking the normal operation of each instrument, it entered the horizontal loading stage. After the column is loaded, the maximum initial out-of-plane deflection of the measured steel plate is about 3 mm.

When the specimen is loaded to 0.375% of the interlaminar displacement, there is a little out-of-plane buckling deformation on the basis of the initial imperfection in one floor of steel plate and no obvious change in other parts. When the interstory displacement reaches 0.5%, the buckling of one-story embedded wallboard increases obviously and the drum sound appears in the process of push-pull alternation. When the interlaminar displacement reaches 0.75%, the buckling of the first floor of the embedded steel plate continues to increase. Under the action of the main tensile stress, the steel plate forms an obvious tension band along the diagonal direction of the steel plate, as shown in Figure 6(a), while the second floor of the embedded steel plate also begins to buckle. When the interstory displacement reaches 1%, the tension band of the two-floor wallboard begins to form, as shown in Figure 6(b); the out-of-plane buckling of the embedded wallboard becomes more and more serious, but the waveform remains unchanged; at this time, there is no obvious deformation in other parts of the specimen; and there is no obvious bolt slip in the discontinuous cover-plate connection joints connected with the ground beam.

When the interfloor displacement is loaded to 1.5%, the wall slips slightly at the corner bolts connected with the ground beam; cracks appear on the eastern side of the weld between the first wall plate and the fishtail plate, as shown in the Figure 7(a), which shows that there is no obvious deformation of the frame beam at this time. When the specimens are loaded to the second circle of 2% interstory displacement, the bending deformation of the two-floor frame beams begins to be visible to the naked eye, and that of the first-floor frame beams begins to bend, but the deformation pattern is different from that of the second-floor frame beams. The buckling deformation is shown in Figure 7(c), while the top beam is only equipped with the embedded wallboard in the lower floor, so U-shaped bending deformation will be formed during hysteretic loading, as shown in Figure 7(c). The buckling sound of the wallboard increases with the increase of out-of-plane buckling when push-pull alternates.

When the specimen is loaded to 4% of the interfloor displacement, the out-of-plane buckling of the wallboard is very serious, first floor of the wallboard begins to appear cracks caused by the alternating tension bands of the wallboard, as shown in Figure 7(b) The welded joints between beams and columns are not damaged, and the bending failure of the top beam is serious, as shown in Figure 7(e), but the deformation of the first floor of the frame beam is much smaller than that of the top beam, as shown in Figure 7(d). The displacement of the load level has exceeded the maximum distance of the rolling guide between the vertical jack and the reaction frame and the test is stopped.

3.3.2. Specimen A. In accordance with the requirements of JGJ 82-2011 of "Technical Regulations for High Strength Bolt Connection of Steel Structures" [15], the final bolt of specimen A is screwed after the axial force of the column is applied and a torque wrench is used to make the bolt reach the pretightening force specified in the design. After the final tightening is completed, the maximum initial deflection outside the steel plate surface measured is less than 2 mm.

When the specimen is loaded to 0.375% of the interlaminar displacement, there is a little out-of-plane buckling deformation on the basis of the initial imperfection in one floor of the steel plate and no obvious change in other parts. When the interstory displacement reaches 0.5%, the buckling of one-story embedded wallboard increases obviously and the drum sound appears in the process of push-pull alternation, which shows that in the first two loading stages, the experimental phenomena of specimen A and specimen B are almost the same and only slightly different in the wall plate buckling waveform. When the interstory displacement reaches 0.75%, the buckling of one-story embedded wallboard continues to develop and the tension band begins to form. It can be seen from the mark of the corner of the lower-story embedded steel plate that the embedded steel plate is at the corner and the bolt slipped slightly. When the interstory displacement reaches 1%, the tension band of the two-floor built-in wallboard also appears obviously; at the corner bolt of the first floor, the wallboard sliding continues to develop and the frame beam begins to show small deformation visible to the naked eye.

When the specimen is loaded to 1.5% of the interstory displacement, the embedded wallboard on the second floor begins to slip at the corner bolts connected with the middle beam and the deformation of the frame beam is also developing continuously. After loading 2% of the interstory

(a) (b)

FIGURE 6: Tension field of specimen B. (a) Interfloor displacement of 0.75%. (b) Interfloor displacement of 1%.

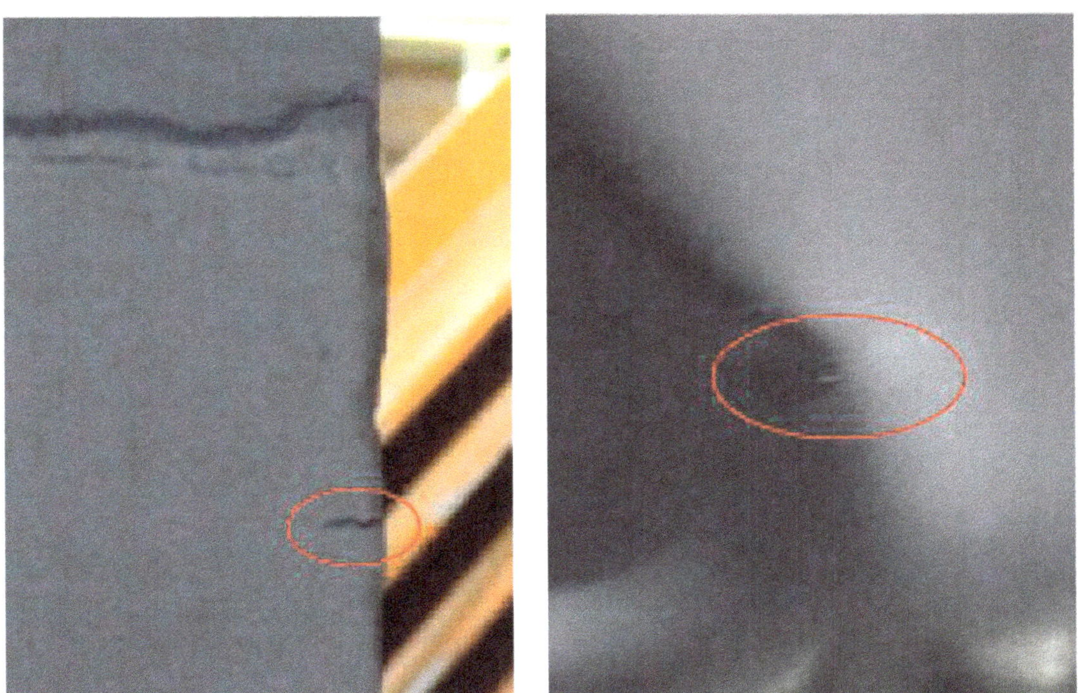

(a) (b)

FIGURE 7: Continued.

(c)

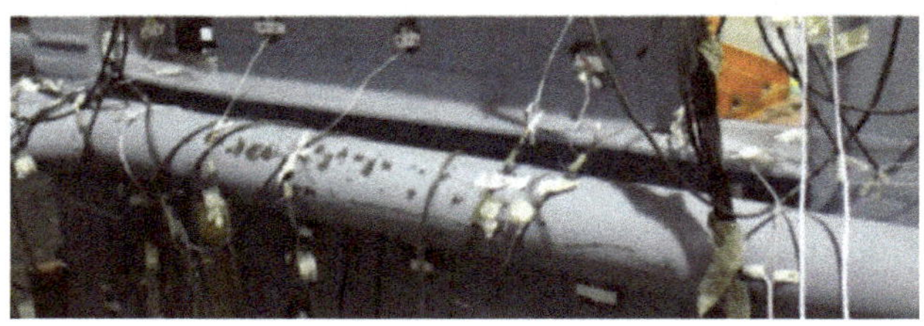

(d)

(e)

FIGURE 7: Failure characteristics of specimen B. (a) Weld crack. (b) Plate tear. (c) Global deformation. (d) Deformation of beam on first floor. (e) Deformation of frame top beam.

displacement, the inverted Z-shaped bending deformation of the frame beam of specimen A is similar to that of specimen B and the U-shaped bending deformation of the top beam is similar, as shown in Figure 8(a). When push-pull alternates, the drum sound of the wall panel increases with the increase of out-of-plane buckling. When the specimen is loaded to 3% of the interstory displacement, the corner slip is quite serious, as shown in Figure 8(d), but there is no relative displacement between the connecting cover plate and the frame beam; the deformation of the frame beam is getting larger and larger, and the paint skin of the web of the frame beam begins to fall off; there is no obvious out-of-plane deformation between the connecting cover plate and the bottom plate at the joint of the embedded wall plate. When the specimen is loaded to 4% of the interstory displacement, there are no cracks similar to specimen B in the embedded steel plates of the two floors of the specimen; the welded joints between the beam and the column are not damaged, and the bending failure of the top beam is serious, as shown in Figure 8(b), but the deformation of the first-floor frame beam is much smaller than that of the top beam, as shown in Figure 8(c). Because the displacement of the next loading stage has exceeded the maximum distance of the rolling guide between the vertical jack and the reaction frame, in order to ensure safety, the test stopped.

DCPC joints are strengthened by the angle steel cover plate; the test results show that the high-strength bolt connection on the steel beam is almost perfect after the test, and there is no obvious deformation in the connection area between the steel beam and the embedded wallboard. On the contrary, the plastic bending deformation of the steel beam in the unconnected area is obvious. From the final deformation morphology of the specimens, it can be seen that the two groups of specimens have similar deformation of frame beams, but the final deformation range of specimen A frame beams is smaller than that of

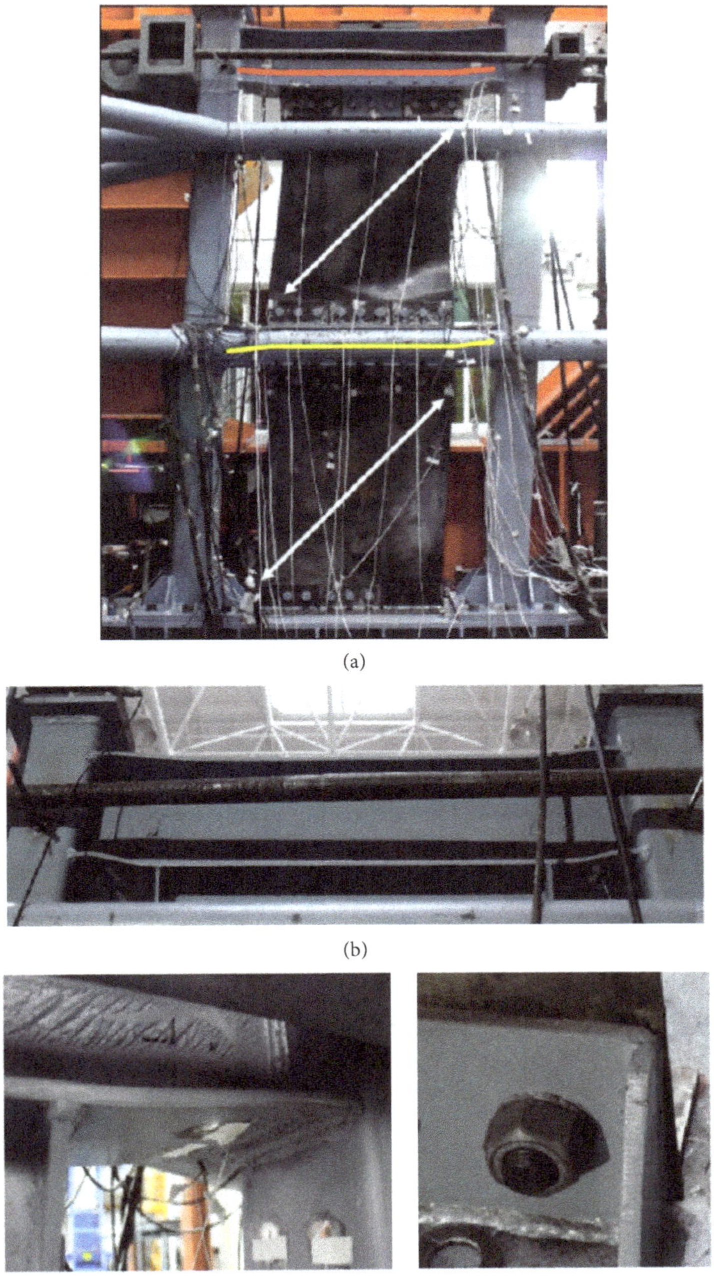

(a)

(b)

(c) (d)

FIGURE 8: Failure characteristics of specimen A. (a) Specimen deformation. (b) Deformation of frame top beam. (c) Deformation of beam on first floor. (d) Slip of connection angle bolt.

specimen B. It is proved that the DCPC joint can ensure good load transfer performance of the steel plate shear wall connection area.

4. Test Results and Analysis

4.1. Hysteresis Curve. Hysteretic curve is the load-displacement curve of specimens under low-cycle reciprocating loads, which comprehensively reflects the rate of deterioration under cyclic loading and is also the main basis for the elastic-plastic dynamic response of the structure [16]. The load-displacement hysteresis curves of the two specimens are shown in Figure 9.

When the actuator is pushed to the west column, the displacement of the structure is greater than zero, which is called positive loading; when the actuator is pulled to the east column, the displacement of the structure is less than zero, which is called reverse loading. Figure 9 shows that the reading of the column top displacement meter is 110 mm and the whole vertex displacement angle is 1/25 at the end of

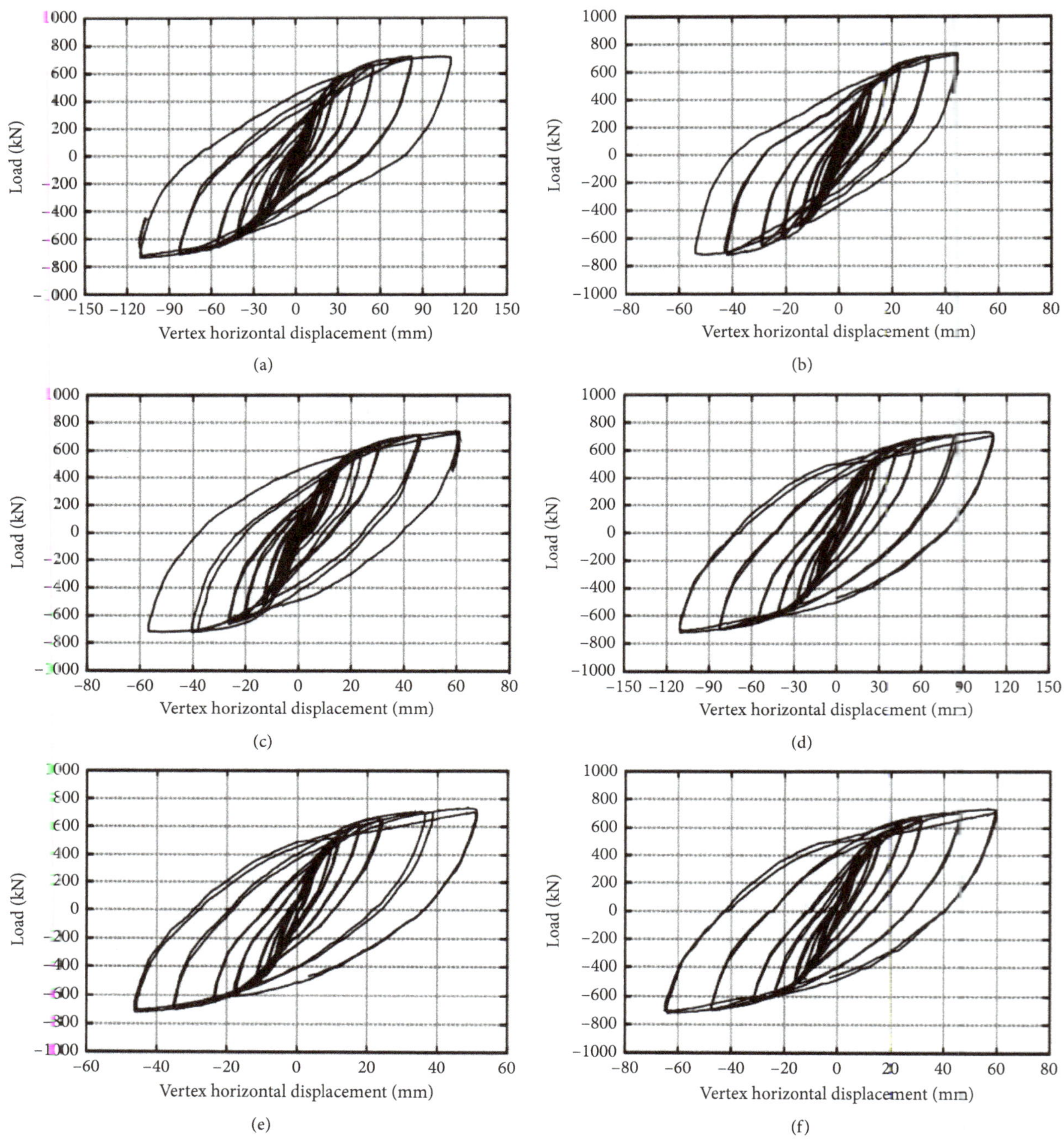

FIGURE 9: Hysteretic curves. (a) Global of specimen B. (b) First floor of specimen B. (c) Second floor of specimen B. (d) Global of specimen A. (e) First floor of specimen A. (f) Second floor of specimen A.

the test. The hysteresis curve shows the following characteristics.

Compared with Figures 9(a) and 9(d), the force and displacement of the two specimens are basically linear at the elastic stage, the hysteretic loop encloses a small area, and the unloading stiffness of the specimens is basically the same as the elastic stiffness. After the specimen yields, the out-of-plane buckling deformation of the embedded wallboard increases continuously, the tension band develops continuously, and the plastic part of the specimen increases gradually. After that, the hysteresis loops of the specimens gradually expand, the envelope area of the ring increases, and the stiffness of each loading stage begins to degenerate. With the increase of loading times, the unloading stiffness decreases slightly compared with the elastic stiffness. When the horizontal load is unloaded to zero, the residual displacement of the specimen increases with the increase of the loading stage. The hysteretic loops

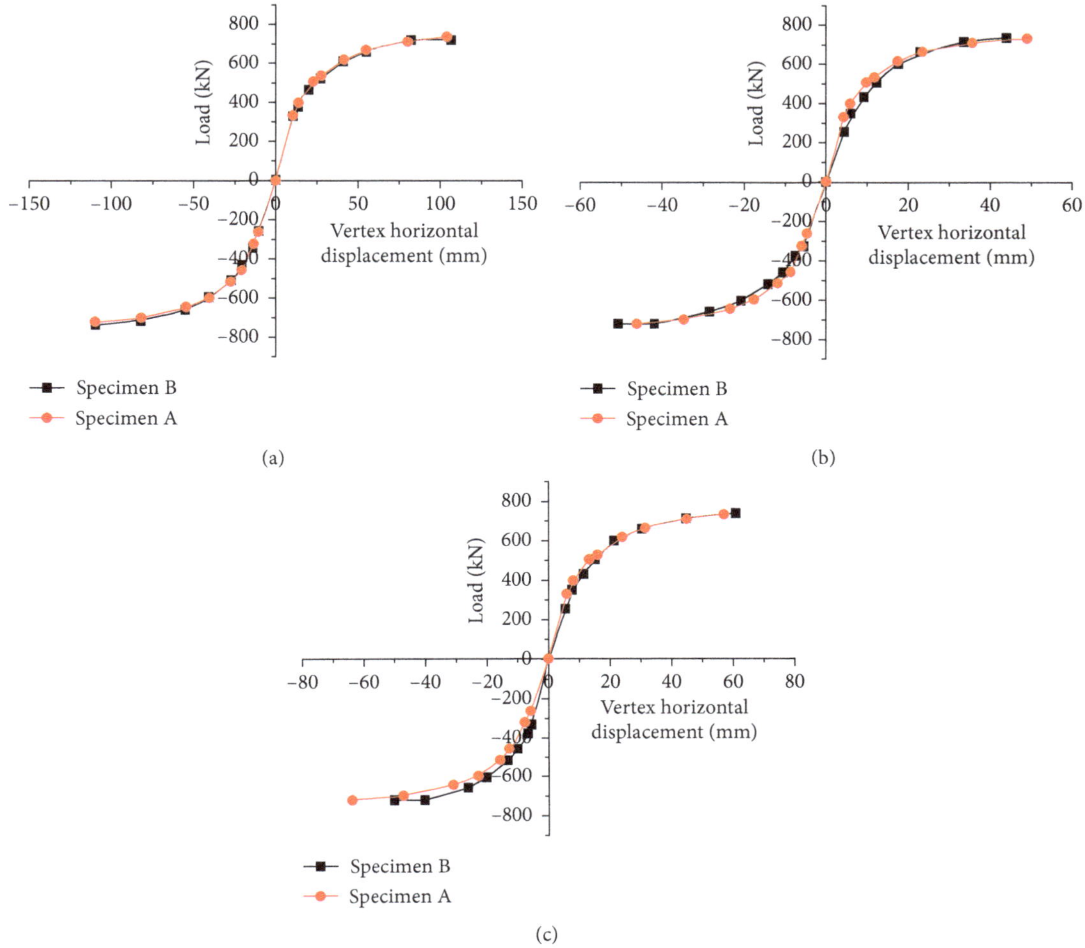

FIGURE 10: Skeleton curves of the specimen. (a) Whole. (b) First floor. (c) Second floor.

of the two specimens were not pinched, but the specimen A was fuller.

From Figures 9(b) and 9(c), it can be seen that from the seventh loading stage, the load-displacement curve on the first floor of specimen B appears asymmetric; under the same horizontal load, the negative interfloor displacement of the first floor is obviously larger than that of the square interfloor displacement. This is due to the tearing of the welding seam of a wall plate connection at this time, which results in the partial failure of the tension band of the negative direction steel plate, thus causing the stiffness asymmetry in the positive and negative directions of the specimen. However, the bearing capacity of the structure did not decrease sharply because of cracks, which indicates that the embedded wall panel has a higher degree of affluence. Up to the end of loading, there was no crack in the second floor of weld, so the hysteretic curve of the second floor was symmetrical. In the later stage of loading, because the top beam is only affected by the tension band of the unilateral embedded wall panel, the two-floor stiffness degenerates rapidly; the upper and lower tension bands of the middle beam can offset each other, the deformation of the first-floor beam is obviously smaller than that of the second-floor top beam, and the stiffness degeneration is slower than that of the first-floor beam, so the displacement of the first floor is gradually smaller than that of the second floor after loading. As can be seen from Figures 9(e) and 9(f), the first and second hysteretic curves of specimen A with DCPC node are full and symmetrical. By comparison with Figures 9(b) and 9(c), the hysteretic curves of each floor of the two specimens are in good agreement.

4.2. Skeleton Curve and Ductility. From Figure 10(a), the overall skeleton curves of the two specimens basically coincide, which shows that the seismic performance of the steel plate shear wall with DPCP joint on both sides is almost the same as that of the traditional steel plate shear wall with welding on both sides.

Figures 10(b) and 10(c) are the contrast diagrams of the skeleton curves of the first and second floors of the two specimens, respectively. By comparison, the skeleton curves of each floor of the two specimens are also in good agreement. Because of cracks in the first floor of specimen B joint weld during loading, the positive and negative directions of the first-floor skeleton curve of the specimen are asymmetrical. However, due to the high

Table 3: Test results of specimen B.

Floor number	Loading direction	K_i (kN/mm)	Yield			Peak			Limit			Displacement ductility coefficient
			Py (kN)	Δy (mm)	$\Delta y/h$ (%)	Pm (kN)	Δm (mm)	$\Delta m/h$ (%)	Pu (kN)	Δu (mm)	$\Delta u/h$ (%)	
Global	Pull direction	25.1	346.6	13.75	0.50	736.75	110.24	4.01	736.75	110.24	4.01	8.02
	Push direction	30.3	376.08	13.78	0.50	720.44	82.39	3.00	718.32	106.69	3.88	7.74
First floor	Pull direction	57.3	346.17	6.14	0.47	736.13	43.90	3.38	736.13	43.90	3.38	7.15
	Push direction	62.3	377.44	7.69	0.59	720.44	41.96	3.23	719.54	50.82	3.91	6.61
Second floor	Pull direction	44.2	345.39	7.68	0.53	736.75	60.84	4.20	736.75	60.84	4.20	7.92
	Push direction	61.2	377.60	6.51	0.45	719.54	40.26	2.78	719.08	50.20	3.46	7.72

Table 4: Test results of specimen A.

Floor number	Loading direction	K_i (kN/mm)	Yield			Peak			Limit			Displacement ductility coefficient
			Py (kN)	Δy (mm)	$\Delta y/h$ (%)	Pm (kN)	Δm (mm)	$\Delta m/h$ (%)	Pu (kN)	Δu (mm)	$\Delta u/h$ (%)	
Global	Pull direction	25.2	321.71	13.64	0.50	719.54	110.13	4.00	719.54	110.13	4.00	8.07
	Push direction	31.8	396.92	13.79	0.50	732.22	104.46	3.80	732.22	104.46	3.80	7.58
First floor	Pull direction	57.1	321.71	5.96	0.46	719.54	46.13	3.55	719.54	46.13	3.55	7.74
	Push direction	75.4	397.38	5.75	0.44	732.22	49.03	3.77	732.22	49.03	3.77	8.53
Second floor	Pull direction	45.7	321.71	7.68	0.53	719.54	64.00	4.41	719.54	64.00	4.41	8.33
	Push direction	55.1	396.92	8.02	0.55	732.22	57.09	3.94	732.22	57.09	3.94	7.12

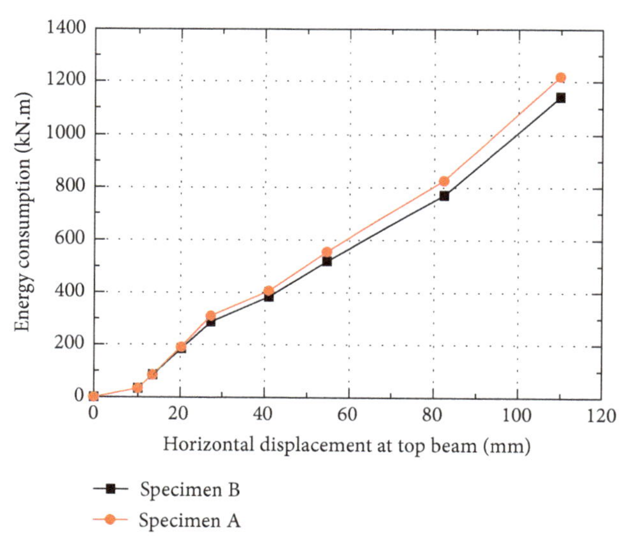

Figure 11: Energy dissipation capacity.

wealthiness of the built-in wallboard, cracks have no significant impact on the seismic performance of the whole structure.

The test results of specimen B and A are given in Tables 3 and 4, respectively. By comparison, the initial stiffness of the two specimens is basically the same and the

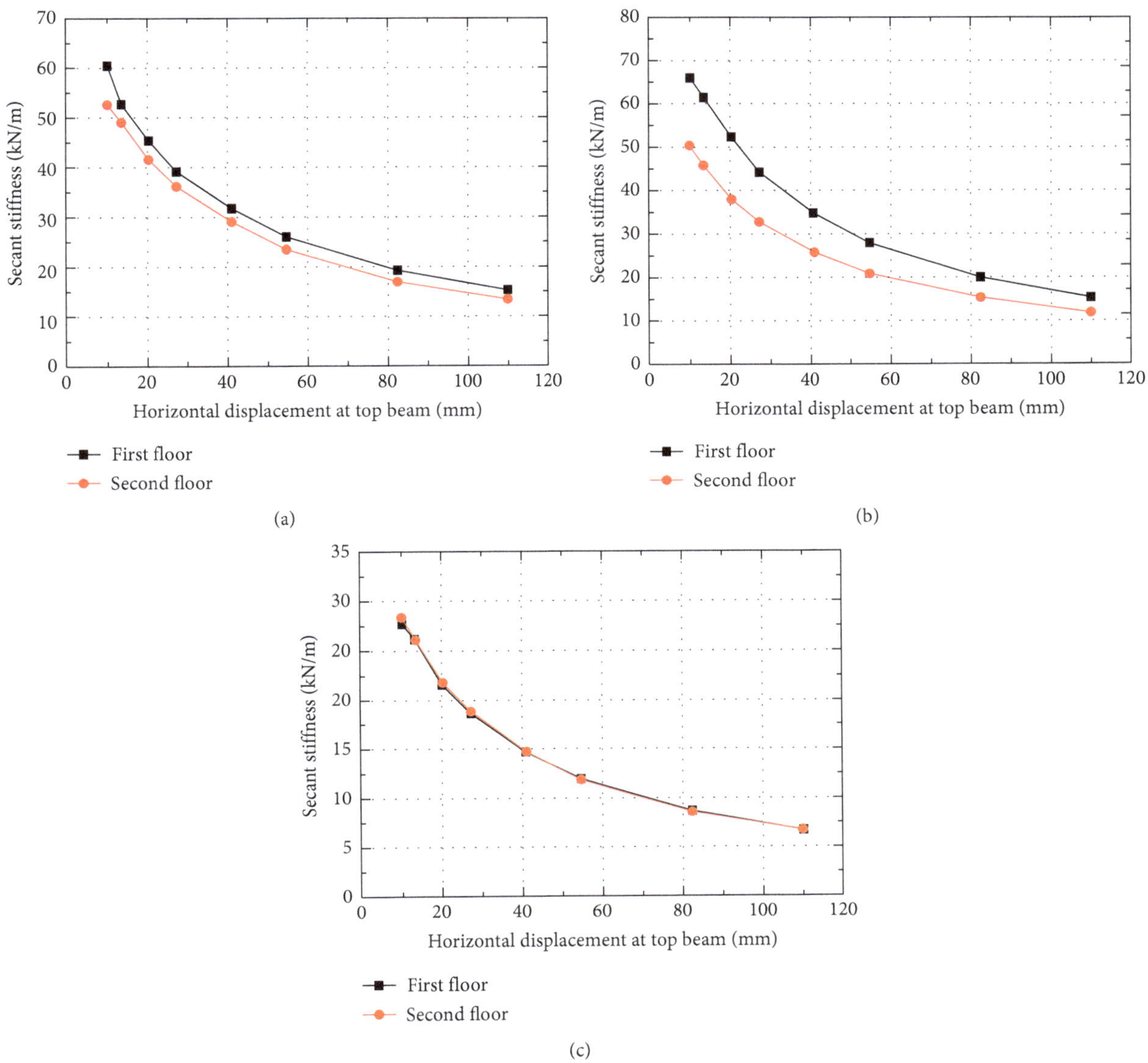

FIGURE 12: Stiffness degeneration. (a) Interlaminar stiffness of specimen B. (b) Interlaminar stiffness of specimen A. (c) Integral stiffness.

ductility coefficients of push and pull directions are about 8.0, which proves that both groups of specimens have good ductility.

4.3. Energy Consumption Capacity. As can be seen from Figure 11, the energy consumption of the two specimens increases with the increase of displacement until the end of loading. In the first three loading stages, the energy consumption of the two specimens is low and almost the same. As the fourth loading stage enters, the energy consumption of the specimen A exceeds than that of the specimen B and the gap increases with the increase of displacement. This is due to the large bolt holes in the embedded wallboard of specimen A. At the fourth loading stage, the wallboard begins to slip at the corner connecting bolts and the slip of the wallboard provides a part of energy dissipation. With the increase of displacement, the sliding area of wallboard increases with the development of tension band, so the energy consumption provided by sliding increases.

4.4. Stiffness Degradation. In order to understand the stiffness variation characteristics of specimens under reciprocating loads, the ratio of the sum of absolute values of load and displacement in two directions is taken as the secant stiffness of specimens at different stages to describe the stiffness degradation of specimens. The secant stiffness of the specimen is shown in Figure 12.

From Figure 12(a), it can be seen that the stiffness of the first floor of specimen B is larger than that of the second floor, and it degenerates synchronously during loading. This is because the stiffening ribs are arranged at the foot of the first floor of the specimen, so the stiffness of the first floor is larger. From Figure 12(b), it can be seen that the stiffness of the first floor of specimen A is also larger than that of the second floor and the degradation rate of the stiffness of the first floor is faster than that of the second floor. This is because in the process of loading, the slip of one-story embedded wallboard at the corner connecting bolt is obviously larger than that of the second floor. Compared with Figures 12(a) and 12(b), the degradation rate of stiffness of specimen A is faster in the initial stage of loading than that of specimen B, but slower in the later stage, which makes the final residual stiffness of each floor equal. Figure 12(c) shows the overall stiffness degradation curves of two specimens, and the two curves are in good agreement. The initial stiffness of specimen A is slightly higher than that of specimen B, which is due to the field assembly of specimen A, and effectively reduces the initial defects that may occur in the transportation of the specimen.

5. Conclusion

Low cyclic loading tests were carried out on the welded steel frame-steel plate shear wall structure and assembled steel frame-steel plate shear wall structure with discontinuous cover-plate connections. The following conclusions are drawn:

(1) The assembled steel frame-steel plate shear wall structure with DCPC has full shuttle-shaped hysteretic curve, stable hysteretic performance, and good ductility.

(2) The assembled steel frame-steel plate shear wall structure with DCPC is connected on both sides, with the increase of displacement, the wallboard first buckles and forms a tension band. Thereafter, the additional bending moment caused by the tension band causes the bending failure of the frame beam. Up to the end of the test, there was no obvious damage to the frame column, which indicated that the structure itself conformed to the seismic design concept of "strong frame, weak wallboard, strong column, and weak beam."

(3) The seismic performance of the assembled steel frame-steel plate shear wall structure with DCPC is comparable to that of the welded steel frame-steel plate shear wall structure, and the energy dissipation capacity of the former is better than that of the latter.

(4) The bolt hole sliding energy dissipation is adopted in DCPC joints, which effectively protects the frame beam and ensures the integrity of the components in the joint area and provides good postearthquake repair function.

(5) The assembled steel plate shear wall with DCPC effectively reduces the number of connecting bolts and is easy to assemble and convenient for opening and stiffness adjustment.

Conflicts of Interest

The authors declare that they have no conflicts of interest.

Acknowledgments

This work was supported by the National Key R&D Program (2017YFC0701614).

References

[1] X. T. Wang, Z. Chao, G. Q. Jiang et al., "Experimental research on seismic behavior of concrete-filled square steel tubular column frame-thin steel plate shear walls with middle opening," *Journal of Building Structures*, vol. 36, no. 8, pp. 16–23, 2015.

[2] A. L. Zhang, X. Zhang, X. C. Liu et al., "Experimental study on seismic behavior of steel frame-fabricated thin steel plate shear wall with two sides connection," *Engineering Mechanics*, vol. 35, no. 9, pp. 54–63, 2018.

[3] Y. H. Wang, W. L. Cao, and G. Li, "The seismic behaviour of the materials used in the steel-plate shear walls with concrete filled steel tube columns," *Advanced Materials Research*, vol. 194-196, pp. 2014–2017, 2011.

[4] H.-S. Hu, J.-G. Nie, J.-S. Fan, M.-X. Tao, Y.-H. Wang, and S.-Y. Li, "Seismic behavior of CFST-enhanced steel plate-reinforced concrete shear walls," *Journal of Constructional Steel Research*, vol. 119, pp. 176–189, 2016.

[5] S.-Y. Seo, H.-D. Yun, and Y.-S. Chun, "Hysteretic behavior of conventionally reinforced concrete coupling beams in reinforced concrete coupled shear wall," *International Journal of Concrete Structures and Materials*, vol. 11, no. 4, pp. 599–616, 2017.

[6] H. B. Shim, J. H. Lee, and H. S. Park, "Shear force of headed shear studs, high-strength shear studs and steel plate embedded in concrete: an experimental study," *Materials Research Innovations*, vol. 19, no. sup8, pp. S8–S145, 2015.

[7] W. Hou, B. Chen, Z. Guo et al., "Experimental study on seismic behavior of embedded steel plate reinforced concrete coupling beams," *China Civil Engineering Journal*, vol. 50, no. 2, pp. 9–18, 2017.

[8] Y. Wu, D. Kang, L. Gao et al., "Seismic behavior of bolted endplate connection between steel reinforced concrete (SRC) wall and SRC beam for use in high-rise buildings," *International Journal of Civil Engineering*, vol. 16, no. 2, pp. 1–12, 2018.

[9] Y. Wu, J. Fu, Y.-B. Lan, T. au, and Y.-B Yang, "seismic performance of endplate connections between steel reinforced concrete walls and steel beams," *Structural Engineering International*, vol. 28, no. 2, pp. 208–217, 2018.

[10] F.-X. Ding, G.-A. Yin, L.-P. Wang, D. Hu, and G.-Q. Chen, "Seismic performance of a non-through-core concrete between concrete-filled steel tubular columns and reinforced concrete beams," *Thin-Walled Structures*, vol 110, pp. 14–26, 2017.

[11] B. Rong, R. Liu, G. Feng, C. You, and F. Apostolos, "Flexural capacity formula of diaphragm-through joints of concrete-filled square steel tubular columns," *Transactions of Tianjin University*, vol. 23, no. 3, pp. 258–266, 2017.

[12] M. A. Rusli, A. Murtopo, I. Siswanto, and M. F Siswanto, "Full height rectangular opening castellated steel beam with diagonal stiffener," *Applied Mechanics and Materials*, vol. 881, pp. 150–157, 2018.

[13] S. Karthika, P. Lakshimikandhan, and G. Dhinakaran, "Load deflection characteristics of sustainable infilled concrete wall panels," *Arabian Journal for Science and Engineering*, vol. 41, no. 2, pp. 451–460, 2016.

[14] N. Balasubramani and R. Thenmozhi, "Behaviour and strength of innovative steel concrete columns with SCC," *Advanced Materials Research*, vol. 984-985, pp. 684–692, 2014.

[15] JGJ 82-2011, *Technical Specification for High Strength Bolt Connection of Steel structure*, China Architecture& Building Press, Beijing, China, 2011.

[16] K. Wang, S.-F. Yuan, Z.-X. Zhi, G.-L. Shi, and D.-F. Cao, "Experimental study on hysteretic behavior of composite frames with concrete-encased CFST columns," *Journal of Constructional Steel Research*, vol. 123, pp. 110–120, 2016.

Permissions

All chapters in this book were first published in AMSE, by Hindawi Publishing Corporation; hereby published with permission under the Creative Commons Attribution License or equivalent. Every chapter published in this book has been scrutinized by our experts. Their significance has been extensively debated. The topics covered herein carry significant findings which will fuel the growth of the discipline. They may even be implemented as practical applications or may be referred to as a beginning point for another development.

The contributors of this book come from diverse backgrounds, making this book a truly international effort. This book will bring forth new frontiers with its revolutionizing research information and detailed analysis of the nascent developments around the world.

We would like to thank all the contributing authors for lending their expertise to make the book truly unique. They have played a crucial role in the development of this book. Without their invaluable contributions this book wouldn't have been possible. They have made vital efforts to compile up to date information on the varied aspects of this subject to make this book a valuable addition to the collection of many professionals and students.

This book was conceptualized with the vision of imparting up-to-date information and advanced data in this field. To ensure the same, a matchless editorial board was set up. Every individual on the board went through rigorous rounds of assessment to prove their worth. After which they invested a large part of their time researching and compiling the most relevant data for our readers.

The editorial board has been involved in producing this book since its inception. They have spent rigorous hours researching and exploring the diverse topics which have resulted in the successful publishing of this book. They have passed on their knowledge of decades through this book. To expedite this challenging task, the publisher supported the team at every step. A small team of assistant editors was also appointed to further simplify the editing procedure and attain best results for the readers.

Apart from the editorial board, the designing team has also invested a significant amount of their time in understanding the subject and creating the most relevant covers. They scrutinized every image to scout for the most suitable representation of the subject and create an appropriate cover for the book.

The publishing team has been an ardent support to the editorial, designing and production team. Their endless efforts to recruit the best for this project, has resulted in the accomplishment of this book. They are a veteran in the field of academics and their pool of knowledge is as vast as their experience in printing. Their expertise and guidance has proved useful at every step. Their uncompromising quality standards have made this book an exceptional effort. Their encouragement from time to time has been an inspiration for everyone.

The publisher and the editorial board hope that this book will prove to be a valuable piece of knowledge for researchers, students, practitioners and scholars across the globe.

List of Contributors

M. Rodríguez-Martín
Department of Cartographic and Terrain Engineering, University of Salamanca, Polytechnic School of Avila, Hornos Caleros 50, 05003 Ávila, Spain
Faculty of Sciences and Arts, Catholic University of Avila (UCAV), C/Canteros s/n, 05005 Ávila, Spain

S. Lagüela
Department of Cartographic and Terrain Engineering, University of Salamanca, Polytechnic School of Avila, Hornos Caleros 50, 05003 Ávila, Spain
Applied Geotechnologies Research Group, University of Vigo, Campus Lagoas-Marcosende, Rúa Maxwell s/n, 36310 Vigo, Spain

D. González-Aguilera and P. Rodríguez-Gonzálvez
Department of Cartographic and Terrain Engineering, University of Salamanca, Polytechnic School of Avila, Hornos Caleros 50, 05003 Ávila, Spain

Zhengning Li, Fuan Wei, Peiqing La, Hongding Wang and Yupeng Wei
State Key Laboratory of Advanced Processing and Recycling of Nonferrous Metals, Lanzhou University of Technology, Lanzhou 730050, China

Dongjun Chen
Chongqing Jiaotong University, Chongqing 400074, China
CSCEC Xinjiang Construction & Engineering (Group) Co., Ltd., Urumchi 830000, China

Benniu Zhang, Xingxing Li, Chong Tu, Can Yuan, Wenzhao Li, Zhixiang Zhou and Zongbao Liang
Chongqing Jiaotong University, Chongqing 400074, China

Martin Vild and Miroslav Bajer
Faculty of Civil Engineering, Institute of Metal and Timber Structures, Brno University of Technology, Veveří 331/95, 602 00 Brno, Czech Republic

Li-hui Dai, Cheng Wu, Feng-jiang An and Sha-sha Liao
State Key Laboratory Explosion Science and Technology, Beijing Institute of Technology, Beijing 100081, China

Adel T. Abbas and Essam A. Al-Bahkali
Department of Mechanical Engineering, King Saud University, Riyadh 11421, Saudi Arabia

Karim Hamza
Department of Mechanical Engineering, University of Michigan, Ann Arbor, MI 48105, USA

Mohamed F. Aly
Department of Mechanical Design, American University in Cairo, New Cairo, Cairo 11835, Egypt

Maider Muro and Garikoitz Artola
Metallurgy Research Centre IK4 AZTERLAN, Aliendalde Auzunea 6, 48200 Durango, Spain

Anton Gorriño and Carlos Angulo
Department of Mechanical Engineering, University of the Basque Country (UPV/EHU), Plaza Ingeniero Torres Quevedo 1, 48013 Bilbao, Spain

Tianqi Li, Yingying Zhang and Lei Gao
School of Mechanical Engineering, Liaoning Shihua University, Fushun 113001, Liaoning, China

Yunhao Zhang
School of Aeronautical Manufacturing Engineering, Nanchang Hangkong University, Nanchang 330000, Jiangxi, China

Wiriyakorn Phanitwong, Arkarapon Sontamino and Sutasn Thipprakmas
Department of Tool and Materials Engineering, King Mongkut's University of Technology Thonburi, Bangkok 10140, Thailand

Waleed Elghazaly and Saied Elghazaly
Steel Technology Lab, Central Metallurgical R&D Institute, Cairo, Egypt

R. Rashad
Mechanical Design, Production Engineering, Cairo University, Cairo, Egypt

Sayed Elmohr
Siegwart Co., Cairo, Egypt

Dezheng Liu, Yan Li, Haisheng Liu and Zhongren Wang
Department of Mechanical Engineering, Hubei University of Arts and Science, Xiangyang, Hubei 441053, China

Yu Wang
Shanghai Institute of Applied Physics, Chinese Academy of Sciences, Shanghai 201800, China

Adel T. Abbas, Mohanad Alata, Magdy M. El Rayes and Ehab A. El Danaf
Department of Mechanical Engineering, College of Engineering, King Saud University, Riyadh 11421, Saudi Arabia

Adham E. Ragab
Department of Industrial Engineering, College of Engineering, King Saud University, Riyadh 11421, Saudi Arabia

Xiao-liang Xu, Zhou-dao Lu, Ling-zhi Li and Chang-jiu Jiang
Research Institute of Structural Engineering and Disaster Reduction, College of Civil Engineering, Tongji University, 1239 Siping Road, Shanghai 200092, China

O. Lupicka and B. Warcholinski
Faculty of Technology and Education, Koszalin University of Technology, Sniadeckich 2, 75-453 Koszalin, Poland

Young-Ho Kim
Institute of R&D, Ji Seung Consultant Co., Ltd., Seoul 06128, Republic of Korea

Jae-Yoon Kang and Hyun-Bon Koo
Korea Institute of Civil Engineering and Building Technology, Goyang 10223, Republic of Korea

Dae-Jin Kim
Department of Architectural Engineering, Kyung Hee University, Yongin 17104, Republic of Korea

Lihua Xu, Biao Li, Yin Chi, Changning Li, Biao Huang and Yuchuan Shi
School of Civil Engineering, Wuhan University, Wuhan, China

Zülküf Demir and Rifat Yakut
Mechanical and Manufacturing Engineering of Technology, Faculty of Batman University, 72100 Batman, Turkey

D. M. Martinez de la Escalera
Facultad de Ingeniería, Universidad Autónoma de Campeche, Campus V, Av. Ing. Humberto Lanz, 24085 Campeche, CAM, Mexico

J. J. Ramos-Hernandez
Instituto de Ciencias Físicas, Universidad Nacional Autónoma de México, Avenida Universidad s/n, 62210 Cuernavaca, MOR, Mexico

E. Porcayo-Palafox, J. Porcayo-Calderon and J. G. Gonzalez-Rodriguez
CIICAp, Universidad Autónoma del Estado de Morelos, Avenida Universidad 1001, 62209 Cuernavaca, MOR, Mexico

L. Martinez-Gomez
Instituto de Ciencias Físicas, Universidad Nacional Autónoma de México, Avenida Universidad s/n, 62210 Cuernavaca, MOR, Mexico
Corrosion y Protección (CyP), Buffon 46, 11590 México City, Mexico

Tae-Hyu Ha
POSCO, Seoul, Republic of Korea

Bong-Ho Cho
Department of Architecture, Ajou University, Suwon, Republic of Korea

Hongjin Kim
Department of Architectural Engineering, Kyungpook National University, Daegu, Republic of Korea

Dae-Jin Kim
Department of Architectural Engineering, Kyung Hee University, Yongin, Republic of Korea

Jae Yuel Oh, Jungmin Lee and Kang Su Kim
Department of Architectural Engineering, University of Seoul, 163 Seoulsiripdae-ro, Dongdaemun-gu, Seoul 02504, Republic of Korea

Deuck Hang Lee
Department of Civil and Environmental Engineering, University of Illinois at Urbana-Champaign, 205 N. Mathews Ave., Urbana, IL 61801, USA

Sung-Bae Kim
Department of Structural Design, The Naeun Structural Engineering, 22 Yangpyeong-ro, Yeongdeungpo-gu, Seoul 07205, Republic of Korea

Index